The Analysis
of Biological Data

The Analysis of Biological Data

Michael C. Whitlock and Dolph Schluter

ROBERTS AND COMPANY PUBLISHERS
Greenwood Village, Colorado

Roberts and Company Publishers
4950 South Yosemite Street, F2 #197
Greenwood Village, Colorado 80111 USA
Internet: www.roberts-publishers.com
Telephone: (303) 221-3325
Facsimile: (303) 221-3326
Internet: www.roberts-publishers.com

Publisher: Ben Roberts
Development Editor and Copyeditor: John Murdzek
Proofreader: Gunder Hefta
Art Studio: Lineworks, Inc.
Text Designer: Mark Ong, Side By Side Studios
Cover Photographer: Christopher Marley/Form and Pheromone
Cover Designer: Mark Ong, Side By Side Studios
Permissions Coordinator: Laura Gabbard Roberts
Compositor: Side by Side Studios

ISBN: 978-0-9815194-0-1

Library of Congress Cataloging-in-Publication Data

Whitlock, Michael.
 The analysis of biological data / Michael Whitlock and Dolph Schluter.
    p. cm.
 ISBN 978-0-9815194-0-1
 1. Biometry--Textbooks. I. Schluter, Dolph. II. Title.
 QH323.5.W48 2009
 570.1'5195--dc22

                          2008010898

10 9 8 7 6 5

To Sally and Wilson, Andrea and Maggie

Contents in brief

Contents

PART **5**   **MODERN STATISTICAL METHODS**

Preface

Modern biologists need the powerful tools of data analysis. As a result, an increasing number of universities offer, or even require, a basic data analysis course for all their biology students. We have been teaching such a course at the University of British Columbia for the last two decades. Over this period, we have sought a textbook that covered the material we needed in a first course at just the right level. We found that most texts were too technical and encyclopedic, or else they didn't go far enough, missing methods that were crucial to the practice of modern biology. We wanted a book that had a strong emphasis on intuitive understanding to convey meaning, rather than an over-reliance on formulas. We wanted to teach by example, and the examples needed to be interesting. Most importantly, we needed a biology book, addressing topics important to biologists handling real data.

We couldn't find the book that we needed, so we decided to write this one to fill the gap. We include several unusual features that we have discovered to be helpful for effectively reaching our audience:

Interesting biology examples. Our teaching has shown us that biology students learn data analysis best in the context of interesting examples drawn from the medical and biological literature. Statistics is a means to an end, a tool to learn about nature. By emphasizing what we can learn about biology, the power and value of statistics becomes plain. Plus, it's just more fun for everyone concerned.

Every chapter has several biological examples of key concepts, and each example is prefaced by a substantial description of the biological setting. The examples are illustrated with photos of the real organisms, so that students can look at what they're learning about. The emphasis on real and interesting examples carries into the problem sets; for each chapter, there are dozens of questions based on real data about biological issues.

Intuitive explanations of key concepts. Statistical reasoning requires a lot of new ways of thinking. Students can get lost in the barrage of new jargon and multitudinous tests. We have found that starting from an intuitive foundation, away from all the details, is extremely valuable. We take an intuitive approach to basic questions: What's a good sample? What's a confidence interval? Why do an experiment? The first several chapters establish this basic knowledge, and the rest of the book builds on it.

Practical data analysis. As its title suggests, this book focuses on data rather than the mathematical foundations of statistics. We teach how to make good graphical displays, and we emphasize that a good graph is the beginning point of any good data analysis. We give equal time to estimation and hypothesis testing, and we avoid treating the P-value as an end in itself. The book does not demand a knowledge of mathematics beyond simple algebra. We focus on practicality over nuance, on biological usefulness over theoretical hand-wringing. We teach not only the "right" way of doing something, but also highlight some of the pitfalls that might be encountered.

We know that a computer will be available for most calculations, so we focus on the concepts of biological data analysis and how statistics can help extract scientific insight from data. With the power of modern computers at hand, the challenge in analyzing data becomes knowing what method to use and why.[1] We imagine and hope that every course using this book will have a component encouraging students to use computer statistical packages. We are also aware that the diversity of such packages is immense, and so we have not tied the book to any particular program.

Practical experimental design. A biologist cannot do good statistics—or good science—without a practical understanding of experimental design. Unlike most books, we discuss basic topics in experimental design, such as controls, randomization, pseudoreplication, and blocking, and we do it in a practical, intuitive way.

Up-to-date on the basics. Believe it or not, the best confidence interval for the proportion is not the one you probably learned as an undergraduate. Nonparametric statistics do not effectively test for differences in means (or medians, for that matter) without some fairly strong assumptions that we normally hear little about. With these and many other topics, we have brought the coverage of basic, everyday topics in statistics up to date.

Coverage of modern topics. Modern biology uses a larger toolkit than a generation ago. In this book, we go beyond most introductory books by establishing the conceptual principles of important topics, such as likelihood, nonlinear regression, randomization, meta-analysis, and the bootstrap.

Useful summaries. Near the end of each chapter is a short, clear summary of the key concepts, and most chapters end with Quick Formula Summaries that put most equations in one easy-to-find place.

Interleaves. Between chapters are short essays that we call interleaves. These interleaves cover a variety of conceptual and common-sense topics that are crucial for the interpretation of statistical results in scientific research. Several of them focus on

---

[1] "A computer lets you make more mistakes faster than any invention in human history—with the possible exceptions of handguns and tequila." —Mitch Ratcliffe, in *Technology Review,* 1992

ways that science can go wrong when concepts are misapplied—and how to account for such mistakes. Although the interleaves are set outside the boundaries of the chapters, they complement the material in the core chapters, and we strongly recommend that they not be skipped.

After five years of writing, you hold the result in your hands. We think *The Analysis of Biological Data* provides a good background in data analysis for biologists, covering a broad range of topics in a practical and intuitive way. It works for our classes; we hope that it works for yours, too.

Organization of the book

The Analysis of Biological Data is divided into five blocks, each with a handful of chapters. We recommend starting with the first block, because it introduces many basic concepts that are used throughout the book. These early chapters are meant to be read in their entirety.

After the first block, most chapters progress from the most general topics at the start to more specialized topics by the end. Each chapter is structured so that a basic understanding of the topic may be obtained from the earliest sections. For example, in the chapter on analysis of variance (Chapter 15), the basics are taught in the first two sections; reading Sections 15.1 and 15.2 gives roughly the same material that most introductory statistics texts provide about this method. Sections 15.3–15.6 explain additional twists and other interesting applications.

The last block of chapters (Chapters 18–21) is mainly for the adventurous and the curious. These chapters introduce several topics, such as likelihood, bootstrapping, and meta-analysis, that are commonly encountered in the biological and medical literature but that are not often mentioned in an introductory course. These chapters introduce the basic principles of each topic, how the methods work, and point to where you might look to find out more.

A basic course could be taught by using only Chapters 1–17 and, within this subset of chapters, by stopping after Sections 5.6, 7.3, 8.4, 9.3, 12.6, 13.6, 15.2, 16.4, and 17.5 in their respective chapters. We suggest that all courses highlight the topics covered in the interleaves.

Each chapter ends with a series of problems that are designed to test students' understanding of the concepts and the practical application of statistics. The problems are divided into Practice Problems and Assignment Problems. Short answers to all Practice Problems are provided in the back of the book; answers to the Assignment Problems are available to instructors only from the publisher. For a copy, contact Ben Roberts at bwr@roberts-publishers.com or (303) 221-3325. Other teaching resources for the book are available online at http://www.roberts-publishers.com/whitlock/teaching.

A word about the data

The data used in this book are real, with a few well-marked exceptions. For the most part, these data were obtained directly from published papers. In some cases, we contacted the authors of articles who generously provided the raw data for our use. Often, when raw data were not provided in the original paper, we resorted to obtaining data points by digitizing graphical depictions, such as scatter plots and histograms. Inevitably, the numbers we extracted differ slightly from the original numbers because of measurement error. In rare cases, we generated data by computer that matched the statistical summaries in the paper. In all cases, the results we present are consistent with the conclusions of the original papers.

Acknowledgments

This book would not have been possible for the two of us alone. Many other people contributed to it in substantial ways. The clarity and accuracy of its contents were improved by the careful attention of a lot of generous readers, including Arianne Albert, Brad Anholt, Cecile Ane, Eric Baack, James Bryant, Martin Buntinas, C. Ray Chandler, Christiana Drake, Jonathan Dushoff, Steven George, Aleeza Gerstein, George Gilchrist, Brett Goodwin, Mike Hickerson, Darren Irwin, Nusrat Jahan, Philip Johns, Istvan Karsai, Robert Keen, John Kelly, Rex Kenner, Ben Kerr, Joseph G. Kunkel, Todd Livdahl, Brian C. McCarthy, Eli Minkoff, Robert Montgomerie, Spencer Muse, Courtney Murren, Claudia Neuhauser, Patrick C. Phillips, Jay Pitocchelli, James Robinson, Simon Robson, Michael Rosenberg, Noah Rosenberg, Nathan Rank, Bruce Rannala, Mark Rizzardi, Michael Russell, Ronald W. Russell, Andrew Schaffner, Andrea Schluter, William Thomas, Michael Travisano, Thomas Valone, Bruce Walsh, Grace A. Wyngaard, and Sam Yeaman. Many of these people read multiple chapters, and they all improved the clarity and accuracy of the book. Sally Otto and Allan Stewart-Oaten earned our undying gratitude by reading and commenting on the entire book. Of course, any errors that remain are our own fault; we didn't always take everyone's advice, even perhaps when we should have. If we have forgotten anyone, you have our thanks even if our memories are poor.

We owe a debt to the students of BIOL 300 at the University of British Columbia, who class-tested this book over the last several years. The book also benefited by class testing at several colleges and universities, in courses by Brad Anholt (University of Victoria), Eric Baack (Luther College), George Gilchrist (College of William and Mary), Mike Hickerson (Queens College, City University of New York), Nusrat Jahan (James Madison University), Susan Lehman (Brock University), Jean Richardson (Brock University), Simon Robson (James Cook University), and Grace A. Wyngaard (James Madison University). George Gilchrist and his students gave us a very detailed and extremely helpful set of comments at a crucial stage of the book. The following students from UBC and other institutions uncovered significant errors in earlier versions of the book: Carol Baskauf, Jessica Beaubier, Edward Cheung, Lorena Cheung, Stephanie Cheung, Denise Choi, Peter Dunn, Samrad Ghavimi, Steve Green, Inderjit Grewal, Sarah Hamanishi, Gurpreet Khaira, Jung Min Kim, Arleigh Lambert, Alexander Leung, Mira Li, Flora Liu, Dianna Louie, Johnston Mak, Giovanni Marchetti, Sarah Neumann, Ruth Ogbamichael, Marion Pearson, Trevor Schofield, Meredith Soon, Erin Stacey, Andrew Tierman, Michelle Uzelac,

John Wakeley, Hillary Ward, Chris Wong, Irene Yu, Anush Zakaryan, Paul Zhou, and Jon-Paul Zacharias. We give special thanks to Nick Cox, who kindly read a previous printing of this book with extraordinary care.

A number of researchers kindly sent us their original data, including Matt Arnegard, Audrey Barker-Plotkin, Butch Brodie, Pamela Colosimo, Kevin Fowler, Chris Harley, Luke Harmon, Andrew Hendry, Peter Keightley, Fredrik Liljeros, Jean Thierry-Mieg, Jeffrey S. Mogil, Patrik Nosil, Margarita Ramos, Rick Relyea, Jake Socha, Brian Starzomski, Richard Svanback, Andrew Trites, Jason Weir, Jack Werren, and Martin Wikelski.

The book was edited and copyedited by editor extraordinaire John Murdzek: his humor may be warped, but his editorial direction is straight as can be. John worked overtime getting this book out on a tight schedule, even finding the typo in "*Photinus ignites.*" Tom Webster turned our graphs and illustrations into beautiful art, and Laura Roberts spent many an hour tracking down the wonderful photos, and Jeff Whitlock very generously provided many of the beautiful photos in this book. Mark Ong and his team at Side by Side Studios did a fantastic job of turning the manuscript into a book. Eric Baack has our special appreciation for slaving over the problem sets to create the answer keys. Gunder Hefta and Aleeza Gerstein corrected numerous errrors with their careful proofreading. Finally, Ben Roberts deserves our greatest thanks, for all of his support and vision in making this book happen, and especially for Clause 24.

The book was started while MCW was supported by the Peter Wall Institute for Advanced Studies at UBC as a Distinguished-Scholar-in-Residence, and the majority of the final stages of the book were written while he was a Sabbatical Scholar at the National Evolutionary Synthesis Center in North Carolina (NSF #EF-0423641). DS began working on the book while a visiting professor in Developmental Biology at Stanford University. The scholarly support and environment provided by each of these institutions was exceptional—and greatly appreciated.

Finally, we would like to give great thanks to all of the people that have taught us the most over the years. MCW would like to thank Dave McCauley, Mike Wade, Nick Barton, Ben Pierce, Kevin Fowler, Patrick Phillips, Sally Otto, and Betty Whitlock.

About the authors

Michael Whitlock is an evolutionary biologist and population geneticist. He is a Professor of Zoology at the University of British Columbia, where he has taught statistics to biology students since 1995. He is currently Editor-in-Chief of *The American Naturalist.*

Dolph Schluter is Professor and Canada Research Chair in the Zoology Department and Biodiversity Research Center at the University of British Columbia. He is known for his research on the ecology and evolution of Galapagos finches and threespine stickleback. He is a fellow of the Royal Societies of Canada and London.

1

Statistics and samples

1.1 What is statistics?

Biologists study the properties of living things. Measuring these properties is a challenge, though, because no two individuals from the same biological population are ever exactly alike. We can't measure everyone in the population, either, so we are constrained by time and funding to limit our measurements to a *sample* of individuals drawn from the population. Sampling brings uncertainty to the project because, by chance, properties of the sample are *not* the same as the true values in the population. Thus, measurements made from a sample are affected by who happened to get sampled and who did not.

Statistics is a technology that describes and measures aspects of nature from samples. Most importantly, statistics lets us *quantify the uncertainty* of these measures—that is, statistics makes it possible to determine the likely magnitude of their departure from the truth.

Statistics is about **estimation**, the process of inferring an unknown quantity of a target population using sample data. Properly applied, the tools for estimation allow us to approximate almost everything about populations using only samples. Examples range from the average flying speed of bumblebees, to the risks of exposure to cell phones, to the variation in beak size of finches on a remote Galápagos island. We can also estimate the proportion of people with a particular disease that die per year and the fraction who recover when treated.

Most importantly, we can assess differences between groups and relationships between variables. For example, we can test the effects of different drugs on the possibility of recovery, we can measure the association between the lengths of fingers and whether individuals play the piano, and we can test whether the survival of women and children during shipwrecks differs from that of men.

> *Estimation* is the process of inferring an unknown quantity of a population using sample data.

All of these quantities describing populations—namely, averages, proportions, measures of variation, and measures of relationship—are called **parameters**. Statistics tells us how best to estimate these parameters using our measurements of a sample. The parameter is the truth, and the estimate is an approximation of the truth, subject to error. If we were able to measure every possible member of the population, we could know the parameter without error, but this is rarely possible. Instead, we use estimates based on incomplete data to approximate this true value. With the right statistical tools, we can determine just how good our approximations are.

> A *parameter* is a quantity describing a population, whereas an *estimate* is a related quantity calculated from a sample.

Statistics is also about **hypothesis testing**, the process of determining how well a "null" hypothesis about a population quantity fits a sample of data. The **null hypothesis** is a specific claim regarding the population quantity. It is made for the purposes of argument and often embodies the skeptical point of view. Examples are "this new drug is no improvement on its predecessor," and "metabolic rate increases with body mass according to Kleiber's law, by the 3/4 power." Biological data usually get more interesting and informative if the null hypothesis is found to be inadequate.

Statistics has become an essential tool in almost every area of biology—as indispensable as the PCR machine, calipers, binoculars, and the microscope. This book presents the ideas and methods needed to use statistics effectively, so that we can improve our understanding of nature.

Chapter 1 begins with an overview of samples—how they should be gathered and the conclusions that can be drawn from them. We also discuss the types of variables that can be measured from samples, introducing terms that will be used throughout the book.

1.2 **Sampling populations**

Our ability to obtain reliable measures of population characteristics—and to assess the uncertainty of these measures—depends critically on how we sample populations. It is often at this early step in an investigation that the fate of a study is sealed, for better or worse, as Example 1.2 demonstrates.

| Example 1.2 | **Raining cats** |

In an article published in the *Journal of the American Veterinary Medical Association,* Whitney and Mehlhaff (1987) presented results on the injury rates of cats that had plummeted from buildings in New York City according to the number of floors they had fallen. Fear not: no experimental scientist tossed cats from different altitudes to obtain the data for this study. Rather, the cats had fallen (or jumped) of their own accord. The researchers were merely recording the fates of the cats that ended up at the veterinary hospital for repair. The damage caused by such falls was dubbed Feline High-Rise Syndrome, or FHRS.[1]

Not surprisingly, cats that fell five floors fared worse than those dropping only two, and those falling seven or eight floors tended to suffer even more (see Figure 1.2-1). But the astonishing result was that things got better after that. On average, the number of injuries was reduced in cats that fell more than nine floors. This was true in every injury category. Their injury rates approached that of cats that had fallen only two floors! One cat fell 32 floors and walked away with only a chipped tooth.

This effect cannot be attributed to the ability of cats to right themselves so as to land on their feet—a cat needs less than one story to do that. The authors of the article put forth a more surprising explanation. They proposed that after a cat attains terminal velocity, which happens after it has dropped six or seven floors, the falling cat relaxes, and this change to its muscles cushions the impact when the cat finally meets the pavement.

---

[1] "The diagnosis of high-rise syndrome is not difficult. Typically, the cat is found outdoors, several stories below, and a nearby window or patio door is open" (Ruben 2006).

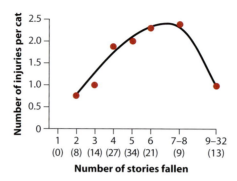

Figure 1.2-1 A graph plotting the average number of injuries sustained per cat according to the number of stories fallen. Numbers in parentheses indicate sample sizes. Modified from Diamond (1988).

Remarkable as these results seem, aspects of the sampling procedure raise questions. A clue to the problem is provided by the sample size for each height. This is the number of cats that fell a particular number of floors, and is indicated along the horizontal axis of Figure 1.2-1. No cats fell just one floor, and the number of cats falling increases with each floor from the second floor to the fifth. Yet, surely, every building in New York having at least five floors has a fourth floor, too, with open windows no less inviting. What can explain this curious trend?

Our strong suspicion is that not all falling cats were taken to the vet and that the chance of a cat making it to the vet was affected by the number of stories it had fallen. Perhaps most cats that tumble out of a first- or second-floor window suffer only indignity, which is untreatable. Any cat appearing to suffer no physical damage from a fall of even a few stories may likewise skip a trip to the vet. At the other extreme, a cat plunging 20 stories might also avoid a trip to the vet, heading to the nearest pet cemetery instead.

This example illustrates the kinds of questions of interpretation that arise if samples are biased. If the sample of cats delivered to the vet clinic is, as we suspect, a distorted subset of all the cats that fell, then the measures of injury rate and injury severity will also be distorted. We cannot say whether this bias is enough to cause the surprising downturn in injuries at a high number of stories fallen. At the very least, though, we can say that, if the chances of a cat making it to the vet depends on the number of stories fallen, the relationship between injury rate and number of floors fallen will be distorted.

Good samples are a foundation of good science. In the rest of this section we give an overview of the concept of sampling, what we are trying to accomplish when we take a sample, and the inferences that are possible when researchers get it right.

Populations and samples

The first step in collecting any biological data is to decide on the target population. A **population** is the entire collection of individuals or units that a researcher is interested in. Ordinarily, a population is composed of a large number of individuals; in most cases, we assume that the population is effectively infinitely large. Examples of populations include

- ▶ all cats that have fallen from buildings in New York City,
- ▶ all the genes in the human genome,
- ▶ all individuals of voting age in Australia,
- ▶ all paradise flying snakes, *Chrysopelea paradisi*, in Borneo, and
- ▶ all children in Vancouver, Canada, suffering from asthma.

A **sample** is a much smaller set of individuals selected from the population.[2] The researcher uses this sample to draw conclusions that, hopefully, apply to the whole population. Examples include

- ▶ the fallen cats brought to one veterinary clinic in New York City,
- ▶ a selection of 20 human genes,
- ▶ a pub full of Australian voters,
- ▶ eight paradise tree snakes caught by researchers in Borneo, and
- ▶ a selection of 50 children in Vancouver, Canada, suffering from asthma.

A *population* is all the individuals or units of interest, whereas a *sample* is a subset of units taken from the population.

In the above examples, the basic unit of sampling is literally a single individual. Sometimes, however, the basic unit of study is a *group* of individuals, in which case a sample consists of a *set* of such units. Examples of units include a single family, a colony of microbes, a plot of ground in a field, an aquarium of fish, and a cage of mice. Scientists use several terms to indicate the sampling unit, such as "unit," "individual," "subject," or "replicate."

Properties of good samples

Estimates based on samples are doomed to depart somewhat from the true population characteristics simply by chance. This chance difference from the truth is called **sampling error**. The spread of estimates resulting from sampling error indicates the **precision** of an estimate. The lower the sampling error, the higher the precision. Larger samples are less affected by chance and so, all else being equal, larger samples will have lower sampling error and higher precision.

---

[2]   In biology, a "blood sample" or a "tissue sample" might refer to a substance taken from a single individual. In statistics, we reserve the word "sample" to refer to a subset of individuals drawn from a population.

Sampling error is the chance difference between an estimate and the population parameter being estimated.

Ideally, our estimate is **accurate** (or **unbiased**), meaning that the average of estimates is centered on the true population value. If samples are not properly taken, measurements made on them might systematically underestimate (or overestimate) the population parameter. This is a second kind of error called **bias**.

Bias is a systematic discrepancy between estimates and the true population characteristic.

The major goal of sampling is to minimize sampling error and bias in estimates. Figure 1.2-2 illustrates these goals by analogy with shooting at a target. Each point represents an estimate of the population bull's-eye (i.e., of the true characteristic). Ideally, all the estimates are tightly grouped, indicating low sampling error, and they are centered on the bull's-eye, indicating low bias. Estimates are precise if they are tightly grouped and highly repeatable, with different samples giving similar answers. Estimates are accurate if they are centered on the bull's-eye. Estimates are imprecise, on the other hand, if they are spread out, and they are biased (inaccurate) if they are displaced systematically to one side of the bull's-eye. The shots (estimates) on the upper right-hand target in Figure 1.2-2 are widely spread out but centered on the bull's-eye, so we say that the estimates are accurate but imprecise. The shots on the lower left-hand target are tightly grouped but not near the bull's-eye, so we say that they are precise but inaccurate. Both precision and accuracy are important, because a lack of either means that an estimate is likely to differ greatly from the truth.

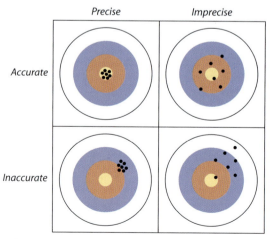

Figure 1.2-2 Analogy between estimation and target shooting. An accurate estimate is centered around the bull's-eye, whereas a precise estimate has low spread.

A final goal of sampling is to allow the precision of an estimate to be quantified. There are several quantities available to measure precision, which we discuss in Chapter 4.

The sample of cats in Example 1.2 falls short in achieving some of these goals. If uninjured and dead cats do not make it to the pet hospital, then estimates of injury rate are biased. Injury rates for cats falling only two or three floors are likely to be overestimated, whereas injury rates for cats falling many stories might be underestimated.

Random sampling

The common assumption of the methods presented in this book is that the data come from a **random sample**. A random sample is a sample from a population that fulfills two criteria.

First, every unit in the population must have an **equal chance** of being included in the sample. This is not as easy as it sounds. A botanist estimating plant growth might be more likely to find the taller individual plants or to collect those closer to the road. Some members of animal or human populations may be difficult to collect because they are shy of traps, never answer the phone, ignore questionnaires, or live at greater depths or distances than other members. These hard-to-sample individuals might differ in their characteristics from those of the rest of the population, so underrepresenting them in samples would lead to bias.

Second, the selection of units must be **independent**. In other words, the selection of any one member of the population must in no way influence the selection of any other member. This, too, is not easy to ensure. Imagine, for example, that a sample of adults is chosen for a survey of consumer preferences. Because of the effort required to contact and visit each household to conduct an interview, the lazy researcher is tempted to record the preferences of multiple adults in each household and add their responses to those of other adults in the sample. This violates the criterion of independence, because the selection of one individual has increased the probability that another individual from the same household will also be selected. This will skew the data if individuals from the same household have preferences more similar to one another than is obtained from individuals randomly chosen from the population at large. With non-independent sampling, our sample size is effectively smaller than we think. This, in turn, will cause us to miscalculate the precision of the estimates.

> In a *random sample*, each member of a population has an equal and independent chance of being selected.

In general, the surest way to minimize bias and allow sampling error to be quantified is to obtain a random sample.

Random sampling minimizes bias and makes it possible to measure the amount of sampling error.

How to take a random sample

Obtaining a random sample is easy in principle but can be challenging in practice. A random sample can be obtained by using the following step-by-step procedure:

1. Create a list of every unit in the population of interest, and give each unit a number between 1 and the total population size.
2. Decide on the number of units to be sampled (call this number n).
3. Using a random-number generator, generate n random integers between 1 and the total number of units in the population.
4. Sample the units whose numbers match those produced by the random-number generator.

An example of this process is shown in Figure 1.2-3. In both panels of the figure, we've drawn the locations of all 5699 trees present in 2001 in a carefully mapped tract of Harvard Forest (Barker-Plotkin et al. 2006). Every tree in this population has a unique number between 1 and 5699 to identify it. We used a computerized random-number generator to pick $n = 20$ random integers between 1 and 5699, where 20 is the desired sample size. The 20 random integers, after sorting, are as follows:

156, 167, 232, 246, 826, 1106, 1476, 1968, 2084, 2222, 2223, 2284, 2790, 2898, 3103, 3739, 4315, 4978, 5258, 5500

These 20 randomly chosen trees are identified by red dots in the left panel of Figure 1.2-3.

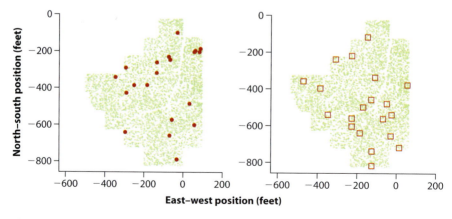

Figure 1.2-3. The locations of all 5699 trees present in the Prospect Hill Tract of Harvard Forest in 2001 (green circles). The red dots in the left panel are a random sample of 20 trees. The squares in the right panel are a random sample of 20 quadrats (each 20 feet on a side).

How realistic is this procedure? Creating a numbered list of every individual member of a population might be feasible for patients recorded in a hospital data base, for children registered in an elementary-school system, or for some other populations for which a registry has been built. The feat is impractical for most plant populations, however, and unimaginable for most populations of animals or microbes. What can be done in such cases?

One answer is that the basic unit of sampling doesn't have to be a single individual—it can be a group, instead. For example, it is easier to use a map to divide a forest tract into many equal-sized blocks or plots, and then to create a numbered list of these plots, than it is to produce a numbered list of every tree in the forest. To illustrate this second approach, we divided the Harvard Forest tract into 836 plots of 400 square feet each. With the aid of a random-number generator, we then identified a random sample of 20 plots, which are identified by the squares in the right panel of Figure 1.2-3.

The trees contained within a random sample of plots do *not* constitute a random sample of trees, for the same reason that all of the adults inhabiting a random sample of households do not constitute a random sample of adults. Trees in the same plot are not sampled independently. The data in this case must be handled carefully. A simple technique is to take the average of the measurements of all of the individuals within a unit as the single independent observation for that unit.

Random numbers should always be generated with the aid of a computer. Haphazard numbers made up by the researcher are not likely to be random (see Example 19.1). Most spreadsheet programs and statistical software packages on the computer include random-number generators.

The sample of convenience

One undesirable alternative to the random sample is the **sample of convenience**, a sample based on individuals that are easily available to the researcher. The researchers must assume (i.e., dream) that a sample of convenience is unbiased and independent like a random sample, but there is no way to guarantee it.

> A *sample of convenience* is a collection of individuals that are easily available to the researcher.

The main problem with the sample of convenience is bias, as the following examples illustrate:

▶ The injury rate of cats that have fallen from high-rise buildings is likely to be underestimated compared with a random sample, if measured only on cats that are brought to a veterinary clinic. Uninjured and fatally injured cats are less likely to make it to the vet and into the sample.

▶ The spectacular collapse of the North Atlantic cod fishery in the last century was caused in part by overestimating cod densities in the sea, which led to excessive allowable catches by fishing boats (Walters and Maguire 1996). Density estimates were too high because they relied heavily on the rates at which the fishing boats were observed to capture cod. However, the fishing boats tended to concentrate in the few remaining areas where cod were still numerous, and they did not randomly sample the entire fishing area.

▶ The Literary Digest Poll was the largest poll in history (questionnaires were sent to 10 million people, of which 2.4 million responded), but it predicted the wrong outcome to the 1936 U.S. federal election (Freedman et al. 1997). This was probably because the list of people to receive questionnaires was obtained from magazine subscriptions, telephone books, automobile registrations, and club memberships. This tended to leave out people in low-income families whose voting preferences were very different from the higher-income people who received questionnaires.

A sample of convenience might also violate the assumption of independence if individuals in the sample are more similar to one another in their characteristics than individuals chosen randomly from the whole population. This is likely if, for example, the sample includes a disproportionate number of individuals who are friends or who are related to one another.

Volunteer bias

Human studies in particular must deal with the possibility of **volunteer bias**, which is a bias resulting from a systematic difference between the pool of volunteers (the **volunteer sample**) and the population to which they belong. The problem arises when the behavior of the subjects affects whether they are sampled.

In a large experiment to test the benefits of a polio vaccine, for example, participating schoolchildren were randomly chosen to receive either the vaccine or a saline solution (serving as the control). The vaccine proved effective, but the rate at which children in the saline group contracted polio was found to be higher than in the general population. Perhaps parents of children who had not been exposed to polio prior to the study, and therefore had no immunity, were more likely to volunteer their children for the study than parents of kids who had been exposed (Brownlee 1955, Bland 2000).

Compared with the rest of the population, volunteers might be

▶ more health conscious and more proactive;
▶ low-income (if volunteers are paid);
▶ more ill, particularly if the therapy involves risk, because individuals who are dying anyway might try anything;
▶ more likely to have time on their hands (e.g., retirees and the unemployed are more likely to answer telephone surveys);

▶ more angry, because people who are upset are sometimes more likely to speak up; or

▶ less prudish, because people with liberal opinions about sex are more likely to speak to surveyors about sex.

Such differences can cause substantial bias in the results of studies. Bias can be minimized, however, by careful handling of the volunteer sample, but the resulting sample is still inferior to a random sample.

Real data in biology

In this book we use real data hard-won from observational or experimental studies in the lab and field and published in the literature. Do the samples on which the studies are based conform to the ideals outlined above? Alas, the answer is often no. Random samples, however much desired, are often not achieved by biologists working in the trenches. Real data are frequently based on samples that are not random, as the falling cats in Example 1.2 demonstrate.

Biologists deal with this problem by acknowledging that the problem exists, by pointing out where biases might arise in their studies,[3] and by carrying out further studies that attempt to control for any sampling problems evident in earlier work.

1.3   Types of data and variables

With a sample in hand, we can begin to measure variables. A **variable** is any characteristic or measurement that differs from individual to individual. Examples include running speed, reproductive rate, and genotype. Estimates (e.g., average running speed of a random sample of 10 lizards) are also variables, because they differ by chance from sample to sample. **Data** are the raw measurements of one or more variables made on a sample of individuals.

> *Variables* are characteristics that differ among individuals.

Categorical and numerical variables

Variables can be categorical or numerical.

Categorical variables describe membership in a category or group. They describe named characteristics of individuals that do not have magnitude on a

---

[3]   We biologists are generally happier to find such flaws in other researchers' data than in our own.

numerical scale. Categorical variables are also called attribute or qualitative variables. Examples of categorical variables include

- ► sex chromosome genotype (e.g., XX, XY, XO, XXY, or XYY),
- ► method of disease transmission (e.g., water, air, animal vector, or direct contact),
- ► predominant language spoken (e.g., English, Mandarin, Spanish, Indonesian, etc.),
- ► life stage (e.g., egg, larva, juvenile, subadult, or adult),
- ► snakebite severity score (e.g., minimal severity, moderate severity, or very severe), and
- ► size class (e.g., small, medium, or large).

A categorical variable is **nominal** if the different categories have no inherent order. Nominal means "name." Sex chromosome genotype, method of disease transmission, and predominant language spoken are nominal variables. In contrast, the values of an **ordinal** categorical variable can be ordered, despite lacking magnitude on the numerical scale. Ordinal means "having an order." Life stage, snakebite severity score, and size class are ordinal categorical variables.

> *Categorical data* are named characteristics of individuals that do not have magnitude on a numerical scale.

A variable is **numerical** when measurements of individuals are quantitative and have magnitude. These variables are numbers. Measurements that are counts, dimensions, angles, rates, and percentages are numerical. Examples of numerical variables include

- ► core body temperature (e.g., degrees Celsius, °C),
- ► territory size (e.g., hectares),
- ► cigarette consumption rate (e.g., average number per day),
- ► age at death (e.g., years),
- ► number of mates, and
- ► number of amino acids in a protein.

> *Numerical data* are quantitative measurements that have magnitude on a numerical scale.

Numerical data are either continuous or discrete. **Continuous** numerical data can take on any real-number value within some range. Between any two values of a continuous variable, an infinite number of other values are possible. In practice, continuous data are rounded to a predetermined number of digits, set for convenience or by the limitations of the instrument used to take the measurements. Core body temperature, territory size, and cigarette consumption rate are continuous variables.

In contrast, **discrete** numerical data, such as counts, come in indivisible units. Number of mates and number of amino acids in a protein are discrete numerical measurements. Number of cigarettes consumed on a specific day is a discrete variable, but the rate of cigarette consumption is a continuous variable when calculated as an average number per day over a large number of days. In practice, discrete numerical variables are often analyzed as though they were continuous, if there is a large number of possible values.

Just because a variable is indexed by a number does not mean it is a numerical variable. Numbers might also be used to name categories (e.g., family 1, family 2, etc.). Numerical data can be reduced to categorical data by grouping, though the result contains less information (e.g., "above average" and "below average").

Explanatory and response variables

One major use of statistics is to relate one variable to another, by examining associations between variables and differences between groups. Measuring an association is equivalent to measuring a difference, statistically speaking. Showing that "the proportion of survivors *differs* between treatment categories" is the same as showing that the variables "survival" and "treatment" are *associated*.

Often in our analysis we try to predict one of the variables, deemed the **response variable**, from a second variable, called the **explanatory variable**. For example, if we wanted to examine the possibility that high blood pressure leads to an increase in the risk of strokes, then we would treat blood pressure as the explanatory variable and record whether an individual had a stroke as the response variable. In the falling-cats example, the number of stories fallen is the explanatory variable, and the number of injuries sustained is the response variable.

Sometimes you will hear variables referred to as "independent" and "dependent." These are the same as explanatory and response variables, respectively. Strictly speaking, if one variable depends on the other, then neither is independent, so we prefer to use "explanatory" and "response" throughout this book.

1.4 Frequency distributions and probability distributions

Different individuals in a sample will have different measurements. We can see this variability with a frequency distribution. The **frequency** of a specific measurement in a sample is the number of observations having a particular value of the measurement. The **frequency distribution** shows how often each value of the variable occurs in the sample.

> The *frequency distribution* describes the number of times each value of a variable occurs in a sample.

Figure 1.4-1 shows the frequency distribution for the measured beak depths of a sample of 100 finches from a Galápagos Island population.[4]

The large-beaked ground finch on the Galápagos Islands.

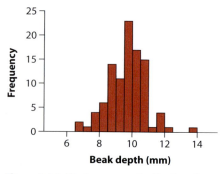

Figure 1.4-1 The frequency distribution of beak depths in a sample of 100 finches from a Galápagos Island population (Boag and Grant 1984). The vertical axis indicates the frequency, the number of observations in each 0.5-mm interval of beak depth.

We use the frequency distribution of a sample to inform us about the distribution of the variable (beak depth) in the population from which it came. Looking at a frequency distribution gives us an intuitive understanding of the variable. For example, we can see which values of beak depth are common and which are rare, we can get an idea of the average beak depth, and we can start to understand how variable beak depth is among the finches living on the island.

The distribution of a variable in the whole population is called its **probability distribution**. There are several theoretical probability distributions that can be used to approximate the frequency distributions occurring in real life. Which one you pick depends on what you want to approximate. We'll learn about a few theoretical probability distributions as we proceed through the book.

For a continuous variable like beak depth, the distribution in the population is often approximated by a theoretical probability distribution known as the **normal distribution**. The normal distribution drawn in Figure 1.4-2, for example, approximates the probability distribution of beak depths in the finch population from which the sample of 100 birds was drawn.

---

[4]   Beak depth is the height of the beak near its base.

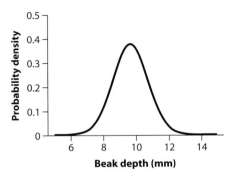

Figure 1.4-2 A normal distribution. This probability distribution is often used to approximate the distribution of a variable in the population from which a sample has been drawn.

The normal distribution is the familiar "bell curve." It is the most important probability distribution in all of statistics. You'll learn a lot more about it in Chapter 10. Most of the methods presented in this book for analyzing data depend on the normal distribution in some way.

1.5 Types of studies

Data in biology are obtained from either an experimental study or an observational study. In an **experimental study**, the researcher assigns different treatment groups or values of an explanatory variable randomly to the individual units of study. A classic example is the clinical trial, where different treatments are assigned randomly to patients in order to compare responses. In an **observational study**, on the other hand, nature assigns treatment groups or values of an explanatory variable to individuals. The researcher has no control over which units fall into which groups.

Studies of the health consequences of cigarette smoking in humans are all observational studies, because it is ethically impossible to assign human beings to smoking and nonsmoking treatments to assess the effects of smoking. The individuals in the sample have made the decision themselves about whether to smoke. The only experimental studies of the health consequences of smoking have been carried out on non-human subjects, such as mice, where researchers can assign individuals randomly to smoking and nonsmoking groups.

> A study is *experimental* if the researcher assigns treatments randomly to individuals, whereas a study is *observational* if the assignment of treatments is *not* made by the researcher.

The distinction between experimental studies and observational studies is that experimental studies can determine cause-and-effect relationships between variables, whereas observational studies can only point to associations. An association between smoking and lung cancer might be due to the effects of smoking *per se*, or perhaps to an underlying predisposition to lung cancer in those individuals prone to smoking. It is difficult to distinguish these alternatives with observational studies alone. For this reason, experimental studies of the health hazards of smoking in non-human animals have helped make the case that cigarette smoking is dangerous to human health. But experimental studies are not always possible, even on animals. Smoking in humans, for example, is also associated with a higher suicide rate (Hemmingsson and Kriebel 2003). Is this association caused by the effects of smoking, or is it caused by the effects of some other variable?

Just because a study was carried out in the laboratory does not mean that the study is an experimental study in the sense described here. A complex laboratory apparatus and careful conditions may be necessary to obtain measurements of interest, but such a study is still observational if the assignment of treatments or groups is out of the control of the researcher.

1.6   Summary

▶ Statistics is a technology for measuring aspects of populations from samples and for quantifying the uncertainty of the measurements.

▶ Much of statistics is about estimation, which infers an unknown quantity of a population using sample data.

▶ Statistics also allows hypothesis testing, a method to determine how well a null hypothesis about a population characteristic, established for the purposes of argument, fits the sample data.

▶ Sampling error is the chance difference between an estimate describing a sample and that parameter of the whole population. Bias is a systematic discrepancy between an estimate and the population quantity.

▶ The goals of sampling are to increase the accuracy and precision of estimates and to ensure that it is possible to quantify precision.

▶ In a random sample, every individual in a population has the same chance of being selected and the selection of individuals is independent.

▶ A sample of convenience is a collection of individuals easily available to a researcher, but it is not usually a random sample.

▶ Volunteer bias is a systematic discrepancy in a quantity between the pool of volunteers and the population.

▶ Variables are measurements that differ among individuals.

▶ Variables are either categorical or numerical. A categorical variable describes which category an individual belongs to, while a numerical variable is expressed as a number.

▶ In studies of association between variables, one variable is often designated as the response variable (i.e., the variable being predicted) and the other variable is designated as the explanatory variable (i.e., the variable being used to predict the response).

▶ In experimental studies, the researcher is able to assign subjects randomly to different treatments or groups. In observational studies, the assignment of individuals to treatments is not controlled by the researcher.

PRACTICE PROBLEMS

Answers to the practice problems are provided at the end of the book, starting on page 613.

1. Which of the following categorical variables are nominal? Which are ordinal?
   a. Letter grades
   b. First letter of students' last names
   c. Brand of antidepressant
   d. Body mass index (e.g., underweight, normal, overweight, or obese)

2. Refer to Example 1.2 on the study of cats falling from buildings in New York City.
   a. Is this an observational or an experimental study? Explain your answer.
   b. Of the variables measured in Figure 1.2-1, which is the explanatory variable?
   c. Which is the response variable?

3. Which of the following numerical variables are continuous? Which are discrete?
   a. Number of injuries sustained in a fall
   b. Fraction of birds in a large population infected with avian flu virus
   c. Number of crimes committed
   d. Logarithm of body mass

4. A researcher carried out a study on the peppered moth to examine how the proportion of melanic (black) individuals changed over time in England. She measured all specimens of the moth that were collected and deposited in museums and large private collections, and she also recorded the year in which each moth was collected. This enabled her to plot the proportion of melanic moths year by year.
   a. Can the specimens from a given year be considered a random sample from the moth population?
   b. If not a random sample, what type of sample is it?
   c. What type of error might be introduced by the sampling method when estimating the proportion of melanic moths?

5. What feature of an estimate—precision or accuracy—is most strongly affected when individuals differing in the variable of interest do not have an equal chance of being selected?

6. On February 25, 2004, CNN conducted a poll on their website, inviting visitors to the site to answer the following question: *Should the U.S. Constitution be amended to ban same-sex marriages?* Of the 402,485 votes cast, 168,448 answered "yes" and 234,037 answered "no."
   a. Is the variable measured in this survey categorical or numerical?
   b. Can the 402,485 votes be regarded as a random sample of people across the U.S.? Why or why not?
   c. The fraction of people answering "yes" was 168,448/402,485 = 0.42. What type

of bias might affect the accuracy of this estimate?

7. In a study of stress levels in U.S. army recruits stationed in Iraq, researchers obtained a complete list of the names of recruits in Iraq at the time of the study. They listed the recruits alphabetically and then numbered them consecutively. One hundred random numbers between 1 and the total number of recruits were then generated using a random-number generator on the computer. The 100 recruits whose numbers corresponded to those generated by the computer were interviewed for the study.
   a. What is the population of interest in this study?
   b. The 100 recruits interviewed were randomly sampled as described. Are the measurements still affected by sampling error? Explain.
   c. What are the main advantages of random sampling in this example?

8. An important quantity in conservation biology is the number of plant and animal species inhabiting a given area to be set aside for protection. To survey the community of small mammals inhabiting Kruger National Park, a large series of live traps were placed randomly throughout the park for one week in the main dry season of 2004. Traps were set each evening and checked the following morning. Individuals caught were identified, tagged (so that new captures could be distinguished from recaptures), and released. At the end of the survey, the total number of small mammal species in the park, M, was estimated by m, the total number of species captured in the survey.
   a. What is the population being estimated in the survey?
   b. Is the sample of individuals captured in the traps likely to be a random sample? Why or why not? In your answer, address the two criteria that define a sample as random.
   c. Is the number of species in the sample (m) likely to be an unbiased estimate of M, the total number of small mammal species in the park? Why or why not?

9. Survival of sparrows over their first year of life was measured on a large university campus. The nests of 20 breeding pairs were located by careful search, representing less than 1% of all the breeding pairs on campus. Chicks in each of these nests were given a unique set of color bands on their legs so that they could be identified individually once they had fledged from the nests. A total of 60 chicks were banded, of which only four survived to the following year. The researchers estimated the survival rate in the population as 4/60. Do the 60 chicks constitute a random sample? What consequences might this have for the estimate of survival rate?

ASSIGNMENT PROBLEMS

10. Identify which of the following variables are discrete and which are continuous:
    a. Number of warts on a toad
    b. Survival time after poisoning
    c. Temperature of porridge
    d. Number of bread crumbs in ten meters of trail
    e. Length of wolves' canines

11. A study was carried out in women to determine whether the psychological consequences of having an abortion differ from those experienced by women who have lost their fetus by other causes at the same stage of pregnancy. Was this an observational study or an experimental study? Why?

12. Identify whether the following variables are numerical or categorical. If numerical, state whether the variable is discrete or continuous. If categorical, state whether the variable is nominal or ordinal.
    a. Number of sexual partners in a year
    b. Petal area of rose flowers

c. Key on the musical scale
d. Heart beats per minute of a Tour de France cyclist, averaged over the duration of the race
e. Stage of fruit ripeness (e.g., underripe, ripe, or over ripe)
f. Angle of flower orientation relative to position of the sun
g. Tree species
h. Year of birth
i. Gender
j. Birth weight

13. A researcher dissected the retinas of 20 randomly sampled fish belonging to each of two subspecies of guppy in Venezuela. Using a sophisticated laboratory apparatus, the researcher was able to compare the two groups of fish in the wavelengths of visible light to which the cones of the retina were most sensitive.

a. Is this an experimental study or an observational study? Why?
b. What are the two variables being associated in this study?
c. Which is the explanatory variable and which is the response variable?

14. A random sample of 500 households was identified in a major North American city using the municipal voter registration list. Five hundred questionnaires went out, directed at one adult in each household, which asked a series of questions about attitudes regarding the municipal recycling program. Eighty of the 500 surveys were filled out and returned to the researchers.

a. Can the 80 households that returned questionnaires be regarded as a random sample of households? Explain?
b. What type of bias might affect the survey outcome?

15. Current telephone polls carried out to estimate voter or consumer preferences do not make calls to cell phones. The reason? Persons carrying cell phones might be driving at the time of the call (they might also have to pay for the call).

a. How might the strategy of leaving out cell phones affect the goal of obtaining a random sample of voters or consumers?
b. Which criterion of random sampling is most likely to be violated by the problems you identified in part (a): equal chance of being selected, or the independence of the selection of individuals?

16. The average age of piñon pine trees in the coast ranges of California was investigated by placing a 10-hectare plot randomly on a distribution map of the species in California using a computer. Researchers then recorded the location of the random plot, found it in the field, and flagged it using compass and measuring tape. They then proceeded to measure the age of every piñon pine tree within the 10-hectare plot. The average age within the plot was used to estimate the average age of the whole California population.

a. What is the population of interest in this study?
b. Were the trees sampled randomly from this population? Why or why not?

17. In a study of heart rate in ocean-diving birds, researchers harnessed 10 randomly sampled wild-caught cormorants to a laboratory contraption that monitored vital signs. (The bird portrayed at the opening of the chapter is a double-crested cormorant.) Each cormorant was subjected to six artificial "dives" over the following week (one dive per day). A dive consisted of rapidly immersing the head of the bird in water by tipping the harness. In this way, a sample of 60 measurements of heart rate in diving birds was obtained. Do these 60 measurements represent a random sample? Why or why not?

Biology and the history of statistics

The formal study of probability began in the mid-17th century when Blaise Pascal and Pierre de Fermat started to describe the mathematical rules for determining the best gambling strategies. Gambling and insurance continued to motivate the development of probability for the next couple of centuries.

The application of probability to data did not happen until much later. The importance of variation in the natural world, and by extension to samples from that world, was made obvious by Charles Darwin's theory of natural selection. Darwin highlighted the importance of variation in biological systems as the raw material of evolution. Early followers of Darwin saw the need for quantitative descriptions of variation and the need to incorporate the effects of sampling error in biology. This led to the development of modern statistics. In many ways, therefore, modern statistics was an offshoot of evolutionary biology.

One of the first pioneers in statistical data analysis was Francis Galton, who began to apply probability thinking to the study of all sorts of data. Galton was a real polymath, thinking about nearly everything and collecting and analyzing data at every chance. He said, "Whenever you can, count."[1] He invented the study of fingerprints, he tested whether prayer increased the life span of preachers compared with others of the middle class, and quantified the heritable nature of many important traits. He once

[1] Quoted in Newman, J. R. *The World of Mathematics*; Simon & Schuster: New York, 1956.

recorded his idea of the attractiveness of women seen from the window of a train headed from London to Glasgow, finding that "attractiveness" (at least according to Galton) declined as a function of distance from London. Galton is best known, though, for his twin interests in evolution (he was the first cousin of Darwin) and data analysis. He invented the idea of regression, which we will learn more about in Chapter 17. Galton was also responsible for establishing a lab that brought more researchers into the study of both statistics and biology, including Karl Pearson and Ronald Fisher.

Biological questions motivated the development of most of the basic statistical tools, and biologists helped to develop those tools.

Karl Pearson, like Galton, was interested in many spheres of knowledge. Pearson was motivated by biological data—in particular, by heredity and evolution. Pearson was responsible for our most often-used measure of the correlation between numerical variables. In fact, the correlation coefficient that we will learn about in Chapter 16 is often referred to as Pearson's correlation coefficient. He also made many contributions to the study of regression analysis and invented the χ^2 contingency test (Chapter 9). He also invented the term *standard deviation* (Chapter 3).

Sir Ronald Fisher

Last, but definitely not least, Ronald Fisher was one of the great geniuses of the 20th century. Fisher is well known in evolutionary biology as one of the three founders of the field of theoretical population genetics. Among his many contributions are the demonstration that Mendelian inheritance is compatible with the continuous variation we see in many traits, the accepted theory for why most animals conceive equal numbers of male and female offspring, and a great deal of the mathematical machinery we use to describe the process of evolution. But his contributions did not end there. He is probably also the most important figure in the history of statistics. He developed the analysis of variance (Chapter 15), likelihood (Chapter 20), the *P*-value (Chapter 6), randomized experiments (Chapter 14), multiple regression, and many other tools used in data analysis. His mathematical knowledge was made practical by a life-long association with biologists, particularly agricultural scientists at the Rothamsted Experimental Station in England. Fisher solved problems associated with the analysis of real data as he encountered them, and he generalized them for application to many other related questions. Moreover, Fisher developed experimental designs that would give more information than might otherwise be possible.

What this short hagiography is intended to demonstrate is that the early history of statistics is tightly bound up with biology. Biological questions motivated the development of most of the basic statistical tools, and biologists helped to develop those tools. Biology and statistics naturally go hand in hand.

2

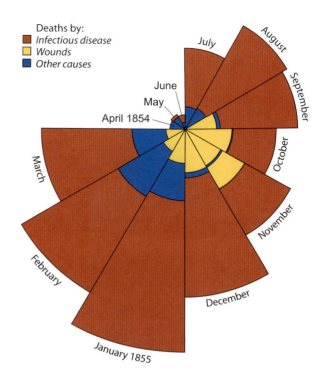

July

August

September

June

May

April 1854

October

March

November

February

December

January 1855

Displaying data

The human eye is a natural pattern detector, adept at spotting trends and exceptions in visual displays. For this reason, biologists spend hours creating and examining visual summaries of their data—graphs and, to a lesser extent, tables—in a way that can be readily examined and interpreted. Graphs enable visual comparisons of measurements between groups and expose relationships between different variables. They are also the principal means of communicating results to a wider audience.

Florence Nightingale (1858) was one of the first to put graphs to good use. In her famous wedge diagrams, such as the one shown in the figure above, she visualized the causes of death of British troops during the Crimean War. The number of cases is indicated by the area of a wedge, and the cause of death by color. The diagrams showed convincingly that disease was the main cause of soldier deaths during the wars, not wounds or other causes. With these graphs, she successfully campaigned for military and public-health measures that saved many lives.

Good graphs are a prerequisite for good data analysis, revealing general patterns in the data that bare numbers cannot. Therefore, the first step in any data analysis or statistical procedure is to graph the data and look at it. We'll follow this process throughout the book. Humans are a visual species, with brains evolved to process visual information. Take advantage of millions of years of evolution, and look at graphical representations of your data first.

In this chapter, we give a tour of basic methods for constructing graphs and tables and when to use them. We also review the types of data, whether numerical or categorical, and the basic goals of visual analysis: to display frequencies, to illustrate associations between variables, and to show differences between groups. We provide general guidance for common situations without attempting to cover every type of display. More specialized displays will appear later in the book, where they are most relevant.

2.1 Displaying frequency distributions

Recall from Chapter 1 that the *frequency* of occurrence of a specific measurement in a sample is the number of observations having that particular measurement. Moreover, the *frequency distribution* of a variable shows the frequencies of all its values in the data.

Relative frequency is the proportion of observations having a given measurement, calculated as the frequency divided by the total number of observations. The **relative frequency distribution** shows the proportion of occurrences of each value in the data set.

> The *relative frequency distribution* describes the fraction of occurrences of each value of a variable.

There are many methods for displaying frequency distributions, depending on whether the variable is categorical or numerical.

Frequency tables and bar graphs for categorical data

Let's start with displays for categorical variables. A **frequency table** is a text display of the number of occurrences of each category in the data set. A **bar graph** uses the height of rectangular bars to visualize the frequency (or relative frequency) of occurrence of each category.

> A *bar graph* uses the height of rectangular bars to display the frequency distribution (or relative frequency distribution) of a categorical variable.

Example 2.1A illustrates both kinds of displays.

| Example 2.1A | **Causes of teenage deaths** |

Teenagers die from many causes, but some causes are more lethal than others. Table 2.1-1 lists causes of death of American teenagers between the ages of 15 and 19 in 1999, and the number of teenagers (frequency) dying from each cause (Anderson 2001). Only the top 10 causes are listed individually.

Table 2.1-1 is a frequency table that shows the frequency distribution of the causes of teenage deaths (a nominal variable). Here, alternative values of the variable "cause of death" are listed in a single column, and frequencies of occurrence are listed next to them in a second column. The categories have no intrinsic order, but comparing the rel-

Table 2.1-1   Frequency table showing the 10 most common causes of death in Americans between 15 and 19 years of age in 1999.

Cause of death	Frequency
Accidents	6,688
Homicide	2,093
Suicide	1,615
Malignant tumor	745
Heart disease	463
Congenital abnormalities	222
Chronic respiratory disease	107
Influenza and pneumonia	73
Cerebrovascular diseases	67
Other tumor	52
All other causes	1,653
Total	13,778

ative importance of each cause is made easier by arranging the categories in order of their importance, from the most frequent at the top to the least frequent at the bottom.

This display shows clearly that accidents were the main cause of death to American teenagers, occurring at a frequency three times that of homicide, the next most common cause. Together, homicide and suicide (the third most common cause of death) exceeded any single agent of disease (such as malignant tumor, heart disease, or congenital abnormalities). In all, accidental and violent ends dominated other categories.

These patterns stand out even more vividly in the bar graph shown in Figure 2.1-1, in which frequency is depicted by the height of rectangular bars. Exact frequencies can no longer be discerned, but, to compensate, we gain a clear picture of how steeply the numbers drop in the less frequent categories. Not everyone dies in the same way, but some ways are much more common than others.

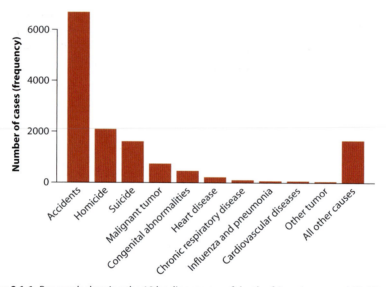

Figure 2.1-1 Bar graph showing the 10 leading causes of death of Americans aged 15–19 years in 1999. Total number of cases: $n = 13{,}778$. The frequencies are taken from Table 2.1-1.

Construction rules for bar graphs

The top edge of each bar conveys all the information about frequency, but the eye responds mainly to the area of the bars. For this reason, it doesn't much matter whether frequency or relative frequency is indicated along the y-axis of the graph. It is important, though, that the baseline is at zero—otherwise, the area and the height of the bars are misleading.

To be effective, a bar graph should obey some standard rules. For an ordinal categorical variable, such as snakebite severity score, the values along the horizontal axis should be in the natural order (e.g., minimally severe, moderately severe, and very severe). When the variable is nominal, as in Figure 2.1-1, ordering categories by frequency of occurrence aids in the visual presentation of the information. The bars should be of equal width, and each bar must rise from a baseline of zero, with the result that bar area and height are proportional to frequency. A base frequency other than zero, or unequal bar width, would give the eye a distorted view of the relative importance of each category. The bars should stand apart, not be fused together, and it is a good idea to provide the total number of observations (*n*) in the figure legend.

Frequency tables and histograms for numerical data

A frequency distribution for a numerical variable can be displayed either in a frequency table or in a **histogram**. A histogram uses the area of rectangular bars to display frequency. The data values are split into consecutive intervals, or "bins," usually of equal width, and the frequency of observations falling into each bin is presented.

> A *histogram* uses the area of rectangular bars to display the frequency distribution (or relative frequency distribution) of a numerical variable.

We discuss how histograms are made in greater detail using the data in Example 2.1B.

Example 2.1B   ### Abundance of desert bird species

How many species are common in nature and how many are rare? One way to address this question is to construct a frequency distribution of species abundance. The data in Table 2.1-2 are from a survey of the breeding birds of Organ Pipe Cactus National Monument in southern Arizona, USA. The measurements were extracted from the North American Breeding Bird Survey, a continent-wide data set of estimated bird numbers (Sauer et al. 2003).

We treated each species in the survey as a sampling unit and took the abundance of each species as its measurement. The range of abundance values was divided into 13 intervals of equal width (0–50, 50–100, and so on), and the number of species falling

Table 2.1-2   The abundance of each species of bird encountered during four surveys in Organ Pipe Cactus National Monument.

Species	Abundance	Species	Abundance
Greater roadrunner	1	Turkey vulture	23
Black-chinned hummingbird	1	Violet-green swallow	23
Western kingbird	1	Lesser nighthawk	25
Great-tailed grackle	1	Scott's oriole	28
Bronzed cowbird	1	Purple martin	33
Great horned owl	2	Black-throated sparrow	33
Costa's hummingbird	2	Brown-headed cowbird	59
Canyon wren	2	Black vulture	64
Canyon towhee	2	Lucy's warbler	67
Harris's hawk	3	Gilded flicker	77
Loggerhead shrike	3	Brown-crested flycatcher	128
Hooded oriole	4	Mourning dove	135
Northern mockingbird	5	Gambel's quail	148
American kestrel	7	Black-tailed gnatcatcher	152
Rock dove	7	Ash-throated flycatcher	173
Bell's vireo	10	Curve-billed thrasher	173
Common raven	12	Cactus wren	230
Northern cardinal	13	Verdin	282
House sparrow	14	House finch	297
Ladder-backed woodpecker	15	Gila woodpecker	300
Red-tailed hawk	16	White-winged dove	625
Phainopepla	18		

into each abundance interval was counted and presented in a frequency table to help see patterns (Table 2.1-3).

The shape of the frequency distribution is more readily obvious in a histogram of these same data (Figure 2.1-2). Here, frequency (number of species) in each abundance interval is indicated by bar area.

The frequency table and histogram of the bird abundance data reveal that the majority of bird species have low abundance. Frequency falls steeply with increasing abundance.[1] The white-winged dove (pictured in Example 2.1B) is exceptionally abundant at Organ Pipe Cactus National Monument, accounting for a large fraction of all individual birds encountered in the survey.

---

[1]   This pattern is a remarkably general one in nature, common to many types of organisms. Only a few species are common, whereas most species are rare.

Table 2.1-3 Frequency distribution of bird species abundance at Organ Pipe Cactus National Monument.[a]

Abundance	Frequency (Number of species)
0–50	28
50–100	4
100–150	3
150–200	3
200–250	1
250–300	2
300–350	1
350–400	0
400–450	0
450–500	0
500–550	0
550–600	0
600–650	1
Total	43

[a] *Source:* Data are from Table 2.1-2.

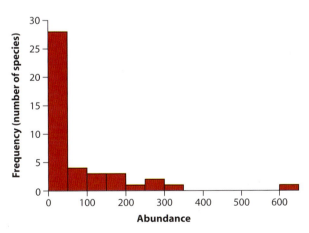

Figure 2.1-2 Histogram illustrating the frequency distribution of bird species abundance at Organ Pipe Cactus National Monument. Total number of bird species: $n = 43$.

Describing the shape of a histogram

The histogram reveals the shape of a frequency distribution. Some of the most common shapes are displayed in Figure 2.1-3. Any interval of the frequency distribution that is noticeably more frequent than surrounding intervals is called a peak. The **mode** is the interval corresponding to the highest peak. For example, a bell-shaped frequency distribution has a single peak (the mode) in the center of the range of observations. A frequency distribution having two distinct peaks is said to be **bimodal**.

> The *mode* is the interval corresponding to the highest peak in the frequency distribution.

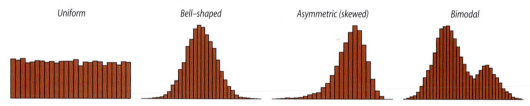

Uniform   Bell–shaped   Asymmetric (skewed)   Bimodal

Figure 2.1-3 Some possible shapes of frequency distributions.

A frequency distribution is **symmetric** if the pattern of frequencies on the left half of the histogram is the mirror image of the pattern on the right half. The uniform distribution and the bell-shaped distribution in Figure 2.1-3 are symmetric. If a frequency distribution is not symmetric, we say that it is **skewed**. The asymmetric and bimodal distributions in Figure 2.1-3 are skewed. The abundance data for desert bird species are skewed right (Figure 2.1-2), having a long tail extending to the right. The asymmetric distribution in Figure 2.1-3 is skewed left.[2]

> *Skew* refers to asymmetry in the shape of a frequency distribution for a numerical variable.

---

[2]   The nomenclature of skew seems backwards to many people. Focus on the sharp tail of the distribution extending to the left in the third distribution in Figure 2.1-3. We say it is skewed left because it seems to have a "skewer" sticking out to the left, like the skewer through a shish kebab.

Extreme data points lying well away from the rest of the data are called **outliers**. The histogram of desert bird abundance (Figure 2.1-2) includes one extreme observation (the white-winged dove) that falls well outside of the range of abundance of other bird species. The white-winged dove, therefore, is an outlier. Outliers are common in biological data. They can result from mistakes in recording the data or, as in the case of the white-winged dove, they may represent real features of nature. Outliers should always be investigated. They should be removed from the data set only if they can be shown to be errors.

> An *outlier* is an observation well outside the range of values of other observations in the data set.

Interval width can affect histogram shape

How wide should the intervals be when drawing a histogram? The choice of interval width must be made carefully because it can affect the conclusions drawn from the graph, as we show with Example 2.1C.

Example 2.1 C   **How many peaks?**

Sockeye salmon (*Oncorhynchus nerka*) spend years in the Pacific Ocean before returning to their natal rivers to reproduce and then die. Some individuals, however, spend more years at sea than other individuals before returning, which means that, in any given year, the salmon breeding in a river might include different age groups. As a result, reproducing adults vary in body  size. Figure 2.1-4 shows three different histograms that depict the body mass of 228 female sockeye salmon sampled from Pick Creek, Alaska, in 1996 (Hendry et al. 1999).

The leftmost histogram of Figure 2.1-4 was drawn using a narrow interval width. The result is a somewhat bumpy frequency distribution that suggests the existence of two or even more peaks. The rightmost histogram uses a wide interval. The result is a smoother frequency distribution that masks the second of the two dominant peaks. The middle histogram uses an intermediate interval that shows two distinct body-size groups. The fluctuations from interval to interval within size groups are less noticeable.

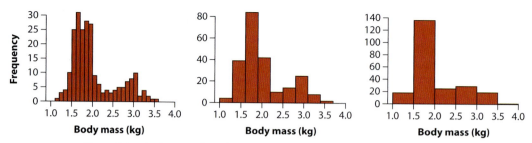

Figure 2.1-4 Body mass of 228 female sockeye salmon sampled from Pick Creek in Alaska (Hendry et al. 1999). The same data are shown in each case, but the interval widths are different: 0.1 kg (left), 0.3 kg (middle), and 0.5 kg (right).

To choose the ideal interval width we must decide whether the two distinct body-size groups are "real," in which case the histogram should show both, or whether a bimodal shape is an artifact produced by too few observations.[3]

Construction rules for histograms

Histograms should follow a few simple design rules. Each bar must rise from a baseline of zero, so that the area of each bar is proportional to frequency. Unlike bar graphs, there is no separation between adjacent histogram bars. This reinforces the perception of a numerical scale, with bars grading from one into the next. In this text we follow convention by placing an observation whose value is exactly at the boundary of two successive intervals into the higher interval. For example, the Gila woodpecker, with a total of 300 individuals observed (Table 2.1-2), is recorded in the interval 300–350, not in the interval 250–300.

There are no strict rules about the number of intervals that should be used in frequency tables and histograms. Some computer programs use Sturges's rule of thumb, in which the number of intervals is $1 + \ln(n)/\ln(2)$, where n is the number of observations and ln is the natural logarithm. The resulting number is then rounded up to the higher integer (Venables and Ripley 2002). Many regard this rule as overly conservative, and in this book we tend to use a few more intervals than Sturges. The number of intervals should be chosen to best show patterns and exceptions in the data, and this requires good judgment rather than strict rules. Computers allow you to try several alternatives to help you determine the best option.

When breaking the data into intervals for the histogram, use readable numbers for breakpoints —for example, break at 0.5 rather than 0.483. Finally, it is a good idea to provide the total number of individuals in the accompanying legend.

---

[3]   The two distinct body-size classes in this salmon population correspond to two age groups.

2.2   Quantiles of a frequency distribution

All information about the shape of the frequency distribution of a numerical variable is contained in its percentiles and the closely related quantiles.

Percentiles and quantiles

The Xth **percentile** of a distribution is the value below which $X\%$ of the individuals lie; this is also the $X/100$ **quantile.** For example, the measurement that splits a frequency distribution into equal halves—that is, where half the measurements are smaller and half the measurements are larger—is the 50th percentile.[4] The 50th percentile is also the 0.50 quantile, the value below which 0.50 of the individual values lie. The 10th percentile (i.e., the 0.10 quantile), on the other hand, is the measurement where 10% of the observations are less than or equal to it, and the other 90% exceed it.

> The *percentile* of a measurement specifies the percentage of observations less than or equal to it; the remaining observations exceed it.

Cumulative frequency distribution

The **cumulative frequency distribution** is a graph of all the quantiles of a numerical variable. To draw Figure 2.2-1, which shows the cumulative frequency distribution of the abundances of desert bird species from Example 2.1-2, all of the measurements were first sorted from smallest to largest. Next, the fraction of observations less than or equal to each data value were calculated. This fraction, which is called the cumulative relative frequency, is plotted by the height of the curve above the corresponding data value. Finally, these points were connected with straight lines to form an ascending curve. The result is an irregular sequence of "steps" from the smallest data value to the largest data value. Each step is flat, but the curve jumps up by $1/n$ at every observed measurement, where n is the number of observations, to a maximum of 1. There may be multiple jumps at one measurement if multiple data points have the same measurement. The curve provides lots of details because every observation is illustrated.[5]

---

[4]   The 50th percentile is known as the median, a quantity explored more fully in Chapter 3.

[5]   Some prefer to drop the steps and instead draw a series of straight line segments connecting the points. An example is shown later in Figure 2.4-2.

Cumulative relative frequency at a given measurement is the fraction of observations less than or equal to that measurement.

The curve in Figure 2.2-1 confirms that most bird species at Organ Pipe Cactus National Monument had low abundance. Half the species (corresponding to a cumulative relative frequency of 0.50) had abundances below 18, which is a small portion of the full range of abundance measurements observed. Three-fourths of the species (those whose cumulative relative frequency was 0.75 and below) had abundance below 128, which is still less than one-fourth of the range of abundance measurements.

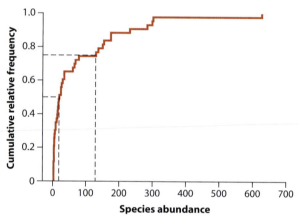

Figure 2.2-1 The cumulative frequency distribution of the abundance of bird species at Organ Pipe Cactus National Monument (solid curve). Horizontal dotted lines indicate the cumulative relative frequencies 0.50 (lower) and 0.75 (upper); vertical lines indicate corresponding quantiles of bird species abundance (18 and 128). The data are $n = 43$ bird species, obtained from Table 2.1-2.

Because the eye responds so readily to area, the histogram is the better choice for displaying the shape of a single frequency distribution. Histograms are also a better method for detecting outliers, which can be overlooked in a cumulative distribution function. However, with practice, the cumulative frequency distribution can be very useful, especially for comparing frequency distributions between multiple groups.

2.3   Associations between categorical variables

The full power of data visualization is realized when it is used to investigate associations between variables and differences between groups. The methods that we use to graphically display relationships depend on the types of variables involved—that is, whether the variables are categorical or numerical. The present section shows a few ways to exhibit relationships between two categorical variables, based upon Example 2.3.

| **Example 2.3** | **Reproductive effort and avian malaria** |

Is reproduction hazardous? If not, then it is difficult to explain why adults in many organisms hold back on the number of offspring they appear capable of raising. Oppliger et al. (1996) investigated the impact of reproductive effort on the susceptibility to malaria[6] in wild great tits (*Parus major*) breeding in nest boxes. They divided 65 nesting females into two treatment groups. In one group of 30 females, each bird had two eggs stolen from her nest, causing the female to lay an additional egg. The extra effort required might increase stress on these females. The remaining 35 females were left alone, establishing the control group. A blood sample was taken from each female 14 days after her eggs hatched to test for infection by avian malaria, a common parasite.

Contingency tables

The relationship between experimental treatment and the incidence of malaria is displayed in Table 2.3-1. Table 2.3-1 is a **contingency table**, a frequency table for two (or more) categorical variables. It is called a contingency table because it shows how

Table 2.3-1 Contingency table showing the incidence of malaria in female great tits in relation to experimental treatment.

	Experimental treatment group		
	Control group	**Egg-removal group**	**Row total**
Malaria	7	15	22
No malaria	28	15	43
Column total	35	30	65

---

6   Malaria is a common cause of death in humans, but avian forms of the disease are even more prevalent in many bird species.

values for one of the variables (the incidence of malaria, in this case) is contingent upon the value of a second variable (the experimental treatment group).

Each subject (bird) is represented exactly once in the four "cells" of Table 2.3-1. A cell is one combination of categories of the row and column variables in the table. The explanatory variable (i.e., experimental treatment) is displayed in the columns, whereas the response variable, the variable being predicted (i.e., incidence of malaria), is displayed in the rows. The frequency of subjects in each treatment group is given in the column totals, and the frequency of subjects with and without malaria is given in the row totals.

According to Table 2.3-1, malaria was detected in 15 of the 30 birds subjected to egg removal, but in only seven of the 35 control birds. The result suggests that the stress of egg removal, or the effort involved in producing one extra egg, increases female susceptibility to avian malaria.

> A *contingency table* gives the frequency of occurrence of all combinations of two (or more) categorical variables.

Table 2.3-1 is an example of a 2 × 2 ("two-by-two") contingency table, because it displays the frequency of occurrence of all combinations of two variables, each having exactly two categories. Larger tables are possible if the variables have more than two categories.

Grouped bar graph

The **grouped bar graph** uses heights of rectangles to graph the frequency of occurrence of all combinations of two (or more) categorical variables. Figure 2.3-1 shows the grouped bar graph for the avian malaria experiments. Grouped bar graphs follow all of the conventions of bar graphs, except that different groups (e.g., malaria and no malaria) are indicated by different colors or shades. The space between bars from dif-

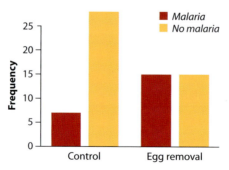

Figure 2.3-1 Grouped bar graph for reproductive effort and avian malaria in great tits. The data are from Table 2.3-1, where *n* = 65 birds.

ferent groups of the explanatory variable (e.g., control and egg removal) should be wider than the space between bars between groups of the response variable (e.g., malaria and no malaria). In the control group, most birds had no malaria (the yellow bar is much taller than the red bar), whereas in the experimental group, the frequency of subjects with and without malaria was equal (Figure 2.3-1).

> A *grouped bar graph* uses the height of rectangular bars to display the frequency distributions (or relative frequency distributions) of two or more categorical variables.

Mosaic plot

A **mosaic plot** is similar to a grouped bar plot except that the bars within groups are stacked on top of one another rather than placed side by side. The mosaic plot for the great tit malaria data is shown in Figure 2.3-2. The bar area in a mosaic plot represents the relative frequency (i.e, the proportion) within each group. In Figure 2.3-2, for example, most of the individuals in the control group did *not* get malaria, so the area of the red bar (no malaria) is greater than the area of the yellow bar (malaria). The width of each vertical stack, moreover, is proportional to the number of observations in that group. In Figure 2.3-2, the wider stack for the control group reflects the greater total number of individuals in this treatment (35) compared with the number in the egg-removal treatment (30).

What this all means, then, is that the area of each box is proportional to the relative frequency of that combination of variables in the whole data set. Thus, the mosaic plot for the great tit malaria data shows that the incidence of disease was higher in the treatment group (egg removal) than in the control group. Finally, the order of groups along the horizontal axis is predetermined in the case of ordinal data, but the order is arbitrary in the case of nominal data, such as that in Figure 2.3-2.

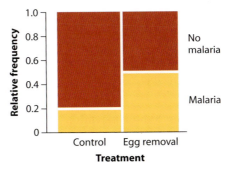

Figure 2.3-2 Mosaic plot for reproductive effort and avian malaria in great tits. Yellow indicates birds with malaria, whereas red indicates birds free of malaria. The data are from Table 2.3-1, where $n = 65$ birds.

A mosaic plot provides only relative frequencies, not the absolute frequency of occurrence in each combination of variables. This might be considered a drawback, but keep in mind that the most important goal of graphs is to depict the *pattern* in the data rather than exact figures.

> The *mosaic plot* uses the area of rectangles to display the relative frequency of occurrence of all combinations of two categorical variables.

Of the three methods for presenting the same data—the contingency table, the mosaic plot, and the grouped bar graph—which is best? The answer depends on the circumstances. The main benefit of the grouped bar graph over the contingency table is that bar height or area is easier to compare between groups, visually, than the numbers themselves. This advantage is reduced if the explanatory and response variables both have many categories, increasing the complexity of the graph. The mosaic plot, on the other hand, is simpler to read than the grouped bar graph.

Deciding which type of display is most effective for a given circumstance is best done by trying several methods and choosing among them on the basis of information, clarity, and simplicity.

2.4 Comparing numerical variables between groups

In Section 2.4, we address ways to display differences between groups when one variable is numerical. We illustrate the uses of histograms and cumulative frequency distributions for comparisons between groups.

Comparing histograms between groups

One useful way to compare multiple frequency distributions is to plot a histogram separately for each group. It is easier to compare the groups if the histograms are stacked vertically, making sure that the scale is the same on each *x*-axis. Example 2.4 illustrates the principle.

| Example 2.4 | **Blood responses to high elevation** |

The amount of oxygen obtained in each breath at high altitude can be as low as one-third that obtained at sea level. Studies have begun to examine whether indigenous people living at high elevations have physiological attributes that compensate for the reduced availability of oxygen. A reasonable expectation is that they should have more hemoglobin, the molecule that binds and transports oxygen in the blood. To test this, researchers sampled blood from males in a sea-level population from the USA and from three high-altitude human populations: the high Andes, high-elevation Ethiopia, and Tibet (Beall et al. 2002). Results are shown as histograms in Figure 2.4-1.

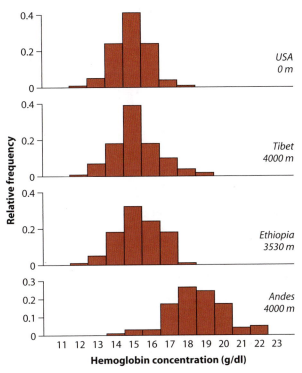

Figure 2.4-1 Grouped histogram showing the hemoglobin concentration in males living at high altitude in three different parts of the world. A fourth population at sea level (USA) is included as a control.

The surprising finding is that only men from the high Andes had elevated hemoglobin concentrations. Men from high-elevation Ethiopia and Tibet were not noticeably different in hemoglobin concentration from the sea-level group.[7] These patterns are readily apparent when the histograms for all groups are on the same scale and are stacked vertically (as in Figure 2.4-1).

_____

7   Mysteriously, oxygen levels in the blood of highland Ethiopian men are just as high as those in men living at sea level (data not shown), despite their concentrations of hemoglobin. The physiological mechanism behind this feat is not yet known.

Comparing cumulative frequencies

We can also look at the association between a numerical variable and a categorical variable by comparing cumulative frequency distributions. This approach illustrates the distributions for several groups in a single compact display. Figure 2.4-2 shows cumulative frequency distributions of the hemoglobin concentration in the four groups of men from Example 2.4.

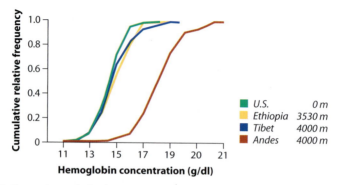

Figure 2.4-2 Grouped cumulative frequency distributions of hemoglobin concentration in males living at high altitude in Ethiopia, Tibet, and the Andes. A fourth, sea-level population (USA) is included as a control. Redrawn from Beall et al. (2002).

Note how the four cumulative frequency distributions in Figure 2.4-2 are similar in shape. The distribution for Andean men is merely shifted to the right compared with other groups. There is only slight overlap in hemoglobin concentrations between the Andean group and the other three groups. The cumulative frequency graphs allow these features to be seen in a compact display.

2.5 Displaying relationships between a pair of numerical variables

In this section, we present three ways to display the relationship between two numerical variables. These are scatter plots, line graphs, and maps.

Scatter plot

A **scatter plot** shows the pattern of association between two numerical variables. Each observation is represented by a point on a graph having two axes. The position along the horizontal axis (the *x*-axis) indicates the measurement of the explanatory variable. The position along the vertical axis (the *y*-axis) indicates the measurement of the response variable. The pattern in the resulting cloud of points indicates whether an association between the two variables is positive (in which case the points tend to run from the lower left to the upper right of the graph), negative (the points run from the upper left to the lower right), or absent (no discernible pattern). A scatter plot can also reveal whether the relationship between two variables follows a straight line or a more complicated curve. Example 2.5A explains how to construct a scatter plot.

| Example 2.5A | **Sins of the father** |

The bright colors and elaborate courtship displays of the males of many species posed a problem for Charles Darwin: how can such elaborate traits evolve? His answer was that they evolved because females are attracted to them when choosing a mate. But why would females choose those kinds of males? One possible answer 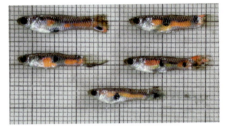 is that females choose fancy males so that their sons will be attractive as well.

A recent laboratory study examined how attractive traits in guppies are passed from father to son (Brooks 2000). The attractiveness of sons (a score representing the rate of visits by females to corralled males, relative to a standard) was compared with their fathers' ornamentation (a composite index of several aspects of male color and brightness). The father's ornamentation is the explanatory variable in the resulting scatter plot of these data (Figure 2.5-1).

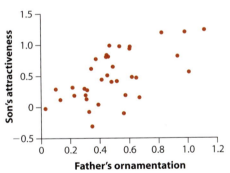

Figure 2.5-1 Scatter plot showing the relationship between the ornamentation of male guppies and the average attractiveness of their sons. Total number of families: $n = 36$.

Each dot in the scatter plot is a father–son pair. The father's ornamentation is the explanatory variable and the son's attractiveness is the response variable. The plot shows a *positive* association between these variables (note how the points tend to run from the lower left to the upper right of the graph). Thus, the sexiest sons come from the most gloriously ornamented fathers, whereas unadorned fathers produce less attractive sons.

> A *scatter plot* is a graphical display of two numerical variables in which each observation is represented as a point on a graph with two axes.

Line graph

A **line graph** is a powerful tool for displaying trends in time or some other ordered series. The graph is like a scatter plot, except that only one *y*-measurement is displayed for every *x*-observation. Adjacent points along the *x*-axis are connected by a line segment. Example 2.5B illustrates the line graph.

Example 2.5B | **Cyclic fluctuations in lynx numbers**

The Hudson's Bay Company controlled the fur trade for centuries across most of northern Canada. During this period, the company kept track of the total number of pelts of fur-bearing animals turned in at its far-flung outposts. Figure 2.5-2 shows a line graph that depicts the number of lynx pelts turned in to fur trading posts in Canada each year from 1752 to 1819 (Poland 1892, cited in Elton and Nicholson 1942).

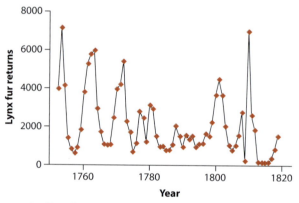

Figure 2.5-2 Line graph of lynx fur returns to the Hudson's Bay Company from 1752 to 1819.

The lines connecting the points in Figure 2.5-2 make the trends in lynx numbers over time more visible. For example, the graph shows that lynx numbers peaked roughly every 10 years. The steepness of the slope of the line segments reflects the speed of change in lynx numbers from one year to the next; notice, for example, that lynx numbers often changed dramatically between adjacent years. This is a famous time series that continues to be studied by ecologists trying to understand the causes of wild fluctuations in animal numbers.

> A *line graph* uses dots connected by line segments to display trends in a measurement over time or other ordered series.

Maps

A **map** is the spatial equivalent of the line graph, displaying a numerical response measurement at multiple locations on a surface. The explanatory variable is again an ordered series, such as points in space. One *y*-measurement is displayed for each point or region of the surface, as shown in Example 2.5C.

Example 2.5C	**The Antarctic ozone hole**

Since the first alarms were raised in the 1970s, the National Aeronautics and Space Administration (NASA) has been measuring atmospheric ozone concentrations around the globe. Measurements by orbiting satellites are taken using backscattered ultraviolet radiance. The data shown in Figure 2.5-3 are the ozone concentrations for October 6, 1987, which were recorded at many points on a fine grid covering the Southern Hemisphere. Points on the grid are colored to represent the ozone concentration above it. The view in Figure 2.5-3 is directly over the South Pole, and the outer limits of the view are about 15 degrees south of the equator (NASA 2007).

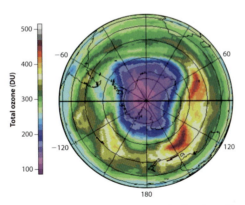

Figure 2.5-3 Map displaying ozone concentrations over the Southern Hemisphere, from the South Pole (center) outward to about 15 degrees south of the equator (outside edge). The measurements are the daily recordings for October 6, 1987, in Dobson Units (DU). Points on the surface of the Southern Hemisphere are colored to represent the ozone concentration above them. Modified from NASA (2007).

The map in Figure 2.5-3 contains an enormous amount of data, yet the pattern is easy to see. The span and depth of the ozone "hole" above the South Pole are clearly evident.

Maps can indicate measurements at points on any surface. The surface can be a spatial grid (such as in the ozone example) or political or geological boundaries on the earth's surface. They can even be used to indicate measurements on representations of any two- or three-dimensional object, such as the brain or the body.

2.6   Principles of effective display

The purposes of graphs and tables are really two-fold. First, displays are a vital tool for analyzing data. Second, displays are used to communicate patterns in data to a wider audience in the form of paper reports, slide shows, or web content. The two purposes are largely coincident because the most revealing displays will be the best for identifying patterns in the data and for communicating these patterns. Both purposes require displays that are clear, honest, and efficient.[8]

Principles of graphical display

The following principles will help to increase the effectiveness of your graphs:

▶ Show the data
▶ Represent magnitudes accurately
▶ Draw graphical elements clearly, minimizing clutter
▶ Make displays easy to interpret
▶ Clearly identify the axes

The benefits of "showing the data" can be seen in the scatter plot in Figure 2.6-1. The measurements were collected as part of an investigation of interactions between

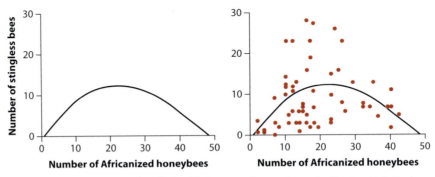

Figure 2.6-1 Curve depicting the relationship between the numbers of native tropical stingless bees and Africanized honey bees on flowering shrubs in French Guiana. The data have been erased in the left panel. Redrawn from Roubik (1978).

---

8   An enlightening discussion about the power of proper graphics can be found in Edward Tufte's 1983 book, *The Visual Display of Quantitative Information.*

the aggressive Africanized honey bee introduced to tropical South America and native stingless bees. A curve through the data suggests that numbers of the two types of bees are positively associated at low values but negatively associated at higher values. A graph presenting this curve alone (*left panel*) raises questions such as: How strong is the association? Is this the best curve? Are outliers affecting the pattern? Including the data points (*right panel*) allows the viewer to make up his or her own mind about the answers to these questions.

"Representing magnitudes accurately" sounds like an easy criterion to meet, but violations are still found in the literature. An example is the upper bar graph in Figure 2.6-2, which depicts government spending on education each year since 1998 in British Columbia. The area of each bar is not proportional to the magnitude of the value displayed. As a result, it exaggerates the differences. The figure falsely suggests that spending increased 20-fold over time, but the real increase is less than 20%. It is

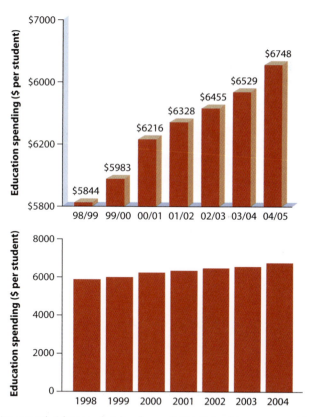

Figure 2.6-2 *Upper graph:* A bar graph, taken from a British Columbia government brochure, indicating education spending per student in different years. *Lower graph:* A revised presentation of the same data, in which the magnitude of the spending is proportional to the area of the bars. The revised graph more clearly portrays the relatively slow increase in the amount of money spent on education through these years. The upper graph is modified from British Columbia Ministry of Education (2004).

more honest to plot the bars[9] with a baseline of zero, as in the lower graph in Figure 2.6-2.

A zero baseline is essential in bar graphs and histograms because the eye instinctively reads bar area as proportional to relative magnitude. Other graphs, such as the line plot, do not always have this same purpose and so are not as constrained. The main purpose of a scatter plot is to depict association, not magnitude or frequency, and the choice of baseline should be made with this goal in mind. In general, use good judgment when deciding the appropriate baseline in a graph. If the goal is to display relative frequencies or magnitudes, then a zero baseline is important.

There's more to criticize about the upper graph of Figure 2.6-2. The bars have been drawn to create a three-dimensional effect, for instance, which is not only unnecessary but distorts the height and area of the bars, again misleading the eye. Such unnecessary and confusing embellishments are known as "chartjunk" (Tufte 1983). Nonessential graphical elements should be stripped away to allow patterns in the data to stand out clearly.

In order to "draw graphical elements clearly, minimizing clutter," try to avoid including too much information in a graph. Consider the map in Figure 2.6-3, for example, which displays two variables in blocks of continental Africa: the diversity of bird species and the number of native human languages. The goal of the study was

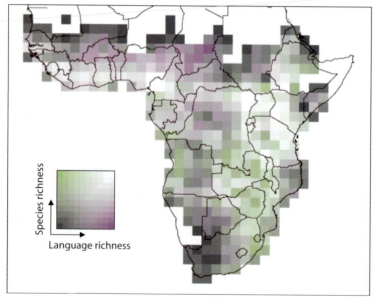

Figure 2.6-3 Map displaying the number of bird species and the number of distinct human languages present in each square of a grid of continental Africa. Reproduced from Moore et al. (2002).

---

[9]   A still better alternative would be to use a line graph rather than a bar plot, since the data are an ordered series of measurements.

to explore the possibility that biological diversity and human cultural diversity have similar underlying causes, such as landscape productivity.

Figure 2.6-3 presents a remarkable amount of spatial data on the two variables and their association, but the complexity of the map is high. We found ourselves returning repeatedly to the figure color guide for help in reading the graph. This complexity made it difficult to see the central result, that language diversity and bird species diversity are positively associated across Africa. Separate maps for species diversity and language diversity, while taking up more space, would have been easier to interpret and would make the result more apparent.

"Making displays (graphs) easy to interpret" and readable means not only "clearly identifying the axes," but also such mundane tasks as choosing appropriate typefaces and colors. Don't always accept the default output of statistical or spreadsheet programs. Label axes clearly using simple unadorned typefaces. Text should be legible even after its size is adjusted to fit the final document. Always provide the units of measurement. Use clearly distinguishable graphical symbols to represent different groups. Up to a fifth of your male audience is red–green color-blind. To make your graphs more visible to this group, make sure that your colors also differ in intensity, and apply redundant coding to distinguish groups—that is, use distinctive shapes or patterns in addition to colors.

A good graph is like a good paragraph. It conveys information clearly, concisely, and without distortion. A good graph requires careful editing; just as in writing, the first draft is rarely as good as the final product.

Follow similar principles in display tables

Tables have diverse purposes. One use is data storage—that is, a place where raw data are kept for reference purposes. Such data tables are often large and are usually inappropriate for communicating general findings. When published, data tables usually appear as appendices or online supplements to printed publications. Table 2.1-2, which lists all of the species of birds observed in Organ Pipe Cactus National Monument, is an example of a data table.

A second type of table has a similar purpose as a graph—that is, to display patterns and exceptions in data. Frequency tables, such as Figure 2.1-3, are examples of display tables. Producing clear, honest, and efficient display tables should follow many of the same principles discussed already for graphs. Numerical detail is less important than the effective communication of results in such tables. Display tables should be compact and present as few significant digits as are necessary to communicate the pattern. Arrange the presentation of numbers to facilitate pattern detection. For example, numbers listed vertically are easier to compare than numbers listed horizontally. Ehrenberg (1977) gives an excellent description of good tables.

When should a graph be used and when should a table? Graphs are the best way to show patterns and exceptions, but they provide little quantitative detail. Tables are best when it is also important to communicate quantitative aspects of the data.

2.7 **Summary**

▶ A frequency table is used to display a frequency distribution for categorical or numerical data.

▶ Bar graphs and histograms are graphical methods for displaying frequency distributions of categorical and numerical variables:

Type of data	Graphical method
Categorical data	Bar graph
Numerical data	Histogram Cumulative frequency distribution

▶ Contingency tables describe the association between two (or more) categorical variables by displaying frequencies of all combinations of categories.

▶ Graphical methods for displaying associations between variables, and differences between groups, include the following:

Types of data	Graphical method
Two categorical variables	Grouped bar graph Mosaic plot
One numerical variable and one categorical variable	Grouped histograms Cumulative frequency distributions Line plot (ordinal categories only)
Two numerical variables	Scatter plot Line plot Map

▶ Graphical displays must be clear, honest, and efficient. Strive to show the data, to represent magnitudes accurately, to minimize clutter, to maximize ease of interpretation, and to use clear graphical elements. Follow the same rules when constructing tables to reveal the true patterns in the data.

PRACTICE PROBLEMS

1. Estimate by eye the relative frequency of the shaded areas in each of the following histograms.

a.

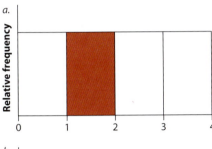

b.

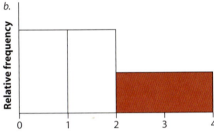

c.

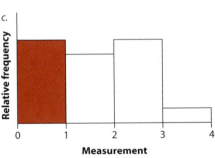

Measurement

2. Using a graphical method from this chapter, draw three frequency distributions: one that is symmetric, one that is skewed, and one that is bimodal.
   a. Identify the mode in each of your frequency distributions.
   b. Is your skewed distribution skewed left or skewed right?
   c. Is your bimodal distribution skewed or symmetric?

3. In the southern elephant seal, males defend harems that may contain hundreds of reproductively active females. Modig (1996) recorded the numbers of females in harems in a population on South Georgia Island. The histograms of the data (below, drawn from data in Modig 1996) are unusual because the rarer, larger harems have been divided into wider intervals. In the upper histogram, bar *height* indicates the relative frequency of harems in the interval. In the lower histogram, bar height is adjusted such that bar *area* indicates relative frequency. Which histogram is correct? Why?

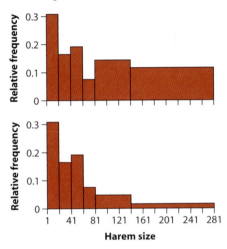

Harem size

4. Draw scatter plots for invented data that illustrate the following patterns:
   a. Two numerical variables that are positively associated.
   b. Two numerical variables that are negatively associated.
   c. Two numerical variables whose relationship is nonlinear.

5. A study by Miller et al. (2004) compared the survival of two kinds of Lake Superior rainbow trout fry (babies). Four thousand fry were from a government hatchery on the lake, whereas 4000 more fry came from wild trout. All 8000 fry were released into a stream flowing into the lake, where they remained for one year. After

one year, the researchers found 78 survivors. Of these, 27 were hatchery fish and 51 were wild. Display these results in the most appropriate table. Identify the type of table you used.

6. The following data are the occurrences in 2002 of the different taxa in the list of endangered and threatened species under the U.S. Endangered Species Act (U.S. Fish and Wildlife Service 2001). The taxa are listed in no particular order in the table.

Taxon	Number of species
Birds	92
Clams	70
Reptiles	36
Fish	115
Crustaceans	21
Mammals	74
Snails	32
Plants	745
Amphibians	22
Insects	44
Arachnids	12

a. Rewrite the table, but list the taxa in a more revealing order. Explain your reasons behind the ordering you choose.
b. What kind of table did you construct in part (a)?
c. Choosing the most appropriate graphical method, display the number of species in each taxon. What kind of graph did you choose? Why?
d. Should the baseline for the number of species in your graph in part (c) be 0 or 12, the smallest number in the data set for 2002? Why?

7. Can environmental factors influence the incidence of schizophrenia? A recent project measured the incidence of the disease among children born in a region of eastern China. One hundred ninety-two of 13,748 babies born in the midst of a severe famine in the region in 1960 later developed schizophrenia. This compared with 483 schizophrenics out of 59,088 births in 1956, before the famine, and 695 out of 83,536 births in 1965, after the famine (St. Claire et al. 2005).

a. What two variables are compared in this example?
b. Are the variables numerical or categorical? If numerical, are they continuous or discrete; if categorical, are they nominal or ordinal?
c. Effectively display the findings in a table. What kind of table did you use?
d. Calculate the relative frequency (proportion) of children born in each of the three years that later developed schizophrenia. Plot these proportions in a line graph. What pattern is revealed?

8. Human diseases differ in their virulence, which is defined as their ability to cause harm. Scientists are interested in determining what features of different diseases make some more dangerous to their hosts than others. The graph below depicts the frequency distribution of virulence measurements, on a log scale, of a sample of human diseases (modified from Ewald 1993). Diseases that spread from one victim to another by direct contact between people are shown in the upper graph. Those transmitted from person to person by insect vectors are shown in the lower graph.

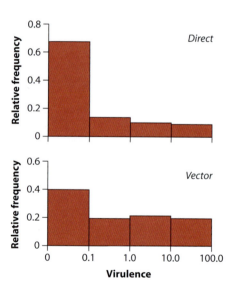

a. Identify the type of graph displayed.
b. What are the two groups being compared in this graph?
c. What variable is being compared between the two groups? Is it numerical or categorical?
d. Explain the units on the vertical (*y*) axis.
e. What is the main result depicted by this graph?

9. Examine the figure below, which indicates the date of first occurrence of rabies in raccoons in the townships of Connecticut, measured by the number of months following 1 March 1991 (modified from Smith et al. 2002).

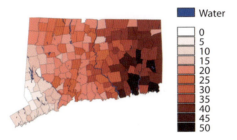

Water
0
5
10
15
20
25
30
35
40
45
50

a. Identify the type of graph shown.
b. What is the response variable?
c. What is the explanatory variable?

10. Could the space shuttle *Challenger* disaster of January 28, 1986, have been predicted? On the night before the accident, experts convened to discuss the implications of an anticipated 31°F temperature the morning of the launch—far lower than any previous launch. They discussed the following raw data, which give the temperature of previous launches and the number of O-ring failures during each launch (Dalal et al. 1989, Tufte 1997). O-rings seal the joints of the rocket motors that boost the shuttle into orbit. A gas leak through one of these joints, caused by O-ring failure, was responsible for the *Challenger* explosion.

a. Using an appropriate graphical method, depict the relationship between the number of O-ring failures and the launch temperature. Which variable is the explanatory variable, and which is the response variable? Why?
b. Examine your plot and decide whether an association is present between the number of O-ring failures and the launch temperature. If the variables are associated, is the association positive or negative?
c. Based on your answer in part (b), what effect might a low launch temperature have on the expected number of O-ring failures?

Launch temperature (°F)	Number of O-ring failures
66	0
70	1
69	0
68	0
67	0
72	0
73	0
70	0
57	1
63	1
70	1
78	0

Launch temperature (°F)	Number of O-ring failures
67	0
53	3
67	0
75	0
70	0
81	0
76	0
79	0
75	2
76	0
58	1

11. Refer to the data in the previous problem.

   a. Display the frequency distribution of launch temperatures using an appropriate graph. Describe the shape of the frequency distribution.

   b. What is the mode of your figure in part (a)?

12. The graph below is taken from a study of married women who had been raised by adoptive parents (Bereczkei et al. 2004). It shows the facial resemblance between the women and their husbands (first bar), between their husbands and the women's adoptive fathers (second bar), and between their husbands and the women's adoptive mothers (third bar). Facial resemblance of a given woman to her husband was scored by 242 "judges," each of whom was given a photograph of the woman, photos of three other women, and a photo of her husband. The judges were asked to decide from the photos which of the four women was the wife of the husband based on facial similarity. Her resemblance was scored as the percentage of judges who chose correctly. If there was no resemblance between a given woman and her husband, then by chance only one in four judges (25%) should have chosen correctly. Resemblance of the woman's husband to the wife's adoptive father and to the wife's adoptive mother was measured in the same way.

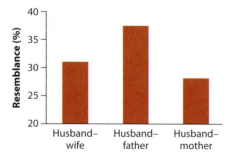

   a. Describe the essential findings displayed in the figure.

   b. Identify the most significant weakness of the figure as a display of these results.

   c. Is this graph displaying a frequency distribution? Explain.

13. Each of the following graphs illustrates an association between two variables. For each graph identify (1) the type of graph, (2) the explanatory and response variables, and (3) the type of data (whether numerical or categorical) for each variable.

   a. Observed fruiting of individual plants in a population of *Campanula americana* according to the number of fruits produced previously (Richardson and Stephenson 1991):

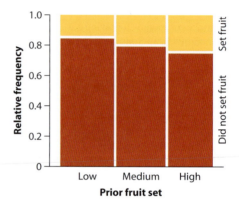

   b. The maximum density of wood produced at the end of the growing season in white spruce trees in Alaska in different years (modified from Barber et al. 2000):

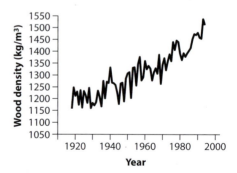

   c. Percent changes in the range sizes of different species of native butterflies (*red*), birds (*blue*), and plants (*green*) of Britain over the past two to four decades (modified from Thomas et al. 2004):

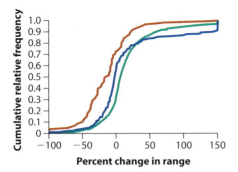

ASSIGNMENT PROBLEMS

14. The accompanying graph depicts a frequency distribution of beak widths of 1017 black-bellied seedcrackers, *Pyrenestes ostrinus,* a finch from West Africa (Smith 1993).

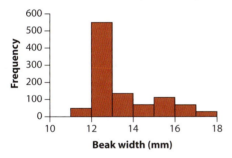

a. What is the mode of the frequency distribution?

b. Estimate by eye the fraction of birds whose measurements are in the interval representing the mode.

c. There is a hint of a second peak in the frequency distribution between 15 and 16 mm. What strategy would you recommend be used to explore more fully the possibility of a second peak?

d. What name is given to a frequency distribution having two distinct peaks?

15. When its numbers increase following favorable environmental conditions, the desert locust, *Schistocerca gregaria*, undergoes a dramatic Jekyll-and-Hyde transformation. From a solitary, cryptic form, it develops into a conspicuously colored, gregarious form that swarms by the billions, ravaging crops. A laboratory experiment found that mechanical stimulation—that is, simulating locusts bumping into one another—was the most effective trigger of the switch. The graph below summarizes the degree

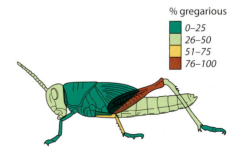

of gregariousness resulting from mechanical stimulation of different parts of the body (from Simpson et al. 2001).

    a. Touching which part of the locust's anatomy results in the greatest response in gregariousness? Touching which part resulted in the smallest response?

    b. Identify the type of graph displayed.

16. The *Cambridge Study in Delinquent Development* was undertaken in north London (UK) to investigate the links between criminal behavior in young men and the socioeconomic factors of their upbringing (Farrington 1994). A cohort of 395 boys was followed for about 20 years, starting at the age of eight or nine. All of the boys attended six schools located near the research office. The following table shows the total number of criminal convictions by the boys between the start and end of the study.

    a. What type of table is this?

    b. How many variables are presented in this table?

    c. How many boys had exactly two convictions by the end of the study?

    d. What fraction of boys had no convictions?

    e. Display the frequency distribution in a graph. Which type of graph is most appropriate? Why?

Number of convictions	Frequency
0	265
1	49
2	21
3	19
4	10
5	10
6	2
7	2
8	4
9	2
10	1
11	4
12	3
13	1
14	2
	Total: 395

    f. Describe the shape of the frequency distribution. Is it skewed or is it symmetric? Is it unimodal or bimodal? Where is the mode in number of criminal convictions? Are there outliers in the number of convictions?

    g. Does the sample of boys used in this study represent a random sample of British boys? Why or why not?

17. The following data are from the *Cambridge Study in Delinquent Development* (see the previous problem). They examine the relationship between the occurrence of convictions by the end of the study and the family income of each boy when growing up. Three categories described income level: inadequate, adequate, and comfortable.

	Income level		
	Inadequate	**Adequate**	**Comfortable**
No convictions	47	128	90
Convicted	43	57	30

    a. What type of table is this?

    b. Display these same data in a mosaic plot.

    c. What type of variable is "income level"? How should this affect the arrangement of groups in your mosaic plot in part (b)?

    d. By viewing the table above and the graph in part (b), describe any apparent association between family income and later convictions.

    e. In answering part (d), which method (the table or the graph) better revealed the association between conviction status and income level? Explain.

18. Male fireflies of the species *Photinus ignitus* attract females with pulses of light. Flashes of longer duration seem to attract the most females. During mating, the male transfers a spermatophore to the female. Besides containing sperm, the spermatophore is rich in protein that is distributed by the female to her fertilized eggs. The data below are measurements of spermatophore mass (in mg) of 35 males (Cratsley and Lewis 2003).

0.047, 0.037, 0.041, 0.045, 0.039, 0.064, 0.064, 0.065, 0.079, 0.07, 0.066, 0.059, 0.075, 0.079, 0.09, 0.069, 0.066, 0.078, 0.066, 0.066, 0.055, 0.046, 0.056, 0.067, 0.075, 0.048, 0.077, 0.081, 0.066, 0.172, 0.08, 0.078, 0.048, 0.096, 0.097

a. Create a graph depicting the frequency distribution of the 35 mass measurements.
b. What type of graph did you choose in part (a)? Why?
c. Describe the shape of the frequency distribution. What are its main features?
d. What term would be used to describe the largest measurement in the frequency distribution?

19. Swordfish have a unique "heater organ" that maintains elevated eye and brain temperatures when hunting in deep, cold water, but its function is unclear. The graph below illustrates the results of a study by Fritsches et al. (2005) that measured how the ability of swordfish retinas to

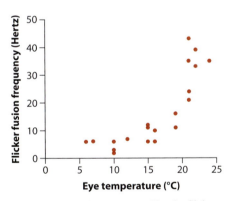

detect rapid motion, measured by the flicker fusion frequency, changes with eye temperature.
a. What types of variables are displayed?
b. What type of graph is this?
c. Describe the association between the two variables. Is the relationship between flicker fusion frequency and temperature positive or negative? Is the relationship linear or nonlinear?
d. The 20 points in the graph were obtained from measurements of six swordfish. Can we treat the 20 measurements as a random sample? Why or why not?

20. The following graph displays the net number of species listed under the U.S. Endangered Species Act between 1980 and 2002 (U.S. Fish and Wildlife Service 2001):

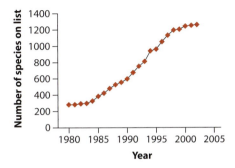

a. What type of graph is this?
b. What does the steepness of each line segment indicate?
c. Explain what the graph tells us about the relationship between the number of species listed and time.

21. Spot the flaw. Examine the following figure, which displays the frequency distribution of similarity values (the percentage of amino acids that are the same) between equivalent (homologous) proteins in humans and pufferfish of the genus, *Fugu* (modified from Aparicio et al. 2002).

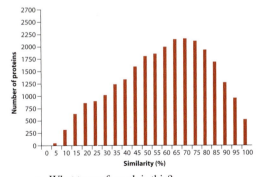

a. What type of graph is this?
b. Identify the main flaw in the construction of this figure.
c. What are the main results displayed in the figure?
d. Describe the shape of the frequency distribution shown.
e. What is the mode in the frequency distribution?

22. The following graph shows the population growth rates of the 204 countries recognized by the United Nations. Growth rate is measured as the average annual percent change in the total human population between 2000 and 2004 (United Nations Statistics Division 2004).

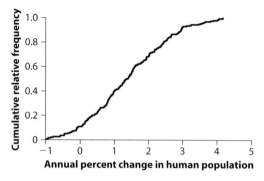

a. Identify the type of graph depicted.
b. Explain the quantity along the y-axis.
c. Approximately what percentage of countries had a negative change in population?
d. Identify by eye the 0.10, 0.50, and 0.90 quantiles of change in population size.
e. Identify by eye the 60th percentile of change in population size.

23. The following data give the photosynthetic capacity of nine individual females of the neotropical tree *Ocotea tenera*, according to the number of fruits produced in the previous reproductive season (Wheelwright and Logan 2004). The goal of the study was to investigate how reproductive effort in females of these trees impacts subsequent growth and photosynthesis.

Number of fruits produced previously	Photosynthetic capacity (μmol O$_2$/m²/s)
10	13.0
14	11.9
5	11.5
24	10.6
50	11.1
37	9.4
89	9.3
162	9.1
149	7.3

a. Graph the association between these two variables using the most appropriate method. Identify the type of graph you used.
b. Which variable is the explanatory variable in your graph? Why?
c. Describe the association between the two variables in words, as revealed by your graph.

24. Each of the following graphs illustrates an association between two variables. For each graph, identify (1) the type of graph, (2) the explanatory and response variables, and (3) the type of data (whether numerical or categorical) for each variable.

a. Taste sensitivity to phenylthiocarbamide (PTC) in a sample of human subjects grouped according to their genotype at the PTC gene—namely, *AA*, *Aa* or *aa* (Kim et al. 2003):

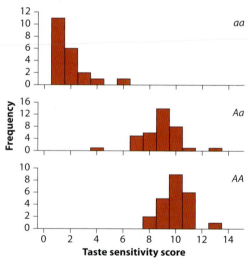

b. Migratory activity (hours of nighttime restlessness) of young captive Blackcaps (*Sylvia atricapilla*) compared with the migratory activity of their parents (Berthold and Pulido 1994):

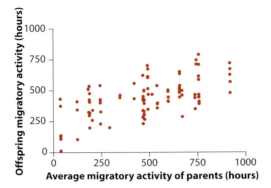

c. Density of fine roots in Monterey pines (*Pinus radiata*) planted in three different years of study (redrawn from Moir and Bachelard 1969, with permission):

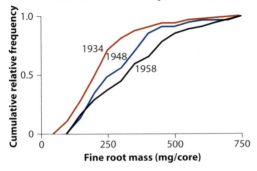

a. What is the main result displayed in this figure?

b. Identify the main weaknesses of this figure.

c. Redraw the figure using the most appropriate method discussed in this chapter. What type of graph did you use?

26. When a courting male of the small Indonesian fish *Telmatherina sarasinorum* spawns with a female, other males sometimes sneak in and release sperm, too. The result is that not all of the female's eggs are fertilized by the courting male. Gray et al. (2007) noticed that courting males occasionally cannibalize fertilized eggs immediately after spawning. Egg eating took place by 61 of 450 courting males who fathered the entire batch; the remaining 389 males did not cannibalize eggs. In contrast, 18 of 35 courting males ate eggs when a single sneaking male also participated in the spawning event. Finally, 16 of 20 males ate eggs when two or more sneaking males were present.

a. Display these results in a table that best shows the association between cannibalism and the number of sneaking males. Identify the type of table you used.

b. Illustrate the same results using a graphical technique instead. Identify the type of graph you used.

d. The frequency of injection-heroin users that share or do not share needles according to their known HIV infection status (Wood et al. 2001):

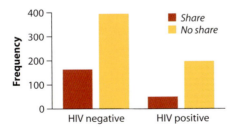

25. Examine the following figure, which displays the percentage of adults over 18 with a "body mass index" greater than 25 in different years (modified from T*he Economist* 2006, with permission). Body mass index is a measure of weight relative to height.

3

Describing data

Descriptive statistics, or summary statistics, are quantities that capture important features of frequency distributions. Whereas graphs reveal shapes and patterns in the data, descriptive statistics provide hard numbers. The most important descriptive statistics for numerical data are those measuring the **location** of a frequency distribution and its **spread**. The location tells us something about the average or typical individual—where the observations are centered. The spread tells us how variable the measurements are from individual to individual—how widely scattered the observations are around the center. The **proportion** is the most important descriptive statistic for a categorical variable, measuring the fraction of observations in a given category.

The importance of calculating the location of a distribution seems obvious. How else do we address questions like "Which species is larger?" or "Which treatment group yielded the greatest response?" The importance of describing distribution spread is less obvious but no less crucial, at least in biology. In some fields of science, variability around a central value is instrument noise or measurement error, but in biology much of the variability signifies real differences among individuals. Different individuals respond differently to treatments, and this variability begs measurement. Measuring variability also gives us perspective. We can ask, "How large are the differences between groups compared with variations within groups?" Biologists also appreciate variation as the stuff of evolution—we wouldn't be here without variation.

In this chapter, we review the most common statistics to measure the location and spread of a frequency distribution and to calculate a proportion. We introduce the use of mathematical symbols to represent values of a variable, and we show formulas to calculate each summary statistic.

3.1   Arithmetic mean and standard deviation

The arithmetic mean is the most common metric to describe the location of a frequency distribution. It is the average of a set of measurements. The standard deviation is the most commonly used measure of distribution spread. Example 3.1 illustrates the basic calculations for means and standard deviations.

| Example 3.1 | **Gliding snakes** |

When a paradise tree snake (*Chrysopelea paradisi*) flings itself from a treetop, it flattens its body everywhere except for the region around the heart. As it gains downward speed, the snake forms a tight horizontal S shape and then begins to undulate widely from side to side. Lift is thereby created, causing the snake to glide away from the source tree. By orienting the head and anterior part of the body, the snake can change direction during a glide to avoid trees, reach a preferred landing site, and even chase aerial prey.

To better understand how lift is generated, Socha (2002) videotaped the glides of eight snakes leaping from a 10-m tower.[1] Among the measurements taken was the rate of side-to-side undulation on each snake. Undulation rates of the eight snakes, measured in Hertz (cycles per second), were as follows:

0.9, 1.4, 1.2, 1.2, 1.3, 2.0, 1.4, 1.6

A histogram of these data is shown in Figure 3.1-1. The frequency distribution has a single peak between 1.2 and 1.4 Hz.

---

[1]   See films of these snakes flying at http://www.flyingsnake.org/video/video.html.

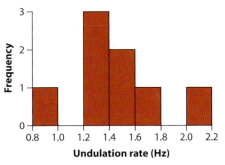

Figure 3.1-1 Undulation rate of gliding paradise tree snakes. $n = 8$ snakes.

The sample mean

The **sample mean** is the average of the measurements in the sample, the sum of all the observations divided by the number of observations. To show its calculation, we use the symbol Y to refer to the variable and Y_i to represent the measurement of individual i. For the gliding snake data, i takes on values between 1 and 8, because there are eight snakes. Thus, $Y_1 = 0.9$, $Y_2 = 1.4$, $Y_3 = 1.2$, $Y_4 = 1.2$, and so on.[2]

The sample mean, symbolized as $\overline{Y}$ (and pronounced "Y-bar"), is calculated as

$$\overline{Y} = \frac{\sum_{i=1}^{n} Y_i}{n},$$

where n is the number of observations. The symbol Σ (uppercase Greek letter "sigma") indicates a sum. The "$i=1$" under the Σ and the "n" over it indicate that we are summing over all values of i between 1 and n, inclusive

$$\sum_{i=1}^{n} Y_i = Y_1 + Y_2 + Y_3 + \cdots + Y_n.$$

When it is clear that i refers to individuals 1, 2, 3, . . ., n, the formula is often written more succinctly as

$$\overline{Y} = \frac{\sum Y_i}{n}.$$

Applying this formula to the snake data yields the mean undulation rate:

$$\overline{Y} = \frac{0.9 + 1.2 + 1.2 + 2.0 + 1.6 + 1.3 + 1.4 + 1.4}{8} = 1.375 \text{ Hz.}$$

---

[2] We have adopted the simple convention of using upper case letters (e.g., Y) when referring to both variable names and data, and prefer to distinguish the two by context. This is a departure from mathematical convention, which reserves upper case exclusively for random variables.

Based on the histogram in Figure 3.1-1, we see that the value of the sample mean is close to the middle of the distribution. Note that the sample mean has the same units as the observations used to calculate it. In the Quick Formula Summary (Section 3.6), we review how the sample mean is affected when the units of the observations are changed, such as by adding a constant or multiplying by a constant.

> The *sample mean* is the sum of all the observations in a sample divided by n, the number of observations.

Variance and standard deviation

The **standard deviation** is a commonly used measure of the spread of a distribution. It measures how far from the mean the observations typically are. The standard deviation is large if most observations are far from the mean, and it is small if most measurements lie close to the mean.

The standard deviation is calculated from the **variance**, another measure of spread. The standard deviation is simply the square root of the variance. The standard deviation is a more intuitive measure of the spread of a distribution, but the variance has mathematical properties that make it useful sometimes as well. The standard deviation from a sample is usually represented by the symbol s, and the sample variance is written as s^2.

To calculate the variance from a sample of data, we must first compute the deviations. A deviation is the difference between a measurement and the mean $(Y_i - \bar{Y})$. Deviations for the measurements of snake undulation rate are listed in Table 3.1-1.

The best measure of the spread of this distribution isn't just the average of the deviations $(Y_i - \bar{Y})$, because this average is always zero (the negative deviations cancel the positive deviations). Instead, we need to average the *squared* deviations (the third column in Table 3.1-1) to find the variance:

Table 3.1-1   Quantities needed to calculate the standard deviation and variance of snake undulation rate $(\bar{Y} = 1.375\text{Hz})$.

Observations (Y_i)	Deviations $(Y_i - \bar{Y})$	Squared deviations $(Y_i - \bar{Y})^2$
0.9	−0.475	0.225625
1.2	−0.175	0.030625
1.2	−0.175	0.030625
1.3	−0.075	0.005625
1.4	0.025	0.000625
1.4	0.025	0.000625
1.6	0.225	0.050625
2.0	0.625	0.390625
Sum	0.000	0.735

$$s^2 = \frac{\Sigma (Y_i - \bar{Y})^2}{n - 1}.$$

By squaring each number, deviations above and below the mean contribute equally[3] to the variance. The summation in the numerator (top part) of the formula, $\Sigma (Y_i - \bar{Y})^2$, is called the **sum of squares** of Y. Note that the denominator (bottom part) is $n - 1$ instead of n, the total number of observations. Dividing by $n - 1$ gives a more accurate estimate of the population variance.[4] We provide a shortcut formula for the variance in the Quick Formula Summary (Section 3.6).

For the snake undulation data, the variance (rounded to hundredths) is

$$s^2 = \frac{0.735}{7} = 0.11 \text{ Hz}^2.$$

The variance has units equal to the square of the units of the original data. To obtain the standard deviation, we take the square root of the variance:

$$s = \sqrt{\frac{\Sigma (Y_i - \bar{Y})^2}{n - 1}}.$$

For the snake undulation data,

$$s = \sqrt{\frac{0.735}{7}} = 0.324037 \text{ Hz}.$$

The standard deviation is never negative and has the same units as the observations from which it was calculated.

> The *standard deviation* is a common measure of the spread of a distribution. It indicates just how different measurements typically are from the mean.

The standard deviation has a straightforward connection to the frequency distribution. If the frequency distribution is bell-shaped, as in Figure 2.1-3, then about two-thirds of the observations will lie within one standard deviation of the mean, and about 95% will lie within two standard deviations. In other words, about 67% of the data will fall between and $\bar{Y} - s$, and $\bar{Y} + s$, about 95% will fall between $\bar{Y} - 2s$ and $\bar{Y} + 2s$. For an in-depth discussion of standard deviation, see Chapter 10.

This straightforward connection between the standard deviation and the frequency distribution diminishes when the frequency distribution deviates from the bell-shaped

---

[3] We could have averaged the absolute values of the deviations instead, to yield the mean absolute deviation. Averaging the square of the deviations is more common because the result, the variance, has many more useful mathematical properties.

[4] The reason is that the average squared deviation of the observations around the sample mean is slightly smaller than the average of the squared deviations around the population mean (the quantity being estimated by the sample mean). Dividing by $n - 1$ instead of n corrects for this bias.

(normal) distribution. In such cases, the standard deviation is less informative about where the data lie in relation to the mean. This point is explored in greater detail in Section 3.3.

In the Quick Formula Summary (Section 3.6), we review how the sample variance and standard deviation are affected when the units of the observations are changed, such as by adding a constant or multiplying by a constant.

Rounding means, standard deviations, and other quantities

To avoid rounding errors when carrying out calculations of means, standard deviations, and other descriptive statistics, always retain as many significant digits as your calculator or computer can provide. Intermediate results written down on a page should also retain as many digits as feasible. Final results, however, should be rounded before being presented.

There are no strict rules on the number of significant digits that should be retained when rounding. A common strategy, which we adopt here, is to round descriptive statistics to one decimal place more than the measurements themselves. For example, the undulation rates in snakes were measured to a single decimal place (tenths). We therefore present descriptive statistics with two decimals (hundredths). The mean rate of undulation for the eight snakes, calculated as 1.375 Hz, would be communicated as

$$\overline{Y} = 1.38 \text{ Hz}.$$

Similarly, the standard deviation, calculated as 0.324037 Hz, would be reported as

$$s = 0.32 \text{ Hz}.$$

Note that even though we report the rounded value of the mean as $\overline{Y} = 1.38$, we used the more exact value, $\overline{Y} = 1.375$, in the calculation of s to avoid rounding errors.

Coefficient of variation

For many traits, standard deviation and mean change together when organisms of different sizes are compared. Elephants have greater mass than mice and also more variability in mass. For many purposes, we care more about the relative variation among individuals. A gain of 10 g for an elephant is inconsequential, but it would double the mass of a mouse. On the other hand, an elephant that is 10% larger than the elephant mean may have something in common with a mouse that is 10% larger than the mouse mean. For these reasons, it is sometimes useful to express the standard deviation relative to the mean. The **coefficient of variation** (CV) calculates the standard deviation as a percentage of the mean:

$$CV = 100\% \, \frac{s}{\overline{Y}}.$$

A higher CV means that there is more variability, whereas a lower CV means that individuals are more consistently the same. For the snake undulation data, the coefficient of variation is

$$CV = 100\% \, \frac{0.324}{1.375} = 24\%.$$

The coefficient of variation only makes sense when all of the measurements are greater than or equal to zero.

> The *coefficient of variation* is the standard deviation expressed as a percentage of the mean.

The coefficient of variation can also be used to compare the variability of traits that do not have the same units. If we wanted to ask, "What is more variable in elephants, body mass or lifespan?," then the standard deviation is not very informative, because mass is measured in kilograms and lifespan is measured in years. The coefficient of variation would allow us to make this comparison.

Calculating mean and standard deviation from a frequency table

Sometimes the data include many tied observations and are given in a frequency table. The frequency table in Table 3.1-2, for example, lists the number of criminal

Table 3.1-2 Number of criminal convictions of a cohort of 395 boys.

Number of convictions	Frequency
0	265
1	49
2	21
3	19
4	10
5	10
6	2
7	2
8	4
9	2
10	1
11	4
12	3
13	1
14	2
Total	395

convictions of a cohort of 395 boys (Farrington 1994; see Assignment Problem 16 in Chapter 2).

To calculate the mean and standard deviation of the number of convictions, notice first of all that the sample size is *not* 15, the number of rows in Table 3.1-2, but 395, the frequency total:

$$n = 265 + 49 + 21 + 19 + \ldots + 2 = 395.$$

Calculating the mean thus requires that the measurement of "0" be represented 265 times, the number "1" be represented 49 times, and so on. The sum of the measurements is thus

$$\sum Y_i = (265 \times 0) + (49 \times 1) + (21 \times 2) + (19 \times 3)$$
$$+ \ldots + (2 \times 14) = 445.$$

The mean of these data is then

$$\bar{Y} = \frac{445}{395} = 1.126582,$$

which we round to $\bar{Y} = 1.1$ when presenting the results.

The calculation of standard deviation must also take into account the number of individuals with each value. The sum of the squared deviations is

$$\sum (Y_i - \bar{Y})^2 = 265(0 - \bar{Y})^2 + 49(1 - \bar{Y})^2 + 21(2 - \bar{Y})^2 + \ldots$$
$$+ 2(14 - \bar{Y})^2 = 2377.671.$$

The standard deviation for these data is therefore

$$s = \sqrt{\frac{2377.671}{395 - 1}} = 2.4566,$$

which we present as $s = 2.5$.

These calculations assume that all the data are presented in the table. It would not work, however, for frequency tables in which the data are grouped into intervals, such as Table 2.1-3.

3.2   Median and interquartile range

After the sample mean, the *median* is the next most common metric used to describe the location of a frequency distribution. It is often displayed in a box plot alongside the *interquartile range*, another measure of the spread of the distribution. We define and demonstrate these concepts with the help of Example 3.2.

Example 3.2   **I'd give my right arm for a female**

Male spiders in the genus *Tidarren* are tiny, weighing only about 1% as much as females. They also have disproportionately large pedipalps, copulatory organs that make up about

10% of a male's mass. (See the adjacent photo; the pedipalps are indicated by arrows.) Males load the pedipalps
with sperm and then search for females to inseminate.
Astonishingly, male *Tidarren* spiders voluntarily amputate
one of their two organs, right or left, just before sexual
maturity. Why do they do this? Perhaps speed is important
to males searching for females, and amputation increases
running performance. To test this hypothesis, Ramos et al. (2004) used video to measure the
running speed of males on strands of spider silk. The data are presented in Table 3.2-1.

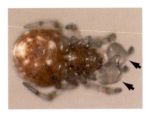

**Table 3.2-1 Running speed (cm/s) of male *Tidarren* spiders before and after
voluntary amputation of a pedipalp.**

Spider	Speed before	Speed after	Spider	Speed before	Speed after
1	1.25	2.40	9	2.98	3.70
2	2.94	3.50	10	3.55	4.70
3	2.38	4.49	11	2.84	4.94
4	3.09	3.17	12	1.64	5.06
5	3.41	5.26	13	3.22	3.22
6	3.00	3.22	14	2.87	3.52
7	2.31	2.32	15	2.37	5.45
8	2.93	3.31	16	1.91	3.40

The median

The **median** is the middle observation in a set of data, the measurement that partitions the ordered measurements into two halves. In other words, the median is the
50th percentile and the 0.5 quantile. To calculate the median, first sort the sample
observations from smallest to largest. The sorted measurements of running speed of
male spiders before amputation (Table 3.2-1) are 1.25, 1.64, 1.91, 2.31, 2.37, 2.38,
2.84, 2.87, 2.93, 2.94, 2.98, 3.00, 3.09, 3.22, 3.41, 3.55 in cm/s. Let $Y_{(i)}$ refer to the ith
sorted observation, so $Y_{(1)}$ is 1.25, $Y_{(2)}$ is 1.64, $Y_{(3)}$ is 1.91, and so on. If the number of
observations (n) is odd, then the median is the middle observation:

$$\text{Median} = Y_{([n+1]/2)}.$$

If the number of observations is even, as in the spider data, then the median is the
average of the middle pair:

$$\text{Median} = [Y_{(n/2)} + Y_{(n/2+1)}]/2.$$

Thus, $n/2 = 8$, $Y_{(8)} = 2.87$, and $Y_{(9)} = 2.93$ for the spider data (before amputation).
The median is the average of these two numbers:

$$\text{Median} = (2.87 + 2.93)/2 = 2.90 \text{ cm/s}.$$

The *median* is the middle measurement of a set of observations.

The interquartile range

Quartiles are values that partition the data into quarters. The first quartile (the 25th percentile) is the middle value of the measurements lying below the median. The second quartile is the median. The third quartile (the 75th percentile) is the middle value of the measurements larger than the median. The **interquartile range** is the span of the middle half of the data, from the first quartile to the third quartile:

Interquartile range = third quartile − first quartile.

Figure 3.2-1 shows the meaning of the median, first quartile, third quartile, and interquartile range for the spider data set (before amputation).

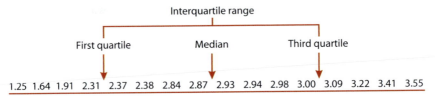

Figure 3.2-1 The first quartile, median, and third quartile break the data set into four equal portions. The median is the middle value, and the first and third quartiles are the middles of the first and second halves of the data. The interquartile range is the span of the middle half of the data.

The first step to calculating the interquartile range is to break the data into halves, leaving out one observation equal to the median if n is odd.[5]

The smaller half of the before-amputation data is

1.25, 1.64, 1.91, 2.31, 2.37, 2.38, 2.84, 2.87.

The first quartile is the median of these observations:

First quartile = (2.31 + 2.37)/2 = 2.34.

The larger half of the before-amputation measurements is

2.93, 2.94, 2.98, 3.00, 3.09, 3.22, 3.41, 3.55,

so the third quartile is the median of these numbers:

Third quartile = (3.00 + 3.09)/2 = 3.045.

---

[5] Don't be surprised if your computer program gives slightly different values from ours for the quartiles and the interquartile range. The method given here is simple to calculate, but it does not give the most accurate estimates of the population quantities. Several improved methods are available (Hyndman and Fan 1996).

The interquartile range is then

$$\text{Interquartile range} = 3.045 - 2.34 = 0.705 \text{ cm/s}.$$

> The *interquartile range* is the difference between the first and third quartiles of the data.

The box plot

A **box plot** displays the median and interquartile range, along with other quantities of the frequency distribution. Figure 3.2-2 shows a box plot for the spider running-speeds data, both before and after amputation.

Each box in Figure 3.2-2 is bounded at the lower edge by the first quartile and at the upper edge by the third quartile. The horizontal line dividing each box is the median. Two lines, called whiskers, extend outward from the box at each end. The whiskers stop at the smallest and largest "non-extreme" values in the data. "Extreme" values are defined as those lying farther from the box edge than 1.5 times the interquartile range. Extreme values are plotted as isolated dots past the ends of the whiskers.[6] There is one extreme value in the box plots shown in Figure 3.2-2, the smallest measurement for running speed before amputation.

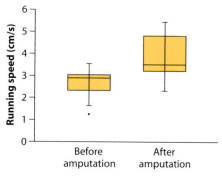

Figure 3.2-2 Box plot of the running speed of 16 male spiders before and after self-amputation of a pedipalp.

A box plot indicates where the bulk of the measurements lie. A histogram can do the same with greater detail, but a box plot picks out a few of the most important features of the frequency distribution. Because it is compact, a box plot is especially useful when comparing more than one distribution.

---

[6] Some computer programs extend whiskers all the way to the most extreme values on each end and do not indicate extreme values with isolated dots.

A *box plot* is a graph that uses lines and a rectangle to display the median, quartiles, extreme measurements, and range of the data.

3.3 How measures of location and spread compare

Which measure of location (the sample mean or the median) is most revealing about the center of a distribution of measurements? And which measure of spread (the standard deviation or the interquartile range) best describes how widely the observations are scattered about the center? To answer these questions, we must compare the behavior of the mean and median, and of the standard deviation and interquartile range, with frequency distributions of different shapes. These alternative measures of location and of spread yield similar information when the frequency distribution is symmetric and unimodal, but the mean and standard deviation become less informative than the median and interquartile range when the data include extreme observations. We compare these measures using Example 3.3.

Example 3.3   **Disarming fish**

The marine threespine stickleback is a small coastal fish named for its defensive armor. It has three sharp spines down its back, two pelvic spines underneath, and a series of lateral bony plates down both sides. The armor seems to reduce mortality from predatory fish and diving birds. In contrast, in lakes and streams, where predators are fewer, stickleback populations have reduced armor. (See the photo in the margin for examples of both types. Bony tissue has been stained red to make it more visible.)

Colosimo et al. (2004) measured the grandchildren of a cross made between a marine and a freshwater stickleback. The study found that much of the difference in number of plates is caused by a single gene, ectodysplasin.[7] Fish inheriting two copies of the gene from the marine grandparent, called *MM* fish, had many plates (the top histogram in Figure 3.3-1). Fish inheriting both copies of the gene from the freshwater grandparent (*mm*) had few plates (the bottom histogram in Figure 3.3-1). Fish having one copy from each grandparent (*Mm*) had any of a wide range of plate numbers (the middle histogram in Figure 3.3-1).

---

[7]   Mutations in the same gene in humans cause the loss of hair, teeth, and sweat glands.

Mean versus median

The mean and median of the three distributions in Figure 3.3-1 are compared in Table 3.3-1. The two measures of location give similar values in the case of the *MM* and *mm* genotypes, whose distributions are fairly symmetric. The mean is smaller than the median in the case of the *Mm* fish, however, whose distribution is strongly asymmetric.

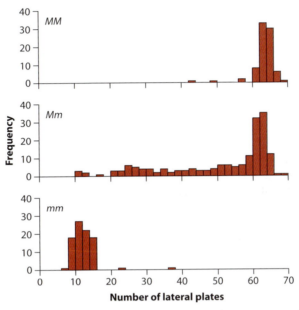

Figure 3.3-1 Frequency distributions of lateral plate number in three genotypes of stickleback, *MM, Mm,* and *mm,* descended from a cross between marine and freshwater grandparents. Plates are counted as the total number down the left and right sides of the fish. The total number of fish: 82 (*MM*), 174 (*Mm*), and 88 (*mm*).

Table 3.3-1 Descriptive statistics for the number of lateral plates of the three genotypes of threespine sticklebacks[8] discussed in Example 3.3.

Genotype	*n*	$\overline{Y}$	Median	*s*	Interquartile range
MM	82	62.8	63	3.4	2
Mm	174	50.4	59	15.1	21
mm	88	11.7	11	3.6	3

---

[8] When listing descriptive statistics in tables, put the same quantity calculated on different groups into one column. Numbers stacked in a single column are easier to compare than numbers placed side by side in a row. Exchange the columns and rows in Table 3.3-1, and you'll see what we mean.

Why are the median and mean different from one another when the distribution is asymmetric? The answer, shown in Figure 3.3-2, is that the median is the middle measurement of a distribution, whereas the mean is the "center of gravity."

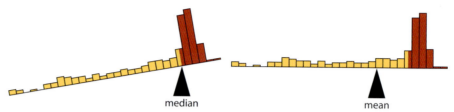

median   mean

Figure 3.3-2 Comparison between the median and the mean using the frequency distribution for the *Mm* genotype (middle panel of Figure 3.3-1). The median is the middle measurement of the distribution (different colors represent the two halves of the distribution). The mean is the center of gravity, the point at which the frequency distribution would be balanced (if observations had weight).

The balancing act illustrated in Figure 3.3-2 suggests that the mean is sensitive to extreme observations. To demonstrate, imagine taking the four smallest observations of the *MM* genotype (top panel in Figure 3.3-1) and moving them far to the left. The median would be completely unaffected, but the mean would shift leftward to a point near the edge of the range of most observations (Figure 3.3-3).

> Median and mean measure different aspects of the location of a distribution. The median is the middle value of the data, whereas the mean is its center of gravity.

Thus, the mean is displaced from the location of the "typical" measurement when the frequency distribution is strongly skewed, particularly when there are extreme observations. The mean is still useful as a description of the population as a whole, but it no longer indicates where most of the observations are located. The median is less sensitive to extreme observations, and hence the median is the more informative descriptor of the typical observation in such instances.

Standard deviation versus interquartile range

Because it is calculated from the square of the deviations, the standard deviation is even more sensitive to extreme observations than is the mean. When the four smallest observations of the *MM* genotype are shifted far to the left, such that the smallest is set to zero (Figure 3.3-3), the standard deviation jumps from 3.4 to 12.0, whereas the interquartile range is not affected. For this reason, the interquartile range is a better indicator of the spread of the main part of a distribution than the standard deviation when the data are strongly skewed to one side or the other, especially when there are extreme observations. On the other hand, the standard deviation is a better indication of the spread among all of the data points.

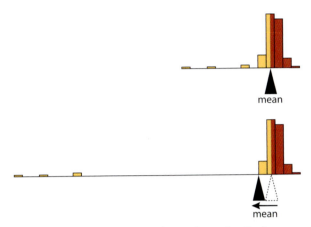

Figure 3.3-3 Sensitivity of the mean to extreme observations using the frequency distribution of the *MM* genotypes (see the upper panel in Figure 3.3-1). The two different colors represent the two halves of the distribution. When the four smallest observations of the *MM* genotype are shifted far to the left (lower panel), the mean is displaced downward, to the edge of the range of the bulk of the observations. The median, on the other hand, which is located where the two colors meet, is unaffected by the shift.

3.4 Proportions

The proportion is the most important descriptive statistic for a categorical variable.

Calculating a proportion

The **proportion** of observations in a given category, symbolized $\hat{p}$, is calculated as

$$\hat{p} = \frac{\text{Number in category}}{n},$$

where the numerator is the number of observations in the category of interest, and n is the total number of observations in all categories combined.[9]

For example, of the 344 individual sticklebacks in Example 3.3, 82 had genotype *MM*, 88 were *mm*, and 174 were *Mm* (Table 3.3-1). The proportion of *MM* fish is

$$\hat{p} = \frac{82}{344} = 0.238.$$

The other proportions are calculated similarly, and all three proportions are listed in Table 3.4-1.

---

[9]   The "hat" in $\hat{p}$ is used to indicate an estimate of the true proportion p.

Table 3.4-1 The number of fish of each genotype from a cross between a marine stickleback and a freshwater stickleback (Example 3.3). The sum of the proportions as written does not add precisely to one because of rounding.

Genotype	Frequency	Proportion
MM	82	0.24
Mm	174	0.51
mm	88	0.26
Total	344	1.00

The proportion is like a sample mean

The proportion $\hat{p}$ has properties in common with the arithmetic mean. To see this, let's create a new numerical variable Y for the stickleback study. Give individual fish i the value $Y_i = 1$ if it has the *MM* genotype and give it the value $Y_i = 0$ otherwise. The sum of all the ones and zeroes, $\sum Y_i$, is the frequency of fish having genotype *MM*. The mean of the ones and zeroes is

$$\bar{Y} = \frac{\sum Y_i}{n} = \frac{82}{344} = 0.238,$$

which is just $\hat{p}$, the proportion of observations in the first category. If we imagine the Y-measurements to have weight, then the proportion is their center of gravity (Figure 3.4-1).

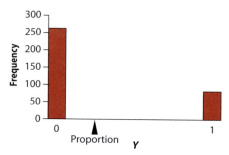

Figure 3.4-1 The distribution of Y, where $Y = 1$ if a stickleback is genotype *MM* and 0 otherwise. The mean of Y is the proportion of *MM* individuals in the sample (0.238).

3.5 Summary

▶ The location of a distribution for a numerical variable can be measured by its mean or by its median. The mean gives the center of gravity of the distribution and is calculated as the sum of all measurements divided by the number of measurements. The median gives the middle value.

▶ The standard deviation measures the spread of a distribution for a numerical variable. It is a measure of the typical distance between observations and the mean. The variance is the square of the standard deviation.

▶ The quartiles break the ordered observations into four equal parts. The interquartile range, the difference between the first and third quartiles, is another measure of the spread of a frequency distribution.

▶ The mean and median yield similar information when the frequency distribution of the measurements is symmetric and unimodal. The mean and standard deviation become less informative about the location and spread of typical observations than the median and interquartile range when the data include extreme observations.

▶ A box plot is a graphical method used to display important features of a frequency distribution. It shows the median, the first and third quartiles, the range of values, and any extreme values.

▶ The proportion is the most important descriptive statistic for a categorical variable. It is calculated by dividing the number of observations in the category of interest by n, the total number of observations in all categories combined.

3.6 Quick Formula Summary

Table of formulas for descriptive statistics

Quantity	Formula
Sample size	n
Mean	$\bar{Y} = \dfrac{\Sigma Y}{n}$
Variance	$s^2 = \dfrac{\Sigma (Y_i - \bar{Y})^2}{n - 1}$
shortcut formula:	$s^2 = \dfrac{\Sigma (Y_i^2) - n\bar{Y}^2}{n - 1}$
Standard deviation	$s = \sqrt{\dfrac{\Sigma (Y_i - \bar{Y})^2}{n - 1}}$
shortcut formula:	$s = \sqrt{\dfrac{\Sigma (Y_i^2) - n\bar{Y}^2}{n - 1}}$
Sum of squares	$\Sigma (Y_i - \bar{Y})^2 = \Sigma (Y_i^2) - n\bar{Y}^2$

Coefficient of variation	$CV = 100\% \dfrac{s}{\bar{Y}}$

Median

$Y_{([n+1]/2)}$　　　(if n is odd)

$(Y_{(n/2)} + Y_{(n/2+1)})/2$　　　(if n is even),

where $Y_{(1)}, Y_{(2)}, \ldots, Y_{(n)}$ are the ordered observations

Proportion

$$\hat{p} = \frac{\text{Number in category}}{n}$$

Effect of arithmetic operations on descriptive statistics

The table below lists the effect on the descriptive statistics of adding or multiplying all the measurements by a constant. The rules listed in the table are useful when converting measurements from one system of units to another, such as English to metric or degrees Celsius to degrees Fahrenheit.

For example, if the mean temperature is 27°C with a standard deviation of 3°C, then measurements in Celsius can be converted to equivalent measurements in Fahrenheit by the transformation

$$°F = 9/5°C + 32.$$

Thus, the new mean temperature is 9/5 (27) +32 = 80.6°F, and the standard deviation is 9/5 (3) = 5.4°F.

Quantity	Symbol	Effect of adding a constant c to all the measurements, $Y' = Y + c$	Effect of multiplying all the measurements by a constant c, $Y' = cY$
Mean	$\bar{Y}$	$\bar{Y}' = \bar{Y} + c$	$\bar{Y}' = c\bar{Y}$
Standard deviation	s	$s' = s$	$s' = cs$
Variance	s^2	$s'^2 = s^2$	$s'^2 = c^2 s^2$
Median	M	$M' = M + c$	$M' = cM$
Interquartile range	IQR	$IQR' = IQR$	$IQR' = c\,IQR$

PRACTICE PROBLEMS

1. A review of the performance of hospital gynecologists in two regions of England measured the outcomes of patient admissions under each doctor's care (Harley et al. 2005). One measurement taken was the percentage of patient admissions made up of women under 25 years old who were sterilized. We are interested in describing what constitutes a typical rate of sterilization, so that the behavior of atypical doctors can be better scrutinized. The frequency distribution of this measurement for all doctors is plotted to the right.

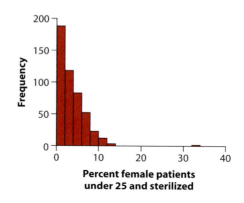

a. Explain what the vertical axis measures.

b. What would be the best choice to describe the location of this frequency distribution, the mean or the median, if our goal was to describe the typical individual? Why?

2. The data displayed in the plot below are from a nearly complete record of body masses (in grams, then converted to log base-10) of the world's native mammals (Smith et al. 2003). The data were divided into three groups: those surviving from the last ice age to the present day ($n = 4061$), those who went extinct around the end of the last ice age ($n = 227$), and those driven extinct within the last 300 years (recent; $n = 44$).

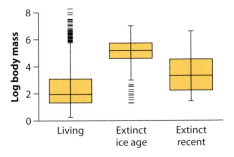

a. What type of graph is this?

b. What does the horizontal line in the center of each rectangle represent?

c. What are the horizontal lines at the top and bottom edges of each rectangle supposed to represent?

d. What are the dashes lying outside of the rectangles?

e. What are the vertical lines extending above and below each rectangle?

f. Compare the locations of the three body size distributions. How do they differ?

g. Compare the shapes of the three frequency distributions. Which are relatively symmetric and which are asymmetric? Explain your reasoning.

h. Which group's frequency distribution has the lowest spread? Explain your reasoning.

i. What has been the likely effect of ice-age and recent extinctions on the median body size of mammals?

3. Mehl et al. (2007) wired 396 volunteers with electronically activated recorders that allowed the researchers to calculate the number of words each individual spoke, on average, per 17-hour waking day. They found that the mean number of words spoken was only very slightly higher for the 210 women (16,215 words) than the 186 men (15,669) in the sample. The frequency distribution of the number of words spoken by all individuals of each sex is shown below (modified from Mehl et al. 2007).

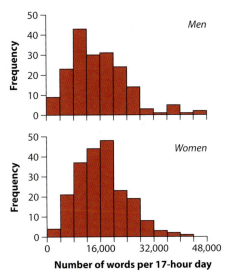

a. What type of graph is shown?

b. What are the explanatory and response variables in the figure?

c. What is the mode of the frequency distribution of each sex?

d. Which sex likely has the higher median number of words spoken per day?

e. Which sex had the highest variance in number of words spoken per day?

4. The following data are measurements of body mass, in grams, of finches captured in mist nets during a survey of various habitats in Kenya, East Africa (Schluter 1988).

Crimson-rumped waxbill   8, 8, 8, 8, 8, 8, 8, 6, 7, 7, 7, 8, 8, 8, 7, 7, 7

Cutthroat finch   16, 16, 16, 12, 16, 15, 15, 17, 15, 16, 15, 16

| White-browed | 40, 43, 37, 38, 43, 33, 35, 37, |
| *sparrow weaver* | 36, 42, 36, 36, 39, 37, 34, 41 |

a. Calculate the mean body mass of each of these three finch species. Which species is largest, and which is smallest? (*Here and forever after, provide units with your answer.*)

b. Which species has the greatest standard deviation in body mass? Which has the least?

c. Calculate the coefficient of variation (CV) in mass for each finch species. How different are the coefficients between the species? Compare the difference in CVs with the differences in standard deviation calculated in part (b).

d. Given the following measurements of beak length, in mm, of the 16 white-browed sparrow weavers, which measurement is more variable in this species (relative to the mean), body mass or beak length?

10.6, 10.8, 10.9, 11.3, 10.9, 10.1, 10.7, 10.7, 10.9, 11.4, 10.8, 11.2, 10.7, 10.0, 10.1, 10.7

5. Assignment Problem 22 in Chapter 2 presents a cumulative distribution function of the population growth rates of 204 countries recognized by the United Nations.

a. Draw a box plot using the information from the graph in that problem.

b. Label three features of this box plot.

6. The spider data in Example 3.2 consist of *pairs* of measurements made on the same subjects. One measurement is running speed before amputation and the second is running speed after amputation. Calculate a new variable called "change in speed," defined as the speed after amputation minus the speed before amputation.

a. What are the units of the new variable?

b. Draw a box plot for the change in running speed. Use the method outlined in Section 3.2 to calculate the quartiles.

c. Based on your drawing in part (b), is the frequency distribution of the change in running speed symmetric or asymmetric? Explain how you decided this.

d. What is the quantity measured by the span of the box in part (b)?

e. Calculate the mean change in running speed. Is it the same as the median? Why or why not?

f. Calculate the variance of the change in running speed.

g. What fraction of observations fall within one standard deviation above and below the mean?

7. If you were to add a constant k to each measurement in Practice Problem 6, what effect would this have on

a. the mean?

b. the standard deviation?

8. If you were to convert all of the observations of change in running speed in Practice Problem 6 from cm/s into mm/s, how would this change

a. the mean?

b. the standard deviation?

c. the median?

d. the interquartile range?

e. the coefficient of variation?

f. the variance?

9. Niderkorn's (1872; from Pounder 1995) measurements on 114 human corpses provided the first quantitative study on the development of rigor

mortis.[10] The data in the following table give the number of bodies achieving rigor mortis in each hour after death, recorded in one-hour intervals.

Hours	Number of bodies
1	0
2	2
3	14
4	31
5	14
6	20
7	11
8	7
9	4
10	7
11	1
12	1
13	2
Total	114

a. Calculate the mean number of hours after death that it took for rigor mortis to set in.
b. Calculate the standard deviation in the number of hours until rigor mortis.
c. What fraction of observations lie within one standard deviation of the mean (i.e., between $(\bar{Y} - s$ and $\bar{Y} + s)$?
d. Calculate the median number of hours until rigor mortis sets in. What is the likely explanation for the difference between the median and the mean?

ASSIGNMENT PROBLEMS

10. The following data are measurements of the number of young cod that "recruited" (grew to the catchable size) to the North Sea population in different years (Beaugrand et al. 2003). The measurements are listed below in $\log_{10}$ units (i.e., the measurement 5.0 represents $10^{5.0}$ recruits) and arranged from low to high rather than by year.

    5.0, 5.1, 5.2, 5.2, 5.2, 5.2, 5.2, 5.2, 5.3, 5.3, 5.3, 5.3, 5.4, 5.4, 5.4, 5.4, 5.5, 5.5, 5.5, 5.5, 5.5, 5.5, 5.5, 5.6, 5.6, 5.6, 5.7, 5.7, 5.7, 5.7, 5.7, 5.8, 5.8, 5.8, 5.9, 5.9, 6.0, 6.0

    a. What was the mean number of recruits (in $\log_{10}$ units) over this period?
    b. What was the standard deviation in the number of recruits (in $\log_{10}$ units)?

    c. In what fraction of years did the number of recruits fall within two standard deviations of the mean?

11. The gene for the vasopressin receptor $V1a$ is expressed at higher levels in the forebrain of monogamous vole species than promiscuous vole species.[11] Can expression of this gene influence monogamy? To test this, Lim et al. (2004) experimentally enhanced $V1a$ expression in the forebrain of 11 males of the meadow vole, a solitary promiscuous species. The percentage of time each male spent huddling with the female provided to him (an index for monogamy) was recorded. The same measurements were taken in 20 control males left untreated.

[10] Rigor mortis is the muscular stiffening that occurs after death. It is caused by linkages forming between actin and myosin in muscle when muscle glycogen is depleted, pH drops, and the level of ATP falls below a critical level.

[11] In monogamous vole species, single males and females form stable mating pairs. In promiscuous voles, no stable pairs form and voles might mate with multiple partners.

Control males: 98, 96, 94, 88, 86, 82, 77, 74, 70, 60, 59, 52, 50, 47, 40, 35, 29, 13, 6, 5

V1a-enhanced males: 100, 97, 96, 97, 93, 89, 88, 84, 77, 67, 61

a. Display these data in a graph.
b. Which group has the higher mean percentage of time spent huddling with females?
c. Which group has the higher standard deviation in percentage of time spent huddling with females?

12. Birds of the Caribbean islands of the Lesser Antilles are descended from immigrants originating from larger islands and the nearby mainland. The data presented here are the approximate dates of immigration, in millions of years, of each of 37 bird species now present on the Lesser Antilles (Ricklefs and Bermingham 2001). The dates were calculated from the difference in mitochondrial DNA sequences between each of the species and its closest living relative on larger islands or the mainland.

0.00, 0.00, 0.04, 0.21, 0.29, 0.54, 0.63, 0.88, 0.96, 1.25, 1.67, 1.75, 1.84, 1.96, 2.01, 2.51, 2.72, 3.30, 3.51, 4.05, 4.85, 6.94, 8.73, 10.57, 11.11, 12.45, 14.00, 17.30, 17.92, 18.05, 18.43, 22.48, 22.48, 23.48, 26.32, 26.45, 28.87

a. Plot the data in a histogram and describe the shape of the frequency distribution.
b. By viewing the graph alone, approximate the mean and median of the distribution. Which should be greater? Explain your reasoning.
c. Calculate the mean and median. Was your intuition in part (b) correct?
d. Calculate the first and third quartiles and the interquartile range.
e. Draw a box plot for these data.

13. The data in the accompanying table are from an ecological study of the entire rainforest community at El Verde in Puerto Rico (Waide and Reagan 1996). Diet breadth is the number of types of food eaten by an animal species. The number of animal species having each diet breadth is shown in the second column. The total number of species listed is $n = 127$.

Diet breadth (number of prey types eaten)	Frequency (number of species)
1	21
2	8
3	9
4	10
5	8
6	3
7	4
8	8
9	4
10	4
11	4
12	2
13	5
14	2
15	1
16	1
17	2
18	1
19	3
20	2
> 20	25
Total	127

a. Calculate the median number of prey types consumed by animal species in the community.
b. What is the interquartile range in the number of prey types? Use the method outlined in Section 3.2 to calculate the quartiles.
c. Can you calculate the mean number of prey types in the diet?

14. Francis Galton (1894) presented the following data on the flight speeds of 3207 "old" homing pigeons traveling at least 90 miles.

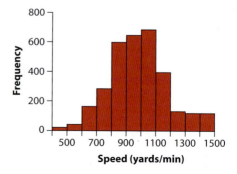

a. What type of graph is this?
b. Examine the graph and visually determine the approximate value of the mean (to the nearest 100 yards per minute). Explain how you obtained your estimate.
c. Examine the graph and visually determine the approximate value of the median (to the nearest 100 yards per minute). Explain how you obtained your estimate.
d. Examine the graph and visually determine the approximate value of the mode (to the nearest 100 yards per minute). Explain how you obtained your estimate.
e. Examine the graph and visually determine the approximate value of the standard deviation (to the nearest 50 yards per minute). Explain how you obtained your estimate.

15. On the basis of your answer to Practice Problem 8, describe how each of the following quantities is changed if every measurement in a data set were multiplied by a constant k:
a. the mean
b. the standard deviation
c. the median
d. the interquartile range
e. the coefficient of variation
f. the variance

16. A study of the endangered saiga antelope (pictured at the beginning of the chapter) recorded the fraction of females in the population that were fecund in each year between 1993 and 2001 (Milner-Gulland et al. 2003). A graph of the data is as follows:

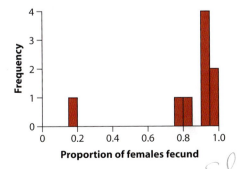

a. Assume that you want to describe the fraction of females that are fecund in a typical year, based on these data. What would be the better choice to describe this fraction, the mean or the median of the measurements? Why?
b. With the same goal in mind, what would be the better choice to describe the spread of measurements around their center, the standard deviation or the interquartile range? Why?

17. Accurate prediction of the timing of death in patients with a terminal illness is important for their care. The following graph compares the survival times of terminally ill cancer patients with the clinical prediction of their survival times (modified from Glare et al. 2003).

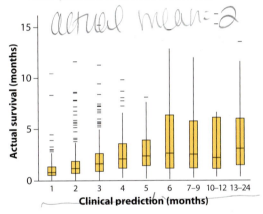

a. Describe in words what features most of the frequency distributions of actual survival times have in common, based on the box plots for each group.
b. Describe the differences in shape of actual survival time distributions between those for one to five months predicted survival times and those for six to 24 months.
c. Describe the trend in median actual survival time with increasing predicted number.
d. The predicted survival times of terminally ill cancer patients tend to overestimate the medians of actual survival times. Are the *means* of actual survival times likely to be closer to, further from, or no different from,

the predicted times than the medians?
Explain.

18. Measurements of lifetime reproductive success
(LRS) of individual wild animals reveal the dis-
parate contributions they make to the next generation. Jensen et al. (2004) estimated LRS of
male and female house sparrows in an island
population in Norway. They measured LRS of
an individual as the total number of "recruits"
produced in its lifetime, where a recruit is an
offspring that survives to breed one year after
birth. Parentage of recruits was determined from
blood samples using DNA techniques. (Assessing parentage from observation alone isn't accurate because pairs aren't entirely faithful in
most bird species.) Their results are tabulated
as follows:

| | Frequency | |
Lifetime reproductive success	Females	Males
0	30	38
1	25	17
2	3	7
3	6	6
4	8	4
5	4	10
6	0	2
7	4	0
8	1	0
> 8	0	0
Total	81	84

a. Which sex has the higher mean lifetime
   reproductive success?
b. Can you think of a reason why sexes might
   have different means in these data?
c. Which sex has the higher variance in reproductive success?

4

Estimating with uncertainty

When biologists carry out a study, their goals are usually more ambitious than a mere description of the resulting data. Rather, data are gathered so that something may be discovered about the larger population from which those individuals came. The descriptive statistics measured on a sample are used to estimate parameters of the population.

For example, the sample mean $\overline{Y}$ is used to estimate the true mean of the population, symbolized using the Greek letter μ ("mu"; pronounced "mew"). The sample standard deviation s is used to estimate the population standard deviation, symbolized by σ (lowercase Greek letter "sigma"). Likewise, the sample proportion $\hat{p}$ (pronounced "p-hat") is an estimate[1] of the population proportion p.

For such estimates of population parameters to be useful, we also need to quantify their uncertainty, or their precision—that is, how far they are likely to be from the tar-

---

[1] We will often use Greek letters (like μ and σ) to describe parameters and roman letters (e.g., $\overline{Y}$ and s) for their estimates. In addition, it is common to put a circumflex, or "hat," (ˆ) over a variable name to show it is an estimate (e.g. $\hat{p}$ is an estimate for p).

get parameter being estimated. Calculating the uncertainty of an estimate is possible when the sample taken is a *random sample*. Recall from Chapter 1 that a random sample is a sample in which all individuals in the population have an equal and independent chance of being selected.

In Chapter 4, we explain the basics of estimation and the uncertainties involved when making generalizations about a population from a sample. We introduce the standard error, a key measure of the precision of an estimate, and we demonstrate how to calculate it in the case of the sample mean. Moreover, we add a brief, conceptual introduction to the confidence interval, another important means of describing the precision of an estimate.

4.1   The sampling distribution of an estimate

Estimation is the process of inferring a population parameter from sample data. The value of an estimate calculated from data is almost never exactly the same as the value of the population parameter being estimated, because sampling is influenced by chance. The crucial question is, "In the face of chance, how much can we trust an estimate?" In other words, what is its *precision*? To answer this question, we need to know something about how the sampling process might affect the estimates we get. We use the **sampling distribution** of the estimate, which is the probability distribution of all the values for an estimate that we *might* have obtained when we sampled the population. We can illustrate the concept of a sampling distribution using samples from a known population, the human genome.

Example 4.1	The length of human genes

The international human genome project was the largest coordinated research effort in the history of biology. It yielded the DNA sequence of all 23 human chromosomes, each consisting of millions of nucleotides chained end to end.[2] These encode the genes whose products—RNA and proteins—shape the growth and development of each individual.

We obtained the lengths of all 20,290 known and predicted genes of the published genome sequence (Hubbard et al. 2005).[3] The length of a gene refers to the total number of nucleotides comprising the coding regions. The frequency distribution of gene lengths

---

[2]   The photo at the beginning of this chapter is a crystal of pure DNA.

[3]   We used the largest known transcript of each gene in release 35 of the human genome (http://www.ensembl.org/ Homo_ sapiens).

in the population of genes is shown in Figure 4.1-1. The figure includes only genes up to 15,000 nucleotides long; in addition, there are 26 longer genes. [4]

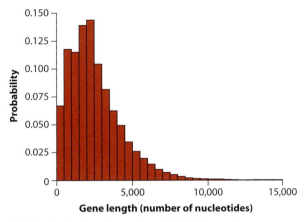

Figure 4.1-1 Probability distribution of gene lengths in the known human genome. The graph is truncated at 15,000 nucleotides; 26 larger genes are too rare to be visible in this histogram.

The histogram in Figure 4.1-1 is like those we have seen before, except that it shows the distribution of lengths in the *population* of genes, not simply those in a *sample* of genes. Because it is the population distribution, the fraction of genes of a given length interval in Figure 4.1-1 represents the *probability* of obtaining a gene of that length when sampling a single gene at random. The probability distribution of gene lengths is skewed, having a long tail extending to the right.

Table 4.1-1 Population mean and standard deviation of gene length in the known human genome.

Name	Parameter	Value (number of nucleotides)
Mean	μ	2622.0
Standard deviation	σ	2036.9

The population mean and standard deviation of gene length in the human genome are listed in Table 4.1-1. These quantities are referred to as *parameters* because they are quantities that describe the population.

In real life, we would not usually know the parameter values of the study population, but in this case we do. We'll take advantage of this to illustrate the process of sampling.

---

[4] The longest human gene, with nearly 100,000 nucleotides, encodes the gigantic protein titin, which is expressed in heart and skeletal muscle. The protein was named for the titans of Greek mythology, giants who ruled the earth until overthrown by the Olympians. Mutations in the titin gene cause heart muscle disease and muscular dystrophy.

Estimating mean gene length with a random sample

To begin, we collected a single random sample of $n = 100$ genes from the known human genome.[5] A histogram of the lengths of the resulting sample of genes is shown in Figure 4.1-2.

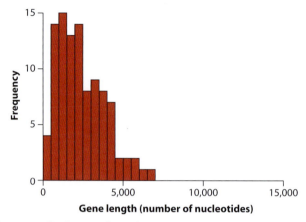

Figure 4.1-2 Frequency distribution of gene lengths in a unique random sample of $n = 100$ genes from the human genome.

The frequency distribution of the random sample (Figure 4.1-2) is not an exact replica of the population distribution (Figure 4.1-1) because of chance. The two distributions nevertheless share important features, including approximate location, spread, and shape. For example, the sample frequency distribution is skewed to the right like the true population distribution.

The sample mean and standard deviation of gene length from the sample of 100 genes are listed in Table 4.1-2. How close are these estimates to the population mean and standard deviation listed in Table 4.1-1? The sample mean is $\bar{Y} = 2411.8$, which is about 200 nucleotides shorter than the population mean of $\mu = 2622.0$. The sample standard deviation ($s = 1463.5$) is also different from the standard deviation of gene length in the population ($\sigma = 2036.9$). We shouldn't be surprised that the sample estimates differ from the parameter (population) values. Such differences are virtually inevitable because of chance in the sampling process.

Table 4.1-2 Mean and standard deviation of gene length Y in our unique random sample of $n = 100$ genes from the human genome.

Name	Statistic	Sample value (number of nucleotides)
Mean	$\bar{Y}$	2411.8
Standard deviation	s	1463.5

---

[5] All genes were listed in a file, one gene per line, with lines numbered from 1 to 20,290. We then used a computer to generate 100 random integers between the values 1 and 20,290, without allowing duplicates. Each random number was then used to draw a gene according to its line number in the list of genes.

The sampling distribution of $\overline{Y}$

We obtained $\overline{Y} = 2411.8$ nucleotides in our single sample, but by chance we might have obtained a different value. For example, on a second random sample of a different 100 genes, we found $\overline{Y} = 2643.5$. Different samples will generate different estimates of the same parameter. If we were able to repeat this sampling an infinite number of times, we could create the probability distribution of our estimate. The probability distribution of values we might obtain for an estimate make up the estimate's **sampling distribution**.

> The *sampling distribution* is the probability distribution of values for an estimate that we might obtain when we sample a population.

The sampling distribution represents the "population" of values for an estimate. It is not a real population, like the squirrels in Muir Woods or all the septuagenarians basking in the Florida sunshine. Rather, the sampling distribution is an imaginary population of values for an estimate. Taking a random sample of n observations from a population and calculating $\overline{Y}$ is equivalent to randomly sampling a *single* value of $\overline{Y}$ from its sampling distribution.

To visualize the sampling distribution for mean gene length, we used the computer to take a vast number of random samples of $n = 100$ genes from the human genome. We calculated the sample mean $\overline{Y}$ each time. The resulting histogram in Figure 4.1-3 shows the values of $\overline{Y}$ that might be obtained when randomly sampling 100 genes, together with their probabilities.

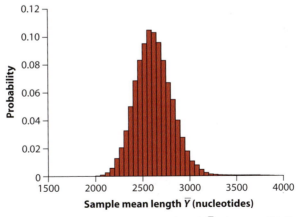

Figure 4.1-3 The sampling distribution of mean gene length, $\overline{Y}$, when $n = 100$. Note the change in scale from Figure 4.1-2.

Figure 4.1-3 makes plain that, although the population mean μ is a constant (2622.0), its estimate $\bar{Y}$ is a variable. Each new sample yields a different $\bar{Y}$ value from the one before. We don't ever see the sampling distribution of $\bar{Y}$ because ordinarily we have only one sample, and therefore only one $\bar{Y}$. Notice that the sampling distribution for $\bar{Y}$ is centered exactly on the true mean, μ. This means that $\bar{Y}$ is an unbiased estimate of μ.[6]

The spread of the sampling distribution of an estimate depends on the sample size. The sampling distribution of $\bar{Y}$ based on $n = 100$ is narrower than that based on $n = 20$, and that based on $n = 500$ is narrower still (Figure 4.1-4). The larger the sample size, the narrower the sampling distribution. And the narrower the sampling distribution, the more precise the estimate will be. Thus, larger samples are desirable whenever possible because they yield more precise estimates. The same is true for the sampling distributions of other estimates of population quantities, not just $\bar{Y}$.

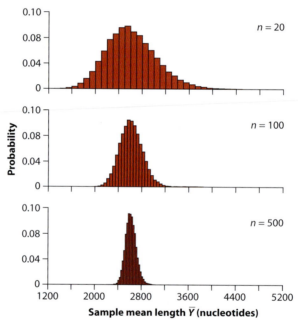

Figure 4.1-4 Comparison of the sampling distributions of mean gene length, $\bar{Y}$, when $n = 20, 100$, and 500.

> Increasing sample size reduces the spread of the sampling distribution of an estimate, increasing precision.

[6]   Notice how the shape of the sampling distribution resembles that of a normal distribution, or "bell curve," more so than the distribution of gene lengths themselves. Most of the methods presented in this book rely on the normal distribution to approximate the sampling distribution of an estimate.

4.2 Measuring the uncertainty of an estimate

In Section 4.2, we show how the sampling distribution is used to measure the precision of an estimate.

Standard error

The standard deviation of the sampling distribution of an estimate is called the **standard error**. Because it reflects the differences between an estimate and the target parameter, the standard error measures the precision of an estimate. Estimates with smaller standard errors are more precise than those with larger standard errors. The smaller the standard error, the less uncertainty there is about the parameter of interest in the whole population.

> The *standard error* of an estimate is the standard deviation of the estimate's sampling distribution.

The standard error of $\bar{Y}$

The standard error of the sample mean is particularly simple to calculate, so we show it here. We can represent the standard error of the mean with the symbol $\sigma_{\bar{Y}}$. It has a remarkably straightforward relationship with σ, the population standard deviation of Y:

$$\sigma_{\bar{Y}} = \frac{\sigma}{\sqrt{n}}.$$

The standard error decreases with increasing sample size. Table 4.2-1 lists the standard error of the sample mean based on random samples of $n = 20$, 100, and 500 from the known human genome.

Table 4.2-1 Standard error of the sampling distributions of mean gene length $\bar{Y}$ according to sample size. These measure the spread of the three sampling distributions in Figure 4.1-4.

Sample size, n	Standard error, $\sigma_{\bar{Y}}$ (nucleotides)
20	455.5
100	203.7
500	91.1

The standard error of $\bar{Y}$ from data

The trouble with the formula for the standard error of the mean ($\sigma_{\bar{Y}}$) is that we almost never know the value of the population standard deviation (σ) and so we cannot calculate $\sigma_{\bar{Y}}$. The next best thing is to use the sample standard deviation (s) as an estimate of σ, in which case we can approximate $\sigma_{\bar{Y}}$. The estimated standard error of the mean is

$$SE_{\bar{Y}} = \frac{s}{\sqrt{n}}.$$

According to this simple relationship, all we need is one random sample to approximate the spread of the entire sampling distribution for $\bar{Y}$. The quantity $SE_{\bar{Y}}$ is usually called the "standard error of the mean."

> The *standard error of the mean* is estimated from data as the sample standard deviation (s) divided by the square root of the sample size (n).

Calculating $SE_{\bar{Y}}$ is so routine in biology that a sample mean should never be reported without it. For example, if we were submitting the results of our unique random sample of 100 genes in Figure 4.1-2 for publication, we would calculate $SE_{\bar{Y}}$ from the results in Table 4.1-2 as follows:

$$SE_{\bar{Y}} = \frac{s}{\sqrt{n}} = \frac{1463.5}{\sqrt{100}} = 146.3.$$

We would then report the sample mean in the text of the paper as

$$\bar{Y} = 2411.8 \pm 146.3\,(SE).$$

Every estimate, not just the mean, has a sampling distribution with a standard error, including proportions, medians, correlations, differences between means, and so on. In the rest of this book, we will give formulas to calculate standard errors of many kinds of estimates.

4.3 Confidence intervals

The **confidence interval** is another common way to quantify uncertainty about the value of a parameter. It is a range of numbers calculated from the data that is likely to contain, within its span, the unknown value of the target parameter. Here in Section 4.3, we introduce the concept without showing exact calculations. Confidence intervals can be calculated for means, proportions, correlations, differences between means, and other population parameters, as later chapters will demonstrate.

A *confidence interval* is a range of values surrounding the sample estimate that is likely to contain the population parameter.

An example is the **95% confidence interval for the mean**. This confidence
interval is a range likely to contain the value of the true population mean μ. It is calculated from the data and extends above and below the sample estimate $\overline{Y}$. You will
encounter confidence intervals frequently in the biological literature. We'll show you
in Chapter 11 how to calculate an exact confidence interval for the mean, but for now
we give you the result and its interpretation. The 95% confidence interval for the
mean calculated from the unique sample of 100 genes (Table 4.1-2) is

$$2121.4 < \mu < 2702.2.$$

This calculation allows us to say, "We are 95% confident that the true mean lies
between 2121.4 and 2702.2 nucleotides." We do *not* say that "there is a 95% probability that the population mean falls between 2121.4 and 2702.2 nucleotides," which
is a common misinterpretation of the confidence interval (2121.4 and 2702.2 are both
constants, so there's no probability involved).

To better understand the correct interpretation of "95% confidence," imagine that
20 researchers independently take unique random samples of $n = 100$ genes from the
human genome. Each researcher calculates an estimate $\overline{Y}$ and then a 95% confidence
interval for the parameter (the population mean, μ). Each researcher ends up with a
different estimate and a different 95% confidence interval, because their samples
are not the same (Figure 4.3-1). On average, however, 19 out of 20 (95%) of the

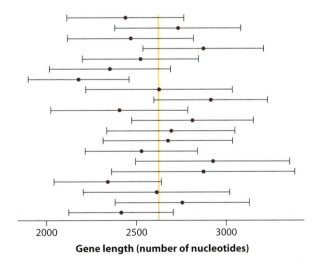

Gene length (number of nucleotides)

Figure 4.3-1 The 95% confidence intervals for the mean calculated from 20 separate random
samples of $n = 100$ genes from the known human genome. Dots indicate the sample means $\overline{Y}$.
The vertical line (colored) represents the population mean, $\mu = 2622.0$. In this example, 19 of 20
intervals included the population mean, whereas one interval did not.

researchers' intervals will contain the value of the population parameter. On average, therefore, *one* out of 20 intervals (5%) will *not* contain the parameter value. None of the researchers will know for sure whether his or her own confidence interval contains the value of the unknown parameter, but all can be "95% confident" that it does.

All numbers falling between the lower and upper bounds of a confidence interval can be regarded as the most plausible values for the parameter, given the data sampled. Values falling outside the confidence interval are less plausible. If the confidence interval is broad, therefore, then the data are not very informative about the location of the population parameter. If the confidence interval is narrow, on the other hand, then we can be confident that the parameter is close to the estimated value.

> The *95% confidence interval* provides a most-plausible range for a parameter. Values lying within the interval are most plausible, whereas those outside are less plausible, based on the data.

The 2SE rule of thumb

A good "quick-and-dirty" approximation to the 95% confidence interval for the population mean is obtained by adding and subtracting two standard errors from the sample mean (the so-called 2SE rule of thumb).

> A rough approximation to the 95% confidence interval for a mean can be found from the sample mean plus and minus two standard errors.

For our unique random sample of 100 genes (Figure 4.1-2), for example, the sample mean for the gene length was $\bar{Y} = 2411.8$ nucleotides, and its standard error was $SE_{\bar{Y}} = 146.3$ nucleotides. Two standard errors below the mean is then

$$\bar{Y} - 2SE_{\bar{Y}} = 2411.8 - (2 \times 146.3) = 2119.2$$

and two standard errors above the mean is

$$\bar{Y} + 2SE_{\bar{Y}} = 2411.8 + (2 \times 146.3) = 2704.4.$$

According to the 2SE rule of thumb, then, the 95% confidence interval for the mean gene length in the population can be approximated as being between 2119.2 and 2704.4 nucleotides. This is not too far off from the more exact confidence interval (i.e., between 2121.4 and 2702.2 nucleotides) that we calculated previously.

4.4 Summary

▶ Estimation is the process of inferring a population parameter from sample data.
▶ All estimates have a sampling distribution, which is the probability distribution of all the possible values of the estimate that might be obtained under random sampling.
▶ The standard error of an estimate is the standard deviation of its sampling distribution. The standard error measures precision. The smaller the standard error, the more precise is the estimate.
▶ The standard error of an estimate declines with increasing sample size.
▶ The confidence interval is a range of values calculated from sample data that is likely to contain within its span the value of the target parameter. On average, 95% confidence intervals calculated from independent random samples will include the value of the parameter 19 times out of 20.
▶ The 2SE rule of thumb (i.e., the sample mean plus or minus two standard errors) provides a rough approximation to the 95% confidence interval for a mean.

4.5 Quick Formula Summary

Standard error of the mean

What is it for? Measuring the precision of the sample estimate $\overline{Y}$ of the population mean μ.

What does it assume? The sample is a random sample.

Estimate: $\mathrm{SE}_{\overline{Y}}$

Parameter: $\sigma_{\overline{Y}}$

Formula: $\mathrm{SE}_{\overline{Y}} = \dfrac{s}{\sqrt{n}}$

where s is the sample standard deviation and n is the sample size. $\mathrm{SE}_{\overline{Y}}$ estimates the population quantity $\sigma_{\overline{Y}} = \dfrac{\sigma}{\sqrt{n}}$, where $\sigma_{\overline{Y}}$ is the standard error of the sample mean, and σ is the standard deviation of Y in the population.

PRACTICE PROBLEMS

1. Examine the times to rigor mortis of the 114 human corpses tabulated in Practice Problem 9 of Chapter 3.
   a. What is the standard error of the mean time to rigor mortis?
   b. The standard error calculated in part (a) measures the spread of what frequency distribution?
   c. What assumption does your calculation in part (a) require?

2. Examine the frequency distribution of gene lengths in the human genome displayed in Figure 4.1-1. Is the population median gene length in the human genome likely to be larger, smaller, or equal to the population mean? Explain.

3. As a general rule, is the spread of the sampling distribution for the sample mean mainly determined by the magnitude of the mean or by the sample size?

4. Seven of the 100 human genes we sampled randomly from the human genome were found to occur on the X chromosome. The sample fraction of genes on the X was thus $\hat{p} = 7/100 = 0.07$. For each of the following statements, specify whether it is true or false:
   a. $\hat{p} = 0.07$ is the fraction of all human genes on the X chromosome.
   b. $\hat{p} = 0.07$ estimates p, the fraction of all human genes on the X chromosome.
   c. $\hat{p}$ has a sampling distribution representing the frequency distribution of values of $\hat{p}$ that we might obtain when we randomly sample 100 genes from the human genome.
   d. The fraction of all human genes on the X chromosome has a sampling distribution.
   e. The standard deviation of the sampling distribution of $\hat{p}$ is the standard error of $\hat{p}$.

5. In a poll of 1,641 people carried out in Canada in November 2005, 73% of people surveyed agreed with the statement that "you don't really expect that politicians will keep their election promises once they are in power" (CBC News 2005).
   a. What is the parameter being estimated?
   b. What is the sample estimate?
   c. What is the sample size?
   d. The poll also reported that "the results are considered accurate within 2.5 percentage points, 19 times out of 20." Explain what this statement likely refers to.

6. The following data are flash durations, in milliseconds, of a sample of 35 male fireflies of the species *Photinus ignitus* (Cratsley and Lewis 2003; see Assignment Problem 18 in Chapter 2).

   79, 80, 82, 83, 86, 85, 86, 86, 88, 87, 89, 89, 90, 92, 94, 92, 94, 96, 95, 95, 95, 96, 98, 98, 98, 101, 103, 106, 108, 109, 112, 113, 118, 116, 119

   a. Estimate the sample mean flash duration. What does this quantity estimate?
   b. Is the estimate in part (a) likely to equal the population parameter? Why or why not?
   c. Calculate a standard error for your sample estimate.
   d. What does the quantity in part (c) measure?
   e. Using an approximate method, calculate a rough 95% confidence interval for the population mean.
   f. Provide an interpretation for the interval you calculated in part (e).

7. Imagine that the results of a study calculated a sample mean of zero, with a narrow 95% confidence interval for the population mean (modified from Borenstein 1997). The most appropriate conclusion is that:
   a. The population mean is likely to be zero or close to zero.
   b. The population mean is probably zero, but there is an equally good chance it is either slightly less than zero or slightly greater than zero.
   c. We can be reasonably certain the mean differs from zero.

ASSIGNMENT PROBLEMS

8. A massive survey of sexual attitudes and behavior in Britain between 1999 and 2001 contacted 16,998 households and interviewed 11,161 respondents aged 16–44 years (one per responding household). The frequency distributions of ages of men and women respondents were the same. The following results were reported on the number of heterosexual partners individuals had had over the previous five-year period (Johnson et al. 2001).

	Sample size, n	Mean	Standard deviation
Men	4620	3.8	6.7
Women	6228	2.4	4.6

  a. What is the standard error of $\bar{Y}$ in men? What is it in women? Assume that the sampling was random.[7]
  b. Which is a better descriptor of the variation among men in the number of sexual partners, the standard deviation or the standard error? Why?
  c. Which is a better descriptor of uncertainty in the estimated mean number of partners in women, the standard deviation or the standard error? Why?
  d. A mysterious result of the study is the discrepancy between the mean number of partners of heterosexual men and women. If each sex obtains its partners from the other sex, then the true mean number of heterosexual partners should be identical. Considering aspects of the study design, suggest an explanation for the discrepancy.

9. Our unique random sample of 100 human genes from the human genome was found to have a median length of 2150 nucleotides. For each of the following statements, specify whether it is true or false.

  a. The median gene length of all human genes is 2150 nucleotides.
  b. The median gene length of all human genes is estimated to be 2150 nucleotides.
  c. The sample median has a sampling distribution with a standard error.
  d. A random sample of 1000 genes would likely yield an estimate of the median closer to the population median than a random sample of 100 genes.

10. The following figure is from the web site of a U.S. national environmental laboratory.[8] It displays sample mean concentrations, with 95% confidence intervals, of three radioactive substances. The text accompanying the figure explained that "the first plotted mean is 2.0 ± 1.1, so there is a 95% chance that the actual result is between 0.9 and 3.1, a 2.5% chance it is less than 0.9, and a 2.5% chance it is greater than 3.1." Is this a correct interpretation of a confidence interval? Explain.

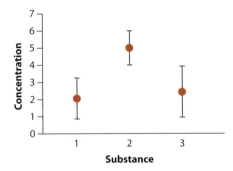

11. One of the great discoveries of biology is that organisms have a class of genes called "regulatory genes" whose only job is to regulate the activity of other genes. How many genes

---

[7] The sampling design was actually more complicated than simply random, a fact which we will ignore. Standard errors and confidence intervals for more complex sampling schemes are not covered in this book.

[8] Redrawn from a figure at http://www.pnl.gov/env /Helpful2.html; accessed January 15, 2006.

does the typical regulatory gene regulate? A recent study of interaction networks in yeast (*S. cerevesiae*) came up with the following data for 109 regulatory genes (Guelzim et al. 2002).

Number of genes regulated	Frequency
1	20
2	10
3	7
4	7
5	8
6	8
7	5
8	2
9	4
10	4
11	3
12	4
13	5
14	1
15	2
16	1
17	3
18	2
19	2
20	3
22	3
25	1
26	1
28	1
29	1
37	1
Total	109

a. What type of graph should be used to display these data?
b. What is the estimated mean number of genes regulated by a regulatory gene in the yeast genome?
c. What is the standard error of the mean?
d. Explain what this standard error measures.
e. What assumption are you making in part (c)?

12. Refer to the previous problem (Assignment Problem 11).
a. Using an approximate method, provide a rough 95% confidence interval for the population parameter.
b. Provide an interpretation of the interval you calculated in part (a).

13. A study was carried out on a domestic goose population to estimate the proportion of individuals infected with flu virus. The estimated proportion was 0.40, with a 95% confidence interval extending from 0.10 to 0.70. Which of the following statements are true?
a. The population proportion is 0.40.
b. The population proportion is likely to be between 0.10 and 0.70.
c. There is a 95% chance that the population proportion is between 0.10 and 0.70.
d. A population proportion of 0.80 is plausible.

Pseudoreplication

Most statistical techniques assume that the data are a random sample from the population, in which each individual has an equal and independent chance of being sampled. Unfortunately, many experiments are conducted in a way that violates the assumption of independence.

How, for example, should we measure whether female birds prefer different types of male birdsong? In some species male birds vary in song complexity, with some males having multiple notes and phrases and other males having a simpler song. A classic way to measure female preferences is to play a tape recording of a song and watch what she does. (Females sometimes have recognizable courtship behaviors that can be used to indicate their interest.) One study[1] recorded the complex song of one male and the relatively simple song of another male, and the researchers played these same songs for 40 different females. The value being measured was how much more often the complex song was preferred by the female compared with the simple song.

A confidence interval was calculated based on the responses of the 40 females, and the result was a very narrow range of plausible values for female preference for song complexity. But what has gone wrong? The measurements of the females' preferences for complexity were not independent. All females listened to the songs of the same two males. All females that listened to the "complex" song were listening to

Pseudoreplication is probably the single most common fault in the design and analysis of ecological field experiments. It is at least equally common in many other areas of research.
—Stuart Hurlbert (1984)

one male, and all females that listened to the "simple" song were listening to another male. In reality, the sample size for complex songs was just $n = 1$, and the sample size for simple songs was $n = 1$ also. In order to understand the effects of song complexity, multiple complex songs would have to be compared to multiple simple songs. As it is, the study only shows that the two males differ in the attractiveness of their songs, not that more complex male songs are more attractive in general.

The bird-song study is an example of pseudoreplication. **Pseudoreplication** occurs in a study whenever individual measurements are not independent but are analyzed as if they are independent of one another. The problem with pseudoreplication is that measurements obtained from individuals not sampled independently might be more similar to one another than measurements made on individuals sampled independently from the population. "Replication" by itself is good—it refers

---

[1] We'll keep the specific example fictional, to avoid singling out one set of authors unfairly, but this is based on many examples in the literature.

to the sampling of multiple independent units from a population, which makes it possible to estimate population characteristics and the precision of those estimates. In general, the larger the level of replication, the greater our confidence in our results. But if we analyze non-independent data points as if they were independent, then we are making a false claim about the amount of replication, hence the "pseudo-" in "pseudoreplication."[2]

Most statistical techniques, including almost everything in this book, assume that each data point is independent of the others. Independence is, after all, built in to the definition of a random sample. When we assume that data points are independent of each other, we give each data point equal credence and weigh its information as heavily as every other point. If two data points are not independent, though, then treating them as independent makes it seem as if we have more information than we really do. We would be treating the data set as if it were larger than it really is, and as a result we would calculate confidence intervals that were too narrow and P-values (see Section 6.2) that were too small.

Imagine that we wanted to know the blood sugar levels of diabetes patients. An overzealous phlebotomist takes 15 samples from each of 10 patients, yielding a total of 150 measurements. How can we treat these 150 data points? If we threw them all together and analyzed them as if we had 150 independent data points, we would commit the sin of pseudoreplication. If the patients were randomly chosen, then we would have only 10 independent data points.

We needn't throw out any of the 15 samples per patient. Rather, for each patient, we could take the average of the 15 measurements and use this as the independent observation. Doing so gives us a more reliable estimate of the blood sugar level of each patient, which can reduce the amount of sampling variation in our estimate compared with an estimate based on only a single measurement per patient. But no matter how many measurements are taken from a patient, the total sample size of this study is still only $n = 10$. By recognizing that the patients, not the measurements, represent our unit of replication, we can analyze the data appropriately.

Pseudoreplication can often be avoided by summarizing the information on each independently sampled individual and using those summaries as the data for the analysis.

Pseudoreplication is often subtle, and it remains a major source of mistakes in the analysis of experiments.

The rate of pseudoreplication has been estimated to be one in every eight field studies in ecology (Hurlbert and White 1993, Heffner et al. 1996). When reading the scientific literature, keep in mind the possibility of pseudoreplication. Be on the lookout for features that group individual data points during the sampling process. Watch out for multiple measurements taken on the same individuals or the same experimental unit. If the number of measurements, and not the number of independent individuals, is counted as the sample size in a statistical analysis, there could be a problem.

Suggested reading

Hurlbert, S. H. 1984. Pseudoreplication and the design of ecological field experiments. *Ecological Monographs* 54: 187–211.

---

[2] The term "pseudoreplication" was coined by Hurlbert (1984) in an extremely readable exposé of the shocking ubiquity of pseudoreplication in biology.

5

Probability

The concept of probability is important in almost every field of science, including biology. Probability is also the backbone of data analysis. In Chapter 4, we learned that we must constantly make statements about probability to quantify the uncertainty of results, and we will see even more ways that this is true throughout the book.

Probability is essential to biology because we almost always look at the natural world by way of a sample, and, as we have seen, chance can play a major role in the properties of samples. In this chapter, we will discuss the basic principles of probability and basic probability calculations. In Chapter 6, we will begin to apply these concepts to data analysis.

5.1 The probability of an event

Imagine that you have 1000 songs on your iPod, each of them recorded exactly once. When you push the "shuffle" button, the iPod plays a song at random from the list of 1000. The probability that the song played is your single favorite song is 1/1000, or 0.001. The probability that the song played is not your favorite song is 999/1000, or 0.999. What exactly do these numbers mean?

The concept of probability rests on the notion of a **random trial** repeated many times. A random trial is a process or experiment that has two or more possible outcomes whose occurrence cannot be predicted. In the iPod example, a random trial consists of pushing the shuffle button once. The specified outcome is "*your favorite song is played*," which is one of 1000 possible outcomes. Other examples of random trials include the following:

▶ Flipping a coin to see if heads or tails comes up,
▶ Rolling a pair of dice to see what the sum of their numbers is,
▶ Randomly sampling an individual from a population of sockeye salmon to see what its weight is, and
▶ Randomly sampling 10 individuals from a population of Dungeness crabs to see what proportion is female.

To describe a probability, we need to define the **event** and the **sample space**. The sample space is the list of all possible outcomes of a random trial. An event is any potential subset of the sample space. For example, there are six possible outcomes if we roll a six-sided die—the numbers one through six. Together these six different numbers represent the sample space. We can define many different events that we might care about, such as "*the result is an even number*," "*the result is a number greater than three*," or even the simple event "*the result is four*."

The **probability** of an event is the *proportion* of all random trials in which the specified event occurs when the same random process is repeated over and over again independently and under the same conditions.[1] If we randomly sample an individual from a population that has 2/3 males, the probability of sampling a male will be 2/3.

> The *probability* of an event is the proportion of times the event would occur if we repeated a random trial over and over again under the same conditions.

A useful shorthand is the following:

Pr[*A*] means "the probability of event *A*."

Thus, if we want to state the probability of "*rolling a four*" with a six-sided die, then we can write

Pr[*rolling a four*] = 1/6.

Because probabilities are proportions, they must always fall between zero and one, inclusive. An event has probability zero if it *never* happens, and an event has probability one if it *always* happens.

---

[1] Believe it or not, the word "probability" has other definitions, even within statistics. In one alternative, "probability" refers to a subjective state of belief by the researcher about the truth. For example, "There's a 95% chance that I turned the gas off before we left for vacation." In this book, though, we define probability only as a proportion.

5.2   Venn diagrams

One useful way to think about the probabilities of events is with a graphical tool called a **Venn diagram**. The area of the diagram represents all possible outcomes of a random trial, and we can represent various events as areas within the diagram. The probability of an event is proportional to the area it occupies in the diagram.

Figure 5.2-1 shows a Venn diagram for one roll of a fair six-sided die. The six possible outcomes fill the diagram, indicating that these are all possible results. The box for each outcome is in this case equally large, showing that these outcomes are equally probable. They each contain 1/6 of the area of the Venn diagram.

We can use Venn diagrams to show more complicated events as well. In Figure 5.2-2, for example, the event "*the result is greater than two*" is shown.

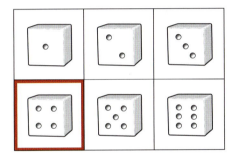   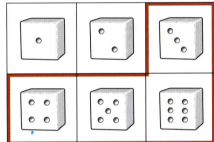

Figure 5.2-1 A Venn diagram for the outcomes of a roll of a six-sided die. The area corresponding to the event "*the result is a four*" is highlighted in red.

Figure 5.2-2 A Venn diagram showing the event "*the result is greater than two*" highlighted in red. The probability of this event is 4/6 = 2/3, equal to the area of the red box.

5.3   Mutually exclusive events

When two events are not simultaneously possible, we say that they are **mutually exclusive**. For example, a single die rolled once cannot yield both a one and a six. The events "*one*" and "*six*" are mutually exclusive.

> Two events are *mutually exclusive* if they cannot both occur simultaneously.

Sometimes physical constraints explain why certain events are mutually exclusive. It is impossible, for example, for more than one number to result from a single roll of a die. Sometimes events are mutually exclusive because they never occur simultaneously in nature. For example, "has teeth" and "has feathers" are mutually exclusive events when we randomly select a single living animal species, because no living animals have both teeth and feathers. If we select a living animal species at random, the probability that it has both teeth and feathers is zero, although plenty of animals have teeth *or* feathers.

In mathematical terms, two events *A* and *B* are mutually exclusive if

$$\Pr[A \text{ and } B] = 0.$$

Here, $\Pr[A \text{ and } B]$ means the probability that both *A* and *B* occur.

5.4   Probability distributions

In Section 1.4, we defined a *probability distribution* as the distribution of a variable in the entire population. The probability distribution of a random trial is the probability of each of the mutually exclusive outcomes of the random trial. Some probability distributions can be described mathematically, while others are just a list of outcomes and their probabilities. The precise meaning of a probability distribution depends on whether the variable is discrete or continuous.

> A *probability distribution* is a list of the probabilities of all mutually exclusive outcomes of a random trial.

Discrete probability distributions

A discrete probability distribution gives the probability of each possible value or outcome of a discrete variable. For example, the probability distribution of outcomes for the single roll of a fair die is given in Figure 5.4-1. In this case, all integers between one and six are equally probable outcomes (probability $= 1/6 = 0.167$). The histogram in Figure 5.4-2 shows the probability distribution for the sum of the

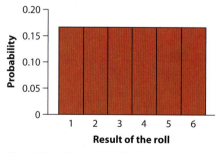

Figure 5.4-1 The probability distribution of outcomes resulting from the role of a single six-sided fair die. The probability of each possible outcome is 1/6 = 0.167.

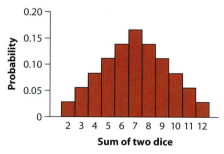

Figure 5.4-2 The probability distribution for the sum of the numbers resulting from rolling two six-sided fair dice.

two numbers resulting from a roll of two dice. Here the different outcomes are *not* equally probable.

By definition, the sum of all probabilities in a probability distribution must add to one.

Continuous probability distributions

Unlike discrete variables, continuous variables can take on any real number value within some range. Between any two values of a continuous variable (call it Y), an infinite number of other values are possible. We describe the distribution with a curve whose height is the **probability density.** The probability density allows us to describe the probability of any range of values.

The normal distribution, first introduced in Section 1.4, is a continuous probability distribution. It is bell-shaped like the curve shown in Figure 5.4-3. We'll see much more of this distribution in Chapter 10.

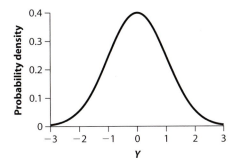

Figure 5.4-3 A normal distribution.

Unlike discrete probability distributions, the height of a continuous probability curve at the value of $Y = 2.4$ does not give the probability of obtaining $Y = 2.4$ when a single measurement is drawn randomly from the population. Because a continuous probability distribution covers an infinite number of possible outcomes, the probability of obtaining any specific outcome is infinitesimally small.

Rather, the height of the curve indicates the density of measurements at $Y = 2.4$. With continuous probability distributions, such as the normal curve, it makes more sense to talk about the probability of obtaining a value of Y within some range. The probability of obtaining a value of Y within some range is indicated by the area under the curve. For example, the probability that a single randomly chosen individual has a measurement lying between the two numbers a and b equals the area under the curve between a and b (Figure 5.4-4).

The area under the curve between a and b is calculated by integrating[2] the probability density function between the values a and b. Integration is the continuous ana-

---

[2] Don't panic! We won't ask you to carry out this integration in this book.

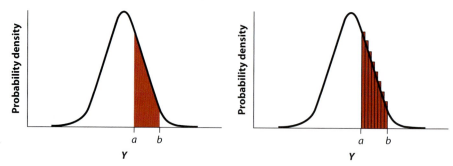

Figure 5.4-4 The probability that a randomly chosen Y-measurement lies between *a* and *b* is the area under the probability density curve between *a* and *b* (left panel). In the right panel we approximate the same area using discrete bars.

log of summation, so integrating the probability density function from *a* to *b* is analogous to adding together the probability of *Y* for all values between *a* and *b* (see the right panel in Figure 5.4-4).

For any probability distribution, the area under the entire curve of a continuous probability density function is always one. Finally, because the probability of any individual *Y*-value is infinitesimally small under a continuous probability density distribution, $\Pr[a \leq Y \leq b]$ is the same as $\Pr[a < Y < b]$.

5.5 Either this or that: adding probabilities

Very often, in life as in statistics, we want to know the probability that we get *either* one event *or* another. In the game of craps,[3] for example, we win if, on the first roll of two dice, the numbers sum to either seven or 11. What is the chance of winning on the first roll?

The addition rule

If the events that we are trying to combine are mutually exclusive, then calculating the probability of one or the other event occurring is both intuitive and easy. The probability of getting either of two mutually exclusive events is simply the sum of the probabilities of each of those events separately. Rolling a seven and rolling an 11 are mutually exclusive events. Therefore, the chance of rolling either a seven or an 11 is the chance of rolling a seven plus the chance of rolling an 11:

$$\Pr[\textit{rolling a 7 or rolling an 11}] = \Pr[\textit{rolling a 7}] + \Pr[\textit{rolling an 11}].$$

---

[3] A game that involves rolling two six-sided dice.

Figure 5.5-1 illustrates the probability of rolling a seven or an 11 with a Venn diagram.

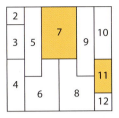

Figure 5.5-1 The probability of rolling either a seven or an 11 is equal to the sum of the probability if rolling a seven plus the probability of rolling an 11, because the two events are mutually exclusive. (This Venn diagram represents the same probability distribution as the histogram in Figure 5.4-2.)

This additive property of the probabilities of mutually exclusive events is called the **addition rule**.

The *addition rule*: If two events A and B are mutually exclusive, then

$$\Pr[A \text{ or } B] = \Pr[A] + \Pr[B].$$

The addition rule extends to more than two events as long as they are all mutually exclusive. Let's say that we want to know the probability of rolling three or more from a single roll of a die. Three or more includes four different outcomes: "*three*," "*four*," "*five*," and "*six*." These four possibilities are mutually exclusive, because you can't roll a four and a five (or any other combination of the four possibilities) at the same time on the same die. Thus, the probability of rolling three or more can be calculated as follows using the addition rule:

$$
\begin{aligned}
\Pr[\text{rolling 3 or more}] &= \Pr[3] + \Pr[4] + \Pr[5] + \Pr[6] \\
&= 1/6 + 1/6 + 1/6 + 1/6 \\
&= 2/3.
\end{aligned}
$$

(To keep the equation short, we used the shorthand notation $\Pr[3]$ to mean the probability of rolling a three, etc.) The Venn diagram for this event was shown in Figure 5.2-2. Notice that the areas (probabilities) of the four mutually exclusive outcomes add up to the area (probability) of the event we were looking for.

The addition rule is about "or" statements. If two events are mutually exclusive and we want to know the probability of being *either* one *or* the other, we can use the addition rule. This property is vital to analyzing data because it allows us to calculate the probabilities of different outcomes of sampling when they are mutually exclusive.

The probabilities of all possible mutually exclusive events add to one

The probabilities of all possible, mutually exclusive outcomes of a random trial must add to one. With a single roll of a fair die, for example, there are six possible mutually exclusive outcomes (the numbers one through six), and each has a probability of 1/6. Therefore, the sum of the probabilities of all outcomes is

$$\Pr[1 \text{ or } 2 \text{ or } 3 \text{ or } 4 \text{ or } 5 \text{ or } 6] = \Pr[1] + \Pr[2] + \Pr[3] + \Pr[4] + \Pr[5] + \Pr[6]$$
$$= 1/6 + 1/6 + 1/6 + 1/6 + 1/6 + 1/6$$
$$= 1.$$

This means that the probability of an event *not* occurring is simply one minus the probability that it occurs. For example, the probability that you do *not* get a two when you roll a fair die is

$$\Pr[\text{not } rolling\ a\ 2] = 1 - \Pr[rolling\ a\ 2] = 5/6.$$

This calculation is much easier than summing the probabilities of all outcomes other than two.

$$\Pr[not\ A] = 1 - \Pr[A].$$

The general addition rule

Not all events, though, are mutually exclusive. It is possible, for example, to roll a one on one die and a six on a different die at the same time. An animal can have both feathers and the ability to fly, although not all animals with feathers can fly and not all flying animals have feathers. If the two events are not mutually exclusive, how do we calculate the probability of either event being true?

In mathematical notation, this generalized addition rule can be written as

$$\Pr[A \text{ or } B] = \Pr[A] + \Pr[B] - \Pr[A \text{ and } B].$$

When events *A* and *B* are mutually exclusive, $\Pr[A \text{ and } B] = 0$, so the generalized addition rule reduces to the addition rule for mutually exclusive events introduced

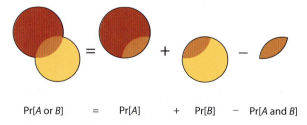

| Pr[A or B] | = | Pr[A] | + | Pr[B] | − | Pr[A and B] |

Figure 5.5-2 The general addition rule. Pr[A and B] is subtracted from Pr[A] + Pr[B] so that the outcomes where both A and B occur (the tan shaded areas) are not counted twice.

previously. The reason why we have to subtract the probability of overlap is illustrated in Figure 5.5-2. If we do *not* subtract the probability of both *A* and *B* occurring, then we will double-count those outcomes where both *A* and *B* occur.

5.6 Independence and the multiplication rule

Science is the study of patterns, and patterns are generated by relationships between events. Men are more likely to be tall, more likely to have a beard, more likely to die young, and more likely to go to prison than women. In other words, height, beardedness, age at death, and criminal conduct are not independent of sex in the human population.

Sometimes, though, the chance of one event does *not* depend on another event. If we roll two dice, for example, the number on one die does not affect the number on the other die. If knowing one event gives us no information about another event, then these two events are independent.

Two events are **independent** if the occurrence of one does not in any way affect the probability that the other will also occur. When rolling the same fair die twice in a row, for example, the probability that the first roll gives a three is 1/6, as we saw previously:

$$\text{Pr}[\textit{first roll is three}] = 1/6.$$

What is the probability that the next roll will also be a three? The probability of rolling a three on the second roll is still 1/6, regardless of whether the first roll was a three or not. Because the outcome of the first roll does not affect the probability of rolling a three on the second roll, we can say that the two events are independent (Figure 5.6-1).

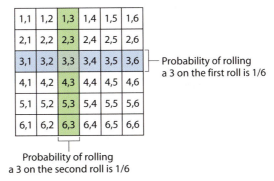

Figure 5.6-1 A Venn diagram for all the possible outcomes of rolling two six-sided dice. The first digit of each pair shows the result of the roll of the first die, and the second number shows the result of the roll of the second die. Rolling a three on the first roll is shown in the blue box. The probability of rolling a three on the second roll is shown in the green box and is the same (1/6) regardless of the result of the first roll.

Two events are *independent* if the occurrence of one does not change the probability that the second will occur.

When the probability of one event depends on the results of another event, then the two events are **dependent**.

Multiplication rule

When two events are independent, then the probability that they both occur is the probability of the first event multiplied by the probability of the second event. This is called the **multiplication rule**. This property of independent events is vital to analyzing data because it allows us to determine whether two or more variables are associated.

The *multiplication rule*: If two events A and B are independent, then

$$\Pr[A \text{ and } B] = \Pr[A] \times \Pr[B].$$

We can see the basis of the multiplication rule in Figure 5.6-1. The area of the Venn diagram that corresponds to *rolling a three on the first die* and *rolling a three on the second die* is the region of overlap between the blue and green areas. Because the two events are independent, the area of this overlap zone is just the probability of being in the blue times the probability of being in the green:

$$\Pr[(\textit{first roll is a three}) \text{ and } (\textit{second roll is a three})]$$
$$= \Pr[\textit{first roll is a three}] \times \Pr[\textit{second roll is a three}]$$
$$= 1/6 \times 1/6$$
$$= 1/36.$$

The multiplication rule pertains to combinations with "and"—that is, that *both* events occur. If we want to know the probability of *this* and *that* occurring, and if the two events are independent, we can multiply the probabilities of each to get the probability of both occurring. Example 5.6A applies the multiplication rule to a study about smoking and high blood pressure.

| Example 5.6A | **Smoking and high blood pressure** |

Both smoking and high blood pressure are risk factors for strokes and other vascular diseases. In the United States, approximately 17% of adults smoke and about 22% have high blood pressure. Research has shown that high blood pressure is not associated with smoking; that is, they seem to be independent of each other (Liang et al. 2001). What is the probability that a randomly chosen American adult has both of these risk factors?

Because these two events are independent, the probability of an individual both *smoking* and *having high blood pressure* is the probability of smoking times the probability of high blood pressure:

$$\text{Pr}[\textit{smoking} \text{ and } \textit{high blood pressure}] = \text{Pr}[\textit{smoking}] \times \text{Pr}[\textit{high blood pressure}]$$
$$= 0.17 \times 0.22$$
$$= 0.037.$$

Therefore, 3.7% of adult Americans will have both of these risk factors for strokes.

This calculation is shown geometrically in the Venn diagram in Figure 5.6-2.

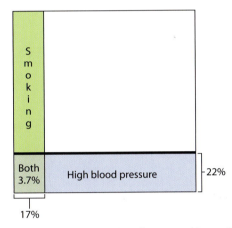

Figure 5.6-2 Venn diagram for the two independent factors smoking and high blood pressure. The probability of having both risk factors is proportional to the area of the rectangle in the bottom left corner.

And versus or

Probability statements involving "and" or "or" statements are common enough, and confusing enough, that it is worth summarizing them together:

▶ The probability of *A* or *B* involves addition. That is, Pr[*A* or *B*] = Pr[*A*] + Pr[*B*] if the two events *A* and *B* are mutually exclusive.
▶ The probability of *A* and *B* involves multiplication. That is, Pr[*A* and *B*] = Pr[*A*] × Pr[*B*] if *A* and *B* are independent.

What may be confusing is that the statement involving "and" requires multiplication, not addition.

Independence of more than two events

The multiplication rule also applies to more than two events, as Example 5.6B demonstrates. If several events are all independent, then the probability of all events occurring is the product of the probabilities that each one occurs.

Example 5.6B   **This thing ate my money!**

Most slot machines in Las Vegas are now electronic. One type of machine gives a payout 9.8% of the times it is played, and each play is independent of the outcome of the previous play.[4] What is the chance that a player with money for eight plays will lose on all eight?

First we need to know the probability of a loss. Loss and win are mutually exclusive outcomes of a single play, so the probability of losing on a play is simply one minus the probability of winning:

$$\text{Pr}[loss] = 1 - \text{Pr}[win] = 1 - 0.098 = 0.902.$$

The probability of losing eight times in a row is then the probability of losing once multiplied times itself eight times:

$$\text{Pr}[losing\ 8\ times] = \text{Pr}[loss\ on\ first\ game] \times \text{Pr}[loss\ on\ second\ game] \times \ldots$$
$$= (\text{Pr}[loss])^8 = (0.902)^8 = 0.438.$$

Thus, there is about a 44% chance of losing money on all eight turns.

Example 5.6C provides a biological application of the independence rule with multiple events.

Example 5.6C   **Mendel's peas**

Like blue eyes in humans, yellow pods in peas is a recessive trait. That is, pea pods are yellow only if both copies of the gene code for yellow. A plant having only one yellow copy and one green copy (a "heterozygote") has green pods just like the pods of plants having two green copies of the gene (a green "homozygote"). Gregor Mendel devised a method to determine whether a green plant was a heterozygote or a homozygote. He crossed the test plant to itself and assessed the pod color of 10 randomly chosen offspring. If all 10 were green, he inferred the plant was a homozygote, but if even one offspring was yellow, the test plant was classified as a heterozygote. What is the chance that his method fails? If the test plant is a homozygote, every offspring is green. If the test plant is a heterozygote, on the other hand, the chance of an offspring being green is 3/4 and the chance of it being yellow is only 1/4. What is the chance that all 10 offspring from a heterozygote test plant are green?

Mendel didn't carry out these calculations, but we can use our rules of probability to figure out the reliability of his approach. The chance that any one of a heterozygote's offspring is green is 3/4. Because the genotype of each offspring is independ-

---

[4]   Contrary to some popular opinion, a machine that has had a long losing streak is no more likely to win in the next play than another machine that won on the previous play. (Consult http://www.WizardofOdds.com for a detailed explanation why.)

ent of the genotypes of other offspring, the probability that all 10 are green can be calculated using the multiplication rule.

$$\Pr[\textit{all 10 green}] = \Pr[\textit{first is green}] \times \Pr[\textit{second is green}] \times \Pr[\textit{third is green}] \times \ldots$$
$$= 3/4 \times 3/4 \times 3/4 \times \ldots = (3/4)^{10} = 0.056.$$

Thus, Mendel likely misidentified about 5.6% of heterozygous individuals. On the other hand, his method correctly identified heterozygotes with probability $(1 - 0.056) = 0.944$.

5.7 Probability trees

A **probability tree** is a diagram that can be used to calculate the probabilities of combinations of events that are the outcomes of multiple random trials. We show how to use probability trees with Example 5.7.

| Example 5.7 | **Sex and birth order** |

Many couples planning a new family would prefer to have at least one child of each sex. The probability that a couple's first child is a boy[5] is 0.512. In the absence of technological intervention, the probability that their second child is a boy is independent of the sex of their first child, and so remains 0.512. Imagine that you are helping a new couple with their planning. If the couple plans to have only two children, what is the probability of getting one child of each sex?

This question requires that we know the probabilities of all mutually exclusive outcomes of two separate variables. The first variable is "*the sex of the first child.*" The second variable is "*the sex of the second child.*" We can start building a probability tree by considering the two variables in sequence. Let's start with the sex of the first child. Two mutually exclusive outcomes are possible—namely, "*boy*" and "*girl*"—which we list vertically, one below the other (see figure at right). We then draw arrows from a single point on the left to both possible outcomes. Along each arrow we write the probability of occurrence of each outcome (0.512 for "*boy*" and 0.488 for "*girl*").

Sex of first child

0.512 → Boy

0.488 → Girl

Now, we list all possible outcomes for the second variable, but we do so separately for each possible outcome of the first variable. For example, for the outcome "*boy*" for the first child, we list both possible outcomes (i.e., "*boy*" and "*girl*") for the sex of

---

[5] This excess of boys is a highly repeatable pattern, measured over tens of millions of births. Many more boys than girls are born. The fraction of males is even higher at conception than at birth, but this fraction declines during pregnancy because male fetuses die at a higher rate than female fetuses. It is thought that sperm bearing a Y chromosome might swim faster—and thus reach the egg sooner—than X-bearing sperm, which would account for the excess of boys.

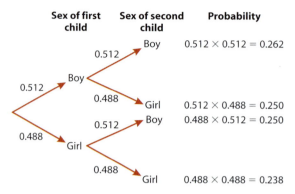

Figure 5.7-1 A probability tree for all possible outcomes of a two-child family.

the second child. Next, we draw arrows originating from the outcome "*boy*" for the first variable to both possible outcomes for the second variable. Then we write the probability of each outcome for the second variable along each arrow. We repeat this process for the case when "*girl*" is the outcome of the first variable. The resulting probability tree is shown in Figure 5.7-1.

At this point, we should check that our outcomes and probabilities are written down correctly. For instance, the probabilities along all arrows originating from a single point must sum to one (within rounding error) because they represent all the mutually exclusive possibilities. If they don't sum to one, we've forgotten to include some outcomes or we've written down the probabilities incorrectly.

With a probability tree, we can calculate the probability of every possible sequence of outcomes of the two variables. A sequence of outcomes is represented by a path along the arrows of the tree that begins at the root at the far left and ends at one of the branch tips on the right. The probability of a given sequence is calculated by multiplying all of the probabilities along the path taken from the root to the tip. For example, the sequence "*boy then girl*" in Figure 5.7-1 has a probability of $0.512 \times 0.488 = 0.250$. On our probability tree, we usually list the probabilities of each sequence of outcomes in a column to the right of the tree tips, as shown in Figure 5.7-1.

Each tree tip defines a unique and mutually exclusive sequence of events. Check Figure 5.7-1 (or any probability tree) to make sure that the probabilities of all possible sequences add to one. If they don't add to one (within rounding error), then something has gone wrong in the construction of the tree.

What is the probability of having one child of each sex in a family of two children? According to the probability tree, two of the four possible sequences result in the birth of one boy and one girl. In the first sequence, the boy is born first, followed by the girl, whereas, in the second sequence, the girl is born first and the boy is born second. These two different sequences are mutually exclusive, and we are looking for the probability of either the first *or* the second sequence. By the addition rule, therefore, the probability of getting exactly one boy and one girl when having two children is the sum of the probabilities of the two alternative sequences leading to this outcome: $0.250 + 0.250 = 0.500$.

We could also use the probability tree in Figure 5.7-1 to calculate probabilities of the following outcomes:

▶ The probability that at least one girl is born,
▶ The probability that at least one boy is born, and
▶ The probability that both children are the same sex.

Calculate these probabilities yourself to test your understanding.[6]

It is not essential to use probability trees when calculating the probabilities of sequences of events, but they are a helpful tool for making sure that you have accounted for all of the possibilities.

5.8 Dependent events

Independent events are mathematically convenient, but when the probability of one event depends on another, things get interesting. Much of science involves identifying variables that are associated.

Sex determination is more interesting in many insects than in humans. In many species, the mother can alter the relative numbers of male and female offspring depending on the local environment. In this case, *sex of offspring* and *environment* are dependent events, as Example 5.8 demonstrates.

Example 5.8	**Is this meat taken?**

The jewel wasp, *Nasonia vitripennis*, is a parasite, laying its eggs on its host, the pupae of flies. The larval *Nasonia* hatch inside the pupal case, feed on the host alive, and grow until they emerge as adults from the now dead, emaciated host. Emerging males and females, possibly brother and sister, mate on the spot.

Nasonia females have a remarkable ability to manipulate the sex of the eggs that they lay.[7] When a female finds a fresh host that has not been previously parasitized, she lays mainly female eggs, producing only the few sons needed to fertilize all her daughters. But if the host has already been parasitized by a previous female, the next female responds by producing a higher proportion of sons.[8] Thus, the state of the host encountered by a female and the sex of an egg laid are dependent variables (Werren 1980).

---

[6]   Answers: Pr[*at least one girl*] = 0.738; Pr[*at least one boy*] = 0.762; Pr[*both same sex*] = 0.500.

[7]   Wasps, like ants and bees, have a very different mechanism than humans for determining the sex of their offspring. All a female has to do to determine the sex of an egg at the time of laying is to control whether or not she fertilizes it with the sperm she has stored. If she fertilizes it, it becomes a female. If not, it's a male.

[8]   The value to her of sons has risen in the second case because there are now plenty of unrelated females to mate with.

Suppose that, when a given *Nasonia* female finds a host, there is a probability of 0.20 that the host already has eggs, laid by a previous female wasp. Presume that the female can detect previous infections without error. If the host is unparasitized, the female lays a male egg with probability 0.05, and a female egg with probability 0.95. If the host already has eggs, then the female lays a male egg with probability 0.90 and a female egg with probability 0.10. Figure 5.8-1 shows a Venn diagram of these probabilities.

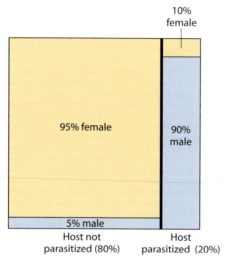

Figure 5.8-1 A Venn diagram showing that the sex of eggs laid by *Nasonia* females depends on the state of the host.

Based on Figure 5.8-1, the events "*host is previously parasitized*" and "*producing a male egg*" are dependent. The probability of laying a male egg changes depending on whether the host has been previously parasitized. Suppose we want to know the probability that a new, randomly chosen egg is male. We can approach this question using a probability tree like the one shown in Figure 5.8-2.

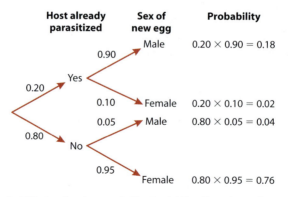

Figure 5.8-2 A probability tree for the sex of offspring laid by *Nasonia* as a function of whether the host has been previously parasitized.

According to the probability tree, there are exactly two paths that yield a male egg. In the first, the host is already parasitized and the mother lays a male egg. This path has probability

Pr[*host already parasitized* and *sex of new egg is male*] = 0.20 × 0.90 = 0.18.

In the second path, the host is not previously parasitized and the female lays a male egg. This second path has probability

Pr[*host not already parasitized* and *sex of new egg is male*]) = 0.80 × 0.05 = 0.04.

The probability of a new egg being male is the sum of the probabilities of these two mutually exclusive paths:

$$Pr[male] = 0.18 + 0.04 = 0.22.$$

The probability of an egg being male in this population is 0.22.

The probability tree shows that the event "*sex of new egg is male*" depends on whether the host encountered by a mother has been previously parasitized. Does this mean that the events "*host already parasitized*" and "*sex of new egg is male*" are not independent? One way to confirm this is via the multiplication rule, which applies only to independent events. The probability that *the host already had been parasitized* and that *the sex of the new egg is male* is 0.18. This is *not* what we would have expected assuming independence, though. If we multiply the probability that the new egg is a male (0.22, as we just calculated) and the probability that a host is already parasitized (0.20), we get 0.22 × 0.20 = 0.044, which is different from the actual probability of these two events (0.18). Based on the definition of independence, then, these two events are *not* independent.

5.9   Conditional probability and Bayes' theorem

If we want to know the chance of an event, we need to take account of all existing information that might affect its outcome. If we want to know the probability that we will see an elephant on our afternoon stroll, for example, we would get a different answer depending on whether our walk was in the Serengeti or downtown Manhattan. The algebra of conditional probability lets us hone our statements about the chances of random events in the context of extra information.

Conditional probability

Conditional probability is the probability of an event given that another event occurs.

> The conditional probability of an event is the probability of that event occurring *given that* a condition is met.

In Example 5.8, the conditional probability that a jewel wasp will lay a male egg is 0.90 *given that* the host that she is laying on already has wasp eggs (i.e., has already been parasitized). Confirm this for yourself by looking at Figure 5.8-2 again. We write conditional probability in the following way:

$$\text{Pr}[\textit{new egg is male} \mid \textit{host is previously parasitized}] = 0.90.$$

More generally, Pr[*event* | *condition*] represents the probability that the event will happen given that the condition is met. The vertical bar in the middle of this expression is a symbol that means "given that" or "when the following condition is met." (Be careful not to confuse it with a division sign.)

The Venn diagram in Figure 5.8-1 illustrates the meaning of this conditional probability. Ninety percent of the area corresponding to "*host parasitized*" represents the cases when the offspring is male, with the remaining 10% being females. The probability of a male is different under the condition "*host parasitized*" than under the condition "*host not parasitized.*"

Conditional probability has many important applications. If we want to know the overall probability of a particular event, we sum its probability across every possible condition, weighted by the probability of that condition. This is known as the **law of total probability**.

According to the *law of total probability,* the probability of an event, *X*, is

$$\text{Pr}[X] = \sum_{\substack{\textit{All values} \\ \textit{of Y}}} \text{Pr}[Y]\text{Pr}[X \mid Y],$$

where *Y* represents all possible mutually exclusive values of the conditions.

One way of thinking about this formula is that it gives the weighted average probability of *X*, averaged over all conditions.

The Venn diagram in Figure 5.8-1 makes it possible to visualize this, too. The probability of being male is obtained by adding the two blue areas, one for the condition when the *host is already parasitized* and the other for when *the host is not already parasitized*. The width of these boxes is proportional to the probability of the condition; the height is proportional to Pr[*male* | *host condition*]. By multiplying the width by the height of each box we find its area (its probability), and by adding all such boxes together we find the total probability of males.

To calculate the probability that a new egg is a male, we must consider two possible conditions: (1) the host is already parasitized and (2) the host is not parasitized. Thus, we'll have two terms on the right side of our equation:

$$\text{Pr}[\textit{egg is male}] = \text{Pr}[\textit{host already parasitized}]$$
$$\text{Pr}[\textit{egg is male} \mid \textit{host already parasitized}]$$
$$+ \text{Pr}[\textit{host not parasitized}]\text{Pr}[\textit{egg is male} \mid \textit{host not parasitized}]$$
$$= (0.20 \times 0.90) + (0.80 \times 0.05) = 0.22.$$

This is the same answer that we got from the probability tree, but now we can see how it can be derived from statements of conditional probability.

The general multiplication rule

With conditional probability statements, we can find the probability of a combination of two events even if they are not independent. When two events are not independent, the probability that both occur can be found by multiplying the probability of one event by the conditional probability of the second event, given that the first has occurred. This is the **general multiplication rule**.

> The *general multiplication rule* finds the probability that both of two events occur, even if the two are dependent:
>
> $$Pr[A \text{ and } B] = Pr[A] \, Pr[B \,|\, A].$$

This rule makes sense, if we think it through. For two events (A and B) to occur, event A must occur. By definition, this happens with probability $Pr[A]$. Now that we know A has occurred, the probability that B also occurred is $Pr[B \,|\, A]$. Multiplying these together gives us the probability of both A and B occurring.

It doesn't matter which event we label A and which we label B. The reverse is also true; that is,

$$Pr[A \text{ and } B] = Pr[B] \, Pr[A \,|\, B].$$

With the jewel wasps, for example, if we wanted to know the probability that *a host had already been parasitized* and that *the mother wasp laid a male egg*, we would multiply the probability that it had been parasitized (0.2) times the probability of a male egg *given that* the egg was already parasitized (0.9), to get 0.18.

If A and B are independent, then having information about A gives no information about B, and therefore $Pr[B \,|\, A] = Pr[B]$. That is, the general multiplication rule reduces to the multiplication rule when the events are independent.

Bayes' theorem

One powerful mathematical relationship about conditional probability is **Bayes' theorem**:[9]

$$Pr[A|B] = \frac{Pr[B|A]\,Pr[A]}{Pr[B]}$$

---

[9]   This theorem is named after its discoverer, the Reverend Thomas Bayes, an 18th century English Presbyterian minister.

Bayes' theorem may seem rather complicated, but it can be derived from the generalized multiplication rule. Because

$$\Pr[A \text{ and } B] = \Pr[B]\,\Pr[A \mid B]$$

and

$$\Pr[A \text{ and } B] = \Pr[A]\,\Pr[B \mid A],$$

it is also true that

$$\Pr[B]\,\Pr[A \mid B] = \Pr[A]\,\Pr[B \mid A].$$

Dividing both sides by $\Pr[B]$ gives Bayes' theorem.

Example 5.9 applies Bayes' theorem to the detection of Down syndrome.

Example 5.9   **Detection of Down syndrome**

Down syndrome (DS) is a chromosomal condition that occurs in about one in 1000 pregnancies. The most accurate test for DS requires amniocentesis, which unfortunately carries a small risk of miscarriage (about 1 in 200). It would be better to have an accurate test of DS without the risks. One such test that has recently come into common use is called the triple test, which screens for levels of three hormones in maternal blood at around 16 weeks of pregnancy.

The triple test is not perfect, however. It does not always correctly identify a fetus with DS (an error called a false negative), and sometimes it incorrectly identifies a fetus with a normal set of chromosomes as DS (an error called a false positive). Under normal conditions, the detection rate of the triple test (i.e., the probability that a fetus with DS will be correctly scored as having DS) is 0.60. The false positive rate (i.e., the probability that a test would say incorrectly that a normal fetus had DS) is 0.05 (Newberger 2000).

Most people's intuition, from looking at these numbers, is that they are acceptable. Based on the probabilities given, the triple test would seem to be right most of the time. But, if the test on a randomly chosen fetus gives a positive result (i.e., it indicates that the fetus has DS), what is the probability that this fetus actually has DS? Make a guess at the answer before we work it through.

To address this question, we need Bayes' theorem. We want to know a conditional probability—the probability that a fetus has DS given that its triple test showed a positive result. In other words, we want to know $\Pr[DS \mid positive\ result]$. Using Bayes' theorem,

$$\Pr[DS \mid positive\ result] = \frac{\Pr[positive\ result \mid DS]\,\Pr[DS]}{\Pr[positive\ result]}.$$

We've been given $\Pr[positive\ result \mid DS]$ and $\Pr[DS]$, the two factors in the numerator, but we haven't been given $\Pr[positive\ result]$, the term in the denominator.

We can figure out the probability of a positive result, though, by using the law of total probability introduced previously in Section 5.9. That is, we can sum over all the possibilities to find the probability of a positive result.

$$
\begin{aligned}
\Pr[\textit{positive result}] &= (\Pr[\textit{positive result} \mid DS]\Pr[DS]) \\
&\quad + (\Pr[\textit{positive result} \mid \textit{no DS}]\Pr[\textit{no DS}]) \\
&= (0.60 \times 0.001) + [0.05 \times (1 - 0.001)] = 0.05055.
\end{aligned}
$$

The probability of something *not* occurring is equal to one minus the probability of it occurring, so the probability that a randomly chosen fetus does *not* have DS is one minus the probability that it has DS. According to Example 5.9, $\Pr[DS] = 0.001$, so $\Pr[\textit{no DS}] = 1 - 0.001 = 0.999$ in the preceding equation.

Now, returning to Bayes' theorem, we can find the answer to our question.

$$
\Pr[DS \mid \textit{positive result}] = \frac{0.60 \times 0.001}{0.05055} = 0.012.
$$

There is a very low probability (i.e., 1.2%) that a fetus with a positive score on the triple test actually has DS!

Many people find it more intuitive to think in terms of numbers rather than probabilities for these kinds of calculations. For every million fetuses tested, 1000 will have DS, and 999,000 will not. Of those 1000, 60% or 600 will test positive. Of the 999,000, 5% or 49,950 will test falsely positive. Out of a million tests, therefore, there are $600 + 49,950 = 50,550$ positive results, only 600 of which are true positives. The 600 true positives divided by the 50,550 total positives is 1.2%, the same answer as we got before. DS babies have a high probability of being detected, but they are a very small fraction of all babies. Thus, the true positive results get swamped by the false positives.

This high false positive ratio is not unusual. Many diagnostic tools have high proportions of false positives among the positive cases. In this case, erring on the side of caution is appropriate because, when the triple test returns a positive result, it can be checked by amniocentesis.

Did you think that the probability of DS with a positive result would be higher? If so, you're not alone. A survey of practicing physicians found that their grasp of conditional probability with false positives was extremely poor (Elstein 1988). In a question about false positive rates, where the correct answer was that 7.5% of patients with a positive test result had breast cancer, 95% of the doctors guessed that the answer was 75%! If these doctors had a better understanding of probability theory, they could avoid overstating the risks of serious disease to their patients, thus reducing unnecessary stress.

5.10 **Summary**

▶ A random trial is a process or experiment that has two or more possible outcomes whose occurrence cannot be predicted.

▶ The probability of an event is the proportion of times the event occurs if we repeat a random trial over and over again under the same conditions.

▶ A probability distribution describes the probabilities of all possible outcomes of a random trial.

▶ Two events (A and B) are mutually exclusive if they cannot both occur (i.e., $\Pr[A$ and $B] = 0$). If A and B are mutually exclusive, then the probability of A or B occurring is the sum of the probability of A occurring and the probability of B occurring (i.e., $\Pr[A$ or $B] = \Pr[A] + \Pr[B]$). This is the addition rule.

▶ The general addition rule gives the probability of either of two events occurring when the events are not mutually exclusive: $\Pr[A$ or $B] = \Pr[A] + \Pr[B] - \Pr[A$ and $B]$. The general addition rule reduces to the addition rule when A and B are mutually exclusive because then $\Pr[A$ and $B] = 0$.

▶ Two events are independent if knowing one outcome gives no information about the other outcome. More formally, A and B are independent if $\Pr[A$ and $B] = \Pr[A]\,\Pr[B]$. This is the multiplication rule.

▶ Probability trees are useful devices for calculating the probabilities of complicated series of events.

▶ If events are not independent, then they are said to be dependent. The probability of two dependent events both occurring is given by the general multiplication rule: $\Pr[A$ and $B] = \Pr[A]\,\Pr[B\,|\,A]$.

▶ The conditional probability of an event is the probability of that event occurring given some condition.

▶ Probability trees and Bayes' theorem are important tools for calculations involving conditional probabilities.

▶ The law of total probability, $\Pr[X] = \sum\limits_{\text{All values of } Y} \Pr[Y]\Pr[X\,|\,Y]$, makes it possible to calculate the probability of an event (X) from all of the conditional probabilities of that event. The law adds up, for all possible conditions (Y), the probability of that condition ($\Pr[Y]$) times the conditional probability of the event assuming that condition ($\Pr[X\,|\,Y]$).

PRACTICE PROBLEMS

1. The following pizza, ordered from the Venn Pizzeria on Bayes Street, is divided into eight slices:

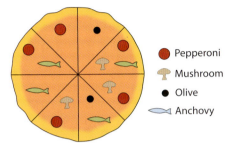

Pepperoni
Mushroom
Olive
Anchovy

The slices might have pepperoni, mushrooms, olives, and/or anchovies. Imagine that, late at night, you grab a slice of pizza totally at random (i.e., there is a 1/8 chance that you grabbed any one of the eight slices). Base your answers to the following questions on the drawing of the pizza.

a. What is the chance that your slice had pepperoni on it?

b. What is the chance that your slice had both pepperoni and anchovies on it?

c. What is the probability that your slice had either pepperoni or anchovies on it?

d. Are pepperoni and anchovies mutually exclusive on the slices from this pizza?

e. Are olives and mushrooms mutually exclusive on the slices from this pizza?

f. Are getting mushrooms and getting anchovies independent when choosing slices from this pizza?

g. If I pick a slice from this pizza and tell you that it has olives on it, what is the chance that it also has anchovies?

h. If I pick a slice from this pizza and tell you that it has anchovies on it, what is the chance that it also has olives?

i. Seven of your friends each choose a slice at random and eat them without telling you what toppings they had. What is the chance that the last slice left has olives on it?

j. You choose two slices at random from this pizza. What's the chance that they both have olives on them? (Be careful—after removing the first slice, the probability of choosing one of the remaining slices changes.)

k. What's the probability that a randomly chosen slice does *not* have pepperoni on it?

l. Draw a pizza for which mushrooms, olives, anchovies, and pepperoni are all mutually exclusive.

2. On any given hunting trip, the probability that a lion will find suitable prey is 0.80. If it finds prey, the probability that it successfully captures it is 0.10. What is the probability that a lion captures prey on a given hunting trip?

3. Cavities in trees are important nesting sites for a wide variety of wildlife, including the white-breasted nuthatch shown on the first page of this chapter. Cavities in trees are much more common in old-growth forests than in recently logged forests. A recent survey in Missouri found that 45 out of 273 trees in an old-growth area had cavities, while the rest did not (Fan et al. 2005). What is the probability that a randomly chosen tree in this area has a cavity?

4. The following bar graph gives the relative frequency of letters in texts from the English language. Such charts are useful for deciphering simple codes.[10]

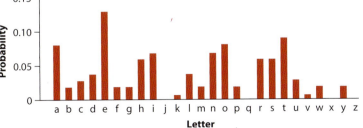

10 See http://www.simonsingh.net/The_Black_Chamber/frequencyanalysis.html.

a. If a letter were chosen at random from a book written in normal English, what is the probability that it is a vowel (i.e., A, E, I, O, or U)?

b. What is the probability that five letters chosen independently and at random from an English text would spell out (in order) "S-T-A-T-S"?

c. What is the probability that two letters chosen at random from an English text are both E's?

5. A large population of giant pandas has five alleles at one gene (labeled A_1, A_2, A_3, A_4, and A_5). The relative frequencies with which these alleles occur in the population are 0.1, 0.15, 0.6, 0.05, and 0.1, respectively. In this randomly mating population, the two alleles present in any individual are independently sampled from the population as a whole.

a. What is the probability that a single allele chosen at random from this population is either A_1 or A_4?

b. What is the probability that an individual has two A_1 alleles?

c. What is the probability that an individual is not $A_1 A_1$?

d. What is the probability, if you drew two individuals at random from this population that neither of them would have an $A_1 A_1$ genotype?

e. What is the probability, if you drew two individuals at random from this population, that at least one of them would have an $A_1 A_1$ genotype?

f. What is the probability that three randomly chosen individuals would have no A_2 or A_3 alleles? (Remember that each individual has two alleles.)

6. Draw a probability tree that shows the probability of rolling a sum of seven when two fair, six-sided dice are rolled (see Figure 5.4-2).

7. After graduating from your university with a biology degree, you are interviewed for a lucrative job as a snake handler in a circus sideshow. As part of your audition, you must pick up two rattlesnakes from a pit. The pit contains eight snakes, three of which have been defanged and are assumed to be harmless, but the other five are definitely still dangerous. Unfortunately, budget cuts have eliminated the herpetology course from the curriculum, so you have no way of telling in advance which snakes are dangerous and which are not. You pick up one snake with your left hand and another snake with your right.

a. What is the probability that you picked up *no* dangerous snakes?

b. Assume that each dangerous snake has an equal probability of biting you if it were picked up. This probability is 0.8. The defanged snakes do not bite. What is the chance that, in picking up your two snakes, you are bitten at least once?

c. Still assuming that the defanged snakes do not bite and the dangerous snakes have a probability of 0.8 of biting, assume that you picked up only one snake and it did not bite you. What is the probability that this snake was defanged?

8. Five different researchers independently take a random sample from the same population and calculate a 95% confidence interval for the same parameter.

a. What is the probability that all five researchers have calculated an interval that includes the true value of the parameter?

b. What is the probability that at least one does *not* include the true parameter value?

9. Schrödinger's cat lives under constant threat of death from the random release of a deadly poison. The probability of release of the poison is 1% per day, and the release is independent on successive days.

a. What is the probability that the cat will survive seven days?

b. What is the probability that the cat will survive a year (365 days)?

c. What is the probability that the cat will die by the end of a year?

10. Seeds fall on a landscape that contains 70% barren rock. The remainder of the landscape is suitable habitat for the seeds to germinate.

Imagine that the location where a seed falls in this landscape is random with respect to the site's suitability for germination.

a. What is the probability that a single seed lands on a suitable site for germination?

b. If two seeds fall independently, what is the probability that both fall on suitable habitats?

c. Imagine that three seeds are put randomly and independently onto this landscape. Use a probability tree to find the probability that exactly two of these three seeds land on suitable habitat.

11. The first test used in screening for HIV infection was the ELISA test, which measured levels of HIV antibodies in a blood sample. In a particular population of 5000 tested individuals, 20 had HIV but tested negative (false negatives), 980 had HIV and tested positive, eight did not have HIV but tested positive (false positives), and 3992 did not have HIV and tested negative.

a. What was the false-positive rate?

b. What was the false-negative rate?

c. If a randomly sampled individual from this population tests positive, what is the probability that he or she has HIV?

12. On a particular slot machine, the probability of winning per play is 9%, and wins are independent from play to play. If you played six times in a row, what is the probability of winning (W) and losing (L) *in the following orders*:

a. WWLWWW

b. WWWWWL

c. LWWWWW

d. WLWLWL

e. WWWLLL

f. WWWWWW

13. In a contest like the women's tennis finals at Wimbledon, the overall winner is determined by which person first wins two sets. Each set continues until there is a winner and loser; there are no ties. If one woman wins the first two sets, the match is finished at two sets. The maximum possible number of sets in a match is three.

a. Imagine that the two women are evenly matched, so that the probability of each woman winning any single set is 50%. Use a probability tree to find the probability that a match lasts exactly two sets. What are the probabilities of the match lasting exactly three sets?

b. Imagine that one woman was better than the other, such that the probability of her victory in any set is 55%. Use a probability tree to find the probability that the weaker woman would win the match.

14. Studies have shown that the probability that a man washes his hands after using the restroom is 0.74, and the probability that a woman washes hers is 0.83. (These are real data from a press release of the American Society for Microbiology, 15 September 2003.) A large room contains 40 men and 60 women. Assume that individual men and women are equally likely to use the restroom. What is the probability that the *next* individual who goes to the restroom will wash his or her hands?

15. The histogram in Figure 5.4-2 shows the probability of any result from rolling two six-sided dice. Use a probability tree to find the probability of rolling two dice whose numbers sum to five.

16. If you have ever tried to take a family photo, you know that it is very difficult to get a picture in which no one is blinking. It turns out that the probability of an individual blinking during a photo is about 0.04 (Svenson 2006).

a. If you take a picture of one person, what is the probability that she will *not* be blinking?

b. If you take a picture of 10 people, what is the probability that at least one person is blinking during the photo?

ASSIGNMENT PROBLEMS

17. A collection of 1600 pea plants from one of Mendel's experiments had 900 that were tall plants with green pods, 300 that were tall with yellow pods, 300 that were short with green pods, and 100 that were short with yellow pods.
    a. Are "tall" and "green pods" mutually exclusive for this collection of plants?
    b. Are "tall" and "green pods" independent for this collection of plants?

18. A normal deck of cards has 52 cards, consisting of 13 each of four suits: spades, hearts, diamonds, and clubs. Hearts and diamonds are red, while spades and clubs are black. Each suit has an ace, nine cards numbered 2 through 10, and three "face cards." The face cards are a jack, a queen, and a king. Answer the following questions for a single card drawn at random from a well-shuffled deck of cards.
    a. What is the probability of drawing a king of any suit?
    b. What is the probability of drawing a face card that is also a spade?
    c. What is the probability of drawing a card without a number on it?
    d. What is the probability of drawing a red card? What is the probability of drawing an ace? What is the probability of drawing a red ace? Are these events ("ace" and "red") mutually exclusive? Are they independent?
    e. List two events that are mutually exclusive for a single draw from a deck of cards.
    f. What is the probability of drawing a red king? What is the probability of drawing a face card in hearts? Are these two events mutually exclusive? Are they independent?

19. The human genome is composed of the four DNA nucleotides: A, T, G, and C. Some regions of the human genome are extremely G–C rich (i.e., a high proportion of the DNA nucleotides there are guanine and cytosine). Other regions are relatively A–T rich (i.e., a high proportion of the DNA nucleotides there are adenine and thymine). Imagine that you want to compare nucleotide sequences from two regions of the genome. Sixty percent of the nucleotides in the first region are G–C (30% each of guanine and cytosine) and 40% are A–T (20% each of adenine and thymine). The second region has 25% of each of the four nucleotides.
    a. If you choose a single nucleotide at random from each of the two regions, what is the probability that they are the same nucleotide?
    b. Assume that nucleotides occur independently within regions and that you randomly sample a three-nucleotide sequence from each of the two regions. What is the chance that these two triplets are the same?

20. In Vancouver, British Columbia, the probability of rain during a winter day is 0.58, for a spring day is 0.38, for a summer day is 0.25, and for a fall day is 0.53. Each of these seasons lasts one quarter of the year.
    a. What is the probability of rain on a randomly chosen day in Vancouver?
    b. If you were told that on a particular day it was raining in Vancouver, what would be the probability that this day would be a winter day?

21. When asking survey questions about potentially embarrassing topics, researchers may worry that the answers given are not honest. If you are asked whether you have ever shoplifted, for example, you may give a false answer if you fear that your answer will reflect poorly on you. One technique to encourage honesty involves probability. Each respondent flips a coin twice. If the result of the first toss is heads, then he answers "yes" if the second coin gives heads and "no" if the second coin comes up tails. If the first coin toss gives tails, however, the respondent flips the coin a second time but ignores the result. He then honestly answers the embarrassing yes-or-no shoplifting question. In this way, no one viewing the results can ever know whether a particular individual had shoplifted even if the answer given is a "yes."

Assume that 20% of a given group of people has shoplifted and that all members of this group are asked to report the results of the procedure above.

a. Draw a probability tree that describes the possible results of such a survey.

b. What is the overall probability that a person from this group will have answered "yes"?

22. Imagine that a long stretch of DNA has only adenine, thiamine, cytosine, and guanine in equal proportions. (These make up the nucleotides of the DNA.) What is the probability of randomly drawing 10 adenines in a row in a sample of 10 randomly chosen nucleotides?

23. Developmental biologists have discovered that a set of genes called the Hox genes is responsible for determining the anterior–posterior identity of regions in the developing embryo. The genes are turned on (expressed) in different regions of the body. One surprising thing about the Hox genes is that they usually occur together in a row on the same chromosome and in the same order as the body parts that they control. For example, the fruit fly *Drosophila melanogaster* has eight Hox genes, and they are on the chromosome in exactly the same order as they are expressed in the body from head to tail (Lewis et al. 1978, Negre et al. 2005).

If the eight genes were thrown randomly onto the same chromosome, what is the probability that they would line up in the same order as the segments in which they are expressed?

24. The flour beetle has 10 chromosomes, roughly equal in size, and it also has eight Hox genes (see Assignment Problem 23) (Brown et al. 2002). If the eight Hox genes were randomly distributed throughout the genome of the beetle, what is the probability that all eight would land on the same chromosome?

25. A seed randomly blows around a complex habitat. It may land on any of three different soil types: a high-quality soil that gives a 0.8 chance of seed survival, a medium-quality soil that gives a 0.3 chance of survival, and a low-quality soil that gives only a 0.1 chance of survival. These three soil types (high, medium, and low) are present in the habitat in proportions of 30:20:50, respectively. The probability that a seed lands on a particular soil type is proportional to the frequency of that type in the habitat.

a. Draw a probability tree to determine the probabilities of survival under all possible circumstances.

b. What is the probability of survival of the seed, assuming that it lands?

c. Assume that the seed has a 0.2 chance of dying before it lands in a habitat. What is its overall probability of survival?

26. A bag contains five pebbles. Three of these pebbles are black and two are white. For each of the following questions, the bag starts again in this state.

a. One pebble is randomly chosen from the bag. What is the probability that it is white?

b. One pebble is drawn from the bag and *not replaced*. This first pebble is black. What is

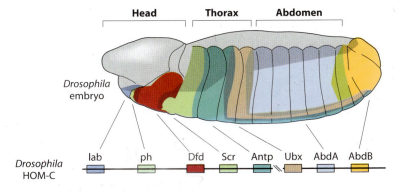

the probability that a second randomly drawn pebble is white?

c. A pebble is drawn randomly from the bag, its color is recorded, and it is replaced in the bag. This process is repeated two more times (for a total of three draws). What is the probability that all three pebbles are white?

d. A pebble is drawn from the bag at random, its color is recorded, and then *it is not replaced in the bag*. A second pebble is drawn at random, and it is also not replaced. Finally, a third pebble is drawn from the bag. What is the probability that all three of these pebbles are white?

e. Why are the answers to parts (c) and (d) different?

27. Blackjack is a game played with an ordinary deck of cards. (See Assignment Problem 18 for a description of such a deck.) "Blackjack" itself means that, out of two cards dealt to a player, one of them is an ace and the other is either a 10, jack, queen, or king. If you are dealt two cards randomly from the same deck, what is the probability that you get blackjack? (Remember that, when a card is dealt, it is removed from the deck.)

28. Ignoring leap years, there are 365 days in a year.

    a. If people are born with equal probability on each of the 365 days, what is the probability that three randomly chosen people have different birthdates?

    b. If people are born with equal probability on each of the 365 days, what is the probability that 10 randomly chosen people have different birthdates?

    c. If, as in fact turns out to be the case, birth rates are higher during some parts of the year than other times, would this increase or decrease the probability that 10 randomly chosen people have different birthdates, compared with your answer in part (b)?

29. During the Manhattan Project, the physicist Enrico Fermi asked Leslie R. Groves, the general in charge, "How do you define a 'great general'?" General Groves replied, "Any general who wins five battles in a row is great." He went on to say that only about 3% of generals are "great." If battles are won entirely at random with a probability of 50% per side, what fraction of generals would be "great" by this definition? How does this compare to the percentage given by the general?

30. What's the probability of being dealt a royal flush (i.e., a 10, jack, queen, king, and ace of the same suit) in a five-card hand from a standard deck of cards? Recall from Assignment Problem 18 that a standard deck has four suits, each with 13 distinct cards, including the five needed for the royal flush. The order in which the cards are dealt is unimportant, and you keep each card as it is dealt (i.e., each card is *not* returned to the deck before the next card is dealt).

31. "After taking 10 mammograms, a patient has a 50% chance of having had at least one false alarm." Given this information (from Elmore et al. 2005), and assuming that false alarms are independent of each other, what is the probability of a false alarm on a single mammogram?

6

Hypothesis testing

Hypothesis testing, like estimation, uses sample data to make inferences about the population from which the sample was taken. Unlike estimation, however, which puts bounds on the value of a population parameter, hypothesis testing asks only whether the parameter differs from a specific "null" expectation. Estimation asks, "How large is the effect?" Hypothesis testing asks, "Is there any effect at all?"

To better understand hypothesis testing, consider the polio vaccine developed by Jonas Salk. In 1954, Salk's vaccine was tested on elementary-school students across the U.S. and Canada. In the study, 401,974 students were divided randomly into two groups: kids in one group received the vaccine, whereas those in the other group (the control group) were injected with sugar water instead. The students were unaware of which group they were in. Of those that received the vaccine, 0.016% developed paralytic polio during the study, whereas 0.057% of the control group developed the disease (Brownlee 1955). The vaccine seemed to reduce the rate of disease by two-thirds, but the difference between groups was quite small, only about four cases per 10,000. Did the vaccine work, or did such a small difference arise purely by chance?

Hypothesis testing uses probability to answer this question. The null hypothesis is that the vaccine didn't work, and that any observed difference between groups happened only by chance. Evaluating the null hypothesis involved calculating the probability of getting as big a difference between groups as that observed if the vaccine had no effect. This probability turned out to be very small. Even though the rate of disease was not hugely different between the vaccine and control groups, the Salk vaccine trial was so large (over 400,000 subjects) that it was able to demonstrate a real difference. Thus, the "null" hypothesis was rejected. The vaccine had an effect, sparing many kids from disease, which was borne out by the success of the vaccine in the ensuing decades.

Hypothesis testing quantifies how unusual the data are, assuming that the null hypothesis is true. If the data are too different from that expected by the null hypothesis, then the null hypothesis is rejected.

> *Hypothesis testing* compares data to the expectations of a specific null hypothesis. If the data are too unusual, assuming that the null hypothesis is true, then the null hypothesis is rejected.

In this chapter, we illustrate the basics of hypothesis testing in the simplest possible setting: a test about a proportion in a single population. Our goal is to present the main concepts with a minimum of calculation. The rest of this book will present many specific methods of hypothesis testing.

6.1   Making and using hypotheses

Formal hypothesis testing begins with clear statements of two hypotheses—the null and alternative hypotheses—about a population quantity. The null hypothesis is the default, whereas the alternative hypothesis usually includes every other possibility except that stated in the null hypothesis.

Null hypothesis

The **null hypothesis** is a specific claim about the value of a population parameter. It is made for the purposes of argument and often embodies the skeptical point of view. Often, the null hypothesis is that the population parameter of interest is zero (i.e., no effect, no preference, no correlation, or no difference). In general, the null hypothesis is a statement that would be interesting to *reject*. For example, if we can reject the statement, "Medication X does not extend the lifespan of patients," then we have learned something useful—that people do in fact live longer when taking medication X.

The null hypothesis, which we can abbreviate as H_0 (pronounced "H-naught" or "H-zero"), is always *specific*; it identifies one particular value for the parameter being studied. In a study to test hypotheses about the impact of drift-net fishing on dolphins, for example, a valid null hypothesis could be the following:

H_0: The density of dolphins *is the same* in areas with and without drift-net fishing.

A clinical trial designed to compare the effects of the antidepressant medication Sertraline (Zoloft) and the older, tricyclic medication Amitriptyline would state the null hypothesis as

H_0: The antidepressant effects of Sertraline *do not differ* from those of Amitriptyline.

In other cases, the null hypothesis might represent an expectation from theory or from prior knowledge. For example, the following are valid null hypotheses:

H_0: Brown-eyed parents, each of whom had one parent with blue eyes, have brown- and blue-eyed children in *a 3:1 ratio.*

H_0: The mean body temperature of healthy humans *is 98.6°F.*

> The *null hypothesis* is a specific statement about a population parameter made for the purposes of argument. A good null hypothesis is a statement that would be interesting to reject.

Alternative hypothesis

Every null hypothesis is paired with an **alternative hypothesis** (abbreviated H_A) that usually represents all other possible parameter values except that stated in the null hypothesis. The alternative hypothesis typically includes possibilities that are biologically more interesting than that stated in the null hypothesis. For this reason the alternative hypothesis is often, but not always, the statement that the researcher hopes is true.

The following are some alternative hypotheses that go with the null hypotheses stated previously:

▶ H_A: The density of dolphins *differs* between areas with and without drift-net fishing.

▶ H_A: The antidepressant effects of Sertaline *differ* from those of Amitriptyline.

▶ H_A: Brown-eyed parents, each of whom had one parent with blue eyes, have brown- and blue-eyed children at *something other than a 3:1 ratio.*

▶ H_A: The mean body temperature of healthy humans *is not* 98.6°F.

> The *alternative hypothesis* includes all other possible values for the population parameter besides the value stated in the null hypothesis.

In contrast to the null hypothesis, the alternative hypothesis is nonspecific. Every possible value for a population characteristic or contrast is included, except that specified by the null hypothesis.

To reject or not to reject

Crucially, null and alternative hypotheses do not have equal standing. The null hypothesis is the only statement being tested with the data. If the data are consistent with the null hypothesis, then we say we have failed to reject it (we never "accept" the null hypothesis). If the data are inconsistent with the null hypothesis, we reject it and accept the alternative hypothesis.

Rejecting H_0 means that we have ruled out the null hypothesized value. It also tells us in which direction the true value likely lies, compared to the null hypothesized value. But rejecting a hypothesis by itself reveals nothing about the magnitude of the population parameter. We use estimation to provide magnitudes.

6.2   Hypothesis testing: an example

To show you the basic concepts and terminology of hypothesis testing, we'll take you through all the steps using a specific example. Our goal is to illuminate the basic process without distraction from the details of the probability calculations. We'll get to plenty of the details in later chapters.

Example 6.2 tests a hypothesis about a proportion, but hypothesis testing can address a wide variety of quantities, such as means, variances, differences in means, correlations, and so on. We'll try to emphasize the general over the specific here. Further details of how to test hypotheses about proportions are discussed in Chapter 7.

Example 6.2	**The right hand of toad**

Humans are predominantly right-handed. Do other animals exhibit handedness as well? Bisazza et al. (1996) tested the possibility of handedness in European toads, *Bufo bufo*, by sampling and measuring 18 toads from the wild. We will assume that this was a random sample. The toads were brought to the lab and subjected one at a time to the same indignity: a balloon was

wrapped around each individual's head. The researchers then recorded which forelimb each toad used to remove the balloon. It was found that individual toads tended to use one forelimb more than the other. At this point the question became: do right-handed and left-handed toads occur with equal frequency in the toad population, or is one type more frequent than the other, as in the human population?

Of the 18 toads tested, 14 were right-handed and four were left-handed. Are these results evidence of a predominance of one type of handedness in toads?

Stating the hypotheses

The number of interest is the proportion of toads in the *population* that are right-handed. Let's call this proportion p. The default statement, the null hypothesis, is that the two handedness types are equally frequent in the population, in which case $p = 0.5$.

H_0: Left- and right-handed toads are *equally frequent* in the population (i.e., $p = 0.5$).

This is a specific statement about the state of the toad population, one that would be interesting to prove wrong. If this null hypothesis is wrong, then toads, like humans, on average favor one hand over the other. This establishes the alternative hypothesis:

H_A: Left- and right-handed toads are *not equally frequent* in the population (i.e., $p \neq 0.5$).

The alternative hypothesis is **two-sided**. This just means that the alternative hypothesis allows for two possibilities: that p is greater than 0.5 (in which case right-handed toads outnumber left-handed toads in the population), or that p is less than 0.5 (i.e., left-handed toads predominate). Neither possibility can be ruled out prior to gathering the data, so both should be included in the alternative hypothesis.

> In a *two-sided* (or two-tailed) test, the alternative hypothesis includes values on both sides of the value specified by the null hypothesis.

The phrase "two-sided" comes from the fact that the alternative hypothesis contains values on either side of the value proposed by the null hypothesis. "Two-tailed" refers to the tails of the sampling distribution, where a "tail" is the region at the upper or lower extreme of a distribution.

The test statistic

The **test statistic** is a quantity calculated from the data that is used to evaluate how compatible the results are with those expected under the null hypothesis.

For the toad study, we use the observed number of right-handed toads as our test statistic. On average, if the null hypothesis were correct, we would expect to observe nine right-handed toads out of the 18 sampled (and nine left-handed toads, too). Instead, we observed 14 right-handed toads out of the 18 sampled. Fourteen, then, is the value of our test statistic.

> The *test statistic* is a quantity calculated from the data that is used to evaluate how compatible the data are with the result expected under the null hypothesis.

The null distribution

Unfortunately, data do not always perfectly reflect the truth. Because of the effects of chance during sampling, we don't really expect to see exactly nine right-handed toads when we sample 18 from the population, even if the null hypothesis is true. There is usually a discrepancy, due to chance, between the observed result and that expected under H_0. The mismatch between the data and the expectation under H_0 can be quite large, even when H_0 is true, particularly if there are not many data. To decide whether the data are compatible with the null hypothesis, we must calculate the probability of a mismatch as extreme as that observed, assuming that the null hypothesis is true.

To obtain this probability, we need to determine the sampling distribution of the test statistic *assuming that the null hypothesis is true*. We need to determine what values of the test statistic are possible under H_0 and their associated probabilities. The probability distribution of values for the test statistic, assuming the null hypothesis is true, is called the "sampling distribution under H_0" or, more simply, the **null distribution**.

> The *null distribution* is the sampling distribution of outcomes for a test statistic under the assumption that the null hypothesis is true.

The tricky part is to figure out what the null distribution is for the test statistic. For the moment, let's use the power of a computer to do the calculations. (We'll learn a simpler and more elegant way to calculate this null distribution in Chapter 7.) Sampling 18 toads under H_0 is like tossing 18 coins into the air and counting the number of "heads" that turn up when they land (letting "heads" represent right-handed toads). Tossing coins mimics well the sampling process under this H_0 because the probability of obtaining heads in any one toss is 0.5, which matches the null hypothesis. When we tossed 18 coins a vast number of times with the aid of a computer and

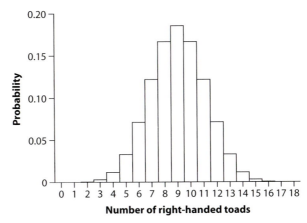

Figure 6.2-1 The null distribution for the test statistic, the number of right-handed toads out of 18 sampled.

counted the number of heads (right-handed toads) each time, we obtained the sampling distribution of outcomes illustrated in Figure 6.2-1. The probabilities themselves are listed in Table 6.2-1.

Table 6.2-1 All possible outcomes for the number of right-handed toads when 18 toads are sampled, and their probabilities under the null hypothesis.

Number of right-handed toads	Probability
0	0.000004
1	0.00007
2	0.0006
3	0.0031
4	0.0117
5	0.0327
6	0.0708
7	0.1214
8	0.1669
9	0.1855
10	0.1669
11	0.1214
12	0.0708
13	0.0327
14	0.0117
15	0.0031
16	0.0006
17	0.00007
18	0.000004
Total	1.0

Based on this null distribution, any number of right-handed toads between 0 and 18 is possible in a random sample of 18 individuals, but some numbers have a much higher probability of occurring than others.

Quantifying uncertainty: the *P*-value

Fourteen right-handed toads out of 18 total is not a perfect match to the expectation of the null hypothesis, but is the mismatch too large to conclude that chance alone is responsible? The usual way of describing the mismatch between data and a null hypothesis is to calculate the chance of getting those data, or data that are even more different from that expected, while assuming the null hypothesis. In other words, we want to know the probability of all results *as unusual or more unusual* than that exhibited by the data. If this probability is small, then the null hypothesis is inconsistent with the data and we would reject the null hypothesis in favor of the alternative hypothesis. If the probability is not small, then we have no reason to doubt the null hypothesis, and we would not reject it.

The probability of obtaining the data (or data that are an even worse match to the null hypothesis), assuming the null hypothesis, is called the **P-value**. If the *P*-value is small, then the null hypothesis is inconsistent with the data and we reject it.[1] Otherwise, we do not reject the null hypothesis.

> The *P-value* is the probability of obtaining the data (or data showing as great or greater difference from the null hypothesis) if the null hypothesis were true.

In practice, we calculate the *P*-value from the null distribution for the test statistic, shown for the toad data in Figure 6.2-2.

According to Figure 6.2-2, 14 or more right-handed toads out of 18 is fairly unusual, assuming the null hypothesis. These values lie at the right tail of the null distribution and have a low probability of occurring if H_0 is true. Equally unusual are zero, one, two, three, or four right-handed toads, which are outcomes at the other tail of the null distribution. Remember that our alternative hypothesis H_A is two-sided: it allows for the possibility that right-handed toads outnumber left-handed toads in the population, and also the possibility that left-handed toads outnumber right-handed toads. Therefore, outcomes from both tails of the distribution that are as unusual as the observed data, or even more unusual, must be accounted for in the calculation of the *P*-value.

---

[1] The *P*-value is usually denoted by an uppercase, italicized *P*, which stems from the word "probability." Don't confuse the *P*-value with the lowercase *p*, used here to indicate the proportion of right-handed toads in the population.

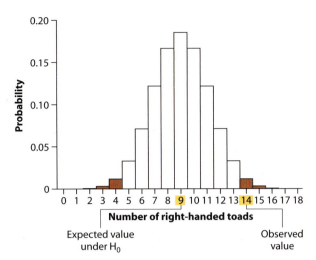

Figure 6.2-2 The null distribution for the number of right-handed toads out of the 18 sampled. Outcomes in red are values as different or more different from the expectation under H_0 than 14, the number observed in the data.

Based on the data in Figure 6.2-2, the probability of 14 or more right-handed toads, assuming the null hypothesis, is

$$\Pr[\textit{14 or more right-handed toads}] = \Pr[14] + \Pr[15] + \Pr[16] + \Pr[17] + \Pr[18]$$
$$= 0.0155,$$

where $\Pr[14]$ is the probability of exactly 14 right-handed toads. We can add the probabilities of 14, 15, 16, 17, and 18 because each outcome is mutually exclusive. This sum is not the P-value, though, because it does not yet include the equally extreme results at the left tail of the null distribution—that is, those outcomes involving a predominance of left-handed toads. The quickest way to include the probabilities of the equally extreme results at the other tail is to take the above sum and multiply by two:

$$P = 2 \times (\Pr[14] + \Pr[15] + \Pr[16] + \Pr[17] + \Pr[18])$$
$$= 2 \times 0.0155$$
$$= 0.031.$$

This number is our P-value. In other words, the probability of an outcome as extreme or more extreme than 14 right-handed toads out of 18 sampled is $P = 0.031$, assuming that the null hypothesis is true.

Statistical significance

Having calculated the P-value, how do we use it? On page 134, we said that if P is "small" we reject the null hypothesis; otherwise, we do not reject H_0. But what value of P is small enough? By convention in most areas of biological research, the boundary between small and not-small P-values is 0.05. That is, if P is less than or equal to 0.05, then we reject the null hypothesis; if $P > 0.05$, we do not reject it.

The P-value for the toad data, $P = 0.031$, is indeed less than 0.05, so we reject the null hypothesis that left-handed and right-handed toads are equally frequent in the toad population. We can conclude from these data that a majority of toads in the population is right-handed.

This decision threshold for P (i.e., $P = 0.05$) is called the **significance level**, which is signified by α (the lowercase Greek letter alpha). In biology, the most widely used significance level is $\alpha = 0.05$, but you will encounter some studies that use a different value for α. After $\alpha = 0.05$, the next most commonly used significance level is $\alpha = 0.01$. In Section 6.3, we explain the consequences of choosing a significance level and why $\alpha = 0.05$ is the most common choice.[2]

> The *significance level, α,* is a probability used as a criterion for rejecting the null hypothesis. If the P-value for a test is less than or equal to α, then the null hypothesis is rejected. If the P-value is greater than α, then the null hypothesis is *not* rejected.

Reporting the results

When writing up your results in a research paper or laboratory report, always include the following information in the summary of the results of a statistical test:

► the value of the test statistic
► the sample size
► the P-value.

Leaving out any of these three values (e.g., presenting only the bare P-value), makes it difficult for the reader to determine how you obtained it. When writing up the results of the toad study, we would need to indicate that 14 out of 18 toads were right-handed (which in this case gives both the test statistic and the sample size) and that $P = 0.031$.

In addition, it is always useful to provide confidence intervals, or at least the standard errors, for the parameters of interest.

---

[2]   At the same time, let's not get carried away. A P-value of 0.051 is not much different from a P-value of 0.049. The boundary of 0.05 should be seen as a guide to interpretation, not as a clear boundary between truth and fiction.

6.3 Errors in hypothesis testing

The most unsettling aspect of hypothesis testing is the possibility of errors. Rejecting H_0 does not necessarily mean that the null hypothesis is false. Similarly, failing to reject H_0 does not necessarily mean that the null hypothesis is true. This is because chance inevitably affects samples. Some uncertainty can be quantified, though, if the data are a random sample, so making rational decisions is possible.

Type I and Type II errors

There are two kinds of errors in hypothesis testing, prosaically named Type I and Type II.

Rejecting a true null hypothesis is a **Type I error**. Failing to reject a false null hypothesis is a **Type II error**. Both types of error are summarized in Table 6.3-1.

Table 6.3-1 Types of error in hypothesis testing.

	Reality	
Decision	**H_0 true**	**H_0 false**
Reject H_0	Type I error	Correct
Do not reject H_0	Correct	Type II error

The significance level, α, gives us the probability of committing a Type I error. If we go along with convention and use a significance level of $\alpha = 0.05$, then we reject H_0 whenever P is less than or equal to 0.05. This means that, if the null hypothesis were true, we would reject it mistakenly one time in 20. Biologists typically regard this as an acceptable error rate.

> *Type I error* is rejecting a true null hypothesis. The significance level α sets the probability of committing a Type I error.

We could reduce our Type I error rate if we wanted to, simply by using a smaller significance level than 0.05. For example, a more cautious approach would be to use $\alpha = 0.01$ instead of 0.05—that is, reject the null hypothesis only if P is less than or equal to 0.01. This would have the beneficial effect of reducing the probability of committing a Type I error down to 0.01 (i.e., one time in 100). Unfortunately, this has the side effect of increasing the chance of committing a Type II error. Reducing α makes the null hypothesis more difficult to reject when true, but it also makes the null hypothesis more difficult to reject when *false*. This is why the convention is to use $\alpha = 0.05$.

> *Type II error* is failing to reject a false null hypothesis.

If a null hypothesis is false, we need to reject it to get the right answer. Failure to reject a false null hypothesis is a Type II error. Because the Salk vaccine really did reduce the probability of catching polio, another study that by chance found the vaccine had no effect would have committed a Type II error.

A study that has a low probability of Type II error is said to have high **power**. Power is the probability that a random sample taken from a population will, when analyzed, lead to rejection of a false null hypothesis. All else being equal, a study is better if it has more power.

> The *power* of a test is the probability that a random sample will lead to rejection of a false null hypothesis.

Power is difficult to quantify, because the probability of rejecting a null hypothesis depends on how different the truth is from the null hypothesis. Detecting a small effect is more difficult than detecting a large effect. Because we never know how large the true value is, we cannot predict with any confidence how much power a study really has.

If we can guess the magnitude of the deviation from the null hypothesis, however, we can usually estimate the power of a study. A study has more power if the sample size is large, if the true discrepancy from the null hypothesis is large, and if the variability in the population is low. We discuss how to calculate power when we study experimental design in Chapter 14.

6.4   When the null hypothesis is not rejected

Example 6.4 represents a study in which the null hypothesis is *not* rejected. We discuss how to interpret such a "non-significant" result.

| Example 6.4 | **The genetics of mirror-image flowers** |

Individuals of most plant species are hermaphrodites (with both male and female sexual organs) and are therefore prone to the worst kind of inbreeding: having sex with themselves. The mud plantain, *Heteranthera multiflora*, has a simple mechanism to avoid "selfing." The female sexual organ (the style) deflects to the left in some individuals and to the right in others (see the pair of flower images adjacent). The male sexual organ (the anther) is on the opposite side. Bees visiting a left-handed plant are dusted with pollen on their right side, which then is deposited on the styles of only right-handed plants visited later. To investigate the genetics of this variation, Jesson and Barrett (2002) crossed pure strains

of left- and right-handed flowers, yielding only right-handed plants in the next generation. These right-handed plants were then crossed to each other. The expectation under a simple model of inheritance would be that their offspring should consist of left- and right-handed individuals in a 1:3 ratio. Of 27 offspring measured from one such cross, six were left-handed and 21 were right-handed. Do these data support the simple genetic model?

The null hypothesis states the expectation of the simple genetic model:

H_0: Left- and right-handed offspring occur at a 1:3 ratio (i.e., the proportion of left-handed individuals in the offspring population is $p = 1/4$).

The alternative hypothesis covers every other possibility:

H_A: Left- and right-handed offspring do not occur at a 1:3 ratio (i.e., $p \neq 1/4$).

The test

As with the study of handedness in toads (Example 6.2), the test in this flower study is two-sided: under the alternative hypothesis, the proportion of left-handed offspring may be less than 1/4 or it may be greater than 1/4. Neither possibility can be ruled out before gathering the data, so both possibilities must be included.

The number of left-handed offspring (out of 27) is the test statistic, although it doesn't much matter whether we use the right- or left-handed flowers, as long as the hypothesis statements reflect our choice. Under the null hypothesis, the expected number of left-handed offspring is $27 \times 1/4 = 6.75$. This expected frequency is a long-run average, because we don't really expect to find 0.75 of a left-handed flower.

To get the null distribution for the number of right-handed offspring, we again used the computer to take a vast number of random samples of 27 individuals from a fictitious population in which 1/4 of the individuals were left-handed. The resulting null distribution is illustrated in Figure 6.4-1.

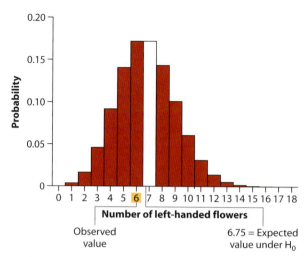

Figure 6.4-1 The null distribution for the number of left-handed individuals out of 27 sampled. Red bars on the left indicate outcomes less than or equal to six, the observed number of left-handed individuals. Red bars on the right indicate values on the right tail of the null distribution that are also as unusual or more unusual than the data.

The observed number of left-handed flowers (six) is less than the expected number from the null hypothesis (6.75), so we begin the calculation of the *P*-value by determining the probability of obtaining *six or fewer* left-handed offspring, assuming that the null distribution is true:

$$\Pr[\textit{number of left-handed flowers} \leq 6] = \Pr[6] + \Pr[5] + \ldots + \Pr[0].$$

Summing these probabilities (given here by the height of the bars in Figure 6.4-1) yields

$$\Pr[\textit{number of left-handed flowers} \leq 6] = 0.471.$$

This sum yields only the probability under the left tail of the null distribution (Figure 6.4-1). Because the test is two-sided, we also need to account for outcomes at the other tail of the distribution that are as unusual or more unusual than the outcome observed. The easiest way to obtain *P* is to multiply the above sum by two, yielding

$$P = 2 \Pr[\textit{number of left-handed flowers} \leq 6]$$
$$= 0.942.$$

The *P*-value is quite high, and there is a high probability of getting data like these when the null hypothesis is true. The *P*-value is not less than or equal to the conventional significance level $\alpha = 0.05$, so the null hypothesis is *not rejected*.

Interpreting a non-significant result

What does failure to reject H_0 mean? Can we conclude that the null hypothesis is true? Sadly, we can't, because it is always possible that the true value of the proportion p differs from 1/4 by a small or even moderate amount. Instead, it's possible that H_0 was not rejected either because of chance or because the power of the test was limited by the sample size of 27.

How, then, do we interpret the result? A valid conclusion is that the data are *compatible* or *consistent* with the null hypothesis; in other words, the data are compatible with the simple genetic model of inheritance of handedness in the mud plantain. There is no need or justification to build more complex genetic models of inheritance for flower handedness. A time may come when a new study—one with a larger sample size and more power—convincingly rejects the null hypothesis. If so, then a new genetic model will need to be developed. In the meantime, no evidence warrants a more complex scenario. This attitude treats the null hypothesis for what it is: the default until data show otherwise.

Keep in mind that a study should never be terminated having only the results of a hypothesis test to show for it. Drawing conclusions about populations from data also requires that we estimate magnitudes and put bounds on these estimates (Chapter 4). These bounds will help to identify the most plausible set of parameter values given the data. If we fail to reject a null hypothesis, but the confidence interval is large, then we know that we do not yet have enough information. But if the confidence interval is narrow and tightly bounded around the null hypothesis, then any real deviation from H_0 is likely to be either small or nonexistent.

6.5   One-sided tests

The studies of handedness in toads (Example 6.2) and flowers (Example 6.4) required two-sided tests, but **one-sided tests** are justified in some circumstances, instead. In a one-sided test, the alternative hypothesis includes values for the population parameter exclusively to one side of the value stated in the null hypothesis.

> In a *one-sided* (or one-tailed) test, the alternative hypothesis includes parameter values on only one side of the value specified by the null hypothesis. H_0 is rejected only if the data depart from it in the direction stated by H_A.

For example, imagine a study designed to test whether daughters resemble their fathers. In each trial of the experiment a subject examines a photo of one girl and photos of two adult men, one of whom is the girl's father. The subject must guess which man is the father. If there is no daughter–father resemblance, then the probability that the subject guesses correctly is only 1/2:

H_0: Subjects pick the father correctly one time in two ($p = 1/2$).

The only reasonable alternative hypothesis is that daughters indeed resemble their fathers, in which case the probability that the subject guesses correctly should *exceed* 1/2:

H_A: Subjects pick the father correctly *more than* one time in two ($p > 1/2$).

The test is "one-sided" because the alternative hypothesis includes values for the parameter on only one side of the value stated in the null hypothesis. This is justified if the parameter values on the other side of the value stated in the null hypothesis are inconceivable for any reason other than chance. It is not really conceivable that daughters would on average resemble their fathers *less* than other randomly chosen men.

The alternative hypothesis for a one-tailed test must be chosen before looking at the data. Data will always be on one side or another of the null hypothesis, so the data themselves cannot be used to predict which direction the deviation might be.

Now let's imagine that the study was carried out using 18 independent trials (which would require 18 different sets of photographs and subjects) and that 13 out of 18 subjects successfully guessed the father of the daughter. Under the null hypothesis we would expect only $18 \times 1/2 = 9$ correct guesses on average. The null distribution, obtained by taking a vast number of random samples of 18 subjects from a fictitious population in which 1/2 guess correctly, is that shown in Figure 6.2-1. The actual probabilities are those presented previously in Table 6.2-1.

Calculating a *P*-value for a one-sided test is different from the procedure used in a two-sided test. This is because the only outcomes that would cause us to reject H_0 and accept H_A are those with an unusually large number of correct guesses (the direction stated in the alternative hypothesis). Therefore, we only examine the right tail of the null distribution (Figure 6.5-1).

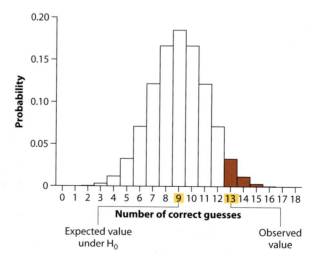

Figure 6.5-1 The null distribution for the number of subjects that correctly guessed the father of the daughter from photographs. Bars filled in red on the right tail correspond to values greater than or equal to 13, the observed number of correct guesses.

If 13 subjects guessed the father correctly, then the P-value is

$$P = \Pr[\textit{number of correct guesses} \geqslant 13]$$
$$= \Pr[13] + \Pr[14] + \Pr[15] + \ldots + \Pr[18]$$
$$= 0.048,$$

assuming that H_0 is true. There is no need to multiply this number by two, as in the case of the two-sided tests presented earlier, because we are only accounting for the probability in one tail of the distribution.

To reiterate: the appropriate tail of the null distribution to use when calculating P is given by the alternative hypothesis. Had we observed only four correct guesses out of 18, the P-value would still be calculated using the probabilities in the right tail of the null distribution, even though four correct guesses is smaller than the null expectation:

$$P = \Pr[\textit{number of correct guesses} \geqslant 4]$$
$$= 0.996.$$

One-sided tests should be used sparingly because the decision whether to use a one-sided or a two-sided alternative hypothesis is usually less clear-cut than the daughter–father resemblance study, and is therefore subjective. For example, what if we carried out a subsequent study to test whether daughters, when they marry, choose husbands that resemble their fathers? The null hypothesis is that there is no resemblance, but what is the alternative hypothesis? Should it be one-sided (husbands resemble fathers) or two-sided (husbands resemble fathers or husbands are very unlike fathers)? One researcher may have a clear theoretical basis for predicting a deviation from the null hypothesis in one direction, but a second researcher may have a different prediction. More importantly, even when we predict that a result may go in a particular direction, we may still be tempted to interpret a result in the tail of the opposite direction as a significant deviation from the null hypothesis. Two-tailed tests keep us honest.

For these reasons, we recommend using two-sided tests except in very special circumstances, and we adopt that policy in this book.

6.6 Hypothesis testing versus confidence intervals

The confidence interval puts bounds on the most-plausible values for a population parameter based on the data in a random sample (Chapter 4). Would confidence intervals and hypothesis tests on the same data, then, give the same answer? In other words, if the parameter value stated in the null hypothesis fell outside the 95% confidence interval estimated for the parameter, would H_0 be rejected by a hypothesis test at $\alpha = 0.05$? And, if the parameter value stated in the null hypothesis fell inside the 95% confidence interval, would a test at $\alpha = 0.05$ fail to reject H_0?

The answer is almost always "yes."[3] A 95% confidence interval usually contains all values of the parameter that would not be rejected as a null hypothesis if it were tested with the same data at $\alpha = 0.05$.

Why, then, don't we just skip hypothesis testing altogether? The confidence interval seems to do everything hypothesis testing can do, and it has the added benefit of being much more revealing about the actual magnitude of the parameter.

Hypothesis testing is used all the time in biology. And while it is fair to say that confidence intervals are under-utilized in biological research, hypothesis testing is informative, too. Its main use is to decide whether sufficient evidence has been presented to support a scientific claim (Frick 1996). The kinds of claims addressed by hypothesis testing are largely qualitative, such as "this new drug is effective" or "this pollutant harms fish." Such claims are embodied by the alternative hypothesis, with "sufficient evidence" defined as $P \leq 0.05$ (or some other significance level). For this reason, both hypothesis testing and confidence intervals are a big part of statistical analyses and of this book.

6.7 **Summary**

▶ Hypothesis testing uses data to decide whether a parameter equals the value stated in a null hypothesis. If the data are too unusual, assuming the null hypothesis is true, then we reject the null hypothesis.

▶ The null hypothesis (H_0) is a specific claim about a parameter. The null hypothesis is the default hypothesis, the one assumed to be true unless the data lead us to reject it. A good null hypothesis would be interesting if rejected.

▶ The alternative hypothesis (H_A) includes all values for the parameter other than that stated in the null hypothesis.

▶ Rejecting H_0 does not provide much information about the magnitude of the parameter.

▶ The test statistic is a quantity calculated from data used to evaluate how compatible the data are with the null hypothesis.

▶ The null distribution is the sampling distribution of the test statistic under the assumption that the null hypothesis is true.

▶ The P-value is the probability of obtaining the data (or data showing as great or greater difference from the null hypothesis) if the null hypothesis were true. If P is less than or equal to α, then H_0 is rejected.

▶ The threshold α is called the significance level of a test. Typically, α is set to 0.05.

▶ The P-value is not the probability that the null hypothesis is true or false.

▶ A Type I error is rejecting a true null hypothesis. A Type II error is failing to reject a false null hypothesis:

---

[3]   In a few cases, the most powerful test method and the best method for calculating a confidence interval for the parameter use different approximations or slightly different sampling distributions, so they would not yield the identical answer in every instance.

Decision	Reality	
	H_0 **true**	H_0 **false**
Reject H_0	Type I error	(no error)
Do not reject H_0	(no error)	Type II error

Reality

▶ The probability of making a Type I error is set by the significance level, α. If $\alpha = 0.05$, then the probability of making a Type I error is 0.05.

▶ The power of a test is the probability that a random sample, when analyzed, leads to rejection of a false null hypothesis.

▶ Increasing sample size increases the power of a test.

▶ In a two-sided test, the alternative hypothesis includes parameter values on both sides of the parameter value stated by the null hypothesis. In a one-sided test, the alternative hypothesis includes parameter values on only one side of the parameter value stated by the null hypothesis.

▶ Most hypothesis tests are two-sided. One-sided tests should be restricted to rare instances in which a parameter value on one side of the null value is inconceivable.

PRACTICE PROBLEMS

1. Assume that a null hypothesis is *true*. Say whether each of the following statements is true:
   a. Calculating the *P*-value assumes that the sample is a random sample.
   b. If you reject H_0 with a test, you will be making a Type II error.

2. Do people have powers of extrasensory perception (ESP)? Some people claim to have such abilities or to know someone who has them. Other people disbelieve such claims. Imagine that you were to set up an experiment to test the existence of ESP. Each trial in your experiment involves one person privately rolling a fair six-sided die and holding the image of the face of the die that turned up firmly in her mind. In another room, a second person attempts to identify the result correctly.
   a. What would be the null hypothesis for your test?
   b. What would be the alternative hypothesis?

3. Identify whether each of the following statements is more appropriate as the null hypothesis or as the alternative hypothesis in a test:

   a. Hypothesis: The number of hours preschool children spend watching television affects how they behave with other children when at day care.
   b. Hypothesis: Most genetic mutations are deleterious.
   c. Hypothesis: A diet of fast foods has no effect on liver function.
   d. Hypothesis: Cigarette smoking influences risk of suicide.
   e. Hypothesis: Growth rates of forest trees are unaffected by increases in carbon dioxide levels in the atmosphere.

4. What effect does reducing the value of the significance level from 0.05 to 0.01 have on
   a. the probability of committing a Type I error?
   b. the probability of committing a Type II error?
   c. the power of a test?
   d. the sample size?

5. Assume a random sample. What effect does increasing the sample size have on

a. the probability of committing a Type I error?
b. the probability of committing a Type II error?
c. the power of a test?
d. the significance level?

6. In the toad experiment (Example 6.2), what would the P-value have been if
   a. 15 toads were right-handed and the rest were left-handed?
   b. 13 toads were right-handed and the rest were left-handed?
   c. 10 toads were right-handed and the rest were left-handed?
   d. seven toads were right-handed and the rest were left-handed?

7. The MathWorld web page on hypothesis testing declares that "Hypothesis testing is the use of statistics to determine the probability that a given hypothesis is true" (Weisstein 2007). Is this statement true or false? Explain.

8. Why do we "fail to reject H_0" rather than "accept H_0" after a test in which the P-value is calculated to be greater than α?

9. An imaginary researcher examined the 18 largest mammal species in Asia that occur on both the mainland and on islands. Sixteen of the mammal species were smaller on islands than on the mainland. The remaining two species were larger on the islands than the mainland. With these data, test whether Asian large mammals differ in size between islands and the mainland in a consistent direction. Proceed through all steps of the hypothesis testing procedure. Use the sampling distribution in Table 6.2-1 to calculate your P-value. Apply the conventional significance level, $\alpha = 0.05$.

10. A clinical trial was carried out to test whether a new treatment has an effect on the rate of recovery of patients suffering from a debilitating disease. The null hypothesis "H_0: the treatment has no effect" was rejected with a P-value of 0.04. The researchers used a significance level of 5%. State whether each of the following conclusions is correct. If not, explain why.
    a. The treatment has only a small effect.
    b. The treatment has some effect.
    c. The probability of committing a Type I error is 0.04.
    d. The probability of committing a Type II error is 0.04.
    e. The null hypothesis would not have been rejected if the significance level was set to $\alpha = 0.01$.

ASSIGNMENT PROBLEMS

11. Identify whether each of the following statements is more appropriate as a null hypothesis or an alternative hypothesis:
    a. Hypothesis: The number of hours that grade-school children spend doing homework predicts their future success on standardized tests.
    b. Hypothesis: King cheetahs on average run the same speed as standard spotted cheetahs.
    c. Hypothesis: The mean length of African elephant tusks has changed over the last 100 years.
    d. Hypothesis: The risk of facial clefts is equal for babies born to mothers who take folic acid supplements compared with those from mothers who do not.
    e. Hypothesis: Caffeine intake during pregnancy affects mean birth weight.

12. State the most appropriate null and alternative hypothesis for each of the following experiments or observational studies:
    a. A test of whether cigarette smoking causes lung cancer.
    b. An experiment to test whether mean herbivore damage to a genetically modified crop plant differs from that in the related unmodified crop.
    c. A test of whether industrial effluents from a factory into the Mississippi River are affecting fish densities downstream.

d. A test of whether municipal safe-injection sites for drug addicts influences the rate of HIV transmission.

13. Assume that a null hypothesis is *true*. Which one of the following statements is true?
   a. A study with a larger sample is more likely than a smaller study to get the result that $P < 0.05$.
   b. A study with a larger sample is less likely than a smaller study to get the result that $P < 0.05$.
   c. A study with a larger sample is equally likely compared to a smaller study to get the result that $P < 0.05$.

14. Assume that a null hypothesis is *false*. Which one of the following statements is true?
   a. A study with a larger sample is more likely than a smaller study to get the result that $P < 0.05$.
   b. A study with a larger sample is less likely than a smaller study to get the result that $P < 0.05$.
   c. A study with a larger sample is equally likely compared to a smaller study to get the result that $P < 0.05$.

15. Imagine an experiment with guppies in which females from a wild population are presented with a choice of two males. One of the males is from her own population, whereas the other is from a second population in a nearby stream. The goal of the study is to test whether females have a preference. A total of 18 females are tested, all randomly sampled from the population. Males are also randomly sampled. The same males are never used more than once, to ensure the independence of the experimental trials. Twelve females choose the male from her own population, whereas six females choose the male from the other population.
   a. What are the null and alternative hypotheses for the test?
   b. Why should the test be two-sided?
   c. Using the sampling distribution under H_0 given in Table 6.2-1, calculate the P-value for the test.
   d. What is the interpretation of the P-value in part (c)?

e. What is the best estimate, from the data, of the population parameter, p?

16. Imagine that two researchers independently carry out clinical trials to test the same null hypothesis, that COX-2 selective inhibitors (which are used to treat arthritis) have no effect on the risk of cardiac arrest. They use the same population for their study, but one experimenter uses a sample size of 60 subjects, whereas the other uses a sample size of 100. Assume that all other aspects of the studies, including significance levels, are the same between the two studies.
   a. Which study has the higher probability of a Type II error, the 60-subject study or the 100-subject study?
   b. Which study has higher power?
   c. Which study has the higher probability of a Type I error?
   d. Should the tests be one-tailed or two-tailed? Explain.

17. A newspaper report of a study, in which the sex ratio of children born to families in a native community of Ontario was found to deviate significantly from the continental average ratio, claimed that "there was only a 1-percent probability that the results were due to chance."[4] What do you think this statement refers to? Rewrite this statement so that it is correct.

18. A group of researchers tested whether snakes tend to choose a warm resting site when both a warm site and a cool site are presented to them. Their hypotheses were H_0: Snakes do not prefer the warmer site. H_A: Snakes prefer the warmer site. They carried out the experiment and with their data calculated a one-tailed P-value of $P = 0.03$. They rejected their null hypothesis and concluded that snakes prefer the warmer sites.
   a. Is a one-tailed test appropriate here? Explain.
   b. What would their hypothesis statements have been had they used a two-tailed test instead?
   c. What would their P-value have been had they used a two-tailed test instead?

---

4   *Globe and Mail*, November 15, 2005.

Why statistical significance is not the same as biological importance

In the early 20th century, the word "significant" had one dominant meaning. If you said that something was "significant" you meant that it *signified* or showed something—that it had or conveyed a meaning. When R. A. Fisher said that a result was significant, he meant that the data showed some difference from the null hypothesis. In other words, we were able to learn something from those data. This sense of the word "significant" has persisted in the scientific literature. When discussing data, a "statistically significant" result means that a null hypothesis has been rejected.

But languages, like species, evolve. Over the 20th century, the word "significant" came to mean *important*. Today, outside of statistics, when we say that something is significant, we usually mean that it has value and import—that it ought to be paid attention to. This creates ambiguity when the term is applied now to scientific findings. When a newspaper article describes a new scientific result as "a significant new finding" or "a significant advance in our knowledge," for example, is that a statistical statement or a value judgment? Sometimes the meaning is blurred.

A statistically significant result is not the same as a biologically important result. An important result in biology refers to one whose effect is large enough to matter in some way. The mean lengths of the third molars differ between species of early mammals, but

Full of sound and fury, Signifying nothing.

—Shakespeare, *Macbeth*

this doesn't by itself mean that the difference is vital. To address the importance of the difference, we need to know its magnitude and how it might matter.

The problem is that extremely small, biologically uninteresting effects can be statistically significant, as long as the sample size is sufficiently large. For example, automobile accidents increase during full moons, and this result is statistically significant (Templer et al. 1982). Such results attracted media attention because they bring to mind stories of werewolves and vampire legends. But the size of the effect is about 1%, far too small to make it worth changing our driving habits or public policy. In fact, almost any null hypothesis can be disproved with a large enough sample. Are there really populations out there *exactly* equal to each other in every way? We should care about the differences only if they are large enough to matter.

We already have some sense that statistically significant does not mean important when we read of scientific studies that showed such earth-shattering facts as "teenagers like to listen to music sometimes," "driving fast makes the ride feel more bumpy," and "spiders scare some people." Each of these results was backed by hard data and statistics, but each surprised no one.

On the other hand, a result can be important even if it is not statistically significant. Sometimes new data suggest a pattern that, if true, is very important, provoking further study. For example, the first studies testing whether administering streptokinase prevents strokes did not reject the null hypothesis that this drug had no effect on mortality rate. But they were small studies that showed a suggestive pattern. As a result, further larger studies were conducted, and streptokinase was eventually shown to be effective in reducing mortality rates from strokes. This is why we don't "accept the null hypothesis"—a small study on a small effect will have a low probability of rejecting a false null hypothesis.

At the other extreme, sometimes it is important to show the lack of an effect. A large-scale study of the efficacy of hormone replacement therapy (HRT) showed no statistically significant evidence for a benefit of HRT to post-menopausal women. Moreover, confidence intervals showed that any plausible effect was small. HRT had been in wide use with substan-

tial money being invested and with some known deleterious side effects. Knowing that it had little benefit saved a great deal of resources and prevented many of the side effects. This result was not "statistically significant," but it was very important medically.

When presenting data, we should always report the estimated magnitude of the effect with a confidence interval, not just the P-value. The confidence interval gives us a plausible range for the size of the effect, and if this interval includes values with greatly varying interpretations, we know that we have to revisit the question with further data. We should look at a graphical presentation of the data to gauge the magnitude of the effect. The importance of a result depends on the value of the question and the size of the effect. Statistical significance tells us merely how confidently we can reject a null hypothesis, but not how big or how important the effect is.

7

Analyzing proportions

What proportion of people with Lou Gehrig's disease will live 10 years? What proportion of a red wolf population is female? In what fraction of years per decade does global temperature increase? Each of these questions is about a proportion, the fraction of the population that has a particular characteristic of interest. The proportion of individuals sharing some characteristic in a population is also the probability that an individual randomly sampled from that population will have that attribute. A proportion can range from zero to one.

In this chapter, we'll describe how best to estimate a population proportion using a random sample, including a confidence interval. We continue the development, begun in Chapter 4, of one of the major themes of data analysis: estimation with confidence intervals.

We also outline the best way to test hypotheses about a population proportion. In Chapter 6, we used a computer to take a vast number of random samples to obtain the sampling distribution for a proportion under a null hypothesis. Here in Chapter 7 we show a much quicker method to test hypotheses about a population proportion, called the *binomial test*, which provides exact *P*-values.

The key to estimation and hypothesis testing is an understanding of the sampling distribution for a proportion. Therefore, we begin this chapter by exploring the properties of random samples from a population when each individual can be categorized into one of two types.

7.1 The binomial distribution

Consider a measurement made on individuals that divides them into two mutually exclusive groups, such as success or failure, alive or dead, or left-handed or right-handed. In the population, a fixed proportion p of individuals fall into one of the two groups (call it "success") and the remaining individuals fall into the other group (call it "failure"). Calling one of the categories "success" and the other "failure" is a convenience, not a value judgment.[1]

If we take a random sample of n individuals from this population, the sampling distribution for the number of individuals falling into the "success" category is described by the **binomial distribution**. The term *binomial* reveals its meaning: there are only two (*bi-*) possible outcomes and both are named (*-nomial*) categories.

> The *binomial distribution* provides the probability distribution for the number of "successes" in a fixed number of independent trials, when the probability of success is the same in each trial.

Formula for the binomial distribution

The binomial formula gives the probability of X successes in n trials, where the outcome of any single trial is either success or failure. The binomial distribution assumes that

- the number of trials (n) is fixed,
- separate trials are independent, and
- the probability of success (p) is the same in every trial.

Under these conditions, the probability of getting X successes in n trials is

$$\Pr[X \ successes] = \binom{n}{X} p^X (1 - p)^{n-X}.$$

---

[1] For example, if we are measuring the fraction of hurricanes that hit Florida in a given year, then we might call a hurricane hit a "success" and a miss a "failure." We are clearly not cheering for the hurricane; the categories are for convenience only. The terms have their origins in gambling, where success and failure correspond more appropriately to winning and losing, respectively.

The left side of this equation, Pr[X *successes*], means the "probability of X successes," where X is an integer between 0 and n. On the right-hand side, the quantity $\binom{n}{X}$ is called "n choose X." This represents the number of unique ordered sequences of successes and failures that yield exactly X successes in n trials.[2] The term is shorthand for

$$\binom{n}{X} = \frac{n!}{X!(n-X)!},$$

where $n!$ is called "n factorial" and refers to the product

$$n! = n \times (n-1) \times (n-2) \times (n-3) \times \ldots \times 2 \times 1.$$

By definition, $0! = 1$, so $\binom{n}{0}$ and $\binom{n}{n}$ are both equal to 1. Similarly, $X!$ is "X factorial" and $(n-X)!$ is "$(n-X)$ factorial." Factorials get very large very fast. For example, $5! = 120$, but $20! = 2,432,902,008,176,640,000$. Even with a reasonably small number of trials, calculating the binomial coefficient can require a good calculator or computer.[3]

Suppose, for example, that you randomly sample $n = 5$ wasps from a population (representing five random trials), where each wasp has probability $p = 0.2$ of being male. The probability, then, that exactly three of the wasps in your sample are male is

$$\begin{aligned}
\Pr[3 \ males] &= \binom{5}{3}(0.2)^3(1-0.2)^{5-3} \\
&= \frac{(5 \times 4 \times 3 \times 2 \times 1)}{(3 \times 2 \times 1)(2 \times 1)}(0.2)^3(0.8)^2 \\
&= 0.0512.
\end{aligned}$$

The binomial distribution describes the sampling distribution for the *number* of successes in a random sample of n trials, but it also describes the *proportion* of successes. Because the number of trials is fixed at n, the probability that the sample has the proportion X/n successes is the same as the probability that the sample has X successes. For example, the probability that a proportion 0.6 of the five wasps are male is the same as the probability that exactly three are male—namely, 0.0512.

Number of successes in a random sample

Let's look at a complete binomial distribution. Imagine, for example, that we are randomly sampling $n = 27$ individuals from a population in which $p = 0.25$ of the indi-

---

[2] This term is also called the *binomial coefficient*.

[3] Some simple multiplication tricks will keep your calculator from maxing out. For example, $\frac{20!}{18!2!}$ can be rewritten as $\frac{20 \times 19 \times 18!}{18!2!}$, which simplifies to give $\frac{20 \times 19}{2!} = 190$, because $18!$ cancels out.

viduals are "successes" and $1 - p = 0.75$ of the individuals are "failures." What is the probability that the sample contains exactly X successes? This scenario is identical to the one stated by the null hypothesis in Example 6.4 (the study of mirror-image flowers). Recall that we sampled a population of mud plantains in which $p = 0.25$ had left-handed flowers and $1 - p = 0.75$ had right-handed flowers. The process of taking a random sample from this population exactly matches the assumptions of the binomial distribution: n independent random trials, where the probability of success in each trial p is equal in every trial. Hence, the binomial distribution can be used to determine the probability of any given number of successes.

For example, the probability of getting exactly six left-handed flowers with $n = 27$ and $p = 0.25$ is

$$\Pr[6\ \textit{left-handed flowers}] = \binom{27}{6}(0.25)^6(1 - 0.25)^{27-6}.$$

The binomial coefficient, "27 choose 6," is

$$\binom{27}{6} = \frac{27!}{6!(27 - 6)!} = \frac{27!}{6!\ 21!}$$

$$= \frac{27 \times 26 \times 25 \times ... \times 3 \times 2 \times 1}{(6 \times 5 \times 4 \times 3 \times 2 \times 1)(21 \times 20 \times 19 \times ... \times 3 \times 2 \times 1)}$$

$$= 296{,}010.$$

Therefore,

$$\Pr[6\ \textit{left-handed flowers}] = 296{,}010\ (0.25)^6\ (0.75)^{21} = 0.1719$$

In other words, there is about a 17% chance that exactly six of 27 randomly chosen flowers are left-handed, if the proportion of left-handed flowers in the population is 0.25.

We continue in this way to calculate all the probabilities of the sampling distribution (Table 7.1-1). Check some of the probabilities in this table for practice.

The probability distribution given in Table 7.1-1 is plotted in Figure 7.1-1. The distribution in Figure 7.1-1 looks the same as the distribution in Figure 6.4-1, which was obtained by taking a vast number of random samples of $n = 27$ from the population using a computer. In other words, the mathematical formula for the binomial distribution predicts the distribution generated using a computer simulation of the sampling process, but it does so much more rapidly and easily. The binomial distribution also gives *exact* probabilities. As a result, we can use the binomial distribution in place of repeated sampling on the computer to test null hypotheses.

Table 7.1-1 The probability of obtaining X left-handed flowers out of $n = 27$ randomly sampled, if the proportion of left-handed plants in the population is 0.25.

Number of left-handed flowers (X)	Pr[X]	Number of left-handed flowers (X)	Pr[X]
0	4.2×10^{-4}	14	0.0018
1	0.0038	15	5.1×10^{-4}
2	0.0165	16	1.3×10^{-4}
3	0.0459	17	2.8×10^{-5}
4	0.0917	18	5.1×10^{-6}
5	0.1406	19	8.1×10^{-7}
6	0.1719	20	1.1×10^{-7}
7	0.1719	21	1.2×10^{-8}
8	0.1432	22	1.1×10^{-9}
9	0.1008	23	7.9×10^{-11}
10	0.0605	24	4.4×10^{-12}
11	0.0312	25	1.8×10^{-13}
12	0.0138	26	4.5×10^{-15}
13	0.0053	27	5.5×10^{-17}

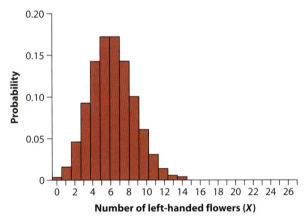

Figure 7.1-1 Plot of the probabilities given in Table 7.1-1 that were calculated from the binomial distribution. The plot shows the probability of obtaining X left-handed flowers out of $n = 27$ randomly sampled, if the proportion of left-handed plants in the population is 0.25.

Sampling distribution of the proportion

If there are X successes out of n trials, then the proportion of successes is $\hat{p}$:

$$\hat{p} = \frac{X}{n}.$$

(We pronounce $\hat{p}$ as "p-hat." Recall that p refers to the proportion in the population, whereas $\hat{p}$ refers to the *sample* proportion.)

We can use the same hypothetical population of flowers, having a true proportion of $p = 0.25$ successes, to illustrate the sampling distribution for the estimate of proportion $\hat{p}$. The panel on the top in Figure 7.1-2 is the sampling distribution when $n = 10$ (a relatively small sample size), whereas the panel on the bottom is the distribution for a larger sample size, $n = 100$. Both are based on binomial distributions, but rather than showing the number of successes X, we have divided X by n to yield $\hat{p}$.

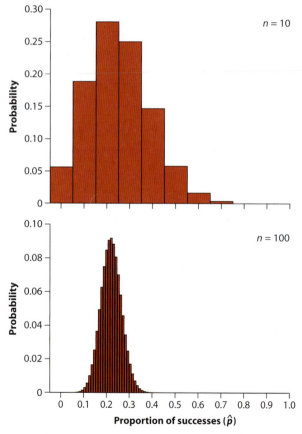

Figure 7.1-2 The sampling distribution for the proportion of successes $\hat{p}$ for sample size $n = 10$ (*top*) and $n = 100$ (*bottom*). In both of these graphs, the population proportion is $p = 0.25$. The distribution is narrower (smaller standard deviation) when n is larger.

The mean of the sampling distribution of $\hat{p}$, the proportion of successes in a random sample of size n, is given by

$$\text{Mean}[\hat{p}] = p.$$

In other words, the proportion of success in random samples is the same *on average* as the proportion of successes in the population. Therefore, $\hat{p}$ is an unbiased estimate of the population proportion—on average, it gives the right answer.

Notice in Figure 7.1-2 how the sample size affects the width of the sampling distribution for $\hat{p}$. When n is large (bottom panel), the sampling distribution is narrow. This effect is quantified in the formula for the standard error of $\hat{p}$. (Remember from Section 4.2 that the standard error of an estimate is the standard deviation of its sampling distribution.) The standard error of $\hat{p}$ (designated by $\sigma_{\hat{p}}$) is

$$\sigma_{\hat{p}} = \sqrt{\frac{p(1-p)}{n}}.$$

The sample size (n) is in the denominator of the standard error equation, so the standard error decreases as the sample size increases. That is why the estimates from samples of size 10 in Figure 7.1-2 (top panel) are more spread out than the estimates based on 100 individuals (bottom panel). Larger samples yield more precise estimates. The improvement in precision as sample size increases is called the **Law of Large Numbers**.

7.2   Testing a proportion: the binomial test

The **binomial test** applies the binomial sampling distribution to hypothesis testing for a proportion. The types of questions that it is suitable for have already been encountered in Chapter 6. The binomial test is used when a variable in a population has two possible states (i.e., "success" and "failure"), and we wish to test whether the relative frequency of successes in the population (p) matches a null expectation (p_0). The hypothesis statements look like this:

H_0: The relative frequency of successes in the population is p_0.
H_A: The relative frequency of successes in the population is not p_0.

The null expectation (p_0) can be any specific proportion between zero and one, inclusive.

> The *binomial test* uses data to test whether a population proportion (p) matches a null expectation (p_0) for the proportion.

Example 7.2 shows how to apply the binomial test to real data.

| Example 7.2 | **Sex and the X** |

A study of 25 genes involved in spermatogenesis (sperm formation) found their locations in the mouse genome. The study was carried out to test a prediction of evolutionary theory that such genes should occur disproportionately often on the X chromosome.[4] As it turned out, 10 of the 25 spermatogenesis genes (40%) were on the X chromosome (Wang et al. 2001). If genes for spermatogenesis 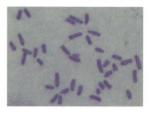 occurred "randomly" throughout the genome, then we would expect only 6.1% of them to fall on the X chromosome because the X chromosome contains 6.1% of the genes in the genome. Do the results support the hypothesis that spermatogenesis genes occur preferentially on the X chromosome?

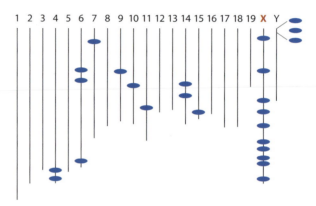

Figure 7.2-1 Cartoon of the mouse genome. Each vertical line represents one of the mouse chromosomes and indicates its length relative to the others. Each mark on a line indicates a single gene involved in spermatogenesis. Note the abundance of these genes on the X chromosome.

The null hypothesis is that the spermatogenesis genes would be on the X chromosome about 6.1% of the time, if they were randomly spread around the genome. To express this in terms of the binomial distribution, let's call the placement of each gene in the sample a "trial," and if the gene is on the X chromosome we'll call it a "success." The null hypothesis is that the probability of success (p) is 0.061. The more interesting alternative hypothesis is that the probability of success (p) is *not* 0.061—that is, either spermatogenesis genes occur *more* frequently than 0.061 or they occur *less* frequently on the X chromosome then expected by chance.

---

[4] New, recessive mutations are expressed only in males if they occur on the X chromosome, because they are not masked by the old allele at a companion chromosome. As a result, recessive mutations that benefit males are more likely to be seen by natural selection if they are on the X chromosome than if they are on other chromosomes.

We can write these hypotheses more formally as follows:

H_0: The probability that a spermatogenesis gene falls on the X chromosome is
  $p = 0.061$.

H_A: The probability that a spermatogenesis gene falls on the X chromosome is
  something other than 0.061 ($p \neq 0.061$).

Note once again the asymmetry of these two hypotheses. The null hypothesis is very specific, while the alternative hypothesis is not specific, referring to every other possibility. Also note that there are two ways to reject the null hypothesis: there can be an excess of spermatogenesis genes on the X chromosome (i.e., $p > 0.061$) or there can be too few (i.e., $p < 0.061$). Too few is not inconceivable, so it should also be included in the alternative hypothesis. Therefore, the test is two-sided.

The next step is to identify the test statistic that will be used to compare the observed result with the null expectation. In the case of the binomial test, the test statistic is the observed number of successes. For the data in Example 7.2, that would be 10 spermatogenesis genes on the X chromosome. The null expectation is $0.061 \times 25 = 1.525$. On average, we expect the fraction 0.061 of the 25 spermatogenesis genes sampled—namely, 1.525—to be located on the X chromosome if H_0 is true. Therefore, we know that in the *sample* more genes were found on the X chromosome than were expected by the null hypothesis.

The question now is whether we are likely to get such an excess by chance alone if the null hypothesis were true. To decide this we need the null distribution, the sampling distribution for the test statistic assuming that the null hypothesis is true. As mentioned previously, the sampling distribution for the number of successes X in a random sample of n individuals from a population having the proportion p of successes is described by the binomial distribution. Under the null hypothesis, the proportion is $p = 0.061$, so, for the above data (where $n = 25$ genes), the null distribution is given by

$$\Pr[X \text{ } successes] = \binom{25}{X}(0.061)^X(1 - 0.061)^{25-X}.$$

This null distribution allows us to calculate the *P*-value, the probability of getting a result as extreme as, or more extreme than, 10 spermatogenesis genes on the X chromosome when the null expectation is 1.525. Because the test is two-tailed, P is the probability of getting 10 or more genes on the X chromosome plus the probability of results as extreme as, or more extreme than, at the other tail of the null distribution, corresponding to too *few* genes on the X chromosome. The easiest way to include all the extreme outcomes is to double the probability of getting 10 or more:

$$P = 2 \Pr[\text{number of successes} \geq 10].$$

The probability of getting exactly 10 out of 25 on the X chromosome, when the probability of being on the X chromosome is 0.061, is

$$\Pr[10 \text{ } successes] = \binom{25}{10}(0.061)^{10}(1 - 0.061)^{15} = 9.07 \times 10^{-7}.$$

Table 7.2-1 Probabilities in the right-hand tail of the binomial distribution with $n = 25$ and $p = 0.061$.

Number of genes on X	Probability under the null hypothesis	Number of genes on X	Probability under the null hypothesis
10	9.1×10^{-7}	18	4.2×10^{-17}
11	8.0×10^{-8}	19	1.0×10^{-18}
12	6.1×10^{-9}	20	2.0×10^{-20}
13	4.0×10^{-10}	21	3.1×10^{-22}
14	2.2×10^{-11}	22	3.6×10^{-24}
15	1.0×10^{-12}	23	3.1×10^{-26}
16	4.3×10^{-14}	24	1.7×10^{-28}
17	1.5×10^{-15}	25	4.3×10^{-31}

This calculation is listed in Table 7.2-1, along with the probability of more than 10 genes on the X chromosome, if the null hypothesis were true.

The probability of getting 10 *or more* spermatogenesis genes on the X chromosome, assuming the null hypothesis is true, is the sum over all of these mutually exclusive possibilities:

$$\Pr[\textit{number of successes} \geq 10] = \Pr[10] + \Pr[11] + \Pr[12] + ... + \Pr[25]$$
$$= 9.9 \times 10^{-7}.$$

The final *P*-value is

$$P = 2\,\Pr[\textit{number of successes} \geq 10] = 2\,(9.9 \times 10^{-7}) = 1.98 \times 10^{-6}.$$

This *P*-value[5] is well below the conventional significance level of $\alpha = 0.05$. The probability of getting a result as extreme as, or more extreme, than the observed result is very low if the null hypothesis were true. Therefore, we reject the null hypothesis and conclude that there *is* a disproportionate number of spermatogenesis genes on the X chromosome. Our best estimate of the proportion of spermatogenesis genes that are located on the mouse X chromosome is

$$\hat{p} = \frac{10}{25} = 0.40,$$

which is much greater than 0.061, the proportion stated in the null hypothesis.

Approximations for the binomial test

The binomial test gives us an exact *P*-value and can be applied in principle to any data that fits into two categories. But calculating *P*-values for the binomial test can be

---

[5] Some computer programs for the binomial test give a slightly different value for *P*, because they calculate the probability of extreme results at both tails differently, whereas we just calculate the probability at one tail and then multiply by two. Multiplying by two is conservative and simple.

tedious without a computer, especially when n is large. There are alternatives that can be used under certain situations that are faster to calculate. They yield only approximate P-values, but they can save a lot of time when appropriate. We discuss two of them in subsequent chapters, but here we just want to let you know that they exist. The first is the χ^2 goodness-of-fit test (Chapter 8), and the second is the normal approximation to the binomial test (Chapter 10).

7.3 Estimating proportions

Here we show you how to measure the precision of an estimate of a population proportion. We explain how to estimate a standard error for a sample proportion and how to calculate a confidence interval for a population proportion. We'll motivate this endeavor with the data from Example 7.3 throughout.

Example 7.3   **Radiologists' missing sons**

Male radiologists have long suspected that they tend to have fewer sons than daughters. What is the proportion of males among the offspring of radiologists? In a sample of 87 offspring of "highly irradiated" radiologists, 30 were male (Hama et al. 2001). Assume that this was a random sample.

The best estimate of the population proportion of male offspring in this population is

$$\hat{p} = \frac{X}{n} = \frac{30}{87} = 0.345.$$

Estimating the standard error of a proportion

As we learned in Section 4.2, the standard deviation of the sampling distribution for an estimate is known as its standard error. We have seen in Section 7.1 that the standard deviation of $\hat{p}$ (and therefore its standard error) is

$$\sigma_{\hat{p}} = \sqrt{\frac{p(1-p)}{n}}.$$

We cannot usually calculate $\sigma_{\hat{p}}$ because we don't know the population parameter p. However, we can estimate this standard error ($\mathrm{SE}_{\hat{p}}$) using the estimate of the proportion:

$$\mathrm{SE}_{\hat{p}} = \sqrt{\frac{\hat{p}(1-\hat{p})}{n-1}}.$$

In the sample of offspring of radiologists, the standard error of the estimate of the proportion that are male is approximated by

$$SE_{\hat{p}} = \sqrt{\frac{\hat{p}(1 - \hat{p})}{n - 1}} = \sqrt{\frac{0.345(1 - 0.345)}{87 - 1}} = 0.051.$$

This value tells us how close, on average, our sample estimate ($\hat{p}$) is likely to be to the population proportion (p).

Confidence intervals for proportions—the Agresti–Coull method

Recall from Section 4.3 that a confidence interval is the range of most plausible values of the parameter we are trying to estimate, based on the data. The 95% confidence interval of a proportion will enclose the true value of the proportion 95% of the time that it is calculated from new data.

There are many methods in the statistical literature for calculating an approximate confidence interval for a proportion. Unfortunately, none of them are exact. We recommend using the **Agresti–Coull method** (Agresti and Coull 1998). With the Agresti–Coull method, we first calculate a number called p':

$$p' = \frac{X + 2}{n + 4}.$$

The confidence interval for a proportion is

$$p' - Z\sqrt{\frac{p'(1 - p')}{n + 4}} < p < p' + Z\sqrt{\frac{p'(1 - p')}{n + 4}},$$

where Z is a constant that depends on the confidence interval to be calculated. For a 95% confidence interval, $Z = 1.96$. For a 99% confidence interval, $Z = 2.58$.[6]

Getting back to the radiologists and their many daughters, recall that our best estimate of the true proportion of sons among the offspring of highly irradiated radiologists is 0.345. We can calculate the 95% confidence interval for this estimate using $X = 30$ and $n = 87$ in the Agresti–Coull method:

$$p' = \frac{X + 2}{n + 4} = \frac{30 + 2}{87 + 4} = 0.352.$$

---

[6] Z is a "standard normal deviate." We discuss it in greater detail in Chapter 10.

Using $Z = 1.96$, the 95% confidence interval is

$$p' - Z \sqrt{\frac{p'(1 - p')}{n + 4}} < p < p' + Z \sqrt{\frac{p'(1 - p')}{n + 4}}$$

$$0.352 - 1.96 \sqrt{\frac{0.352(1 - 0.352)}{87 + 4}} < p < 0.352 + 1.96 \sqrt{\frac{0.352(1 - 0.352)}{87 + 4}}$$

$$0.254 < p < 0.450.$$

This interval does *not* include the value 0.512, which is the proportion of sons typically found in the human population. In other words, 0.512 is not one of the most plausible values for the offspring of radiologists. We can be confident, therefore, that the proportion of sons of radiologists is much lower than the population average, assuming that the data are indeed a random sample. The reason for so few sons is not known.

Confidence intervals for proportions—the Wald method

The most commonly used method to determine a confidence interval for a proportion is called the **Wald method**. In fact, most statistics packages for the computer use the Wald method. We show it here only because it is used so often, but we do not recommend using it because it is not accurate in some commonly encountered situations.

The Wald method brackets the population estimate $\hat{p}$ by a multiple of its standard error. By the Wald method, the 95% confidence interval of the proportion is

$$\hat{p} - 1.96 \, \text{SE}_{\hat{p}} < p < \hat{p} + 1.96 \, \text{SE}_{\hat{p}}.$$

For the radiologist data in Example 7.3, for instance, the 95% confidence interval calculated using the Wald method is $0.244 < p < 0.445$.

Unfortunately, the method is only accurate when n is large and the population p is not close to 0 or 1. A 95% confidence interval should bracket the true population parameter in 95% of samples. Unfortunately, when n is small, or when p is close to 0 or 1, the Wald confidence interval for the proportion contains the true proportion less than 95% of the time.

7.4   Deriving the binomial distribution

We eventually use several probability distributions in this book. Each of these has been mathematically derived from first principles, and often this derivation is quite challenging. The binomial distribution, on the other hand, is relatively straightforward to derive. Here in this section we sketch out how this is done.

Imagine that we randomly sample n individuals from a population and that we take them in order, one at a time, from one to n. We want to calculate the probability of getting X successes. The first step in calculating this probability is to determine the number of sequences of successes and failures that lead to X successes in total.

For most values of X, there are many different sequences of successes and failures that will yield a total of exactly X successes. For example, imagine that we sample five children and we want to know the probability of getting exactly three boys (B) and two girls (G). For a sample this small, it is relatively easy to list all of the possible sequences of $n = 5$ trials that yield three boys:

BBBGG BBGBG BBGGB BGBBG BGBGB

BGGBB GBBBG GBBGB GBGBB GGBBB

There are exactly 10 such sequences.

In general, the number of different sequences of n events yielding exactly X successes is given by the binomial coefficient, $\binom{n}{X}$. In the above example of $X = 3$ boys out of $n = 5$ children, the binomial coefficient is

$$\binom{5}{3} = \frac{5!}{3!(5-3)!} = \frac{5 \times 4 \times 3 \times 2 \times 1}{(3 \times 2 \times 1)(2 \times 1)} = 10.$$

The next step in finding the probability of X successes in n random trials relies on the assumption that separate trials are independent. In this case, each success happens with the same probability (p), and each failure happens with probability $1 - p$. Thus, the probability of any string of successes and failures is the product of these probabilities for each event. Thus, a single sequence that has X successes and $n - X$ failures has probability

$$p^X(1 - p)^{n-X}.$$

Under independence, each sequence of trials yielding exactly X successes has this same probability of occurring.

The last step is to add up the probabilities of all sequences yielding exactly X successes. Because each of the alternative sequences is mutually exclusive and each has the same probability, we find the overall probability of X successes in n trials by multiplying the probability of any given sequence by the number of sequences:

$$\Pr[X\ successes] = \binom{n}{X}p^X(1 - p)^{n-X}.$$

This is the formula for the binomial distribution that we first introduced in Section 7.1 and used thereafter.

7.5 **Summary**

▶ The binomial distribution expresses the probability of getting X successes out of n trials, assuming that each trial is independent and has the same probability (p) of a success.

▶ The best estimate of a population proportion is the sample proportion.

▶ According to the Law of Large numbers, very large samples will have a proportion of successes that is close to the true proportion in the population.

▶ The binomial test compares the observed number of successes in a data set to that expected under a null hypothesis. The null distribution of the number of successes under H_0 is the binomial distribution, and so the binomial formula can be used to calculate the P-value for the test.

▶ A confidence interval for a proportion can be found using the Agresti–Coull method.

7.6 **Quick Formula Summary**

Binomial distribution

Formula: $\Pr[X\ successes] = \binom{n}{X} p^X (1 - p)^{n-X}$, where p is the probability of success, X is the number of successes, and n is the number of trials.

Proportion

Estimate: $\hat{p} = \dfrac{X}{n}$

Standard error: The standard error of a proportion is estimated by $SE_{\hat{p}} = \sqrt{\hat{p}(1 - \hat{p})/(n - 1)}$.

Agresti–Coull confidence interval for proportions

What does it assume? A random sample.

Formula: $\left(p' - Z\sqrt{\dfrac{p'(1 - p')}{n + 4}}\right) < p < \left(p' + Z\sqrt{\dfrac{p'(1 - p')}{n + 4}}\right)$, where $p' = \dfrac{X + 2}{n + 4}$, X is the number of successes in the sample, n is the sample size, and $Z = 1.96$ for a 95% confidence interval (the critical value for a standard normal distribution).

Binomial test

What is it for? Tests whether a population proportion (p) matches a null expectation (p_0) for the proportion.

What does it assume? A random sample.

Test statistic: The observed number of successes, X.

Formula: $P = 2\left(\sum_{i=X}^{n} \Pr[i \; successes]\right)$ for $X/n > p_0$ or $P = 2\left(\sum_{i=0}^{X} \Pr[i \; successes]\right)$ for $X/n < p_0$, where X is the observed number of successes, n is the sample size, and $\Pr[i$ successes] is the probability of obtaining i successes from n trials given by the binomial distribution.

PRACTICE PROBLEMS

1. In 1955, John Wayne played Genghis Khan in a movie called *The Conqueror*. It was not an artistic success. More unfortunately, the movie was filmed downwind of the site of 11 above-ground nuclear bomb tests. Of the 220 people who worked on this movie, 91 had been diagnosed with cancer by the early 1980s, including Wayne, his co-stars, and the director. According to large-scale epidemiological data, only about 14% of people of this age group, on average, should have been stricken with cancer within this time frame.[7] We want to know whether there is evidence for an increased cancer risk of people associated with this film.
   a. What is the best estimate of the probability of a member of the cast or crew getting cancer within the study interval? Assume that this probability is the same for each member of the cast.
   b. What is the standard error of your estimate? What does this quantity measure?
   c. What is the 95% confidence interval for this probability estimate? Does this interval

bracket the typical cancer rate of 14% for people of the same age group? Interpret this result.

2. In the U.S., paper currency often comes into contact with cocaine, either directly, during drug deals or usage, or in counting machines where it wears off from one bill to another. A forensic survey collected fifty $1 bills and measured the cocaine content of the bills. Forty-six of the bills had measurable amounts of cocaine on them (Jenkins 2001).
   a. From these data, what is the best estimate of the proportion of US one-dollar bills that have a measurable cocaine content? Assume that the sample of bills was a random sample.
   b. Assuming that the 50 bills were collected in a random sample, what is a 99% confidence interval for the estimate in part (a)?
   c. What is the correct interpretation of your 99% confidence interval?

[7] See http://www.straightdope.com/classics/a2_016.html.

3. For the following scenarios, state whether the binomial distribution would describe the probability distribution of possible outcomes. If not, explain why not.

a. The number of red cards out of the first five cards drawn from the top of a regular, randomly shuffled deck of cards.

b. The number of red balls out of 10 drawn one by one from a vat of 50 red and blue balls, if the balls are replaced and mixed after each draw.

c. The number of red balls out of 10 drawn one by one from a vat of 50 red and blue balls, if the balls are *not* replaced after each draw.

d. The number of red flowers in a square meter in a field.

e. The number of red-eyed flies among 200 *Drosophila* individuals drawn at random from a large population.

f. The total number of red-eyed flies in five *Drosophila* families, each of 40 individuals, with the families chosen at random from a large population.

4. The mite *Demodex folliculorum* lives in the hair follicles of humans, including the follicles of the eyelashes. (The bluish creatures in the photo below are these mites. The yellowish shaft in the

photo is a human hair.) Having heard that "most people" have these mites living in their skin and eyelashes, we wanted to know what "most" really meant. The only data we could find was in a paper that compared 16 North American women with a skin condition called rosacea to 16 women that did not have this skin condition (call this the "control" group). Of the 16 women in the control group, 15 of the 16 had these mites. All 16 of the women in the rosacea group had mites (Al Am et al. 1997).

a. From these data, a researcher estimated the proportion of people in North America who have these mites as 31/32. What is wrong with this estimate?

b. What is your best estimate for the proportion of North Americans without rosacea who have these mites? Assume random sampling.

c. What is the 95% confidence interval for this estimate?

d. What is the best estimate (with a 95% confidence interval) for the proportion of women with rosacea who have these mites?

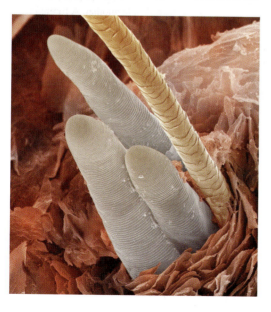

5. In the garden spider *Araneus didadematus*, the female often attempts to eat the male before or after mating, making sex a daunting prospect for the males. (This seemingly bizarre behavior, called sexual cannibalism, is not uncommon in the animal world.) In a series of mating observations, the courting male was captured and eaten by the female in 21 of 52 independent mating trials (Elgar and Nash 1988).

a. Based on the sample, estimate the proportion of males that are eaten by females, and give a 95% confidence interval for this estimate.

b. Is this estimate and confidence interval consistent with a true proportion of 50% capture of males? Is it consistent with 10% capture?

c. If the sample were much larger and the data showed instead that 210 out of 520 matings involved sexual cannibalism, would this change the estimated proportion? Would it change the confidence interval? How? (You don't need to do the calculations—just answer qualitatively.)

6. As our planet warms, as it has done for the last century or so, the change in temperature will have major effects on life. There are basically three possibilities for what might happen to a species: it can evolve to be better adapted to the new temperature, it can move closer to the poles so that it experiences temperatures closer to what it has experienced in the past, or it can go extinct. There have been a large number of studies of the second possibility. A recent study of the range limits of European butterflies found that, of 24 species that had changed their ranges in the last 100 years, 22 of them had moved further north and only two had moved further south (Parmesan et al. 1999). Assume that these 24 are a random sample of butterfly species with altered ranges. Test the hypothesis that the fraction of butterfly species moving north is different from the fraction moving south.

7. Imagine that there were two studies of the prevalence of melanism (solid black coat color) in leopards. Both estimated the proportion of black individuals in a Javanese population of leopards. One estimated that the proportion of black leopards in this population was 52%, with a 95% confidence interval that ranged from 46% to 58%. The other study estimated the proportion to be 64% with a 95% confidence interval that ranged from 35% to 85%.
   a. Which study most likely had a larger sample size?
   b. Which estimate is more believable?
   c. Are the estimated proportions from these two studies more different than we would expect by chance?

8. Mice have litters of several pups at once. While these are still in the mother's uterus, they are arranged in a line, so each fetus that is not on the end is between two of its siblings. It has been shown that female fetuses located between two male fetuses (2M) experience higher testosterone levels than those adjacent to no male fetuses (0M), because the hormone is produced by the males and diffuses across the fetal membranes and through the amniotic fluid to adjacent females. The higher fetal testosterone levels are known to have several effects on the females later in life, including increased aggression levels, growth rates, and even which eye opens first. One study wanted to measure whether fetal testosterone levels affected how attractive these female mice were to male mice as adults (vom Saal and Bronson 1980). Twenty-four male mice were given a choice between a female that was 0M and another female that was 2M. (Both the males and females were randomly chosen from their populations.) Each male was placed on a platform, and he could jump into the cage of

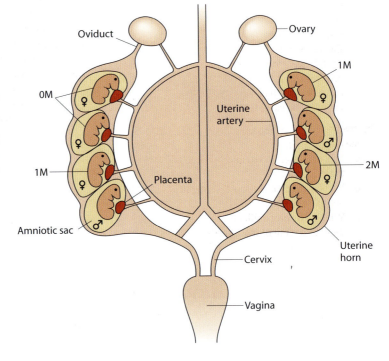

whichever female he preferred. Nineteen of the 24 males chose the 0M female.

a. Is this evidence that the males' choices are affected by the fetal positioning of the females?

b. If the two females presented to each male in a given trial had been sisters, would this have been a better or worse experimental design? Why?

9. A giant vat contains large numbers of two types of bacteria, called strain A and strain B. Assume that 30% of the bacteria in the vat are strain A, and the other 70% are strain B. The vat is well-mixed. Twenty technicians each collect a random sample of 15 cells from the vat and then determine which strain each cell belongs to.

a. What will the proportion of strain A cells be in these 20 samples, on average?

b. Each technician counts the number of strain A bacteria cells in his or her sample. To what distribution should the number of A strain bacteria in samples conform?

c. Each technician counts the *proportion* of strain A cells in his or her sample. What should the standard deviation be among samples in this proportion?

d. Each technician correctly calculates a 95% confidence interval for the proportion of A cells in the vat. On average, what fraction of technicians will construct an interval that includes the value 0.3?

10. Twelve six-sided dice are rolled. Assume all of the dice are fair.

a. What is the expected number of threes out of these 12 rolls?

b. What is the probability of rolling no threes out of the 12 rolls?

c. What is the probability of rolling exactly three threes out of the 12 rolls?

d. On average, what is the sum of the number of spots showing on top of the 12 dice?

e. What is the probability that all the dice show only ones and sixes (in any proportion)?

11. A common perception is that people may be able to delay their deaths from chronic illness until after a special event, like Christmas. Out of 12,028 deaths from cancer in either the week before or after Christmas, 6052 happened in the week before (Young and Hade 2004).

a. What is the best estimate of the proportion of deaths out of this time interval that occurred in the week before Christmas?

b. What is the 95% confidence interval for this estimate?

c. Use this confidence interval to ask, "Are these data consistent with a true value of 50%?" Do these data support the common perception?

12. In 1964, Ehrlich and Raven proposed that plant chemical defenses against attack by herbivores would spur plant diversification. In a test of this idea (Farrell et al. 1991), the number of species in 16 pairs of sister clades[8] whose species differed in their level of chemical protection were counted. In each pair of sister clades, the plants of one clade were defended by a gooey latex or resin that was exuded when the leaf was damaged, whereas the plants of the other clade lacked this defense. In 13 of the 16 pairs, the clade with latex/resin was found to be the more diverse (had more species), whereas in the other three pairs the sister clade lacking latex/resin was found to be the more diverse.

a. With these data, test whether the clade having latex/resin and its sister clade lacking it are equally likely to be more diverse.

b. Is this a controlled experiment or an observational study? Why?

[8] A clade is a group of species all descended from the same ancestor. For example, all the rodents form a clade. Sister clades are two clades that are each other's closest relatives. For example, the rodents and the rabbits are each other's closest relatives, so they are sister clades.

ASSIGNMENT PROBLEMS

13. In the same survey of the cocaine content of currency discussed in Practice Problem 2, heroin was detected on seven of the 50 bills.
    a. What is the best estimate of the proportion of U.S. one-dollar bills that have heroin on them?
    b. What is the 95% confidence interval for this estimate?
    c. If the proportion that you estimated from these data were in fact the true proportion in the population, what would be the probability of getting exactly seven bills with heroin when 50 bills are randomly sampled?

14. We all believe that we see most of what goes on around us, at least the most obvious things. Recently, however, psychologists have identified a phenomenon called "selective looking" which means that, if our attention is drawn to one aspect of what we see, we can miss even seemingly obvious features presented at the same time. In a striking demonstration of this phenomenon, a series of randomly chosen students was shown a video of six people throwing a basketball around, and they were asked to count how many times the people in white shirts threw the ball (Simons and Chabris 1999). In the middle of

Figure provided by Daniel Simons. Simons, D.J., & Chabris, C.F. (1999). "Gorillas in our midst: Sustained inattentional blindness for dynamic events." *Perception*, 28, 1059–1074.

this video,[9] a woman dressed as a gorilla walked through the shot, pausing in the center to thump her chest, and then walked out of the shot. Look at the photo, and you will realize that nothing could be more obvious. Or was it? Of the 12 students watching the video, only five noticed the gorilla.
    a. What is the best estimate from these data of the proportion of students in the population who notice the woman in the gorilla suit?
    b. What is the 95% confidence interval for the proportion of students in the population who notice the woman in the gorilla suit?
    c. What is the best estimate from these data of the proportion of students who *fail to notice* the woman in the gorilla suit?

15. A survey in the UK interviewed 200 shoppers in grocery stores about whether or not they had ever received injuries as a result of food or drink packaging, such as cuts sustained while cleaning up broken glass containers (Winder et al. 2001). Of these 200, 109 had received injuries "over the last few years" (and 27% of those injuries were significant enough to be treated by a doctor or emergency room).
    a. What is the best estimate, and 95% confidence interval, for the proportion of shoppers in this population that have injured themselves with their food or drinks?
    b. Do you think this was a random sample of all UK consumers? What factors might have rendered it a nonrandom sample?

16. The Global Amphibian Assessment in 2005 declared that 1856 out of 5743 of the known species of amphibians worldwide are at risk of extinction. (Approximately 122 species have gone extinct already since 1980.) Amphibians are vulnerable to environmental change, so they

9   You can see the video at http://viscog.beckman.uiuc .edu/grafs/demos/15.html.

are thought to be a bellwether of coming changes for other species.

a. What is the proportion of amphibian species that are at risk of extinction?

b. Would it make sense to calculate a confidence interval for this proportion in the usual way? Why or why not?

17. When one generation reproduces to form the next, the frequencies of alleles in the population can change by chance from generation to generation in a process called random genetic drift. An experiment was carried out using a very large laboratory population of the common fruit fly, *Drosophila melanogaster*, in which there were two eye-color alleles (gene copies) present, one red (the norm for these flies) and the other brown. The frequency of the red allele in this base population was 0.5. The researchers created a new group of flies containing 32 alleles by randomly sampling from this large population (Buri 1956). They created a second new group of 32 alleles by sampling again from the base population. They repeated this procedure until there were a large number of new populations, all containing 32 alleles. The frequency distribution of the proportions of red-eyed alleles in the new groups closely matched a binomial distribution with $n = 32$ and $p = 0.5$.

a. On average, what do we expect the mean proportion of the red-eyed allele to be in the new groups?

b. What should the standard deviation among groups in the proportion of red-eyed alleles be?

c. If a randomly chosen new group has 60% red-eyed alleles, what proportion of its alleles are for brown eyes?

d. What is the probability that a randomly chosen group will have *exactly* 50% red-eyed alleles?

e. What is the probability that a randomly chosen group will have 30 or more red-eyed alleles?

18. In one of the many samples from the *Drosophila* eye-color study first presented in Assignment Problem 17, there were nine out of 32 brown-eyed alleles. In another, there were 26 out of 32 brown-eyed alleles.

a. For each of these samples, separately estimate the proportion of brown-eyed alleles in the overall population of eye-color alleles. Calculate the 95% confidence intervals for these estimates.

b. Do these confidence intervals overlap? How can you reconcile this with the fact that both samples came from the same population?

19. An RSPCA survey of 200 randomly chosen Australian pet owners found that 10 said that they had met their partner through owning the pet.[10] Find the 95% confidence interval for the proportion of Australian pet owners who find love through their pets.

20. One classical experiment on ESP (extrasensory perception) tests for the ability of an individual to show telepathy—that is, to read the mind of another individual. This test uses five cards with different designs, all known to both participants. In a trial, the "sender" sees a randomly chosen card and concentrates on its design. The "receiver" attempts to guess the identity of the card. Each of the five cards is equally likely to be chosen, and only one card is the correct answer at any point.

a. Out of 10 trials, a receiver got four cards correct. What is her success rate? What is her expected rate of success, assuming she is only guessing?

b. Is her higher actual success rate reliable evidence that the receiver has telepathic abilities? Carry out the appropriate hypothesis test.

c. Assume another (extremely hypothetical) individual tried to guess the ESP cards 1000 times and was correct 350 of those times. This is extremely significantly different from

10. http://www.expertguide.com.au/news/default.asp?action=article&ID=273

the chance rate, yet the proportion of her successes is lower than the individual in part (a). Explain this apparent contradiction.

21. In a test of Murphy's Law, pieces of toast were buttered on one side and then dropped. Murphy's Law predicts that they will land butter-side down. Out of 9821 total slices of toast dropped, 6101 landed butter-side down. (Believe it or not, these are real data![11])
    a. What is a 95% confidence interval for the probability of a piece of toast landing butter-side down?
    b. Using the results of part (a), is it reasonable to believe that there is a 50:50 chance of the toast landing butter-side down or butter-side up?

22. Out of 67,410 surgeries tracked in a study in the UK, 2832 were followed by surgical site infections (Coello et al. 2005).
    a. What is a 99% confidence interval for the proportion of surgeries followed by surgical site infection in the UK? Assume that the data are a random sample.
    b. If the study had followed a total of only 674 surgeries from the same population, would the confidence interval have been wider or narrower?

23. Each member of a large genetics class grows 12 pea plants from an independent pea family. Each family is expected to have 3/4 plants with smooth peas and 1/4 plants that have wrinkled peas.
    a. On average, how many wrinkled pea plants will a student see in her 12?
    b. What is the standard deviation of the *proportion* of wrinkled pea plants per student?
    c. What is the variance of the proportion of wrinkled pea plants per student?
    d. Predict what proportion of the students saw exactly two wrinkled pea plants in their sample.

---

[11] http://www.mathsyear2000.org/thesum/issue8/page5.html

Correlation does not require causation

Science is aimed at understanding how the world works. Identifying the causes of events is a key part of the process of science. A first step in that process is the discovery of patterns, which usually involves noticing associations between events. When two events are associated, the possibility is raised that one is the cause of the other.

For example, our ancestors noticed that chewing on willow bark made sufferers of headaches and fevers feel better. We now know that the association between bark-chewing and pain relief is explained by the fact that salicylic acid in the bark blocks the release of prostaglandins in the body, which mediate pain and inflammation. This association led to the invention of aspirin.

An association (or correlation) between variables is a powerful clue that there may be a causal relationship between those variables. Smoking is associated with lung cancer; drinking is associated with fatal automobile accidents; taking streptomycin is associated with reduced bacterial infection. These associations exist because one thing causes the other, and the causal relationship was discovered because someone noticed their correlation.

The problem is that two variables can be correlated without one being the cause of the other. A correlation between two variables can result instead from a common cause. For example, the number of violent crimes tends to increase when ice cream sales increase. Does this mean that violence instills a deep need for ice cream? Or does it mean that squabbles over who ate the last of the Chunky Monkey® escalate to

violence? Perhaps, but the more likely explanation for this association is that they share a common cause: hot weather encourages both ice cream consumption and bad behavior. Ice cream sales and violence are correlated, but this doesn't mean that one causes the other.

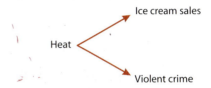

If we plot the mean life expectancy of the people of a country against the number of televisions per person in that country, we see quite a strong relationship (Rossman 1994):

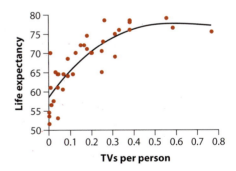

But the magical healing powers of the TV have yet to be demonstrated. Instead, it is likely that both televisions per capita and life expectancy have a common cause in the overall wealth of the citizens of the country.

These examples demonstrate the problems posed by **confounding variables**. A confounding variable is an unmeasured variable that changes in tandem with one or more of the measured variables, which gives the false appearance of a causal relationship between the measured variables. The apparent relationship between violence and ice cream sales is probably the result of the confounding variable *temperature*. *Overall wealth*, or something related to it, is probably the confounding variable in the correlation between the number of televisions and life expectancy.

Even more confusingly, a correlation between two variables may actually result from reverse causation. That is, the variable identified as the effect by the researcher may actually be the cause. For example, studies have repeatedly shown that babies that are breast-fed grow slightly slower than babies that are fed by formula. This has been interpreted as evidence that breast-feeding causes slow growth, but the truth turns out to be the reverse. Babies that grow rapidly are more likely to feed more, to be more demanding, and to be moved off the breast onto formula to give the poor mothers a break. Experimental studies confirm that exclusive breast-feeding has no measurable effect on infant growth (Kramer and Kakuma 2002).

These examples demonstrate the limitations of **observational studies,** which illuminate patterns but are unable to fully disentangle the effects of measured explanatory variables and unmeasured confounding variables. The main

purpose of **experimental studies** is to disentangle such effects. In an experiment, the researcher is able to assign subjects randomly to different treatment groups. Random assignment breaks the association between the confounding variable and the explanatory variable, allowing the causal relationship between treatment and response variables to be assessed.

Sir Ronald Fisher, our hero in many other respects, never believed that smoking caused lung cancer; instead, he thought that smoking may be caused by a genetic predisposition and that this genetic effect might also predispose one to cancer.[1] In other words, he thought that the genotype of an individual was a confounding factor. Fisher himself invented experimental design, which would make it possible to test his claim. In theory, one could assign subjects randomly to smoking and nonsmoking treatments and any underlying correlation with genetics would be broken, because on average, just as many subjects genetically predisposed to cancer would be assigned to both treatments. If Fisher's hypothesis were correct, the smokers would not have an increased cancer rate. Such an experiment is not ethically possible with humans, but it has been done with other species.

Finding correlations and associations between variables is the first step in developing a scientific view of the world. The next step is determining whether these relationships are causal or coincidental. This requires careful experimentation.

---

[1] Fisher was a lifelong smoker who consulted with the Tobacco Manufacturers' Standing Committee at the time he made this claim.

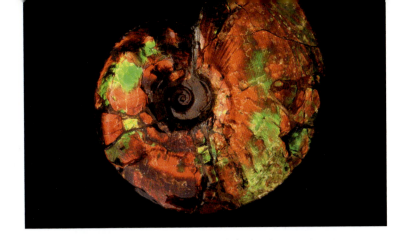

8

Fitting probability models to frequency data

The binomial test, introduced in the preceding chapter, is an example of a "goodness-of-fit test." A **goodness-of-fit test** is a method for comparing an observed frequency distribution with the frequency distribution that would be expected under a simple probability model governing the random occurrence of different outcomes. In Chapter 6, for example, we tested the simplest probability model imaginable: whether left- and right-handed toads occur with equal probability in a population. In Chapter 7, we tested whether the frequency of sperm genes on the X chromosome is proportional to the size of that chromosome. Rejecting the null hypothesis in both cases revealed real patterns in nature.

The binomial test, however, is limited to categorical variables with only two possible outcomes. In this chapter, we introduce a more general goodness-of-fit test that allows us to handle categorical and discrete numerical variables having more than two outcomes. It also allows us to assess the fit of more complex random models. We also show how to test the fit between observed frequency data and the frequencies predicted by simple probability models and how to interpret the results if the null hypothesis is rejected.

8.1   Example of a random model: the proportional model

The **proportional model** is a simple probability model in which the frequency of occurrence of events is proportional to the number of opportunities. We encountered the proportional model in Example 7.2 when we tested whether the frequency of sperm genes on the mouse X chromosome was proportional to the size of that chromosome. Here we will explore the χ^2 goodness-of-fit test on a proportional model by using the data from Example 8.1.

Example 8.1   **No weekend getaway**

The U.S. National Center for Health Statistics records information on each new baby born, such as time and date of birth, weight, and sex (Ventura et al. 2001). One bit of information available from these data is the day of the week on which each baby was born. Under the proportional model, we would expect that babies should be born at the same frequency on all seven days of the week. But is this true? Table 8.1-1 lists the number of babies born on each day of the week for a random sample of 350 births from 1999. Figure 8.1-1 displays these data in a bar graph.

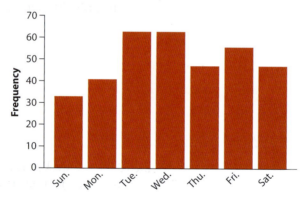

Figure 8.1-1 Bar graph of the day of the week for 350 births in the U.S. in 1999.

The data show a lot of variation in the number of births from one day to the next during the week, from a low of 33 on Sundays to a high of 63 on Tuesdays and Wednesdays. Under the proportional model, which will be the null hypothesis, the number of births on Monday should be proportional to the numbers of Mondays in 1999, except for chance differences. The same should be true for the other days of the week. Does the variation among days evident in Table 8.1-1 represent only chance variation? We can test the fit of the proportional model to the data with a χ^2 goodness-of-fit test.

Table 8.1-1 Day of the week for 350 births in the U.S. in 1999.

Day	Number of births
Sunday	33
Monday	41
Tuesday	63
Wednesday	63
Thursday	47
Friday	56
Saturday	47
Total	350

8.2 χ^2 **goodness-of-fit test**

The χ^2 **goodness-of-fit test** uses a quantity called χ^2 to measure the discrepancy between an observed frequency distribution and the frequencies expected under a simple random model serving as the null hypothesis. (χ is the Greek letter "chi" and is usually pronounced "kye" in English.) The random model is rejected if the discrepancy, χ^2, is too large.

> The χ^2 *goodness-of-fit test* compares frequency data to a model stated by the null hypothesis.

Null and alternative hypotheses

Under the proportional model, each day of the week should have the same probability of a birth (see Example 8.1). This is the simplest possible model, so it's our null hypothesis:

H_0: The probability of birth is the same on every day of the week.
H_A: The probability of birth is *not* the same on every day of the week.

Once again, the null (H_0) and alternative (H_A) hypotheses are statements about the population from which the data are a random sample. The null hypothesis is very specific, describing the expectation under the simple probability model. The alternative hypothesis is not specific, because it includes every other possibility.

Observed and expected frequencies

Because the proportional model is the null hypothesis, we use it to generate the expected frequency of births on each day of the week. We expect the number of births on each day of the week to reflect the number of times each day of the week occurred in 1999. It turns out that in 1999 every day of the week occurred 52 times—except Friday, which occurred 53 times. In Table 8.2-1, we divide these numbers by 365, the total number of days in 1999, yielding proportions.

Table 8.2-1 Expected frequency of births on each day of the week in 1999 under the proportional model.

Day	Number of days in 1999	Proportion of days in 1999	Expected frequency of births
Sunday	52	52/365 = 0.142	49.863
Monday	52	0.142	49.863
Tuesday	52	0.142	49.863
Wednesday	52	0.142	49.863
Thursday	52	0.142	49.863
Friday	53	53/365 = 0.145	50.822
Saturday	52	0.142	49.863
Sum	365	1	350

We can now use these proportions to calculate the expected frequencies of births for each day under the proportional model. For example, there were 350 total births, and under H_0 the fraction 52/365 of them should have occurred on Sundays. The expected frequency for Sunday is therefore

$$Expected = 350 \times \frac{52}{365}$$
$$= 49.863.$$

Note that expected frequencies can have fractional components, even if, in any given case, the number of individuals per category will be an integer. This occurs because the expected frequencies are the average values expected with the null model.

The sum of the *Expected* values should be the same as the sum of the *Observed* values (i.e., 350, except for rounding error). If this isn't the case, you need to check your calculations for errors.

The χ^2 test statistic

The χ^2 statistic measures the discrepancy between the observed and expected frequencies. It is calculated by the following sum:

$$\chi^2 = \sum_i \frac{(Observed_i - Expected_i)^2}{Expected_i}$$

Observed$_i$ is the frequency of individuals observed in the *i*th category, and *Expected$_i$* is the frequency expected in that category under the null hypothesis. The numerator of this quantity is a difference between the data and what was expected, which is squared so that positive and negative deviations are treated equally. When we divide this squared deviation by the expected value, the deviation of the observed and expected is scaled to the expected value.

> The χ^2 statistic measures the discrepancy between observed and expected frequencies.

It's important to notice that the χ^2 calculations use the *absolute* frequencies (i.e., counts) for the observed and expected frequencies, not proportions or *relative* frequencies. Using proportions in the calculation of χ^2 will give the wrong answer.

In the Example 8.1 data set, *i* can take on the values one through seven, where Sunday = 1, Monday = 2, etc. If the observed frequencies in all categories exactly matched the expected frequencies under the null hypothesis, χ^2 would be zero. The larger χ^2 is, the greater the discrepancy between the data and the frequencies expected under the null hypothesis.

To determine χ^2, we must calculate (*Observed$_i$* – *Expected$_i$*)^2/*Expected$_i$* for each day of the week. For Sundays, for example, *i* = 1 and

$$\frac{(Observed_1 - Expected_1)^2}{Expected_1} = \frac{(33 - 49.863)^2}{49.863} = 5.70.$$

Repeating this calculation for the rest of the days, we obtain the values shown in the last column of Table 8.2-2. (Make sure you can obtain these same values for yourself.) Note that this discrepancy is largest for Sundays, which has the largest difference between the number of observed births and the number of expected births.

Adding these up we get

$$\chi^2 = 5.70 + 1.58 + 3.46 + 3.46 + 0.16 + 0.53 + 0.16 = 15.05.$$

The χ^2 statistic is the test statistic for the χ^2 goodness-of-fit test, the quantity measuring the level of agreement between the data and the null hypothesis. All we need now is the sampling distribution of the χ^2 test statistic under H_0. This will allow us to decide whether $\chi^2 = 15.05$ is large enough to warrant rejection of the null hypothesis.

Table 8.2-2 Observed and expected numbers of births on each day of the week under the proportional model.

Day	Observed number of births	Expected number of births	$\dfrac{(Observed - Expected)^2}{Expected}$
Sunday	33	49.863	5.70
Monday	41	49.863	1.58
Tuesday	63	49.863	3.46
Wednesday	63	49.863	3.46
Thursday	47	49.863	0.16
Friday	56	50.822	0.53
Saturday	47	49.863	0.16
Sum	350	350	15.05

The sampling distribution of χ^2 under the null hypothesis

Recall from Chapter 6 that we can determine the sampling distribution for a test statistic under the null hypothesis by computer simulation.[1] This approach is tedious and not recommended, but we ran the simulation for these data just to show you what the approximate null distribution for the χ^2 statistic looks like (see the histogram in Figure 8.2-1).

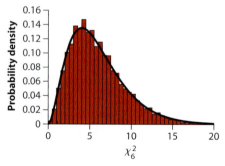

Figure 8.2-1 A histogram showing the approximate sampling distribution of χ^2 values for the data in Example 8.1 under the null hypothesis. The solid black curve shows the theoretical χ^2 probability distribution with six degrees of freedom. The curve provides an excellent approximation of the sampling distribution of the χ^2 test statistic under the null hypothesis.

---

[1]   In each step of the simulation, assign each of 350 births with equal probability to the 365 days of 1999, count the number of births that fall on each of the seven days of the week, and then calculate χ^2 as the discrepancy between the simulated frequencies and the frequencies expected under the proportional model. Repeat this procedure a vast number of times, each time recording the χ^2 value. This procedure yields a close approximation to the sampling distribution of the χ^2 statistic under H_0, provided that the number of repetitions is very large.

Fortunately, there's an easier way to obtain the null sampling distribution for the χ^2 statistic. The null distribution is well approximated by the theoretical χ^2 distribution, which has a known mathematical form. (See the solid curve superimposed on Figure 8.2-1.) Happily, the key features of this theoretical χ^2 distribution (hereafter referred to as "the χ^2 distribution") have been compiled in tables that are easy and quick to use. We demonstrate the use of the tables in the next two subsections of Section 8.2.

The χ^2 distribution is actually a family of distributions, and the particular one that we need for the birth data is specified by a number called the **degrees of freedom** (*df*, for short).

> The number of *degrees of freedom* specifies which of a family of distributions to use as the null distribution.

The degrees of freedom[2] for a χ^2 goodness-of-fit test is calculated using the following formula:

df = (Number of categories) − 1 − (Number of parameters estimated from the data).

Ignore the last term of this formula (i.e., "Number of parameters estimated from the data") for now because it is zero for the birth data. We explain what this term means in Sections 8.5 and 8.6, where we first use it.

The birth data have seven categories (one for each day of the week), so the number of degrees of freedom is

$$df = 7 - 1 = 6.$$

This tells us that we need to compare our χ^2 value calculated from the birth data ($\chi^2 = 15.05$) to the χ^2_6 distribution with six degrees of freedom. (The subscript "6" on χ^2_6 indicates the number of degrees of freedom.)

The χ^2 distribution with six degrees of freedom is shown as the black curve in Figure 8.2-1. Note how similar the χ^2 distribution (the solid line) is to the simulated distribution (the histogram), only smoother. The smallest possible value for χ^2 is zero. On the right, the χ^2 distribution extends to positive infinity. This distribution is the one we will use to calculate the P-value for our test.

---

2   The number of degrees of freedom for the χ^2 goodness-of-fit test is calculated as the number of categories minus the number of constraints imposed on the expected frequencies. In a goodness-of-fit test, the sum of the expected frequencies is constrained to be the same as the total number of observations; we always lose a degree of freedom because of this constraint. Additional constraints are imposed if we use the data to generate other numbers needed to calculate the expected frequencies, as Examples 8.5 and 8.6 will show. The logic behind this calculation is that every constraint causes the expected frequencies to be that much more similar to the observed frequencies, and we must adjust the degrees of freedom to compensate.

Calculating the *P*-value

The *P*-value for a test is the probability of getting a result as extreme as, or more extreme than, the result observed if the null hypothesis were true. For the χ^2 goodness-of-fit test, the *P*-value is the probability of getting a χ^2 value *greater* than the observed χ^2 value calculated from the data ($\chi^2 = 15.05$ for the birth data). Remember that, if the data exactly matched the expectation of the null hypothesis, χ^2 would be zero. Greater deviations from the null expectation result in a higher value of χ^2. As a result, we use only the right tail of the χ^2 distribution to calculate *P*.

The χ^2 distribution is a continuous probability distribution, so probability is measured by the area under the curve, not the height of the curve (Chapter 5). The probability of getting a χ^2 value greater than a single specified value, which is what we need to calculate a *P*-value, is equal to the area under the curve to the right of that value extending to positive infinity.

The data from Example 8.1 yielded $\chi^2 = 15.05$. The probability of getting a χ^2 value of 15.05 or greater is equal to the area under the χ_6^2 curve beyond 15.05, as shown by the region highlighted in red in Figure 8.2-2.

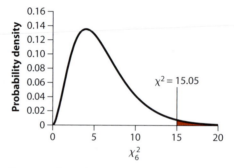

Figure 8.2-2 The χ^2 distribution with six degrees of freedom. The red area shows the probability of getting a χ^2 value greater than or equal to 15.05.

How do we go about finding this area beyond the measured χ^2 value? We have two options available. First, most computer statistical packages will provide the *P*-value directly: $P = 0.0199$. At the standard significance level of $\alpha = 0.05$, such a small *P*-value leads us to reject our null hypothesis. That is, these data provide evidence that births are not "randomly" distributed over the days of the week. The variation in number of births listed in Table 8.1-1 is simply too large to be explained by chance.

The second option for assessing the *P*-value uses critical values, as shown in the next subsection.

Critical values for the χ^2 distribution

The second way to calculate the *P*-value for a χ^2 statistic does not require a computer. This method uses tables of critical values to set bounds on the *P*-value. A **critical**

value is the value of a test statistic that marks the boundary of a specified area in the tail (or tails) of the sampling distribution under H_0. If we want a significance level of $\alpha = 0.05$, for example, we would need to know the value of χ^2 for which the area under the curve to its right is 0.05. This value of χ^2 is called the "critical value corresponding to $\alpha = 0.05$."

> A *critical value* is the value of a test statistic that marks the boundary of a specified area in the tail (or tails) of the sampling distribution under H_0.

Statistical Table A in the back of this book gives critical values for the χ^2 distribution. An excerpt from this table is shown in Table 8.2-3. To read the table, first find the column corresponding to the significance level of interest (we will use the standard $\alpha = 0.05$). Then find the row corresponding to the number of degrees of freedom for the test statistic ($df = 6$ for Example 8.1). The number in the corresponding cell of the table is the critical value $[\chi^2_{6,(0.05)} = 12.59]$. Under the null hypothesis, the probability of obtaining a χ^2 value as extreme as, or more extreme than, 12.59 is 0.05:

$$\Pr[\chi^2_6 \geq 12.59] = 0.05.$$

Table 8.2-3 An excerpt from the table of χ^2 critical values (Statistical Table A).
Numbers down the left side are the number of degrees of freedom (*df*). Numbers across the top are significance levels (α). The critical value for a χ^2 distribution with *df* = 6 and α = 0.05 is 12.59 (indicated in red).

					Significance level α					
df	0.999	0.995	0.99	0.975	0.95	0.05	0.025	0.01	0.005	0.001
1	0.000002	0.00004	0.00016	0.00098	0.00393	3.84	5.02	6.63	7.88	10.83
2	0.002	0.01	0.02	0.05	0.10	5.99	7.38	9.21	10.6	13.82
3	0.02	0.07	0.11	0.22	0.35	7.81	9.35	11.34	12.84	16.27
4	0.09	0.21	0.30	0.48	0.71	9.49	11.14	13.28	14.86	18.47
5	0.21	0.41	0.55	0.83	1.15	11.07	12.83	15.09	16.75	20.52
6	0.38	0.68	0.87	1.24	1.64	**12.59**	14.45	16.81	18.55	22.46
7	0.60	0.99	1.24	1.69	2.17	14.07	16.01	18.48	20.28	24.32
8	0.86	1.34	1.65	2.18	2.73	15.51	17.53	20.09	21.95	26.12

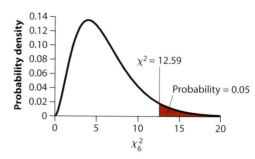

Figure 8.2-3 The χ^2 distribution with six degrees of freedom. The area under the right tail of the curve in red is 5% of the total area under the curve. This is the region to the right of $\chi^2 = 12.59$. Under the null hypothesis, χ_6^2 will be greater than 12.59 with probability 0.05.

Figure 8.2-3 illustrates the area under the curve to the right of 12.59.

Because our observed χ^2 value (15.05) is greater than 12.59 (that is, further out in the right tail of the distribution), χ^2 values of 15.05 or greater occur more rarely under the null hypothesis than 5% of the time. Therefore, our P-value must be less than 0.05,

$$P = \Pr[\chi_6^2 \geq 15.05] < 0.05,$$

so we reject the null hypothesis.

We can use Statistical Table A to get closer to the true P-value. Note that Statistical Table A also includes columns of critical values for other values of α. Our observed χ^2 test statistic (15.05) falls between the critical values for $\alpha = 0.025$ $[\chi_{6,(0.025)}^2 = 14.45]$ and that for $\alpha = 0.01$ $[\chi_{6,(0.01)}^2 = 16.81]$. Our observed test statistic is greater than 14.45 but less than 16.81. Thus, Statistical Table A makes it possible to bound the P-value as

$$0.01 < P < 0.025.$$

This would be reported as just $P < 0.025$ in a scientific paper. Note that this conclusion is consistent with the P-value of 0.0199 calculated by a computer statistics package.

Based on this analysis, we can conclude that births are not equitably distributed over the days of the week. [3]

8.3   Assumptions of the χ^2 goodness-of-fit test

The χ^2 goodness-of-fit test assumes that the individuals in the data set are a random sample from the whole population. This means that each individual was chosen independently of all of the others and that each member of the population was equally

---

[3] This discrepancy is largely due to scheduled C-sections and induced labor, but these obvious causes do not explain the whole effect (Ventura et al. 2001).

likely to find its way into the sample. This is an assumption of every test described in this book.

The sampling distribution of the χ^2 statistic follows a χ^2 distribution only approximately. The approximation is excellent, as long as the following rules[4] are obeyed:

- ▶ None of the categories should have an expected frequency less than one.
- ▶ No more than 20% of the categories should have expected frequencies less than five.

Notice that these restrictions refer to the *expected* frequencies, not to the *observed* frequencies. If these conditions are not met, then the test becomes unreliable.

If one of these conditions is not met, then we have two options. One option, if possible, is to combine some of the categories having small expected frequencies to yield fewer categories having larger expected frequencies (remember to change the degrees of freedom accordingly). We'll see examples of this approach in Section 8.6. A second option is to find an alternative to the χ^2 goodness-of-fit test, perhaps making use of computer simulation (Chapter 19).

8.4   Goodness-of-fit tests when there are only two categories

The χ^2 goodness-of-fit test works even when there are only two categories, so it's a quick substitute for the binomial test (Chapter 7), provided that the assumptions of the χ^2 test are met. The calculations are much quicker than those required for the binomial test, although they are less exact. We demonstrate these calculations using Example 8.4.

Example 8.4 | **Gene content of the human X chromosome**

The sex chromosomes are inherited in a very different pattern from that of the other chromosomes, which is known to affect their evolution in many ways. Are they unusual in other ways? For example, are there as many genes on the human X chromosome as we would expect from its size? The Human Genome Project has found 781 genes on the human X chromosome, out of 20,290 genes found so far in

---

[4]   These restrictions are conservative, because the χ^2 goodness-of-fit test has been shown to work well even with smaller expected values. An alternative rule-of-thumb is that the average expected value should be at least five (Roscoe and Byars 1971).

the entire genome.[5] The X chromosome represents 5.2% of the DNA content of the whole human genome, so under the proportional model we would expect 5.2% of the genes to be on the X chromosome. Is this what we observe?

The null and alternative hypotheses are

H_0: The percentage of human genes on the X chromosome is 5.2%.
H_A: The percentage of human genes on the X chromosome is *not* 5.2%.

Observed frequencies and the frequencies expected under H_0 are listed in Table 8.4-1. The expected number of genes on the X chromosome, under the null hypothesis, is $20{,}290 \times 0.052 = 1{,}055$. We observed only 781, which is substantially fewer. What is the probability of a result as extreme as, or more extreme than, the result observed assuming the null hypothesis?

Table 8.4-1 Numbers of genes on the human X chromosome and on the rest of the genome.

Chromosome	Observed	Expected
X	781	1,055
Not X	19,509	19,235
Total	20,290	20,290

It would be a challenge to calculate the *P*-value using the binomial test. We would need to calculate

$$P = 2 \times \Pr[X \le 781].$$

When the number of trials (genes) is $n = 20{,}290$ and the probability of a gene being on the X chromosome is $p = 0.052$, this number P would be calculated as

$$P = 2 \times (\Pr[X = 0] + \Pr[X = 1] + \Pr[X = 2] + \ldots + \Pr[X = 781]).$$

The tedium of this sum causes the mind to boggle.[6]

It would be much faster to calculate the *P*-value using the χ^2 goodness-of-fit test. This procedure yields

$$\chi^2 = \frac{(781 - 1055)^2}{1055} + \frac{(19509 - 19235)^2}{19235} = 75.1.$$

---

[5] We used release 35 of the human genome, available on the ENSEMBL website in December 2005 (http://www.ensembl.org/Homo_sapiens).

[6] We couldn't resist: the answer is $P = 2.64 \times 10^{-84}$.

This test statistic has two categories and, therefore, only one degree of freedom:

$$df = 2 - 1 = 1.$$

From Statistical Table A, we see that the critical value of χ_1^2 for a significance level $\alpha = 0.05$ is 3.84. Because our observed $\chi^2 = 75.1$ is greater than 3.84, we know that $P < 0.05$, and we reject the null hypothesis. In fact, we can use Statistical Table A to be even more precise. Because our calculated χ^2 is greater than the largest critical value given for one $df\,[\chi_{1,(0.001)}^2 = 10.83]$, we can say that $P < 0.001$. Thus, there are *significantly* fewer genes on the X chromosome in humans than would be expected from its size.

When there are only two categories, the binomial test is the best option when n is small and when the expected frequencies are too low to meet the assumptions of the χ^2 goodness-of-fit test. Even when n is large, though, the binomial test is preferred when a computer is available, because it yields an exact P-value.[7]

8.5 **Fitting the binomial distribution**

The proportional model is not the only probability model that can be tested using a goodness-of-fit approach. Biologists often fit their data to other probability distributions that also represent simple models for how nature behaves. By "model" we mean a mathematical description that mimics how we think a natural process works, or at least how it would work in the absence of complications. For this reason, probability distributions are used as null hypotheses in many branches of biology.

| Example 8.5 | **Designer two-child families?** |

In Chapter 5, we claimed that the sex of consecutive children is independent in humans. For example, having had one boy already does not change the probability that the next child will also be a boy. In the absence of complications, then, we expect the numbers of sons and daughters in families containing two children to match a binomial distribution with $n = 2$ and p equal to the probability of having a son in any single trial. Is this what we see? Rodgers and Doughty (2001) tested this hypothesis using data from the National Longitudinal Survey of Youth, which compiles data on the sex of children in a random sample of families of different sizes.

In this section and the next, we compare data to two important discrete probability distributions, the *binomial distribution* and the *Poisson distribution*. Both describe the probability of getting a certain number of successes from independent trials. The binomial distribution tells us the probability of getting X successes out of n independent

---

[7] This is not always true. Many statistical packages for computers cheat and use an approximation.

trials, while the Poisson distribution tells us the probability of getting X successes in a block of time or space, when each success happens independently.

There are three possible outcomes for families containing exactly two children: zero, one, or two boys. Table 8.5-1 lists the data from 2444 families.

Table 8.5-1 The frequency distribution of the number of boys in families with two children.

Number of boys	Observed number of families
0	530
1	1332
2	582
Total	2444

We can test the fit of the binomial distribution to the data in Table 8.5-1 using the χ^2 goodness-of-fit test. The null and alternative hypotheses are

H_0: The number of boys in two-child families has a binomial distribution.
H_A: The number of boys in two-child families does not have a binomial
      distribution.

Here we are testing the fit of a distribution to the data on multiple families. We are not testing a hypothesis about the mean proportion of boys. When testing the fit to a binomial distribution, we are fitting the results of multiple *sets* of trials, matching a set of frequencies to the expectation of the binomial distribution. This is different from using the binomial distribution to test a null hypothesis about a proportion. In a binomial test we have only one set of trials.

Notice that, in this case, our null hypothesis does not specify p, the probability that an individual offspring is a boy. This complicates our task slightly, because we must first estimate p from the data before we can calculate the expected frequencies.

Here's how we estimate p from the data. There are 4888 children in the study, a value obtained by multiplying the number of families (2444) by the family size (2). The total number of sons is $(2 \times 582) + 1332 = 2496$. Thus, we can estimate the probability of a child being a boy as

$$\hat{p} = 2496/4888 = 0.5106.$$

Next, we use this estimate of p and the binomial distribution with $n = 2$ to calculate the expected frequencies under the null hypothesis. For example, the expected fraction of two-child families having exactly one boy is

$$\Pr[1\ boy] = \binom{2}{1}(0.5106)^1(1 - 0.5106)^1 = 0.49977.$$

Thus, the expected frequency of 2444 two-child families having exactly one boy is

$$Expected\,(1\ boy) = 2444 \times 0.49977 = 1221.4.$$

Table 8.5-2 lists the expected frequencies for all possible outcomes, and Figure 8.5-1 shows expected values alongside the data. Surprisingly, the observed frequencies don't seem to match the frequencies expected under the binomial distribution. There is a shortage of two-child families having either no boys or two boys compared with expectation, and an excess of families having exactly one boy. The differences between observed and expected frequencies are not huge; but can they be explained by chance, or must the null hypothesis be rejected?

Table 8.5-2 Observed and expected number of boys in two-child families.

Number of boys	Observed number of families	Expected number of families
0	530	585.3
1	1332	1221.4
2	582	637.3
Total	2444	2444.0

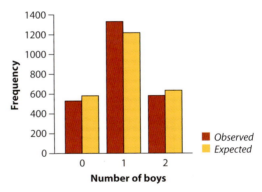

Figure 8.5-1 The observed number of two-child families with a given number of boys (*red*) compared with the frequency expected from a binomial distribution (*gold*). Compared with expected frequencies, there is an excess of two-child families with exactly one boy and a shortage of families with no boys or two boys.

The formula to calculate χ^2, first introduced in Section 8.2, gives us

$$\chi^2 = \frac{(530 - 585.3)^2}{585.3} + \frac{(1332 - 1221.4)^2}{1221.4} + \frac{(582 - 637.3)^2}{637.3} = 20.04.$$

Next, we need to calculate the number of degrees of freedom for our test. There are three categories, which would ordinarily leave us with two degrees of freedom. However, we needed to estimate one parameter from the data to generate the expected frequencies (the probability of boys, p). Using this estimate costs us an extra degree of freedom.[8] As a result, the number of degrees of freedom is

$$df = 3 - 1 - 1 = 1.$$

The critical value for the χ_1^2 distribution having one degree of freedom and a significance level $\alpha = 0.05$ is 3.84 (see Statistical Table A). Because 20.04 is further into the tail of the distribution than 3.84, we know that $P < 0.05$; therefore we reject the null hypothesis. If we probe Statistical Table A further, we find that 20.04 is greater even than the critical value corresponding to $\alpha = 0.001$, so $P < 0.001$. A statistics package on the computer gave us a more exact value, $P = 1.2 \times 10^{-7}$.

These data show that the frequency distribution of the number of boys and girls in two-child families does *not* match the binomial distribution. This means that one of the assumptions of the binomial distribution must not hold in these data. Either the probability of having a son varies from one family to the next, or the individuals within a family are not independent of each other, or both.

What is the reason for the poor fit of the binomial distribution to the number of boys in two-child families? Is the sex of the second child not independent of that of the first, as we've assumed? Are parents manipulating the sex of their children? One likely explanation is that many parents of two-child families having either no boys or two boys are unsatisfied with not having at least one child of each sex and decide to have a third child, thus "removing" their family from the set of two-child families.

8.6   Random in space or time: the Poisson distribution

When the dust settled after the 1980 explosion of Mount St. Helens, spiders were among the first organisms to recolonize the moonscape-like terrain. They dropped out of the air stream and grew fat on insects that arrived in the same way. Let's imagine the frequency distribution of spider landings across the landscape. What would it look like if spider landings were completely "random" in space? The assumptions we need are as follows:

---

[8] Why do we lose another degree of freedom? Setting the parameters of the binomial to equal those estimated from the data would make it easier for the data to fit the null hypothesis, because, whatever the distribution happens to be, we will have forced the expected values to give the same overall sex ratio. We account for this by removing one degree of freedom from our test for every parameter we estimate. This reduction in the degrees of freedom makes it legitimate to estimate parameters from the data to use in calculating the expected values. If we estimate too many parameters, then the number of degrees of freedom would drop to zero and we couldn't do the test.

▶ The probability that a spider lands at a given point on the continuous landscape is the same everywhere (i.e., they aren't more likely to land some places than others).

▶ Whether a spider lands at a given point on the landscape is independent of landings everywhere else (i.e., spiders don't clump together or repel one another).

To count spiders, let's place a large grid across the landscape, breaking it up into equal-sized blocks. (The block size doesn't matter as long as they're large enough to accommodate many potential landing sites.) If both assumptions listed previously are met, then the frequency distribution of the number of spiders landing in blocks will follow a **Poisson distribution**.

> The *Poisson distribution* describes the number of successes in blocks of time or space, when successes happen independently of each other and occur with equal probability at every point in time or space.

The Poisson distribution is a starting place for asking whether events or objects occur randomly in continuous time and space. A Poisson distribution is a reasonable expectation for certain biological counts, such as the number of mutations carried per individual in a population, the number of salmon caught on a given day by sport fishers, or the number of seeds successfully germinated per mother plant. For the biologist, the Poisson distribution is just a *model* for how successes may be distributed in time and space in nature. Life gets interesting when the model doesn't fit the data, because then we learn that one or more of the main assumptions is false, hinting at the existence of interesting biological processes. (For example, some individuals may actually be more prone to mutations than others, some fishers may be better catchers than others, or some plants may produce better quality seeds.)

The alternative to the Poisson distribution is that successes are distributed in some nonrandom way in time or space. Successes can be **clumped**, for example, in which case they occur closer together than expected by chance (see the left panel in Figure 8.6-1), or successes can be **dispersed**, meaning they are spread out more evenly than expected by chance (see the right panel in Figure 8.6-1). A clumped distribution may arise when the presence of one success increases the probability of other successes occurring nearby. Outbreaks of contagious disease, for example, often have a clumped spatial distribution, because individuals catch the disease from their neighbors. A dispersed distribution happens when the presence of one success decreases the probability of another success occurring nearby. Territorial animals are often more dispersed in space than would be expected by chance, for example, because individuals chase each other away. Deviations from the random pattern can therefore help us to identify interesting biological processes that create the patterns.

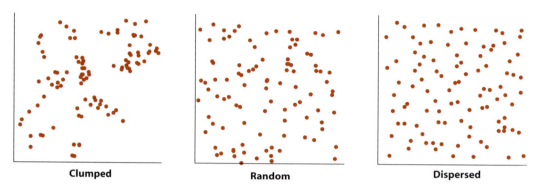

Clumped   **Random**   **Dispersed**

Figure 8.6-1 Distributions of points in space that follow a clumped distribution (*left*), a random distribution (*center*), or a dispersed distribution (*right*). In the "random" distribution, each point has an equal and independent probability of appearing anywhere in the space. If the random graph were divided into equal-sized squares, the number of points per square would follow a Poisson probability distribution.

Formula for the Poisson distribution

The Poisson distribution was derived by Siméon Denis Poisson,[9] a French mathematical physicist. He showed that the probability of X successes occurring in any given block of time or space is

$$\Pr[X \; successes] = \frac{e^{-\mu}\mu^{X}}{X!},$$

where μ is the mean number of independent successes in time or space (expressed as a count per unit time or a count per unit space). Here e, the base of the natural log, is a constant approximately equal to 2.718, and $X!$ is X factorial.

Testing randomness with the Poisson distribution

The main use of the Poisson distribution in biology is to provide a null hypothesis to test whether successes occur "randomly" in time or space. In practice, we usually don't know the exact rate at which successes may occur. So, to make predictions about the probability of different outcomes from a Poisson distribution, we must first estimate the rate from the data.

---

[9]   Poisson is famous for saying, "Life is good for only two things, doing mathematics and teaching mathematics," an opinion no doubt shared by most readers of this book.

| Example 8.6 | **Mass extinctions** |

Do extinctions occur randomly through the long fossil record of Earth's history, or are there periods in which extinction rates are unusually high ("mass extinctions") compared with background rates? The best record of extinctions through Earth's history comes from fossil marine invertebrates, because they have hard shells and therefore tend to preserve well. Table 8.6-1 lists the number of recorded extinctions of marine invertebrate families in 76 blocks of time of similar duration through the fossil record (Raup and Sepkoski 1982).

Table 8.6-1 The frequency of time blocks in the fossil record in which an observed number of marine invertebrate families went extinct.

Number of extinctions (X)	Frequency
0	0
1	13
2	15
3	16
4	7
5	10
6	4
7	2
8	1
9	2
10	1
11	1
12	0
13	0
14	1
15	0
16	2
17	0
18	0
19	0
20	1
> 20	0
Total	76

If the occurrence of family extinctions is "random" in time through the fossil record, then the number of extinctions per block of time should follow a Poisson distribution. Departures from the Poisson distribution could indicate that extinctions tend to be clumped in time and occur in bursts ("mass extinctions"). Another possibility is that extinctions are more evenly spread over time than we would expect if they occurred randomly.

The easiest way to test the randomness of family extinctions is to compare the frequency distribution of extinctions to that expected from a Poisson distribution using a χ^2 goodness-of-fit test. Our hypotheses are

H_0: The number of extinctions per time interval has a Poisson distribution.
H_A: The number of extinctions per time interval does *not* have a Poisson distribution.

To begin the test, we need to estimate μ, the mean number of extinctions per time interval. This is obtained using the sample mean,

$$\overline{X} = \frac{(0 \times 0) + (13 \times 1) + (15 \times 2) + \ldots}{76} = 4.21.$$

(See Section 3.5 to review how to calculate a mean from a frequency table. Remember that there are $n = 76$ separate data points here, not the smaller number indicated by the number of rows in Table 8.6-1.) This sample mean ($\overline{X}$) is used in place of μ in the formula for the Poisson distribution to generate the expected frequencies. We show the calculations next, but first look at the result graphically in Figure 8.6-2.

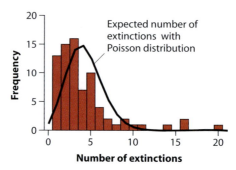

Figure 8.6-2 The frequency distribution of the number of extinctions (*histogram*) compared with the frequencies expected from the Poisson distribution having the same mean (*curve*).

The histogram in Figure 8.6-2 gives the observed frequency distribution of extinctions per time interval, whereas the line graph connects the expected frequencies under the null hypothesis (the Poisson distribution). If you look closely, there is a discrepancy. Compared with the Poisson distribution, the fossil record shows too many time intervals with large numbers of extinctions and too many intervals having very few extinctions, compared with the Poisson distribution. But is the discrepancy between the observed and expected distributions greater than expected by chance? We will use the χ^2 goodness-of-fit test to find out.

In Table 8.6-2 we have tabulated the observed and expected frequencies. The expected frequency for all but the last category of extinctions is computed by applying the formula for the Poisson distribution to get the expected probability, and then multiplying this probability by the total number of intervals (76) to yield the expected frequency. For example, the expected probability of three extinctions in a time interval is

$$\Pr[3 \ extinctions] = \frac{e^{-\mu}\mu^3}{3!} = \frac{e^{-4.21}(4.21)^3}{3!} = 0.1846.$$

Multiplying this result times the total number of intervals in the data set (76), we find the expected number of intervals with three extinctions:

$$Expected[3 \ extinctions] = 76 \times 0.1846 = 14.03.$$

You may want to try to calculate some of the other expected values in Table 8.6-2 for practice.

We have grouped all $X \geq 10$ extinctions into the last category because the expected frequency of larger numbers is getting very small. The expected frequency for the final category is computed by subtracting the sum of all the previous expected values from 76, the total number of time intervals.

Unfortunately, the expected frequencies fail to meet the assumptions of the χ^2 test: one of them is less than one, and more than 20% are less than five. When this happens, we can group categories and try again. For example, combining $X = 0$ and $X = 1$ into a single class and combining all classes with $X \geq 8$ make sense, because the classes grouped are similar. The resulting data, listed in Table 8.6-3, have eight categories.

Table 8.6-2 The observed frequency distribution of extinctions of marine invertebrate families compared with the number expected under the Poisson distribution.

Number of extinctions (X)	Observed frequency of time intervals	Expected frequency of time intervals
0	0	1.13
1	13	4.75
2	15	10.00
3	16	14.03
4	7	14.77
5	10	12.44
6	4	8.72
7	2	5.24
8	1	2.76
9	2	1.29
≥ 10	6	0.86
Total	76	76

Table 8.6-3 The observed and expected frequencies of time intervals with a given number of extinctions of marine invertebrate families.

Number of extinctions (X)	Observed frequency of time intervals	Expected frequency of time intervals
0 or 1	13	5.88
2	15	10.00
3	16	14.03
4	7	14.77
5	10	12.44
6	4	8.72
7	2	5.24
> 8	9	4.91
Total	76	76

Using the standard formula for the χ^2 statistic, we compute

$$\chi^2 = \frac{(13 - 5.88)^2}{5.88} + \frac{(15 - 10.00)^2}{10.00} + \frac{(16 - 14.03)^2}{14.03} + \ldots = 23.93.$$

We have six degrees of freedom for this test, accounting for the one parameter, μ, that we had to estimate from the data:

df = (Number of categories) − 1 − (Number of parameters estimated from the data)
   = 8 − 1 − 1 = 6.

The critical value for $\chi^2_{6,(0.05)}$ is 12.59 (see Statistical Table A). Our χ^2 statistic of 23.93 is further into the tail of the distribution than this critical value is, so our P-value is less than 0.05. More specifically, we can see that $P < 0.001$, because 23.93 is also greater than 22.46, the critical value corresponding to $\alpha = 0.001$. We reject the null hypothesis, therefore, and conclude that extinctions in this fossil record do *not* fit a Poisson distribution.

Comparing the variance to the mean

How can we describe the way that a pattern deviates from the Poisson distribution? One unusual property of the Poisson distribution is that the variance in the number of successes per block of time (the square of the standard deviation) is equal to the mean (μ). If the variance is greater than the mean, then the distribution is clumped. If the variance is less than the mean, then the distribution is dispersed.

For the extinction data, the sample mean number of extinctions is 4.21. The sample variance in the number of extinctions is

$$s^2 = \frac{(0 - 4.21)^2(0) + (1 - 4.21)^2(13) + (2 - 4.21)^2(15) + \ldots}{76 - 1} = 13.72.$$

Because the sample variance (13.72) greatly exceeds the sample mean (4.21), the distribution of extinction events in time is highly "clumped." That is, extinctions tend to occur in bursts (mass extinctions) rather than randomly or evenly in time.

8.7 Summary

▶ The χ^2 goodness-of-fit test compares the frequency distribution of a discrete or categorical variable with the frequencies expected from a probability model.

▶ The χ^2 goodness-of-fit test is more general than the binomial test because it can handle more than two categories. It is also easier to compute, even when there are only two categories.

▶ Goodness-of-fit is measured with the χ^2 test statistic.

▶ The χ^2 test statistic has a null distribution that is approximated by the theoretical χ^2 distribution. The approximation is excellent as long as no expected frequencies are less than one and no more than 20% of the expected frequencies are less than five. It may be necessary to combine categories to meet these criteria.

▶ The theoretical χ^2 distribution is a continuous distribution. Probability is measured by the area under the curve.

▶ The null hypothesis is rejected at significance level α if the observed χ^2 statistic exceeds the critical value of the χ^2 distribution corresponding to α.

▶ Under the proportional probability model, events fall in different categories in proportion to unit size or number of opportunities. Rejecting H_0 implies that the probabilities are not proportional.

▶ If trials are independent, and the probability p of a success is the same for each trial, then the frequency distribution of the number of successes should follow a binomial distribution. Rejecting the null hypothesis that the number of successes follows a binomial distribution implies that trials are not independent, or that the probability of success is not the same for all trials.

▶ The Poisson distribution describes the frequency distribution of successes in blocks of time or space when successes happen independently and with equal probability over time or space. Rejecting a null hypothesis of a Poisson distribution of successes implies that successes are not independent or that the probability of a success occurring is not constant over time or space.

▶ Comparing the variance of the number of successes per block of time or space to the mean number of successes measures the direction of departure from randomness in time or space. If the variance is *greater* than the mean, the successes are clumped; if the variance is *less* than the mean, successes are more evenly distributed than expected by the Poisson distribution.

8.8   Quick Formula Summary

χ^2 Goodness-of-fit test

What is it for? Compares observed frequencies in categories of a single variable to the expected frequencies under a random model.

What does it assume? Random samples. Also that the expected count of each cell is greater than one and that no more than 20% of the cells have expected counts less than five.

Test statistic: χ^2

Distribution under the null hypothesis: χ^2 distributed with df = (Number of categories) $-$ 1 $-$ (Number of parameters estimated from the data).

Formula: $\chi^2 = \sum_i \dfrac{(Observed_i - Expected_i)^2}{Expected_i}.$

Poisson distribution

What is it for? Describes the number of independent events that occur per unit of time or space.

Formula: $\Pr[X \; events] = \dfrac{e^{-\mu}\mu^X}{X!},$

where X is the number of events and μ is the mean number of events per unit time or space.

PRACTICE PROBLEMS

1. Each of the following examples could be addressed with a goodness-of-fit test. From the information given, how many categories and how many degrees of freedom would each test have? Explain your answers.

   a. A die is rolled 50 times to test whether it is fair—that is, whether it has 1/6 chance of coming up on each of its six different sides.

   b. A set of 10 coins is flipped, and the number of heads is recorded. This experiment is repeated with the same coins 1000 times. The test compares the frequency of heads to a binomial distribution with $p = 0.5$.

   c. The scenario is the same as in part (b), except now the question is whether the frequency of heads follows a binomial distribution (p not specified).

   d. A food-protection agency counts the number of insect heads found per 100 grams of wheat flour. The researchers have 500 samples, and they want to know whether the frequency of insect heads in samples follows a Poisson distribution. The samples returned every number from zero to four heads, each more than five times. No sample had more than four heads.

2. Powerball™ is a state-sponsored gambling game in which balls are drawn every Wednesday and Saturday night for large prizes. Data are available for the number of people who buy tickets for each day of the week. Over the course of three years, the number of tickets bought on each day, expressed in millions of tickets, is given in the following table:[10]

Day of the week	Number of tickets sold (in millions)
Sunday	128.0
Monday	275.3
Tuesday	448.9
Wednesday	1063.5
Thursday	244.4
Friday	468.5
Saturday	1060.9

These data provide an opportunity to ask to what extent people procrastinate. A naïve researcher used these data as given to calculate a goodness-of-fit test of a model with equal ticket purchases each day, finding $\chi^2 = 1.38 \times 10^9$. Is this a valid use of the χ^2 goodness-of-fit test? Why or why not?

3. The parasitic nematode *Camallanus oxycephalus* infects many freshwater fish, including shad. The following table gives the number of nematodes per fish (Shaw et al. 1998). Is there evidence that some fish are more or less likely to attract nematodes, or are the nematodes worming their way into the fish at random?

Number of parasites	Frequency
0	103
1	72
2	44
3	14
4	3
5	1
6	1

4. The study of the spatial distribution of vegetation often makes use of random samples of

[10] http://www.dartmouth.edu/~chance/chance_news /recent_news/chance_news_13.02.html

"quadrats," rectangular plots of fixed size placed at random over the sampling region (e.g., a field or forest). The number of plants of each type that are rooted within the quadrats is then counted. In one such study, an investigator counted the number of white pine seedlings growing in eighty 10 m $\times$ 10 m quadrats to test whether the distribution of pine seedlings in the forest was random, clumped, or dispersed. She obtained the following counts:

Number of seedlings	Number of quadrats
0	47
1	6
2	5
3	8
4	6
5	6
6	2
≥ 7	0
Total	80

a. If the null hypothesis of a random distribution of pine seedlings across the forest is correct, to what theoretical probability distribution should the observed frequencies of quadrats containing a given number of seedlings conform?

b. Carry out a formal test of the null hypothesis.

c. If the null hypothesis is rejected in part (b), determine whether the spatial distribution of seedlings is clumped or dispersed.

5. Soccer reaches its apotheosis every four years at the World Cup, attracting worldwide attention and fanatic devotion. The World Cup is widely thought to be the event that determines the best soccer team in the world. If skill were the main determinant of the outcomes of games, then we would expect the scores of the games to be highly variable. On the other hand, if the goals were assigned randomly to each side in each game, with the probability of a goal being the same for each side and game, then we would expect the frequency distribution of goals per side per game to be approximately a Poisson distribution. The data in the following table show the number of goals per side, for all of the games in the 2002 World Cup:

a. What is the mean number of goals per side?
b. Does a Poisson distribution fit these data?

Number of goals	Frequency
0	37
1	47
2	27
3	13
4	2
5	1
6	0
7	0
8	1
> 8	0

6. Sumo wrestling, despite its tradition of honor, has recently been accused of conspiracies between wrestlers to throw matches when a win would benefit one wrestler more than a loss would cost the other. At tournaments, each wrestler has 15 bouts with different opponents, and there is a large difference in the payoff from winning eight out of the 15 bouts, compared with winning just seven. As a result, some wrestlers may be motivated to dishonestly rig the outcome of a bout if one wrestler needs an eighth win and the other has already won eight. This situation might seem impossible to study, but some researchers realized that there might be a statistical signature of fraud if there were an excess of individuals just barely winning eight bouts (Duggan and Leavitt 2002). They therefore collected data on the success rates of 2126 wrestlers. These data are given in the following table, along with the binomial expectation. In these data, the winner and loser of every match are both represented, which means that the probability of winning (p) is 0.5 in every bout.

a. Show the calculations that led to the value in the table for the expected proportion of wrestlers having exactly eight wins out of 15.
b. Does a binomial distribution fit these data?
c. Display in a graph the observed frequency distribution and the expected frequency distribution under the null hypothesis. Interpret

the results in terms of the hypothesis of match-rigging.

d. Assume that there was no cheating. Would you expect these data to match the assumptions of the binomial distribution? If not, is cheating the only interpretation of the discrepancy, or are there other explanations you might propose?

Total wins	Number of wrestlers	Binomial proportion ($p = 0.5$)
0	0	0.00003
1	7	0.00046
2	25	0.00320
3	56	0.01389
4	112	0.04166
5	190	0.09164
6	290	0.15274
7	256	0.19638
8	549	0.19638
9	339	0.15274
10	132	0.09164
11	73	0.04166
12	56	0.01389
13	22	0.00320
14	22	0.00046
15	3	0.00003

7. One thousand coins were each flipped eight times, and the number of heads was recorded for each coin. The results are as follows:

Number of heads	Number of coins
0	6
1	32
2	105
3	186
4	236
5	201
6	98
7	33
8	103

a. Does the distribution of coin flips match the binomial distribution expected with fair coins? (A coin is "fair" if the probability of heads per flip is 0.5.)
b. If the binomial distribution is a poor fit, try to explain in what way the distribution does not match the expectation.

c. Some two-headed coins (which always show heads on every flip) were mixed in with the fair coins. Can you say approximately how many two-headed coins there might have been out of this 1000?

8. Practice Problem 11 from Chapter 7 gave data about death rates from cancer before and after Christmas. Use these data to test whether the holiday affects death rates.

9. Practice Problem 1 from Chapter 7 gave data about the death rates of people working on the movie *The Conqueror*. Test whether the cancer rates in this group were different from the expected rate of 14%.

10. Imagine that a small hospital's emergency room has an average of 20 admissions per Saturday night. If you were a doctor working overtime on such a Saturday night, you might want to know the probability of having a quiet night, one that would let you catch up on some much-needed sleep. Let's call a quiet night one in which five or fewer admissions take place. What is the chance that you get some sleep? Assume that admissions are independent of one another and are just as likely to land in one instant in time as another on a Saturday night.

ASSIGNMENT PROBLEMS

11. If each "success" happens independently of all other successes and with the same probability, what probability distribution is expected for each of the following?
    a. Number of flowers in square-meter blocks in an alpine field
    b. Number of heads out of 10 flips of a coin
    c. Number of bombs landing in city blocks in London in World War II
    d. Daily number of hits on a website
    e. Annual number of elephant attacks on humans in Serengeti National Park
    f. Number of red flowers in sets of 100 flowers in a field of multiple types of flowers

12. The following list gives the number of degrees of freedom and the χ^2 test statistic for several goodness-of-fit tests. Find the *P*-value for each test as specifically as possible from Statistical Table A. If you can, find the *P*-values more exactly using a computer program.

Degrees of freedom	χ^2
1	4.12
4	1.02
2	9.50
10	12.40
1	2.48

13. Windows kill more birds than any other human-related factor. In North America alone, somewhere between 100 million and 1 billion birds die each year by crashing into windows on buildings. This figure represents up to 5% of the total number of birds in the area. One possible solution to this problem is to angle windows downwards slightly, so that they reflect the ground rather than an image of the sky to the flying bird. An experiment was done to compare the number of birds that died as a result of vertical windows, windows angled 20 degrees off vertical, and windows angled 40 degrees off vertical (Klem et al. 2004). The angles were randomly assigned to six identical windows, and the assignments were randomly varied daily for four months. The amount of time that windows were at each of the three angles was the same. Over the course of the experiment, 30 birds were killed by windows in the vertical orientation, 15 were killed by windows set at 20 degrees off vertical, and eight were killed by windows set at 40 degrees off vertical.
    a. Clearly state an appropriate null hypothesis and an alternative hypothesis.

b. What proportion of deaths occurred while the windows were set at a vertical orientation?

c. What statistical test can test the null hypothesis?

d. Carry out the statistical test from part (c). Is there evidence that window angle affects the mortality rates of birds?

e. Why were the windows assigned an angle at random? Why were they changed daily?

14. In the nineteenth century, cavalry were still an important part of the European military complex. While horses have many wonderful qualities, they can be dangerous beasts, especially if poorly treated. The Prussian army kept track of the number of fatalities caused by horse kicks to members of 10 of their cavalry regiments over a 20-year time span. If these fatalities occurred purely by chance, then we would expect the number of deaths by horse kick per regiment per year to follow a Poisson distribution. On the other hand, if some regiments during some years consisted of particularly bad horsemen,[11] then the events would not be independent; we would thus expect a frequency distribution different from the Poisson distribution. The following table shows the data, expressed as the number of fatalities per regiment-year (Bortkiewicz 1898).

Number of deaths (X)	Number of regiment-years
0	109
1	65
2	22
3	3
4	1
> 4	0
Total	200

a. What is the mean number of deaths from horse kicks per regiment-year?

b. Test whether a Poisson distribution fits these data.

15. At a western hospital there were a total of 932 births in 20 consecutive weeks. Of these births, 216 occurred on weekends. Based on these data, are birth rates different on weekends and weekdays?

a. State the null and alternative hypotheses for the test.

b. Name two statistical methods that could be used to test the hypotheses. State the advantages and disadvantages of both.

c. Carry out a hypothesis test to address this question. State your conclusions clearly.

16. Truffles are a great delicacy, sending thousands of mushroom hunters into the forest each fall to find them. A set of plots of equal size in an old-growth forest in Northern California was surveyed to count the number of truffles (Waters et al. 1997). The resulting distribution is presented in the following table. Are truffles randomly located around the forest? If not, are they clumped or dispersed? (The mean number of truffles per plot, calculated from these data, is 0.60.)

Number of truffles per plot	Frequency
0	203
1	39
2	18
3	13
> 3	15

17. A field study of a species of wrasse, a coral reef fish, involved a survey of groups of individuals inhabiting coral outcrops in the northern section of the Great Barrier Reef. Researchers recorded the number and sex of adult individuals on a random sample of outcrops. The following table lists the numbers of males and females observed on 22 outcrops on which exactly six adult fish were found.

Number of males	Number of females	Number of outcrops
0	6	4
1	5	14
2	4	4
> 2	< 4	0
Total		22

[11] Or if they were particularly prone to standing behind their horses . . .

a. Estimate the mean number of males per outcrop with six fish. Provide a standard error for this estimate.

b. Does the number of males on outcrops having six fish have a binomial distribution? Show all steps in carrying out your test.

c. If the number of males on outcrops does not have a binomial distribution, what is the likely statistical explanation (i.e., what assumption of the binomial distribution is likely violated)?

d. Can you suggest a biological explanation for a non-binomial pattern?

18. Hurricanes hit the U.S. often and hard, causing some loss of life and enormous economic costs. They are ranked in severity by the Saffir–Simpson scale, which ranges from category 1 to category 5, with 5 being the worst. In some years, as many as three hurricanes that rate a category 3 or higher hit the U.S. coastline. In other years, no hurricane of this severity hits the U.S. The following table lists the number of years that had zero, one, two, three, or more hurricanes of at least category 3 in severity, over the 100 years of the twentieth century (Blake et al. 2005):

Number of hurricanes category 3 or higher	Number of years
0	50
1	39
2	7
3	4
> 3	0

a. What is the mean number of severe hurricanes to hit the U.S. per year?

b. What model would predict the distribution of hurricanes per year, if they were to hit independently of each other and if the probability of a hurricane were the same in every year?

c. Test the fit of the model from part (b) to the data.

d. Is the annual pattern of hurricanes clumped or dispersed?

19. In snapdragons, variation in flower color is determined by a single gene. *RR* individuals are red, *Rr* (heterozygous) individuals are pink, and *rr* individuals are white. In a cross between heterozygous individuals, the expected ratio of red-flowered:pink-flowered:white-flowered offspring is 1:2:1.

a. The results of such a cross were 10 red-, 21 pink-, and nine white-flowered offspring. Do these results differ significantly (at a 5% level) from the expected frequencies?

b. In another, larger experiment, you count 100 times as many flowers as in the experiment in part (a) and get 1000 red, 2100 pink, and 900 white. Do these results differ significantly from the expected 1:2:1 ratio?

c. Do the proportions observed in the two experiments [i.e., in parts (a) and (b)] differ? Did the results of the two hypothesis tests differ? Why or why not?

Making a plan

Too often, an experimenter does not carefully consider statistical issues until after the study is completed and the data are in hand. Sometimes a flaw in the experimental design becomes obvious only then, when he or she tries to analyze the data. As Fisher once said, "To consult the statistician after an experiment is finished is often merely to ask him to conduct a post mortem examination. He can perhaps say what the experiment died of."[1] To ensure that your experiment is given a statistical clean bill of health and not a toe-tag, it is important to plan the experiment carefully with statistics in mind and to follow that plan throughout the data collection.

Here's a few guidelines to avoid a few common pitfalls. Chapter 14 delves into some of these issues in more detail than is possible here. For now, we list a few sensible procedures to help get you started.

1. *Develop a clear statement of the research question.* This needs to be as specific as possible. Is the question interesting? Has it already been addressed sufficiently in the literature?[2] Identify clear objectives for the experiment.

2. *List the possible outcomes of your experiment.* Once you have a preliminary plan for the treatments you *want* to compare, think of the outcomes you *might* obtain. Can you draw firm conclusions no matter

[1] Indian Statistical Congress, Sankhyā, ca. 1938.
[2] "A month in the laboratory can save an hour in the library."—Westheimer's Discovery

> **To consult the statistician after an experiment is finished is often merely to ask him to conduct a post mortem examination. He can perhaps say what the experiment died of.**
>
> —Fisher

what the outcome? Do these conclusions answer the questions? If not, then modify your design.

3. *Develop an experimental plan.* Write it down. Let it sit for a few days and then review it again.

4. *Keep the design of your experiment as simple as possible.* Do you really need 12 different treatments, or will two suffice? Simplifying the design will make it easier to keep track of your objectives, and it will avoid the need for complex statistical analyses.

5. *Check for common design problems.* Is there replication of treatments? Are these replicates truly independent? Will your sampling method yield random samples? Are there confounding variables that will complicate the interpretation of the results?

6. *Is the sample size large enough?* Avoid getting to the end of an experiment before

discovering that your sample size isn't large enough to demonstrate anything less than an unrealistically large effect. Is the sample size sufficient to produce a confidence interval small enough to permit conclusions, regardless of the size of the treatment effect? Chapter 14 discusses some methods to help in this planning.

7. *Discuss the design with other people.* Many brains think better than one, and others will often see a problem (and hopefully a solution) that wasn't obvious to you. It is better to get that feedback before doing all of the work than to be told after the fact, when it is too late to do anything about it. Science is a social process, so take advantage of the brainpower you have around you.

Careful consideration of these kinds of questions before the experiment starts can save a lot of wasted effort later.

9

Contingency analysis: associations between categorical variables

Biologists are keenly interested in associations between variables and differences between groups. Contingency tables (Section 2.3) display how the frequencies of different values for one variable depend on the value of another variable when both are categorical. In this chapter, we analyze sample data about two categorical variables to infer associations between those variables in populations. We want to determine whether one variable is "contingent" on the other.

Analysis of contingency data can be used to answer questions such as the following:

▶ Do bright and drab butterflies differ in the probability of being eaten?

▶ How much more likely to drink are smokers than nonsmokers?

▶ Are heart attacks less likely among people who take aspirin daily?

In experimental studies, contingency data can determine whether the probability of living or dying differs between medical treatments. We can estimate the differences in these probabilities with odds ratios, which are explained in Section 9.2. We can test these sorts of hypotheses with **contingency analysis**. A contingency analysis allows us to determine whether two (or more) categorical variables are independent. In other words, a contingency analysis allows us to compare different groups to ask whether the proportion of individuals in those groups with some property is the same for all groups.

Contingency analysis estimates and tests for an association between two or more categorical variables.

At the heart of contingency analysis is the investigation of the *independence of variables*. If two variables are independent, then the state of one variable is not related to the probability of the different outcomes of the other variable.

9.1 Associating two categorical variables

An association between two categorical variables implies that the two variables are not independent. During the RMS *Titanic* disaster, for example, women had a lower

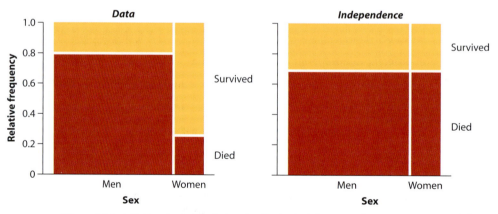

Figure 9.1-1 *Left:* Mosaic plot depicting the death of adult men and women passengers following the shipwreck of the *Titanic*. Survivors are represented by the gold and those that died by the red. The area of each box is proportional to the number of individuals in the sample with those attributes; *n* = 2092 individuals from data in Dawson (1995). *Right:* This is what the mosaic plot would have looked like if death and sex on the *Titanic* were independent. In reality, the probability of death differed between the sexes, so the two variables are not independent.

probability of death than men. Sex and death were not independent; the sex of an individual changed his or her probability of death. The mosaic plot on the left in Figure 9.1-1 shows the relationship between sex and death. If death had been independent of sex, then the probability of death would have been equal for both sexes, and the resulting mosaic plot would look like the one on the right in Figure 9.1-1.

9.2 Estimating association in 2 × 2 tables: odds ratio

The odds ratio measures the magnitude of association between two categorical variables when each variable has only two categories. One of the variables is the response variable—let's call its two categories "success" and "failure." The other variable is the explanatory variable, whose two categories identify the two groups whose probability of success is being compared. The odds ratio compares the proportion of successes and failures between the two groups.

Odds

Consider a variable for which a single random trial yields one of two possible outcomes: success or failure. The probability of success is p and the probability of failure is $1 - p$. The **odds** of success (O) are the probability of success divided by the probability of failure:

$$O = \frac{p}{1 - p}.$$

If $O = 1$ (sometimes written as 1:1 or "the odds are one to one"), then one success occurs for every failure. If the odds are 10 (sometimes written as 10:1) then 10 trials result in success for every one that results in failure.

> The *odds* of success are the probability of success divided by the probability of failure.

The estimate of odds is calculated from a random sample of trials using the observed proportion of successes ($\hat{p}$) as follows:

$$\hat{O} = \frac{\hat{p}}{1 - \hat{p}}.$$

Example 9.2 shows how to estimate odds from sample data.

| **Example 9.2** | **Take two aspirin and call me in the morning?** |

Aspirin, the medicine commonly used for headache and fever, has been shown to reduce the risk of stroke and heart attack in susceptible people. Observational studies have suggested that aspirin may also reduce the risk of cancer. A large, carefully designed experimental study was conducted to test this possibility (Cook et al. 2005). A total of 39,876 women were randomly assigned to one of two different treatments. Of these, 19,934 women received 100 mg of aspirin every other day. The other 19,942 women received a placebo, a chemically inert pill that gives the patient the experience of taking the medication without the chemical effects. The women did not know which treatment they received. The women were monitored for 10 years. During the course of the study, 1438 of the women on aspirin and 1427 of those receiving the placebo were diagnosed with invasive cancer (Table 9.2-1). The results are depicted in a mosaic plot in Figure 9.2-1.

Table 9.2-1 2 × 2 contingency table for the aspirin and cancer experiment.

	Aspirin	**Placebo**
Cancer	1438	1427
No cancer	18,496	18,515

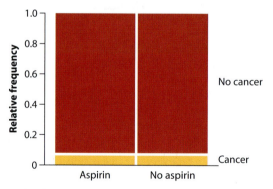

Figure 9.2-1 A mosaic plot showing the results of the study comparing cancer rates in women who took aspirin with women who did not take aspirin; $n = 39{,}876$.

A glance at the mosaic plot suggests that cancer rates did not change much, if at all, as a result of taking aspirin. Let's call *not getting cancer* a "success" and estimate the odds of not getting cancer in the two groups of women.[1] In the aspirin group (group 1), the estimated proportion that did not get cancer is

$$\hat{p}_1 = \frac{18496}{19934} = 0.9279.$$

---

[1]   In medical studies such as this one, the convention is to call the good outcome a success. To calculate odds on survival or getting better, use the probability of survival in the numerator.

We have added the subscript "1" to identify the group. The estimated proportion of women who *did* get cancer is

$$1 - \hat{p}_1 = 1 - 0.9279 = 0.0721.$$

So, the estimated odds of not developing cancer while taking aspirin are

$$\hat{O}_1 = \frac{\hat{p}_1}{1 - \hat{p}_1} = \frac{0.9279}{0.0721} = 12.87.$$

The odds of not getting cancer while on aspirin are 12.87:1, or about 13:1. In common speech, we would say that the odds are 13 to 1 that a woman who took aspirin would not get cancer in the next 10 years.

Similarly, the estimated probability that a woman on the placebo did not develop cancer is

$$\hat{p}_2 = 18515/19942 = 0.9284.$$

So, the odds of a woman on the placebo not getting cancer are

$$\hat{O}_2 = \frac{\hat{p}_2}{1 - \hat{p}_2} = \frac{0.9284}{0.0716} = 12.97,$$

which is also about 13:1. The odds are negligibly better than the odds for women on aspirin.

Odds ratio

We can use the odds ratio to quantify the difference between the odds of women developing cancer on aspirin and on the placebo. The **odds ratio** (*OR*) is just what it sounds like, the ratio of the odds of success between two groups. If O_1 is the odds of success in one group and O_2 is the odds in the other group, then the odds ratio is

$$OR = \frac{O_1}{O_2}.$$

> The *odds ratio* is the odds of success in one group divided by the odds of success in a second group.

If the odds ratio is equal to one, then the odds of success in the response variable is the same for both groups. If the odds ratio is greater than one, then the event has higher odds in the first group than in the second group. Alternatively, if the odds ratio is less than one, then the odds are higher in the second group. The odds ratio is commonly used in medical research, where it is used to measure the improvement in the odds for a response variable resulting from medical intervention compared with a control treatment (the explanatory variable).

For the cancer/aspirin study described in Example 9.2, the estimated odds ratio is given by

$$\widehat{OR} = \frac{\hat{O}_1}{\hat{O}_2} = \frac{12.86}{12.97} = 0.992.$$

(The "hat" on OR in the preceding equation indicates that it is an estimate of the population OR.) The odds of developing cancer while taking aspirin were about the same as the odds while taking the placebo. The estimated odds ratio is less than one, which means that in the data the odds of not getting cancer were slightly lower in the aspirin group than in the placebo group.

Standard error and confidence interval for odds ratio

The sampling distribution for the odds ratio is highly skewed, and so we must convert the odds ratio to its natural log, $\ln(\widehat{OR})$. We can calculate the standard error of the log-odds ratio[2] as

$$SE[\ln(\widehat{OR})] = \sqrt{\frac{1}{a} + \frac{1}{b} + \frac{1}{c} + \frac{1}{d}}.$$

The symbols a, b, c, and d in this equation refer to the observed frequencies in the cells of the contingency table:

	Aspirin	Placebo
Cancer	$a = 1438$	$b = 1427$
No cancer	$c = 18{,}496$	$d = 18{,}515$

An approximate $100(1 - \alpha)\%$ confidence interval for the log-odds ratio is then given by

$$\ln(\widehat{OR}) - Z\, SE[\ln(\widehat{OR})] < \ln(OR) < \ln(\widehat{OR}) + Z\, SE[\ln(\widehat{OR})]$$

where $Z = 1.96$ for a 95% confidence interval and $Z = 2.58$ for a 99% confidence interval.[3] This formula for the confidence interval is an approximation that assumes the sample size is fairly large. To find the confidence interval for the odds ratio, rather than the log odds, we must take the antilog of the upper and lower limits of the interval for the log-odds ratio.

Let's calculate the 95% confidence interval for the aspirin data. We've already calculated $\widehat{OR} = 0.992$, so $\ln(\widehat{OR}) = -0.00803$. The standard error of this estimate of $\ln(\widehat{OR})$ is

---

[2]   This formula will not work when a, b, c, or d is zero.

[3] Z is the critical value for a standard normal distribution. The standard normal distribution is discussed in greater detail in Chapter 10.

$$\text{SE}\left[\ln\left(\widehat{OR}\right)\right] = \sqrt{\frac{1}{a} + \frac{1}{b} + \frac{1}{c} + \frac{1}{d}}$$

$$= \sqrt{\frac{1}{1438} + \frac{1}{1427} + \frac{1}{18496} + \frac{1}{18515}}$$

$$= 0.03878.$$

With this standard error we can calculate the 95% confidence interval. Using $Z = 1.96$ for a 95% confidence interval, we get

$$-0.00803 - 1.96(0.03878) < \ln(OR) < -0.00803 + 1.96(0.03878).$$

The 95% confidence interval for the population log-odds ratio is therefore

$$-0.084 < \ln(OR) < 0.068.$$

To convert this to a confidence interval for the odds ratio, we must take the antilog of the limits of this interval by raising e to the power of each number:

$$e^{-0.084} < OR < e^{0.068}$$

or

$$0.92 < OR < 1.07.$$

The confidence interval for the odds ratio is tightly bounded around 1.00, so the data provide evidence that aspirin has little or no effect on the probability of developing cancer. The data are plausibly consistent with a small beneficial effect, a small deleterious effect, or no effect at all.

9.3 The χ^2 contingency test

The χ^2 **contingency test** is the most commonly used test of association between two categorical variables. It tests the goodness-of-fit to the data of the null model of *independence* of variables.

> The χ^2 contingency test is the most commonly used test of association between two categorical variables.

Example 9.3 illustrates how the method works.

Example 9.3   **The gnarly worm gets the bird**

Many parasites have more than one species of host, so the individual parasite must get from one host to another to complete its life cycle. Trematodes of the species *Euhaplorchis californiensis* use three hosts during their life cycle. Worms mature in birds and lay eggs that pass out of the bird in its feces. The horn snail *Cerithidea californica* eats these eggs, which hatch and grow to another life-stage in the snail, castrating the snail in the process. When an infected snail is eaten by the California killifish *Fundulus parvipinnis*, the parasite develops to the next life stage and encysts in the fish's brain case. Finally, when the killifish is eaten by a bird, the worm becomes a mature adult and starts the cycle over again.

It has been observed that infected fish spend excessive time near the water surface, where they may be more vulnerable to bird predation. This would certainly be to the worm's advantage, as it would increase its chances of being ingested by a bird, its next host. Lafferty and Morris (1996) tested the hypothesis that infection influences risk of predation by birds. A large outdoor tank was stocked with three kinds of killifish: unparasitized, lightly infected, and heavily infected. This tank was left open to foraging by birds, especially great egrets, great blue herons, and snowy egrets. Table 9.3-1 lists the numbers of fish eaten according to their levels of parasitism.

Table 9.3-1 Observed frequencies of fish eaten or not eaten by birds according to trematode infection level.

	Uninfected	Lightly infected	Highly infected	Row total
Eaten by birds	1	10	37	48
Not eaten by birds	49	35	9	93
Column total	50	45	46	141

We can visualize the pattern in the data with a mosaic plot (Figure 9.3-1). Only 2% of the uninfected fish were eaten, while 22% and 80% of the lightly and heavily infected fish, respectively, died from predation.

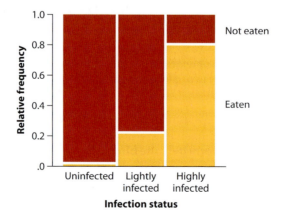

Figure 9.3-1 A mosaic plot for bird predation on killifish having different levels of trematode parasitism. The gold areas represent the relative frequency of fish eaten by birds, and the red areas are fish that escaped bird predation. There are a total of $n = 141$ fish represented in these data.

Hypotheses

We want to test whether the probability of being eaten by birds differs according to infection status. That is, we want to test whether the categorical variables *infection level* and *being eaten* are independent. The null and alternative hypotheses are

H_0: Parasite infection and being eaten are independent.
H_A: Parasite infection and being eaten are *not* independent.

To carry out the χ^2 contingency test, we need to calculate the expected frequencies for each of the cells in Table 9.3-1 under the null hypothesis of independence.

Expected frequencies assuming independence

Recall from Section 5.6 that, if two events are independent, then, by definition, the probability of both occurring is equal to the probability of one event occurring times the probability of the other event occurring (this is the multiplication rule). We use the multiplication rule to calculate the expected proportion of individual fish under each combination of events and then the expected frequencies under the null hypothesis. For example, if infection and being eaten are independent, then

$$\Pr[uninfected \text{ and } eaten] = \Pr[uninfected] \times \Pr[eaten].$$

To calculate the expected fraction of fish both uninfected and eaten, though, we still need to estimate the probability that a fish is uninfected and the chance that a fish was eaten. We can estimate these probabilities from the data in Table 9.3-1. The estimated probability that a fish was uninfected is the total number of uninfected fish in the sample (50) divided by the total number of fish (141):

$$\hat{\Pr}[uninfected] = 50/141 = 0.3546.$$

We've marked the probability estimate with a "hat" ($\hat{\ }$) to indicate that it is an estimate, and not the true value.

We can estimate the probability of being eaten in the same way, by dividing the number of fish eaten (48) by the total number of fish (141):

$$\hat{\Pr}[eaten] = 48/141 = 0.3404.$$

Under the null hypothesis of independence, therefore, the probability of a fish being uninfected *and* eaten is expected to be

$$\hat{\Pr}[uninfected \text{ and } eaten] = 0.3546 \times 0.3404 = 0.1207.$$

This means that the *expected* frequency of fish both uninfected and eaten is this probability (0.1207) times the total number of individuals in the data set (141):

$$Expected[uninfected \text{ and } eaten] = 0.1207 \times 141 = 17.0.$$

Table 9.3-2 Expected frequencies of fish eaten and not eaten by birds, according to trematode infection status.

	Uninfected	Lightly infected	Highly infected	Row totals
Eaten by birds	17.0	15.3	15.7	48
Not eaten by birds	33.0	29.7	30.3	93
Column totals	50	45	46	141

Repeating the preceding procedure for the other cells in Table 9.3-1 gives the expected frequencies of all combinations of outcomes. These values are listed in Table 9.3-2, but you should make sure that you can calculate them on your own.

Note that the row and column totals in Table 9.3-2 match the totals in the actual data (Table 9.3-1). This must be true because we used the proportions in the data themselves to generate our expected frequencies. If the row and column totals are *not* the same for the observed and expected frequencies (within rounding errors), a calculation error has been made.

Note, also, that the expected frequencies don't have to be integers. Remember that we use "expected" in the sense of "on average," so we don't expect an integer.

The χ^2 statistic

The observed frequencies in Table 9.3-1 are quite different from the expected frequencies in Table 9.3-2. We calculate the χ^2 statistic to test whether these discrepancies are greater than expected by chance. Using c to represent the number of columns and r to represent the number of rows,

$$\chi^2 = \sum_{column=1}^{c} \sum_{row=1}^{r} \frac{[Observed(column, row) - Expected(column, row)]^2}{Expected(column, row)}$$

This χ^2 calculation simply adds across all cells of the contingency table. Applied to the data, we get

$$\chi^2 = \frac{(1 - 17.0)^2}{17.0} + \frac{(49 - 33.0)^2}{33.0} + \frac{(10 - 15.3)^2}{15.3} + \frac{(35 - 29.7)^2}{29.7}$$
$$+ \frac{(37 - 15.7)^2}{15.7} + \frac{(9 - 30.3)^2}{30.3}$$
$$= 69.5.$$

Degrees of freedom

The sampling distribution of the χ^2 test statistic under the null hypothesis of independence is approximated by the theoretical χ^2 distribution. To calculate the degrees of freedom for the χ^2 distribution, we count the number of rows (r) and the number of columns (c) in the data table. The number of degrees of freedom, then, is given by

$$df = (r - 1)(c - 1).$$

For Example 9.3, Table 9.3-1 has two rows and three columns, so there are two degrees of freedom:

$$df = (2 - 1)(3 - 1) = 2.$$

P-value and conclusion

The critical value for the χ^2 distribution with $df = 2$ and significance level $\alpha = 0.05$ is 5.99 (see Statistical Table A). Our observed value ($\chi^2 = 69.5$) is further out in the tail of the distribution, much greater than the critical value of 5.99. We therefore reject the null hypothesis that infection level and probability of being eaten are independent in killifish. Instead, the probability of being eaten is contingent upon whether the fish was parasitized. Trematode parasitism in these killifish was associated with higher rates of predation by birds. We reach the same conclusion when we use a computer to calculate the P-value for a χ^2 value of 69.5: $P \approx 10^{-15}$.

How do we explain this result? If differences in infection are truly the cause of the differences in predation risk, then the most likely explanation is that the worms modify fish behavior to increase their chances of being eaten by birds, thus completing the last transition in the worm's life cycle.[4]

A shortcut for calculating the expected frequencies

There is a shortcut formula for calculating the expected frequencies that requires fewer keystrokes on your calculator. The *Expected* cell value for a given row and column is

$$Expected[row\ i, column\ j] = \frac{(Row\ i\ total)(Column\ j\ total)}{Grand\ total}.$$

For the killifish parasite data, for example, the expected frequency for the top left cell in Table 9.3-2 (uninfected fish that were eaten) can be calculated by multiplying the total for its row (50) times the total for its column (48) and dividing by the overall total (141):

$$Expected[row\ 1, column\ 1] = (50 \times 48)/141 = 17.0.$$

---

[4]   Many parasites modify the behavior of their hosts, with the result that their chances of making it to their next host are increased. Liver flukes make ant hosts move to the top of grass blades, where they are more likely to be eaten by grazing cows, the host in which the flukes are able to reproduce. Wire worms make their cricket hosts find and jump into water, where the crickets drown, but where the worm is able to complete its life cycle.

The last cell in a row or column can also be computed by subtraction, because the sum of the expected frequencies for a given row or column is the same as the sum of the observed values. Thus, the expected frequency of the top right cell in Table 9.3-2 is $48 - 17.0 - 15.3 = 15.7$. (The number of cells that we *cannot* calculate by subtraction is the number of degrees of freedom for the test. The expected values of all other cells are fixed and not free to vary.)

This shortcut comes from the definition of independence and the way we estimate the probability of belonging to each row or column:

$$Expected = \Pr[row] \times \Pr[column] \times Grand\ total$$

$$= \left(\frac{Row\ total}{Grand\ total} \right) \times \left(\frac{Column\ total}{Grand\ total} \right) \times Grand\ total.$$

Canceling terms gives the preceding shortcut.

The χ^2 contingency test is a special case of the χ^2 goodness-of-fit test

You may have noticed that, once we specified the expected values, the χ^2 contingency test was very similar to the χ^2 goodness-of-fit test introduced in Chapter 8. This resemblance is no accident, because the χ^2 contingency test is a special application of the more general goodness-of-fit test where the probability model being tested is that of the independence of variables. The number of degrees of freedom for the contingency test obeys the same rules as that for the goodness-of-fit test.[5]

Assumptions of the χ^2 contingency test

The χ^2 contingency test makes the same assumptions as the χ^2 goodness-of-fit test. That is, no more than 20% of the cells can have an expected frequency less than five, and no cell can have an expected frequency less than one.

If these rules are not met, there are at least three options available. First, if the table is bigger than 2×2, then two or more row categories (or two or more column categories) can be combined to produce larger expected frequencies. This should be done carefully, though, so that the resulting categories are still meaningful. (For example, the three categories of infection in a trematode predation experiment could have been

---

[5] Recall, from Section 8.2, that df = (Number of categories) $- 1 -$ (Number of parameters estimated from the data). In a contingency table, the number of categories is $r \times c$. The number of parameters estimated from the data is $(r - 1) + (c - 1)$, reflecting the number of row and column totals needed to generate the expected cell frequencies. After some algebra, this leads to $df = (r - 1)(c - 1)$.

collapsed into two categories—namely, "uninfected" and "infected," if necessary.) Second, if the table is 2 × 2, then the Fisher's exact test should be used instead. The Fisher's exact test is summarized in Section 9.4. Finally, a randomization test may be used instead of the χ^2 test, an approach that we discuss further in Chapter 19.

Correction for continuity

Some statisticians recommend a modified formula to calculate the χ^2 test statistic in the case of a 2 × 2 contingency table. The modification is known as the **Yates correction for continuity**,

$$\chi^2 = \sum_{column=1}^{2} \sum_{row=1}^{2} \frac{\left[\, |Observed(column, row) - Expected(column, row)| - \dfrac{1}{2} \right]^2}{Expected(column, row)}.$$

All the other steps in the Yates corrected test are the same as in an ordinary χ^2 contingency test.

We mention the Yates correction here because you will encounter it in the biological literature. However, we don't recommend that you use it. The correction makes the χ^2 contingency test too conservative (Maxwell 1976). That is, the Yates corrected test overestimates the correct P-value, with the result that the power of the test is reduced—it is less likely to reject a false null hypothesis.

9.4 Fisher's exact test

Fisher's exact test, named after Sir Ronald A. Fisher, provides an exact P-value for a test of association in a 2 × 2 contingency table. The test is an improvement over the χ^2 contingency test in cases where the expected cell frequencies are too low to meet the rules demanded by the χ^2 approximation.

> *Fisher's exact test* examines the independence of two categorical variables, even with small expected values.

The calculation of the P-value in Fisher's exact test is cumbersome and is best done with a computer statistical package. Therefore, we do not detail the calculations here, but instead we focus on what the test can do and when it is appropriate.

| Example 9.4 | **The feeding habits of vampire bats** |

In Costa Rica the common vampire bat, *Desmodus rotundus*, commonly feeds on the blood of domestic cattle. The bat prefers cows to bulls, which suggests that the bats might respond to a hormonal signal. To explore this further, a researcher compared vampire bat attacks on cows in estrous ("in heat") with attacks on cows not in estrous[6] on a particular night (Turner 1975). The results are presented in

Table 9.4-1. Do cows in estrous have a higher chance of being attacked than cows not in estrous?

Table 9.4-1 Numbers of cattle by estrous status and by vampire bat bite status.

	Cows in estrous	Cows not in estrous	Row totals
Bitten by vampire bat	15	6	21
Not bitten by vampire bat	7	322	329
Column totals	22	328	350

The null and alternative hypotheses are

H_0: State of estrous and vampire bat attack are independent.
H_A: State of estrous and vampire bat attack are *not* independent.

We are tempted to analyze these data with a χ^2 contingency test, but, if we calculate the expected values in the usual way, we find that the expected frequency for cows in estrous that were bitten by vampire bats is too low (Table 9.4-2).

Table 9.4-2 The expected frequency values for the vampire bat study.

	Cows in estrous	Cows not in estrous	Row totals
Bitten by vampire bat	1.3	19.7	21
Not bitten by vampire bat	20.7	308.3	329
Column totals	22	328	350

---

[6] His method for doing this was ingenious. It is difficult for humans to distinguish reliably whether cows are in heat, but bulls are very good at it. So, the researcher harnessed paint sponges onto the undersides of the bulls, which each night marked the cows that had been the object of the bull's affections.

According to the null hypothesis, we expect to see only 1.3 cows that were both in heat and bitten by a vampire bat. This means that one out of the four cells (25%) has an expectation less than five, whereas the rule for a χ^2 test is that no more than 20% of cells should have expectations that low.

Because this is a 2×2 contingency table, we can turn to a Fisher's exact test. The null and alternative hypotheses remain the same as in the χ^2 test. Fisher's test proceeds by listing all 2×2 tables that are as extreme or more extreme than the observed table of numbers under the null hypothesis of independence. For example, the following are the more extreme tables in one tail. Because the row and column totals remain the same, we can just change one of the values and adjust the others to match. Focus on the top right corner of each table:

16	5
6	323

17	4
5	324

18	3
4	326

19	2
3	326

20	1
2	327

21	0
1	328

The P-value for Fisher's exact test is the sum of the probabilities of all such extreme tables under the null hypothesis of independence. There are also confidence intervals for odds ratios based on small samples based on the same logic as Fisher's exact test (see Agresti 2002).

We applied the Fisher's exact test to the data in Table 9.4-1, using a statistical program on the computer, and we found that $P < 0.0001$. Thus, we can reject the null hypothesis of independence. Vampire bats evidently preferred the cows in estrous. The reasons for this are not clear.[7]

9.5 G-tests

The G-test is another contingency test often seen in the literature. The G-test is almost the same as the χ^2 test except that the following test statistic is used:

$$\left\{ G = 2 \sum_{column=1}^{c} \sum_{row=1}^{r} Observed(column, row) \ln \left[\frac{Observed(column, row)}{Expected(column, row)} \right] \right\},$$

where "ln" refers to the natural logarithm. Under the null hypothesis of independence, the sampling distribution of the G-statistic is approximately χ^2 with $(r-1)(c-1)$ degrees of freedom.

For the data in Example 9.3 on fish infection and bird predation, for example,

---

[7] It has been speculated that, by drinking the blood of cows in heat, they minimize their intake of hormones in the blood of non-estrous cows that may act as a birth-control pill.

$$G = 2\left(1 \ln\left[\frac{1}{17.0}\right] + 49 \ln\left[\frac{49}{33.0}\right] + 10 \ln\left[\frac{10}{15.3}\right] + 35 \ln\left[\frac{35}{29.7}\right]\right.$$
$$\left. + 37 \ln\left[\frac{37}{15.7}\right] + 9 \ln\left[\frac{9}{30.3}\right]\right)$$

$$= 77.6$$

This G-test statistic would be compared to the χ^2 distribution with $df = (r - 1)(c - 1) = 2$. Again, we would strongly reject the null hypothesis of independence.

The G-test is derived from principles of likelihood (see Chapter 20) and can be applied across a wider range of circumstances than the χ^2 contingency test. While the G-test is preferred by some statisticians, it has been shown to be less accurate for small sample sizes (Agresti 2002), and it is used somewhat less often than the χ^2 contingency test in the biological literature. The G-test has advantages, though, when analyzing complicated experimental designs involving multiple explanatory variables (see Sokal and Rohlf 1995 or Agresti 2002).

9.6 **Summary**

▶ The odds of success are the probability of success occurring divided by the probability of failure.

▶ The odds ratio describes the odds of success in one of two groups divided by the odds of success in the second group. The odds ratio is used to quantify the magnitude of association between two categorical variables, each of which has two categories.

▶ The χ^2 contingency test makes it possible to test the null hypothesis that two categorical variables are independent.

▶ The sampling distribution of the χ^2 statistic under the null hypothesis is approximately χ^2 distributed with $(r - 1)(c - 1)$ degrees of freedom. The χ^2 approximation works well, provided that two rules are met: no more than 20% of the expected frequencies can be less than five, and none can be less than one.

▶ The Fisher's exact test calculates an exact P-value for the test of independence of two variables in a 2×2 table. The test is especially useful when the rules for the χ^2 approximation are not met.

▶ The G-test is an alternative method for contingency analysis.

9.7 Quick Formula Summary

Confidence interval for odds ratio

What does it assume? Random samples.

Formula: $\ln(\widehat{OR}) - Z\,SE[\ln(\widehat{OR})] < \ln(OR) < \ln(\widehat{OR}) + Z\,SE[\ln(\widehat{OR})]$,
where $\ln(\widehat{OR})$, is the natural logarithm of the estimate of odds ratio,

$$OR = \frac{O_1}{O_2},$$

and O_1 and O_2 are the odds for the two groups. $SE[\ln(\widehat{OR})]$ is the standard error of the log-odds ratio,

$$SE[\ln(\widehat{OR})] = \sqrt{\frac{1}{a} + \frac{1}{b} + \frac{1}{c} + \frac{1}{d}},$$

and $Z = 1.96$ for a 95% confidence interval. The confidence interval for OR is found by taking the antilog of the limits of the confidence interval for $\ln(OR)$.

The χ^2 contingency test

What is it for? Testing the null hypothesis of no association between two or more categorical variables.

What does it assume? Random samples; the expected frequency of each cell is greater than one; no more than 20% of the cells have expected frequencies less than five.

Test statistic: χ^2

Sampling distribution under H$_0$: χ^2 distribution with $(r-1)(c-1)$ degrees of freedom, where r and c are the numbers of rows and columns, respectively.

Formula:
$$\chi^2 = \sum_{column=1}^{c} \sum_{row=1}^{r} \frac{[Observed(column, row) - Expected(column, row)]^2}{Expected(column, row)}$$

Fisher's exact test

What is it for? Testing the null hypothesis of no association between two categorical variables, each having two categories. Appropriate with small expected values.

What does it assume? Random samples.

Formula: $P = 2 \sum\limits_{all\ equally\ or\ more\ extreme\ tables} \dfrac{R_1!R_2!C_1!C_2!}{a!b!c!d!n!}$

where R_i and C_i are the row and column totals; a, b, c, and d are the cell values for each of the cells; and n is the total sample size. The summation is over all tables, including the observed table and any tables with the same row and column totals more different from H_0 than the observed table.

G-test

What is it for? Testing the null hypothesis of no association between two or more categorical variables.

What does it assume? Random samples; no more than 20% of cells have expected frequencies less than five.

Test statistic: G

Sampling distribution under H$_0$: χ^2 distribution with $(r - 1)(c - 1)$ degrees of freedom, where r and c are the numbers of rows and columns, respectively.

Formula:

$$\left\{ G = 2 \sum_{column=1}^{c} \sum_{row=1}^{r} Observed(column, row) \ln\left[\frac{Observed(column, row)}{Expected(column, row)} \right] \right\}$$

PRACTICE PROBLEMS

1. The common pigeon found in most American cities is derived from a domesticated European species released in North America. As a result, the pigeons in North America have variations in coloration caused by genes previously selected by pigeon fanciers. An example is the rump, whose feathers are white in wild European pigeons but blue in many pigeons in North America. It has been hypothesized that the white rump of pigeons serves to distract predators like peregrine falcons, and therefore it may be an adaptation to reduce predation. To test this, researchers followed the fates of 203 pigeons, 101 of which had white rumps and 102 of which had blue rumps. Of the white-rumped birds, nine were captured by falcons, while of the blue-rumped birds, 92 were killed (Palleroni 2005).

   a. Do the two kinds of pigeons differ in their rate of capture by falcons? Carry out an appropriate test.

   b. What is the estimated odds ratio for capture by the two groups of pigeons? What is the 95% confidence interval for this odds ratio?

2. Malaria kills more than a million people each year worldwide, making it one of the worst diseases on earth. The disease is caused by an organism called plasmodium, which spreads

between hosts by infected mosquitoes. The more people that each infected mosquito bites, the higher the transmission rates of malaria. Higher transmission rates are bad for human health, but good for the plasmodium. If the plasmodium could somehow cause its mosquito host to bite more people, then it would be able to have more offspring. Is such a thing possible? Does infection by plasmodium cause a mosquitoe to bite more people?

To test this, researchers captured 262 mosquitoes that had human blood in their guts (Koella et al. 1998). They measured two attributes: whether mosquitoes were infected with malaria, and whether they had fed on the blood of more than one person. Multiple blood meals were detected by extracting and testing human DNA from the blood in mosquito guts. Of 173 uninfected mosquitoes, 16 had had multiple blood meals. Of 89 infected mosquitoes, 20 had fed multiple times. Do these data support the idea that infected mosquitoes behave differently than uninfected mosquitoes?

3. In the Australian redback spider, *Latrodectus hasselti*, females are about 50 times larger than males. Many males get eaten by the females during mating. This might sound like a horrible accident, but the possibility exists that males gain some advantage by being cannibalized in this way. Perhaps females are more likely to accept the sperm of a male that she has eaten than of a male that has escaped. Researchers watched the mating behavior of 32 virgin female redback spiders, recording whether she ate her first mate and then whether she rejected advances from a second male later placed in her vicinity (Andrade 1996). The results were as follows:

	1st male eaten	1st male escapes
2nd male accepted	3	22
2nd male rejected	6	1

   a. How does cannibalism affect the odds that the second suitor is accepted?

   b. What method would you use to test the association between these two variables? Why?

4. In many ecosystems, fires are a common and important part of the habitat. Many species have evolved mechanisms for dealing with fire, but for some species the risk posed by the extreme heat is pronounced. Reed frogs, a small species living in West Africa, have been observed hopping away from grass fires long before the heat of the fire had reached the area they were in. This led to the hypothesis that the frogs might hear the fire and respond well before the fire reaches them. To test this hypothesis, researchers played three types of sound to samples of reed frogs and recorded their response (Grafe et al. 2002). Twenty frogs were exposed to the sound of fire, 20 were exposed to the sound of fire played backwards (to control for the range of sound frequencies present in the real sound), and 20 were played equally loud white noise. Of these frogs, 18 hopped away from the sound of fire, six hopped away from the sound of fire played backwards, and none hopped away from the white noise. Is this evidence that reed frogs react consistently to the sound of fire?

5. One of the many problems facing rare species is that individuals may have trouble finding a mate. Even if an individual finds another animal of the same species, it might be the same sex. One possible solution to this problem is for individuals to wait until a possible partner is found before choosing what sex they become. Some fish are famously able to change their sex according to circumstances. One study of a coral reef fish, the goby *Gobiodon erythrospilus*, placed juvenile fish with either an adult male or an adult female (Hobbs et al. 2004).[8] Of the 12 juveniles placed with a male, 11 became female. Of the 10 juveniles placed with an adult female, six became male.

   a. What method can we use to test whether the social context of the juvenile fish affects what sex they become?

   b. Is the effect in the direction predicted?

---

[8] This beautiful fish is shown on the opening page of this chapter.

6. Between 20 and 25 violent acts are portrayed per hour in children's television programming. A study (Johnson et al. 2002) of the possible link between TV viewing and aggression followed the TV viewing habits of children between one and 10 years old. Of these children, 88 watched less than one hour of TV per day, 386 watched 1–3 hours per day, and 233 watched more that three hours per day. Eight years later, researchers evaluated the kids to see if they had a police record or had assaulted another person resulting in injury. The number of aggressive individuals from the three TV watching groups were five, 87, and 67, respectively.

   a. Estimate the proportion of kids in each TV-watching category who subsequently became violent. Give 95% confidence intervals for these estimates.

   b. Is there evidence that childhood TV viewing is associated with future violence? Carry out an appropriate statistical test.

   c. Does this prove that TV watching causes increased aggression in kids? Why or why not?

7. A study by Doll et al. (1994) examined the relationship between moderate intake of alcohol and the risk of heart disease. In all, 410 men (209 "abstainers" and 201 "moderate drinkers") were observed over a period of 10 years, and the number experiencing cardiac arrest over this period was recorded and compared with drinking habits. All men were 40 years of age at the start of the experiment. By the end of the experiment, 12 abstainers had experienced cardiac arrest whereas nine moderate drinkers had experienced cardiac arrest.

   a. Test whether or not the relative frequency of cardiac arrest was different in the two groups of men.

   b. Assume that you were unable to reject the null hypothesis in part (a). Would this imply that drinking has no effect on the risk of cardiac arrest? Why or why not?

8. Postnatal depression affects approximately 8–15% of new mothers. One theory about the onset of postnatal depression is that it may result from the stress of a complicated delivery. If so, then the rates of postnatal depression could be affected by changing the mode of delivery. A study of 10,934 women compared the rates of postnatal depression in mothers who delivered vaginally to those who had voluntary cesarean sections (C-sections) (Patel et al. 2005). Of the 10,545 women who delivered vaginally, 1025 suffered significant postnatal depression. Of the 389 who delivered by voluntary C-section, 48 developed postnatal depression.

   a. Calculate the proportion of women who developed depression after both kinds of childbirth.

   b. Calculate the odds ratio of the risk of developing depression, comparing vaginal birth to C-section. How different are the two procedures for this risk?

   c. Test the hypothesis that the method of childbirth has no effect on the probability of developing postnatal depression.

9. Migraine with aura is a potentially debilitating condition, yet little is known about its causes. A recent study compared 93 people who suffer from chronic migraine with aura to 93 healthy patients (Schwerzmann et al. 2005). The researchers used transesophageal echocardiography to look for cardiac shunts in all of these patients. (A cardiac shunt is a heart defect that causes blood to flow from the right to the left in the heart, causing poor oxygenation.) Forty-four of the migraine patients were found to have a cardiac shunt, while only 16 of the people without migraine symptoms had these heart defects.

   a. What is the odds ratio for the cardiac shunt, comparing the patients with and without migraines.

   b. What is the 95% confidence interval for this odds ratio?

10. A study of 6,839,854 births in the U.S. found a total of 6522 babies were born with a finger defect, either syndactyly (fused fingers), poly-dactyl (extra fingers), or adactyly (fewer than five fingers). Of these babies with finger defects, 5171 were examined in further detail, and

whether the mom smoked was recorded. Of these babies, 4366 had mothers that did *not* smoke while pregnant, and the rest had mothers that did smoke while pregnant. In a sample of 10,342 babies from the same population with normal fingers, 9062 of their mothers did *not* smoke while pregnant, while the remaining 1280 did smoke while pregnant.

a. What is the 95% confidence interval for the proportion of babies born in the U.S. with one of these finger defects? Interpret your result in words.

b. Test whether the mothers' smoking has an effect on the probability of these finger defects.

c. What is the odds ratio for these digital birth defects, as a function of the mother's smoking status? What is the 95% confidence interval for this odds ratio?

ASSIGNMENT PROBLEMS

11. The most common therapy for warts is cryotherapy, in which liquid nitrogen is applied to the area of the wart every few weeks, freezing it for a few seconds until the wart falls off. This isn't as painful as it sounds, but requires a physician's attention and, for some patients, is chilling to contemplate. An alternative therapy, presumably based on the theory that it works to fix everything else, is to put duct tape on the wart and leave it there for a week at a time until the wart goes away. In a randomized trial, 51 patients with warts were randomly assigned to a cryotherapy treatment or a duct-tape treatment (Focht et al. 2002). Of 26 patients treated with duct tape, 22 had their warts completely cured. With the liquid nitrogen treatment, 15 of 25 patients were cured of their warts.

a. Draw a mosaic plot to present these data.

b. Calculate an odds ratio for the two therapies. Which technique appears to work best? What is the 95% confidence interval for this odds ratio? Interpret the result.

c. Test whether duct tape or cryotherapy is a better treatment for warts and interpret the results.

12. In northern Europe, there are two species of flycatcher, the collared flycatcher and the pied flycatcher, that sometimes hybridize (i.e., a male from one species mates with a female from the other). Researchers investigated the sex of the

offspring of such matings to test a pattern found in other birds—namely, that female hybrids are more likely to die before hatching than are male hybrids (Veen et al. 2001). The researchers found that, of 26 hybrid offspring who made it to hatching, 16 were male. This compared with 72 males in 145 "purebred" offspring of matings between members of the same species. Is the proportion of males and females different in the purebred offspring than in the hybrid offspring? Do the appropriate hypothesis test.

13. The same study of pied and collared flycatchers (see Assignment Problem 12) measured the frequencies of different combinations of mated pairs within and between the two species. The numbers of pairs observed are listed in the following table:

	Female collared	Female pied
Male collared	5567	84
Male pied	72	172

a. What fraction of mating pairs involved birds of different species? What is the standard error of your estimate?

b. Test the hypothesis that males and females mate indiscriminately, independently of species.

14. In animals without paternal care, the number of offspring sired by a male increases as the number of females he mates with increases. This fact has driven the evolution of multiple matings in the males of many species. It is less obvious why females mate multiple times, because it would seem that the number of offspring that a female has would be limited by her resources and not by the number of mates, as long as she has at least one mate. To look for advantages of multiple mating, a study of the Gunnison's prairie dog followed females to find out how many times they mated (Hoogland 1998). They then followed the same females to discover whether or not they had given birth later. The results are compiled in the following table:

Number of times female mated:	1	2	3	4	5
Number who gave birth:	81	85	61	17	5
Number who didn't give birth:	6	8	0	0	0

Did the number of times that a female mated affect her probability of giving birth?

a. Calculate expected frequencies for a contingency test.

b. Examine the expected frequencies. Do they meet the assumptions of the χ^2 contingency test? If not, what steps could you take to meet the assumptions and make a test?

c. An appropriate test shows that the number of mates of the female prairie dogs is associated with giving birth. Does this mean that mating with more males increases the probability of giving birth? Can you think of an alternative explanation?

15. Spot the flaw. Since 1953, when Tenzing Norgay and Edmund Hillary reached the summit of Mount Everest, many climbers have attempted to scale the world's two highest mountains, Everest and K2. Norgay and Hillary aided their climb by bringing supplemental oxygen in tanks, and some subsequent groups have attempted to out-do the originals by trying the ascent without supplemental oxygen. In fact, 159 teams comprising 1173 team members have attempted to climb either Everest or K2 between 1978 and 1999. The numbers of individuals who have either survived or died[9] during those attempts is given in the following table (data from Huey and Eguskitza 2000):

	Used supplemental O_2	Did not use supplemental O_2	Row totals
Survived descent:	1045	88	1133
Did not survive descent:	32	8	40
Column totals:	1077	96	1173

A χ^2 contingency test on these data calculated $\chi^2 = 7.694$ with one degree of freedom, which corresponds to $P = 0.0055$. The null hypothesis is that oxygen use has *no* effect on survivorship during these expeditions. What's wrong with this analysis?[10]

16. A recently published study investigated regions of the brain involved in self-recognition (Keenan et al. 2001). Ten subjects were randomly assigned to two groups. Anesthesia was used to inactivate one of the cerebral hemispheres of the brain of all 10 subjects. The left hemisphere was anesthetized in one group, whereas the right hemisphere was inactivated in the other group. Subjects were then shown a picture generated by

---

9   Almost all of the mortality occurred during the descents.

10   The original paper (Huey and Eguskitza 2000) did not make this mistake.

averaging ("morphing") images of the face of a famous celebrity (e.g., Marilyn Monroe) and their own face, and told to remember the picture. After recovery from anesthesia, patients were presented with two pictures and asked to choose the one they had been shown earlier while under anesthesia. The two pictures were the original two images from which the morphed image had been generated (i.e., "self" and "celebrity," but separately this time). All five patients whose left hemisphere had been inactivated chose the picture of "self." Four of the five patients whose right hemisphere had been anesthetized chose the "celebrity" picture, instead (the fifth chose "self"). Explain what test you would use to determine whether the treatment (left- vs. right-hemisphere inactivation) influenced recognition of "self" versus "celebrity," and why you would choose this test (don't necessarily carry out the test, just name it and justify your answer).

17. Vampire bats, as their name implies, feed almost exclusively on blood. A bat must feed every day or it will starve to death, but bats are not always successful at finding a blood meal. Perhaps for this reason, then, they roost during the day in communal groups and sometimes share blood by (and we're not making this up) regurgitative feeding. Researchers measured whether hungry bats were more likely to receive regurgitated blood than were partially fed bats (Wilkinson 1984). Eight bats were captured in the evening before they had fed and were held without feeding until the next morning. As a control, six bats were captured after naturally feeding at night, and they were also held until the following morning. At this time, the bats were returned to their groups. Five of the eight hungry bats were given regurgitated blood meals by group-mates, but none of the six well-fed bats were given a blood meal by other bats. What statistical method would you use to address the question, "Does the probability of being fed by roost-mates vary according to hunger status?"

18. Some people feel that they have good intuition about when others are lying, while others do not feel they have this ability. Are the more "intu-itive" people better able to detect lies? Each of 100 people who thought they had intuitive abilities was shown a video clip of a person stating his or her favorite movie.[11] The person was truthful in some of the clips shown, whereas in others the person was lying. Another 100 people who claimed not to have intuitive abilities were shown similar video clips. Fifty-nine of the 100 "intuitive" subjects correctly identified whether the person in the video was lying, whereas 69 of the 100 "non-intuitive" subjects correctly identified whether the person in the video was lying.

a. Draw a graph that best presents these data.
b. Test whether the success rates of the two groups were different.
c. Are "intuitive" people better at detecting lies than "non-intuitive" people? Calculate an odds ratio and confidence interval for your answer. Interpret your result.

19. *Spot the flaw*. Scottish researchers compared rates of depression between 94 undergraduates who regularly kept diaries and 41 students who did not. They found that people who kept diaries were more likely to have depression than those who did not. The researchers said, "We expected diary-keepers to have some benefit, or be the same, but they were the worst off. You are probably much better off if you don't write anything at all." Why is this an incorrect interpretation of these results?

20. A "Mediterranean diet" (i.e., high in fish, olive oil, red wine, etc.) has recently been touted as a key to a long life. A recent study looked at the death rates of people according to whether their diet had a low component, a medium compo-nent, or a high component of foods that charac-terize a Mediterranean diet (Trichopoulou et al. 2005). In these kinds of studies, it is important to look for other confounding variables, such as smoking, that might be correlated with the main variable under study. For each person in the study, therefore, they also recorded whether the person was a current smoker, a former smoker,

---

[11] http://www.newscientist.com/article.ns?id=dn2054

or had never smoked. We want to know whether there is an association between diet and smoking. The data for men in the study are as follows, where the numbers represent the number of men in each category:

	Low Med. diet	Medium Med. diet	High Med. diet
Never smoked:	2516	2920	2417
Former smoker:	3657	4653	3449
Current smoker:	2012	1627	1294

  a.   Draw a mosaic plot of these data.

  b.   Test whether there is an association between diet and smoking.

21. Some people who have heart attacks do not experience chest pain, although most do. A study of people admitted to emergency rooms with heart attacks compared the death rates of people who had chest pains with those of people who had less typical symptoms (Brieger et al. 2004). Of the 1763 people who had heart attacks *without* chest pain, 229 died, while of the 19,118 people who had heart attacks *with* chest pain, 822 died. Do a hypothesis test to determine whether having chest pains during a heart attack is related to the probability of death.

22. It is well known, and scientifically documented, that yawning is contagious. When we see someone else yawn, or even think about someone yawning, we are very likely to yawn ourselves. (In fact, we predict that you are starting to want to yawn right now.) In a study of yawning contagion, researchers showed subjects one of several pictures, including a picture of a man yawning, the same man smiling, a yawning man with his mouth covered, or a yawning man with his eyes obscured (Provine et al. 1989). Subjects yawned much more often when shown the yawner than the smile, but surprisingly an identical number also yawned when shown the picture with the mouth obscured. This suggests that something besides the mouth is an important trigger. What about the eyes? Seventeen of 30 subjects yawned when confronted with a picture of a

yawning man, while 11 of 30 independent subjects yawned when shown a picture of a yawning man with his eyes covered. Is there evidence in these data that the image of the yawning man's eyes is important in causing contagious yawns?

23. Daycare centers expose children to a wider variety of germs than the children would be exposed to if they stayed at home more often. This has the obvious downside of frequent colds, but it also serves to challenge the immune system of children at a critical stage in their development. A study by Gilham et al. (2005) tested whether exposure of children to a daycare environment affected their probability of later developing acute lymphoblastic leukemia (ALL). They compared 1272 children with ALL to 6238 children without ALL. Of the ALL kids, 1020 had significant social activity outside the home (including daycare) when young. Of the kids without ALL, 5343 had significant social activity outside the home. The rest of both groups lacked regular contact with children who were not in their immediate families.

  a.   Is this an experimental or observational study?

  b.   What are the proportions of children with significant social activity in the two groups (ALL and non-ALL children)?

  c.   What are the odds that a child with ALL had significant social activity?

  d.   What is the odds ratio for the amount of significant social activity, comparing the ALL and non-ALL groups?

  e.   What is the 95% confidence interval for this odds ratio?

  f.   Does this confidence interval lead you to believe that ALL children had different amounts of social activity from the children without the disease? If so, did the ALL children have more or less social activity?

  g.   The researchers interpreted the results of their study in terms of the differing immune-system exposure of the children, but gave several alternative explanations for the pattern. What, if any, are the possible confounding variables?

10

The normal distribution

Measurements of continuous numerical variables abound in biological data. We measure the length and weight of babies, the velocities of swallows, the times between infection and the onset of symptoms, the numbers of cones on pine trees, etc. Typically we take these measurements on a sample of individuals, but we want to be able to make inferences about the continuous variables in the population. For example, "What is a 95% confidence interval for the mean birth weight of American babies?" Or, "Which is faster, on average, an African or a European swallow?" To answer these kinds of questions, we need to understand something about the probability distributions that these data are taken from.

The normal distribution, which we introduced in Section 1.4, is the queen of all probability distributions used in the analysis of biological data. It generates the ubiquitous "bell-shaped curve," which can be used to approximate the frequency distribution of so many biological variables. The normal distribution arguably describes more about nature than any other mathematical function; thus, it takes a preeminent role in biological statistics.

Even more important than its ability to approximate frequency distributions of data, the normal distribution can be used to approximate the *sampling distribution* of estimates, especially of sample means. Many statistical techniques have been developed for dealing with variables that have a normal sampling distribution. The majority of the rest of this book describes these techniques. This chapter describes the normal distribution and explains some of the reasons why it is so important.

10.1 Bell-shaped curves and the normal distribution

Many numerical variables have frequency distributions that are bell-shaped. For example, Figure 10.1-1 shows a histogram of the birth weights of the 4,017,264 singleton[1] births recorded by birth certificate in the U.S. in 1991.

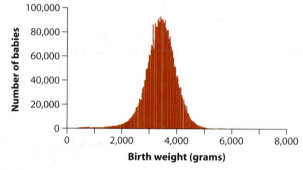

Figure 10.1-1 The frequency distribution of the birth weight of babies born in the U.S. in 1991 (Clemons and Pagano 1999).

Examine the shape of the baby weight distribution. Notice in Figure 10.1-1 that the peak (i.e., the mode) is right at the center of the distribution. If we averaged all of these 4,017,264 data points, we would find the mean to be 3339 grams, which is also right at the mode. The distribution is clearly shaped like a symmetric bell. There are so many data points, and the interval widths in the histogram are so narrow, that the distribution looks almost like a smooth curve. We imagine that if we collected more and more measurements and used even narrower intervals, then the graph would become even smoother.

The theoretical probability distribution describing many bell curves is called the normal distribution. The normal distribution is a continuous probability distribution, which means that it describes the probability distribution of a continuous numerical

---

[1] "Singleton" means that the baby born was not a twin, a triplet, etc.

variable. It is symmetric around its mean. The further values are from the mean, the lower the probability density of observations.

The normal distribution has two parameters to describe its location and spread: the mean and the standard deviation. For example, Figure 10.1-2 shows the normal distribution having the same mean and standard deviation as the baby birth weights. It strongly resembles the frequency distribution of the real data. (The scale on the *y*-axis is different, because the normal distribution shows the probability density, while the data are expressed as counts. To find the expected relative frequency of a particular bin of the histogram, we would integrate the probability density from the lower bound of the bin to the upper bound.)

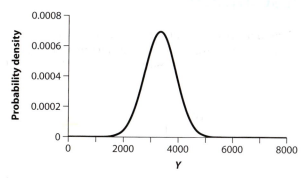

Figure 10.1-2 The normal distribution for a variable *Y* with mean and variance equal to that in the baby birth weight data.

Figure 10.1-3 shows some other examples of data whose frequency distributions resemble the normal curve: the body temperatures of adult humans, the brain sizes of undergraduate students, and the number of bristles on a fly's abdomen. In each case, we have superimposed the normal distribution with the same mean and standard deviation as the data. Bell-shaped frequency distributions appear in nature all the time, and the normal distribution is an excellent approximation to these real distributions. The number of fly bristles is actually a discrete variable, but, with a large number of possible values for this variable, it is still well-approximated by a normal distribution.

Biostatistics makes great use of the normal distribution. The statistical methods in most common use assume that the data come from a normal distribution of measurements. Moreover, as we explain in Section 10.6, the normal distribution can describe some properties of samples from populations that aren't themselves normally distributed.

> The *normal distribution* is a continuous probability distribution describing a bell-shaped curve. It is a good approximation to the frequency distributions of many biological variables.

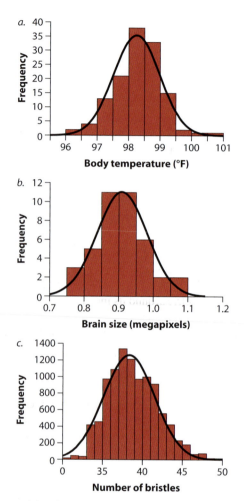

Figure 10.1-3 The normal distribution approximates frequency distributions in nature. *a.* Human body temperature, in degrees Fahrenheit (Shoemaker 1996). *b.* University undergraduate brain size (measured in number of megapixels on an MRI scan) (Willerman et al. 1991). *c.* The number of bristles on the fourth and fifth segments of the abdomens of fruit flies (Falconer and MacKay 1995). The black lines show normal distributions with the same mean and standard deviation as measured in the data.

10.2   The formula for the normal distribution

You may rarely have reason to directly use the formula for the probability density of the normal distribution, but you will often use calculations derived from it, so you need to know what the formula is and what elements it includes. The formula is

$$f(Y) = \frac{1}{\sqrt{2\pi\sigma^2}} \, e^{-\frac{(Y-\mu)^2}{2\sigma^2}}.$$

The value Y can be any real number, ranging from negative infinity to positive infinity, μ is the mean of the distribution, and σ is the standard deviation. The formula also includes the irrational constants $\pi = 3.1415\ldots$ and $e = 2.7182\ldots$.

The mean can take any real value, and the standard deviation can take any positive value. Thus, the "normal distribution" is really an infinite number of distributions, each having its own mean and standard deviation.

10.3 Properties of the normal distribution

The normal distribution has the following features that are worth remembering:

▶ It is a continuous distribution, so probability is measured by the area under the curve rather than the height of the curve.
▶ It is symmetric around its mean.
▶ It has a single mode.
▶ The probability density is highest exactly at the mean.

This final feature, and the symmetry of the normal distribution, imply that the mean, median, and mode are all equal to each other for the normal distribution.

There are some useful rules of thumb about areas under the normal curve. About two-thirds (68.3%, to be more exact) of the area under the normal curve lies within one standard deviation of the mean. In other words, the probability is 0.683 that the measurement of a randomly chosen observation drawn from a normal distribution falls between $\mu - \sigma$ and $\mu + \sigma$, as shown in Figure 10.3-1.

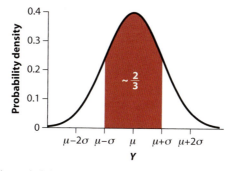

Figure 10.3-1 The probability that a randomly drawn measurement from a normal distribution is within one standard deviation of the mean is approximately 2/3. (More precisely, 0.683 of the observations fall within one standard deviation of the mean.)

Ninety-five percent of the probability of a normal distribution lies within about two standard deviations of the mean (to be more exact, within 1.96 standard deviations). In other words, the probability is 0.95 that the measurement of a randomly chosen observation drawn from a normal distribution falls between $\mu - 1.96\,\sigma$ and $\mu + 1.96\,\sigma$, as shown in Figure 10.3-2.

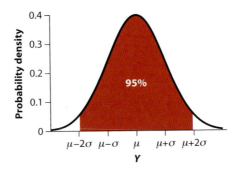

Figure 10.3-2 The probability is 95% that a randomly drawn measurement from the normal distribution is within approximately two standard deviations of the mean (more precisely, exactly 95% lie within 1.96 standard deviations of the mean).

With a normally distributed variable, about two-thirds of individuals are within one standard deviation of the mean, and about 95% are within two standard deviations of the mean.

10.4   The standard normal distribution and statistical tables

A normal distribution with a mean of zero and standard deviation of one is called a **standard normal distribution**.[2] Figure 10.4-1 shows a plot of the standard normal. We will use the symbol Z to indicate a variable having a standard normal distribution.

The *standard normal distribution* is a normal distribution with mean 0 and standard deviation 1.

___
2   "Standard normal" may sound redundant and repetitive (like "bunny rabbit" or "vim and vigor"), but there are an infinite number of normal distributions, depending on what mean and variance are put into the formula, and the standard normal distribution is just one of these.

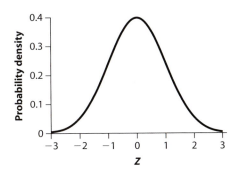

Figure 10.4-1 The standard normal distribution.

Using the standard normal table

Unlike the Poisson and binomial distributions, the probability that a given event occurs when sampling from a normal distribution is difficult to compute by hand, because it requires integration of a complicated function. Instead, we use statistical tables or computers[3] to obtain probabilities under the normal curve. Statistical Table B in the back of this book gives us the probability that a random draw from a standard normal distribution is *above* a given cutoff value. For example, if we drew a number at random from a standard normal distribution, the probability is 0.025 that it would be greater than 1.96. Table 10.4-1, an excerpt from Statistical Table B, shows how we obtained this number.

Table 10.4-1 Probabilities of $Z > a.bc$ under the standard normal curve. The digit before and immediately after the decimal (i.e., $a.b$) are given down the first column, and the second digit after the decimal (i.e., c) is given across the top row. Excerpted from Statistical Table B.

First two digits of $a.bc$	Second digit after decimal (c)									
	0	1	2	3	4	5	6	7	8	9
...										
1.5	0.0668	0.0655	0.0643	0.0630	0.0618	0.0606	0.0594	0.0582	0.0571	0.0559
1.6	0.0548	0.0537	0.0526	0.0516	0.0505	0.0495	0.0485	0.0475	0.0465	0.0455
1.7	0.0446	0.0436	0.0427	0.0418	0.0409	0.0401	0.0392	0.0384	0.0375	0.0367
1.8	0.0359	0.0352	0.0344	0.0336	0.0329	0.0322	0.0314	0.0307	0.0301	0.0294
1.9	0.0287	0.0281	0.0274	0.0268	0.0262	0.0256	0.0250	0.0244	0.0239	0.0233
2.0	0.0228	0.0222	0.0217	0.0212	0.0207	0.0202	0.0197	0.0192	0.0188	0.0183
2.1	0.0179	0.0174	0.0170	0.0166	0.0162	0.0158	0.0154	0.0150	0.0146	0.0143
2.2	0.0139	0.0136	0.0132	0.0129	0.0126	0.0122	0.0119	0.0116	0.0113	0.0110

---

[3]   For example, the Excel function NORMDIST(Z,0,1,TRUE) will return the probability under a standard normal distribution of getting a value less than Z.

Table 10.4-1 has an unusual layout. It provides the probability that Z is greater than a given three-digit cutoff number "*a.bc*," where "*a*," "*b*," and "*c*" refer to digits of the number. To find the probability, begin by finding the first two digits of the cutoff number (*a.b*) in the left-hand column of the table. Then find the second digit of the number after the decimal place (*c*) across the top row. The probability that Z is greater than *a.bc* is given in the corresponding cell for that row and column. To find the probability $\Pr[Z > 1.96]$, for example, we would look down the left-hand column for *a.b* = 1.9, and then go across to the column that corresponds to *c* = 6 to fill in the last decimal. We would find that the probability of a random draw from a standard normal distribution greater than 1.96 is 0.025 (the number outlined in red in Table 10.4-1). This probability is the area under the curve above a standard normal Z of 1.96 (see Figure 10.4-2).

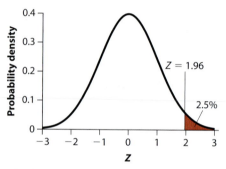

Figure 10.4-2 The area under the standard normal curve above $Z = 1.96$ is 0.025. The area in red is given in Statistical Table B, which is excerpted in Table 10.4-1.

Statistical Table B gives us the probability that Z is greater than a given *positive* number. Probabilities corresponding to *negative* Z values are not included. Recall, however, that all normal distributions are symmetric around their means, and the standard normal distribution is symmetric around $\mu = 0$. This means that

$$\Pr[Z < -\text{number}] = \Pr[Z > \text{number}].$$

Thus, the probability that a random observation from the standard normal distribution is *less than* -1.96 is the same as the probability that an observation is *greater than* 1.96, which is 0.025:

$$\Pr[Z < -1.96] = \Pr[Z > 1.96] = 0.025.$$

The probability that Z lies between a lower bound and an upper bound can be calculated in two steps, as shown in Figure 10.4-3. First, use Statistical Table B to find $\Pr[Z > \textit{lower bound}]$ and $\Pr[Z > \textit{upper bound}]$. Then, calculate the difference between these two probabilities:

$$\Pr[\textit{lower bound} < Z < \textit{upper bound}] = \Pr[Z > \textit{lower bound}] - \Pr[Z > \textit{upper bound}].$$

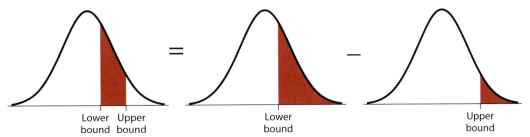

Figure 10.4-3 Calculating the area under the standard normal curve between a lower bound and an upper bound.

Using the standard normal to describe any normal distribution

There are an infinite number of normal distributions, but they are all similar in shape. This allows us to use a simple transformation to obtain probabilities under any normal distribution. To do so, we calculate how many standard deviations a particular value is away from the mean.

$$Z = \frac{Y - \mu}{\sigma}.$$

This standardized Z value is called a **standard normal deviate**.[4] The formula converts Y, which has a normal distribution with mean μ and standard deviation σ, to Z, which has a standard normal distribution. Look at this formula for a second. The numerator, $Y - \mu$, tells us how far away Y is from its mean, measured in the original units. If this value is negative, then Y is less than the mean, and if it is positive, then Y is greater than the mean. If we now divide this value by σ, then we can calculate how far Y is from its mean measured by the number of standard deviations.

The probability of obtaining a measurement that is Z standard deviations from the mean is the same for all normal distributions, including the standard normal distribution. This means that we can use the table of probabilities for the standard normal distribution (Statistical Table B) to find probabilities for *any* normal distribution.

> A *standard normal deviate*, or Z, tells us how many standard deviations a particular value is from the mean.

Example 10.4 shows how the Z-standardization can be used.

---

[4]   Not to be confused with "everyday ordinary pervert." It's not often that you can find a jargon term that seems to be both redundant *and* self-contradictory.

Example 10.4	**One small step for man?**

NASA excludes anyone under 58.5 inches in height and anyone over 76 inches from being an astronaut. In metric units,[5] these values for the lower and upper height restrictions are 148.6 cm and 193.0 cm, respectively. What fraction of the American population is excluded from being an astronaut by these height constraints? The distribution of adult heights within a sex is very well described by a normal distribution,[6] with the mean and standard deviation for adult males in America given by 175.6 cm and 7.1 cm, respectively. For adult American females, the mean height is 162.6 cm with a standard deviation of 6.4 cm.

We can use the standard normal distribution to calculate the proportion of individuals made ineligible for astronaut training as a consequence of their height alone. Let us first do the calculation for males. Let's be very specific about what we are trying to do: we want to know the probability that a male individual has a height that is either less than 148.6 cm or greater than 193.0 cm:

$$\Pr[\text{Height} < 148.6 \text{ or } \text{Height} > 193.0].$$

It becomes much easier to answer the question if we make a sketch of what we are looking for (Figure 10.4-4).

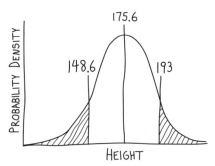

Figure 10.4-4 A sketch can help you keep track of the areas under the curve that you are trying to find. This rough sketch does not show the exact areas, but it properly orients the mean and the values we care about.

The maximum height (193 cm) is above the mean (175.6 cm), and the minimum height (148.6 cm) is below the mean. Based on the drawing in Figure 10.4-4, the outcomes "*Height* < 148.6" and "*Height* > 193.0" are mutually exclusive. By the addi-

---

5   NASA has had infamous difficulty in converting between English and metric units. In 1999, the Mars Climate Orbiter missed Mars and flew off into its own independent orbit around the sun because some components of the navigational system used English units and others used metric. Oops.

6   http://science.ksc.nasa.gov/facts/faq12.html; http://www.tallpages.com/uk/index.php?pag=ukstatist.php

tion rule, therefore, we can determine the fraction of American males whose height excludes them from being an astronaut by summing the two parts:

$$\Pr[Height < 148.6 \text{ or } Height > 193.0] = \Pr[Height < 148.6] + \Pr[Height > 193.0].$$

Let's start with the second part first: what's the probability that an American male is too tall for NASA's restrictions? In other words, what is the probability that an American adult male is taller than 193.0 cm? The first step is to convert $Height = 193.0$ to a standard normal deviate, using the mean (175.6 cm) and standard deviation (7.1 cm) of American male height:

$$Z = \frac{193.0 \text{ cm} - 175.6 \text{ cm}}{7.1 \text{ cm}} = 2.45.$$

That is, a value of 193 cm occurs at 2.45 standard deviations above the mean male height.

What fraction lies above this point? In other words, what is $\Pr[Z > 2.45]$? Using Statistical Table B at the end of this book, we can see that 0.00714 of the area under the standard normal distribution lies more than 2.45 standard deviations above the mean. So, this is our answer to this part of the problem: the fraction 0.00714 of American adult males are taller than 193.0 cm.

Finding the probability of males being too short for NASA's restrictions follows a similar procedure, but with one additional step. Again we convert $Height = 148.6$ (the minimum cutoff) to a standard normal deviate:

$$Z = \frac{148.6 \text{ cm} - 175.6 \text{ cm}}{7.1 \text{ cm}} = -3.80.$$

In other words, 148.6 cm occurs at 3.80 standard deviations below the mean male height. What fraction of American adult males lies *below* this point? What is $\Pr[Z < -3.80]$? Statistical Table B gives us the probability of obtaining a value *greater than* a given number. To find the probability of getting a value *less than* a particular number, remember that the normal distribution is symmetric around its mean. Thus, the probability of being 3.80 standard deviations or more below the mean is the same as the probability of being 3.80 standard deviations or more above the mean:

$$\Pr[Z < -3.80] = \Pr[Z > 3.80].$$

When we look up the probability of being greater than 3.80 in Statistical Table B, we find that $\Pr[Z > 3.80] = 0.00007$. Thus, the fraction 0.00007 of American adult males are shorter than 148.6 cm.

Now, use the addition rule to determine the proportion of American adult men that are excluded from the astronaut program by the height restrictions:

$$\Pr[Height < 148.6 \text{ or } Height > 193.0] = 0.00007 + 0.00714 = 0.00721.$$

The fraction 0.00721 (or 0.72%) of all American adult males are excluded by height. In other words, the NASA height restriction excludes only a small percentage of men.

We can make similar calculations for American adult women. Women's heights follow a normal distribution with mean 162.6 cm and standard deviation 6.4 cm. These values correspond to height restrictions at 2.19 standard deviations below the mean and 4.75 standard deviations above the mean. (Check these numbers for yourself.) Using Statistical Table B, the fraction 0.01426 of the women are too short to meet NASA's guidelines and only about one in a million are too tall. In total, the height restrictions exclude 1.4% of American women from being astronauts.

10.5 The normal distribution of sample means

One of the most important facts about the normal distribution is that it can be used to describe the sampling distribution of some estimates, including the sample mean. The sampling distribution of an estimate lists all the values that we might obtain when we sample a population and describes their probabilities of occurrence (Section 4.1).

> If a variable Y has a normal distribution in a population, then the distribution of sample means $\overline{Y}$ is also normal.

For example, Figure 10.1-2 shows the normal distribution that best matches the distribution of human birth weights in the U.S. This distribution is normal with mean $\mu = 3339$ g and standard deviation $\sigma = 573$ g. Suppose we took a single random sample of 10 babies from this distribution and obtained $\overline{Y} = 3084$ g. This is not equal to the true mean (3339 g) because of chance, or sampling error. The estimate based on any particular sample is influenced by who happened to get sampled and who did not, which is a random process.

Even though all we usually have is a single sample of size n, the estimate has some error, because we might have gotten a different value. The different values for $\overline{Y}$ that we might have obtained, and their associated probabilities, make up the sampling distribution for $\overline{Y}$.

The fact that sample means are normally distributed is a huge time-saver. In Chapter 4, we obtained the sampling distribution for $\overline{Y}$ by going back to the population and taking a vast number of random samples, each of the same size n, calculating $\overline{Y}$ for each sample, and plotting the results. However, we are spared the trouble if Y has a normal distribution in the population, because in this case the distribution of sample means is also normally distributed, as shown in Figure 10.5-1 (we also plot the distribution of individual data points for comparison).

Figure 10.5-1 shows that the mean of the sampling distribution of $\overline{Y}$ is *also* 3339 g, the same as μ, the mean of Y itself. The mean of the sampling distribution of $\overline{Y}$ *always* equals the mean of the original distribution (μ). In other words, the sample mean based on a random sample from a normal distribution gives an unbiased estimate of μ.

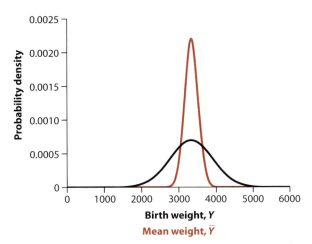

Figure 10.5-1 The normal distribution of sample means ($\bar{Y}$, in red) for samples of size $n = 10$ from the normal distribution describing the population of baby birth weights in Figure 10.1-2 (shown here in black for comparison). The distribution of sample means has the same mean as the individual values, but with a smaller standard deviation.

The standard deviation of the sampling distribution for $\bar{Y}$ is known as the *standard error of the mean* (Section 4.2). It is symbolized as $\sigma_{\bar{Y}}$, and is equal to the standard deviation of Y divided by the square root of the sample size (n):

$$\sigma_{\bar{Y}} = \frac{\sigma}{\sqrt{n}}.$$

This equation is correct even when Y does *not* have a normal distribution.

The standard error describes the typical amount of sampling error when estimating the mean, so it measures the precision of the estimate. Increasing sample size reduces sampling error and hence increases the precision of an estimate, as we saw in Chapter 4. By averaging data from more babies at a time, the sampling distribution based on samples of 100 babies is less "noisy" (i.e., has a smaller sampling error) than the sampling distribution generated from only 10 babies at a time.[7] The sampling distribution of $\bar{Y}$ would be even narrower if we used a larger sample size, as shown in Figure 10.5-2.

In later chapters, we will use the fact that the sampling distribution of $\bar{Y}$ is normal when Y is normal to calculate exact confidence intervals for a population mean and exact P-values when testing hypotheses about means.

---

[7]   This may be the only sense in which it can be said that 100 babies are less noisy than 10.

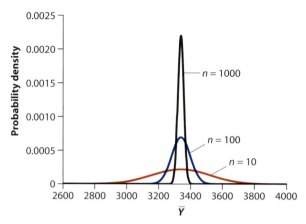

Figure 10.5-2 The distributions of sample means based on sample sizes of $n = 10$ (*red*), $n = 100$ (*blue*), and $n = 1000$ (*black*). (Note the change in scale from the previous figures.)

Calculating probabilities of sample means

The standard normal distribution can be used to calculate the probability of drawing a sample with a mean in a given range, assuming we know the true values of the mean and the standard deviation of the population from which the sample was taken. Given that the distribution of sample means is normal with mean μ and standard deviation $\sigma_{\bar{Y}} = \sigma/\sqrt{n}$, then

$$Z = \frac{\bar{Y} - \mu}{\sigma_{\bar{Y}}}$$

has a standard normal distribution, just like any normally distributed variable.

For example, let's take a random sample of $n = 80$ babies from the population of babies introduced in Section 10.1. In this population, weights are normally distributed with mean $\mu = 3339$ g and standard deviation $\sigma = 573$ g. What is the probability that the sample mean is at least 3370 g? In other words, what is $\Pr[\bar{Y} > 3370]$?

The Z-standardization for $\bar{Y}$ looks like this:

$$Z = \frac{\bar{Y} - \mu}{\sigma_{\bar{Y}}},$$

where $\sigma_{\bar{Y}}$ is the standard deviation of the sampling distribution of $\bar{Y}$, also known as the standard error of the mean. This Z tells us that the sample mean $\bar{Y}$ is Z standard errors above the true mean μ.

For this example, $\sigma_{\bar{Y}}$ is

$$\sigma_{\bar{Y}} = \frac{\sigma}{\sqrt{n}} = \frac{573}{\sqrt{80}} = 64.1 \text{ g}.$$

From this we calculate the Z-score:

$$Z = \frac{\overline{Y} - \mu}{\sigma_{\overline{Y}}} = \frac{3370 - 3339}{64.1} = 0.48.$$

In other words, $\Pr[\overline{Y} > 3370]$ is the same as $\Pr[Z > 0.48]$. Using Statistical Table B, we find that

$$\Pr[Z > 0.48] = 0.314.$$

In other words, $\overline{Y}$ is at least 3370 g in about 31.4% of all random samples from this population with $n = 80$ individuals.

10.6 Central Limit Theorem

We have seen that the means of samples drawn from a normal distribution are themselves normally distributed. One additional reason why the normal distribution is so important in data analysis is that the sampling distribution of sample means $\overline{Y}$ is *approximately* normal even when the distribution of individual data points is *not* normal, provided the sample size is large enough. This astonishing fact is known as the **Central Limit Theorem**. How large a sample is "large enough" depends on the shape of the distribution of observations in the population: the more similar the original distribution is to the normal, the smaller the sample size required to yield a distribution of sample means that is well-approximated by the normal distribution.

> According to the Central Limit Theorem, the sum or mean of a large number of measurements randomly sampled from a non-normal population is approximately normally distributed.

Example 10.6 illustrates the effect that increasing the sample size has on the sampling distribution for the mean, even when the population is highly non-normal.

Example 10.6   **Pushing your buttons**

It is very common in some kinds of psychological studies to measure the response time of a subject following some stimulus. The response time is often measured as the time it takes for the subject to push a button. But how reliable is this measure? To answer this question, an experimenter asked a single subject to press a button as fast as possible after a light went on (Bradley 1980). This single subject repeated the test 2520 times. The distribution of response times is plotted in Figure 10.6-1.

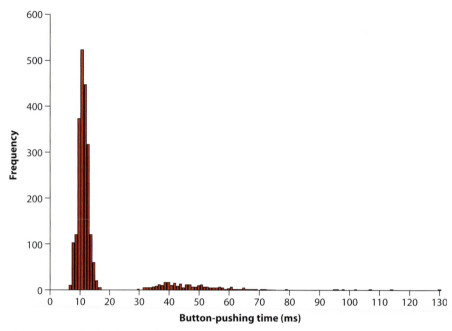

Figure 10.6-1 The distribution of response times (in milliseconds) to push a button after a light went on.

The distribution in Figure 10.6-1 is extremely non-normal because there are multiple peaks and the distribution is highly skewed to the right. The highest peak on the left corresponds to the cluster of response times when the subject managed to hit the button on the first try. The second peak is centered around 45 ms, mostly reflecting trials in which the subject missed the button on his first try and then flailed about before managing to hit it on the second go. This second cluster has greater width than the first. There are also a few larger response times, reflecting responses that took up to four tries to hit the button.

Let's use this distribution to illustrate the Central Limit Theorem by taking samples of increasingly large size and looking at the sampling distributions of sample means. The results are displayed in Figure 10.6-2 for the means of samples of size 1, 2, 8, 64, 128, and 512.

The top left graph in Figure 10.6-2 plots the mean of random samples of size $n = 1$, which simply re-creates the original distribution. The top right graph is a histogram that plots the means of many samples of size $n = 2$. The standard deviation is lower than in the $n = 1$ histogram, and even though two clusters are still present, they are closer together. As n increases (see the $n = 8$ and $n = 64$ histograms in the middle of Figure 10.6-2), the distribution of sample means starts to resemble a normal distribution, until at some point it becomes indistinguishable from the normal distribution around $n = 128$ (the bottom left histogram in Figure 10.6-2).

The most powerful methods available for analyzing data assume that the distribution of sample means (and the distributions of some other estimates, too) follows a

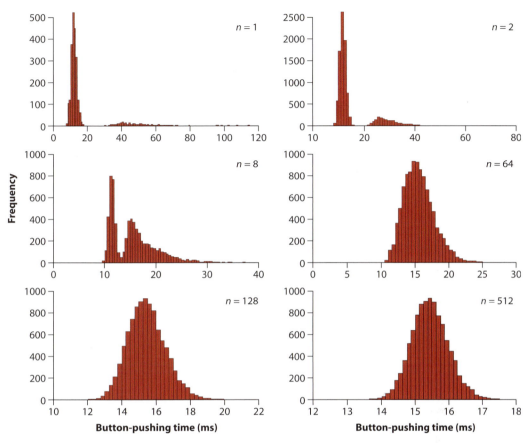

Figure 10.6-2 The frequency distribution of sample means of samples of increasing size. Each histogram displays the means of a large number of repeated samples of size n drawn from the distribution of button-pushing times in Figure 10.6-1. The scale and range of the axes change from graph to graph.

normal distribution. If the sample size is large enough, then the Central Limit Theorem makes it possible to use these methods even when our data are sampled from a population that is *not* normally distributed.

10.7   Normal approximation for the binomial distribution

One frequent application of the Central Limit Theorem is the normal approximation to the binomial distribution. Recall from Section 7.1 that the binomial distribution is a discrete probability distribution. It describes the number of "successes" in n independent trials, where p is the probability of "success" in any one trial.

According to the Central Limit Theorem, a binomial distribution with large n is approximated by a normal distribution. This is because the number of successes is a kind of sum: if each success is labeled as "1" and each failure is labeled as "0," then the count of the number of successes is the sum over all individuals of the 1s and 0s.

The normal approximation to the binomial distribution is useful when you are using a calculator to obtain probabilities for a binomial distribution with a large n—exactly the case when calculating exact binomial probabilities is most time-consuming. This comes up, for example, when carrying out a binomial test on a large sample. With a binomial test, the probability of success p_0 stated in the null hypothesis is often 0.5, in which case n needn't be too large (say 30 or more) in order for the normal approximation to be appropriate.[8] As a rule of thumb, the products np and $n(1-p)$ should both be greater than five in order to use the normal approximation. We use the normal distribution that has the same mean and standard deviation as the binomial distribution—in other words, a mean of np and a standard deviation of $\sqrt{np(1-p)}$.

> When the number of trials n is large, the binomial probability distribution for the number of successes is approximated by a normal distribution having mean np and standard deviation $\sqrt{np(1-p)}$.

Example 10.7 applies the normal approximation to a problem requiring the binomial test.

Example 10.7 ## The only good bug is a dead bug

The brown recluse spider (*Loxosceles reclusa*) often lives in human houses[9] throughout central North America. This spider is a moderate health threat, as its bite causes nasty, slow-healing wounds. Bites are very rarely fatal, but the resulting wounds are so disgusting that you will be very glad we chose to illustrate this example with the spider rather than a photo of the injuries they cause. Information on the spider's diet is useful for developing effective pest management strategies. A diet-preference study (Sandidge 2003) gave each of 41 brown recluse spiders a choice between two crickets, one live and one dead. Thirty-one of the 41 spiders chose the dead cricket over the live one. Does this represent evidence for a diet preference?

---

[8]   The proportion specified by the null hypothesis is what's important, not the proportion in the data.

[9]   One house in Kansas had more than 2000 brown recluse spiders (Sandidge 2003).

The appropriate statistical method for analyzing these data is the binomial test with the following null and alternative hypotheses:

H_0: Brown recluse spiders show no preference for live or dead prey ($p = 0.5$).
H_A: Brown recluse spiders prefer one type of prey over the other ($p \neq 0.5$).

This is a two-tailed test. The outcome of the test will be decided according to the P-value (Section 6.2), the probability of obtaining a result as extreme or more extreme than the observed result (i.e., 31 out of 41 spiders ate the dead prey item), assuming that the null hypothesis were true. This P-value would ordinarily be calculated from the binomial distribution with $n = 41$ and $p = 0.5$:

$$P = 2 \Pr[X \geq 31]$$
$$= 2\,[(\Pr[X = 31] + \Pr[X = 32] + \Pr[X = 33] + \ldots + \Pr[X = 41])],$$

where

$$\Pr[X = i] = \binom{41}{i}(0.5)^i(0.5)^{41-i}.$$

There are 11 different probabilities to calculate in this example, which would be time-consuming to do by hand or by calculator. It is much easier to approximate the answer using the normal distribution. Under the null hypothesis, the mean of the best-fitting normal distribution is

$$\mu = n\,p = (41)(0.5) = 20.5$$

and the standard deviation is

$$\sigma = \sqrt{np(1-p)} = \sqrt{41\,(0.5)\,(0.5)} = 3.20.$$

Now we can use the normal approximation to calculate the probability of getting 31 or more by chance under this distribution, and then multiply by two to yield the P-value:

$$P = 2 \Pr[X \geq 31].$$

We start by converting $X = 31$ into a Z-score using the mean $\mu = 20.5$ and the standard deviation $\sigma = 3.20$:

$$Z = \frac{31 - 20.5}{3.20} = 3.28.$$

Thus, the P-value is calculated using the standard normal distribution as

$$P = 2 \Pr[Z > 3.28].$$

Looking to Statistical Table B, we find that the probability that $Z > 3.28$ is approximately 0.00052. Multiplying this value by two to get the two-tailed P-value, we obtain

$$P = 0.00104.$$

This P-value is less than $\alpha = 0.05$, so we reject the null hypothesis. The spiders indeed show a diet preference for dead prey.

This computation is much simpler than the exact calculation based on the binomial distribution, but how good is the approximation? The exact P-value, calculated using the binomial distribution, is 0.0015, rather than 0.00104. The normal approximation is not exact. It performs best when n is large and when the population proportion p is close to 0.5. Somewhat better approximations can be found by applying a continuity correction. (See the Quick Formula Summary, Section 10.9.) The normal approximation can make some otherwise time-consuming calculations manageable.

10.8 Summary

▶ The normal distribution is a bell-shaped curve, a continuous probability distribution approximating the distribution of many variables in nature.

▶ If a variable has a normal distribution, its mean, median, and mode are all the same. The normal distribution is symmetric around its mean.

▶ To describe a normal distribution, we need to know the mean μ and the standard deviation σ.

▶ Under a normal distribution, about two-thirds (68.3%) of individuals are within one standard deviation of the mean. About 95% are within two standard deviations of the mean.

▶ A standard normal distribution is a normal distribution with mean zero and standard deviation equal to one.

▶ All normal distributions can be converted to the standard normal distribution by computing the number of standard deviations that each measurement is from the mean: $Z = \frac{Y - \mu}{\sigma}$, where Z is called the "standard normal deviate."

▶ Means of random samples drawn from a normal distribution with mean μ and standard deviation σ are also normally distributed, with the same mean μ and with standard deviation $\sigma_{\bar{Y}} = \sigma/\sqrt{n}$, where $\sigma_{\bar{Y}}$ is the standard error of the sample mean.

▶ According to the Central Limit Theorem, the sum (or average) of a large number of observations randomly sampled from a non-normal distribution is approximately normally distributed.

▶ A binomial distribution with large n can be approximated by a normal distribution.

10.9 Quick Formula Summary

Z-standardization

What is it for? Converts values from any normal distribution with known mean μ and known variance σ into standard normal deviates.

What does it assume? The original distribution is normal with known parameters.

Formula: $Z = \dfrac{Y - \mu}{\sigma}$.

Normal approximation to the binomial distribution

What is it for? Approximates the binomial distribution using the normal distribution.

What does it assume? That np and $n(1 - p)$ are both five or greater.

Formula: $\Pr[X \geq Observed] = \Pr\left[Z > \dfrac{Observed - np}{\sqrt{np(1 - p)}}\right]$, where n is the number of trials, p is the probability of success for each trial, and *Observed* is the number of successes in the data.

Continuity correction for the normal approximation to the binomial distribution

What is it for? Improves the match between the normal and binomial distributions. The binomial distribution is discrete, and the normal distribution is continuous. By adding or subtracting 1/2 from the desired value, we can match the normal and binomial distributions better.

Formula: $\Pr[X \geq Observed] = \Pr\left[Z > \dfrac{Observed - \dfrac{1}{2} - np}{\sqrt{np(1 - p)}}\right]$ and

$$\Pr[X \leq Observed] = \Pr\left[Z < \dfrac{Observed + \dfrac{1}{2} - np}{\sqrt{np(1 - p)}}\right],$$

where n is the number of trials, p is the probability of success for each trial, and *Observed* is the number of successes in the data.

PRACTICE PROBLEMS

1. Assume that Z is a number randomly chosen from a standard normal distribution. Use the standard normal table to calculate each of the following probabilities:
   a. $\Pr[Z > 1.34]$ b. $\Pr[Z < -1.34]$
   c. $\Pr[Z > -2.15]$ d. $\Pr[Z < 1.2]$
   e. $\Pr[0.52 < Z < 2.34]$
   f. $\Pr[-2.34 < Z < -0.52]$
   g. $\Pr[Z < -0.93]$ h. $\Pr[-1.57 < Z < 0.32]$

2. Britain's domestic intelligence service MI5 has recently placed an upper limit on the height of its spies, on the assumption that people who are too tall stand out. MI5 wants its agents to blend in with the crowd. To apply to be a spy, you must be no taller than 5 feet 11 inches (180.3 cm) if you are a man, and no taller than 5' 8" (172.7 cm) if you are a woman.
   a. If the mean height of British men is 177.0 cm, with standard deviation 7.1 cm, what

proportion of British men are precluded from being spies by this height restriction? Assume that height follows a normal distribution.

b. The mean height of women in Britain is 163.3 cm, with standard deviation 6.4 cm. Assuming a normal distribution of female height, what fraction of women meet the height standard for application to MI5?

c. Sean Connery, the original James Bond in the movies, is about 183.4 cm tall. By how many standard deviations does he exceed the height limit for spies?

3. Use the three distributions labeled *a*, *b*, and *c* to answer the following questions.

a.

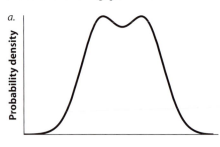

b.

c.

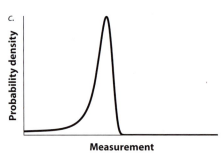

Measurement

a. Which of these distributions is most like the normal distribution? On what basis would you exclude the other two?

b. Which distribution would generate an approximately normal distribution of sample means, calculated from large random samples? Why?

4. The babies born in singleton births in the U.S. have birth weights that are approximately normally distributed with mean 3.339 kg and standard deviation 0.573 kg.

a. What is the probability of a baby being born weighing more than 5 kg?

b. What is the probability of a baby being born weighing between 3 kg and 4 kg?

c. What fraction of babies is more than 1.5 standard deviations from the mean in either direction?

d. What fraction of babies is more than 1.5 kg from the mean in either direction?

e. If you took a random sample of 10 babies, what is the probability that their mean weight $\overline{Y}$ would be greater than 3.5 kg?

5. In each of the following pairs of graphs of normal distributions, which distribution has the highest mean? Which has the highest standard deviation? Pay careful attention to the changes in scale of the *x*-axes.

a.

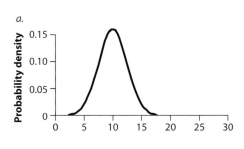

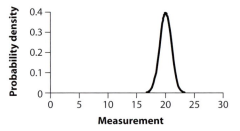

Measurement

b.

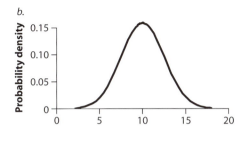

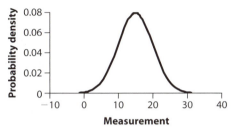

6. In the following graph, the red region accounts for 0.67 of the probability density. Estimate the standard deviation of this normal distribution.

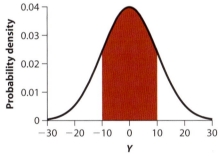

7. Suppose that a variable has a normal distribution, that the mean is 35 mm, and that 20% of the population is larger than 50 mm.
   a. What is the mode of this distribution?
   b. What is the median of this distribution?
   c. Complete the following sentence: "Twenty percent of the distribution is smaller than _____."

8. Use the two distributions labeled I and II below to answer the following questions.

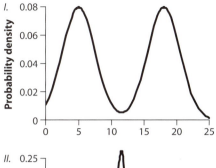

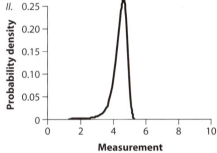

   a. If we drew repeated random samples of individuals from each distribution and calculated the mean for each sample, which distribution would yield a distribution of sample means that most closely followed a normal distribution?
   b. Imagine that we drew the distribution of the sum of 100 random draws from distribution II. What approximate shape would this distribution have?

9. A survey of mitochondrial DNA variation in smelts from an imaginary lake revealed that two haplotypes (genotypes) were present. Thirty percent of individual fish in the population were haplotype A. The remaining 70% were haplotype B. If an experimenter were to sample 400 fish from this population, what is the probability that
   a. at least 130 are haplotype A?
   b. at least 300 are haplotype B?
   c. between 115 and 125 (inclusive) are haplotype A?

10. Ninety-one out of 220 people working as cast and crew of the movie *The Conqueror*, which was filmed in 1955 downwind from nuclear bomb tests, ultimately contracted cancer (see Practice Problem 1 in Chapter 5). Based on age alone, though, only a 14% cancer rate was expected. Test the null hypothesis that the incidence of cancer in this group of people was no different than that expected, using the normal approximation to the binomial distribution.

11. The following table lists the mean and standard deviation of several different normal distributions. For each, a sample of 10 individuals was taken, as well as a sample of 30 individuals. For each sample, calculate the probability that the mean of the sample was greater than the given value of Y.

Mean	Standard deviation	Y	$n = 10$: $Pr[\bar{Y} > Y]$	$n = 30$: $Pr[\bar{Y} > Y]$
14	5	15		
15	3	15.5		
−23	4	−22		
72	50	45		

ASSIGNMENT PROBLEMS

12. The Mercury missions from NASA allowed no astronauts to be taller than 180.3 cm. What fraction of men would that exclude, given a mean height of 175.6 cm and a standard deviation of 7.1 cm?

13. Recall from Example 10.4 that more women (1.4%) than men (0.72%) are excluded from the astronaut program by the height restrictions. What value of minimum height for women would exclude the same total proportion of women as men?

14. Draw a probability distribution that isn't normal. Describe the features of your distribution that identify it as non-normal.

15. In the following graph of a normal distribution, *each* of the two red areas represents 1/6 of the area under the curve. Estimate the following quantities from this graph:

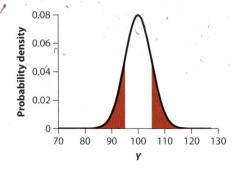

a. The mean
b. The mode
c. The median
d. The standard deviation
e. The variance

16. The proportion of traffic fatalities for each U.S. state resulting from drivers with high alcohol blood levels in 1982 was approximately normally distributed, with mean 0.569 and standard deviation 0.068 (U.S. Department of Transportation Traffic Safety Facts 1999).
a. What proportion of states would you expect to have more than 65% of their traffic fatalities from drunk driving?
b. What proportion of deaths due to drunk driving would you expect to be at the 25th percentile of this distribution?

17. The following table lists the mean and standard deviation of several different normal distributions. For each distribution, calculate the probability of drawing a single value greater than the given value of y and the probability of drawing a value less than y.

Mean	Standard deviation	y	$Pr[Y > y]$	$Pr[Y < y]$
14	5	9		
15	3	18.5		
−23	4	−16		
14,000	5000	9000		

18. In European earwigs, the males sometimes have very long pincers protruding from the end of their abdomens, as shown in the accompanying photo.

Below we have plotted three frequency distributions of sample *means* of pincer lengths (in mm) based on random samples from an earwig population. One of the distributions shows means of samples based on $n = 1$, one shows means of samples based on $n = 2$, and one shows the means of samples based on $n = 8$. Identify which frequency distribution corresponds to each sample size. Explain the basis for your decisions.

19. The crab spider, *Thomisus spectabilis*, sits on flowers and preys upon visiting honeybees, as shown in the photo at the beginning of the chapter. (Remember this the next time you sniff a wild flower.) Do honeybees distinguish between flowers that have crab spiders and flowers that do not? To test this, Heiling et al. (2003) gave 33 bees a choice between two flowers, one of which had a crab spider and the other of which did not. In 24 of the 33 trials, the bees picked the flower that had the spider. In the remaining nine trials, the bees chose the spiderless flower. With these data, carry out the appropriate hypothesis test, using an appropriate approximation.

20. The following table lists the mean and standard deviation of several different normal distributions. For each, a sample of 20 individuals was taken, as well as a sample of 50 individuals. For each sample, calculate the probability that the mean of the sample was less than the given value of Y.

Mean	Standard deviation	Y	$n = 20$: $\Pr[\bar{Y} < Y]$	$n = 50$: $\Pr[\bar{Y} < Y]$
-5	5	-5.2		
10	30	8.0		
55	20	-61.0		
12	3	12.5		

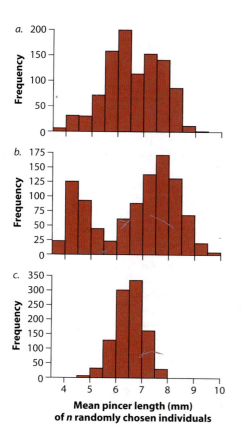

a.

b.

c.

Mean pincer length (mm) of *n* randomly chosen individuals

Controls in medical studies

In 1994, researchers reported on the results of a study describing the effects of a new "wonder" drug (Lanza et al. 1994). The drug, lansoprazole, was intended to treat ulcers, and the study showed that 88% of the people who were treated with this drug got better within four weeks. Surely this was a remarkable achievement!

There was another group of patients that were followed in the study, too, however. At the beginning of the study, the volunteers for the study who had ulcer problems were randomly sorted into two groups. One group received the new medication as planned, but the other group received a chemically inert "sugar pill" instead. The group who received no pharmacologically active medication also improved over the course of the study—in fact, over 40% of the ulcer sufferers in this control group improved over the same four-week period.

In this particular study, the patients treated with the new drug did indeed get better faster than those who were not treated. But the drug was not nearly as effective when compared with the control group as it appeared to be when looked at by itself. Of the 88% who improved after taking the drug, over 40% would have improved even without the medication. How could this be?

I'm addicted to placebos. I could quit but it wouldn't matter.

Steven Wright

There are several reasons. First, for most medical conditions, patients tend to get better over time anyway.[1] Most diseases are neither lethal nor permanent. We tend to go to the doctor when we are feeling at our worst, and therefore the odds are that we would soon start to improve after our worst days anyway, even without a new wonder drug. Second, humans (or at least many of them) like to please others, so there is a tendency to tell doctors that the treatment is more effective than it actually is. Patients in both groups of the study may have described an improvement in their condition, even if this was an exaggeration. Finally, there are benefits to being treated by a physician that go beyond the specifics of a particular drug. A doctor may suggest a new diet or advocate increased rest, for example, and patients in drug trials see doctors more often than they otherwise would and may therefore improve.

One particularly interesting form of benefit from treatment is psychological. In some cases, simply the knowledge of being treated may be sufficient to improve a patient's condition. This is the so-called **placebo effect**, an improvement in a medical condition that results from

---

[1] Hence the old medical adage: "The common cold will go away in a week with proper treatment, but it will take seven days if left untreated."

the psychological effects of medical treatment.[2] A "sugar pill" is a placebo, designed to mimic all the conditions of the medical treatment *except* for the pharmacological effects of the new drug itself.

Placebo effects are well documented for pain relief, but they are more questionable for other kinds of diseases (Hróbjartsson and Gøtzsche 2001). Placebo effects are typically larger for illnesses in which the response variable is subjectively reported by the patient. Placebo effects are smaller or nonexistent in studies that report on more objectively measured variables. The fact that the improvements are subjective does not mean that they are not real; pain, for example, is subjectively experienced but is nonetheless a real condition. In fact, MRI studies have shown that the placebo effect for pain has a neurological basis (Wager et al. 2004).

Even conditions requiring surgical interventions can show improvement without treatment. Studies of human surgery that include a "sham surgery" treatment—that is, surgical incisions without the specific treatment—are rare, for ethical reasons. In many cases, though, sham surgeries have been associated with real improvements in medical conditions. For example, ligation of the mammary artery to treat anginal pain was common in the middle part of the 20th century, with improvement rates of about two-thirds after this surgery. Subsequently, however, two studies showed that improvement rates in patients with this surgery were nearly identical to patients that had received only a sham skin incision (Shapiro and Shapiro 1997).

What we have seen is that patients can improve for a wide variety of reasons, even without specific treatment. It is therefore crucial that medical studies include control treatments, in which a randomized selection of subjects are treated in every way identically to those receiving a new treatment, except for the treatment itself. These controls can take the form of sugar pills, but, by preference, the control patients can receive the currently most effective treatment. In this way, all patients receive care, and if the new drug or treatment is better than the old one, we have evidence in support of switching the most advisable treatment. When we report the results of a study, we want to be able to say that there is an effect, "all else being equal." The goal of careful experimental design is to make "all else equal."

---

[2] "Placebo" is from the Latin for "I will please." In Chaucer's time, this word was used for a sycophant or flatterer who would say whatever would please the listener rather than the truth. The word was then adopted by 19th century physicians to refer to a treatment given to please the patient rather than to cure him or her. Slowly it was realized that there was in fact a therapeutic effect from these placebos and that this had to be controlled for in medical trials.

11

Inference for a normal population

We learned in Chapter 10 that many variables in nature are approximately normally distributed. The majority of methods in this book are geared towards making inference and testing hypotheses about variables that have a normal distribution in the population. In this chapter, we begin the analysis of normally distributed measurements, a theme that will continue for much of the rest of the book.

We start by discussing estimation of the mean, including exact calculations of confidence intervals. Next we describe the simplest hypothesis test on a normal variable, the "one-sample t-test," which lets us ask whether a set of data is consistent with some hypothesized value for the population mean. We finish the chapter by discussing how to make statistical statements about the standard deviation.

11.1 The t-distribution for sample means

To make statistical statements about the precision of an estimate of a mean, we need to describe the probability distribution of $\overline{Y}$ (i.e., its sampling distribution). The sampling distribution is the probability distribution of all the values for an estimate that we might obtain when we sample the population.

Student's *t*-distribution

Recall from Section 10.5 that the sampling distribution for $\overline{Y}$ is a normal distribution if the variable Y is itself normally distributed. As a result, we can use the Z-standardization to calculate probabilities for $\overline{Y}$ under the normal curve:

$$Z = \frac{\overline{Y} - \mu}{\sigma_{\overline{Y}}},$$

where $\sigma_{\overline{Y}}$ is the standard error of the sample mean, the standard deviation of the sampling distribution of $\overline{Y}$.

In general, however, we can't just apply the Z-standardization to $\overline{Y}$-values calculated from real data. This is because to calculate Z we need to know $\sigma_{\overline{Y}}$, yet this is almost never possible. However, we can use the estimate of the standard error:

$$\text{SE}_{\overline{Y}} = \frac{s}{\sqrt{n}},$$

which was first discussed in Section 4.2. The numerator is the sample standard deviation (s), which is just an estimate of the true standard deviation (σ). We will usually call $\text{SE}_{\overline{Y}}$ simply "the standard error of the mean," even though it is really only an estimate of the true standard error.

Substituting $\text{SE}_{\overline{Y}}$ for $\sigma_{\overline{Y}}$ in the formula for Z leads to a related quantity called **Student's *t*,**

$$t = \frac{\overline{Y} - \mu}{\text{SE}_{\overline{Y}}}.$$

This is called "Student's *t*" after its inventor,[1] but we will usually refer to it as simply *t*. The sampling distribution for this statistic is not the normal distribution; instead, *t* has a *t*-distribution. $\text{SE}_{\overline{Y}}$ is not a constant like $\sigma_{\overline{Y}}$ but is a variable, varying by chance from sample to sample (because *s* itself changes from sample to sample). Therefore, the distribution of *t* is not the same as Z. Substituting $\text{SE}_{\overline{Y}}$ for $\sigma_{\overline{Y}}$ adds sampling error to the quantity *t*. As a result, the sampling distribution of *t* is wider than the standard normal distribution. As the sample size increases, *t* becomes more like Z.

> The difference between the sample mean and the true mean ($\overline{Y} - \mu$), divided by the estimated standard error ($\text{SE}_{\overline{Y}}$), has a Student's *t*-distribution with $n - 1$ degrees of freedom.

---

[1] The statistic *t* is called Student's *t* after the nom de plume of the man who first discovered its properties, "Student." (He was more clever at statistics than pseudonyms.) "Student" in reality was William Gosset, an employee of the Guinness Brewing Company in Dublin. Gosset used a pseudonym because Guinness prohibited its employees from publishing, following the unauthorized release of some brewing secrets a few years earlier by another employee. In his honor, Guinness was appointed the official beverage of this book.

The sample size determines the number of *degrees of freedom* (*df*) of the *t*-distribution. The degrees of freedom specify which particular version of the *t*-distribution we need. With a *t*-distribution, we must estimate a parameter from the data—the standard deviation—before calculating *t*. As a result, the number of degrees of freedom for *t*, when applied to inference about one sample, is one less than the number of independent data points:

$$df = n - 1.$$

Consider the *t*-distribution for a sample size of five. With five individuals in a random sample, there are four degrees of freedom. Figure 11.1-1 plots the *t*-distribution with four degrees of freedom in blue; for comparison, the standard normal distribution is shown in red. In most respects, the *t*-distribution is similar to the standard normal distribution. It is symmetric around a mean of zero, is roughly bell-shaped, and has tails that fall off towards plus-infinity and minus-infinity.

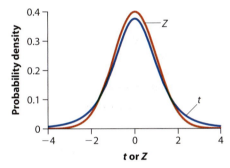

Figure 11.1-1 Student's *t*-distribution with four degrees of freedom (*in blue*), compared with the standard normal distribution (*in red*). The two distributions are similar, though not identical. The tails of the *t*-distribution have more probability than the normal distribution.

The *t*-distribution, however, is fatter in the tails than the standard normal distribution. The difference in the tails is crucial, because it is the tails that matter most when calculating confidence intervals and testing hypotheses.

For example, 95% of the area under the curve of the *t*-distribution with 4 *df* is between −2.78 and 2.78; the remaining 5% lies under the tails outside this range (we explain in the next subsection how we obtained this number). In other words, in 95% of repeated random samples of size $n = 5$ measurements from a normal population, the resulting $\overline{Y}$ will fall within 2.78 estimated standard errors of the true population mean (μ). With the standard normal (*Z*) distribution, on the other hand, 95% of the area under the curve lies between −1.96 and 1.96; the remaining 5% lies beyond these extreme values. The larger range of values of *t*, compared to *Z*, results from the uncertainty about the true value of the standard error.

Finding critical values of the t-distribution

Where did this value of 2.78 come from? The value 2.78 is the "5% critical value" of the t-distribution having $df = 5 - 1 = 4$ degrees of freedom. The 5% refers to the percentage of the area in the tails of the t-distribution (Figure 11.1-2). Every t-distribution has its own critical 5% t-value, depending on the number of degrees of freedom. Once we know the number of degrees of freedom, we can find the critical value using a computer program or using Statistical Table C in the back of this book.

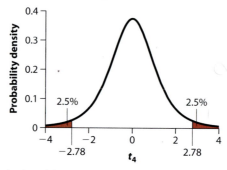

Figure 11.1-2 The critical value of the t-distribution that confines a total of 5% of the area under the curve to its two tails. With four degrees of freedom, this critical value is $t = 2.78$.

We use the symbol $t_{0.05(2), df}$ to indicate the 5% critical t-value of a t-distribution having "df" degrees of freedom. In this notation, the 0.05 stands for the fraction of the area under the curve lying in the tails of the distribution. The "(2)" refers to the fact that the 5% area is divided between the two tails of the t-distribution—that is, 2.5% of the area under the curve lies above $t_{0.05(2), df}$ and 2.5% lies below $-t_{0.05(2), df}$.

We can use the excerpt from Statistical Table C depicted in Table 11.1-1 to show how to find the 5% critical value for a t-distribution with four degrees of freedom. First find the row in the table corresponding to four degrees of freedom. Then find the column that corresponds to $\alpha(2) = 0.05$. The corresponding cell contains the number 2.78, which is the critical value we are looking for.

Table 11.1-1 Critical values of the t-distribution (excerpted from Statistical Table C).

df	$\alpha(2) = 0.20$ $\alpha(1) = 0.10$	$\alpha(2) = 0.10$ $\alpha(1) = 0.05$	$\alpha(2) = 0.05$ $\alpha(1) = 0.025$	$\alpha(2) = 0.02$ $\alpha(1) = 0.01$	$\alpha(2) = 0.01$ $\alpha(1) = 0.005$
1	3.08	6.31	12.71	31.82	63.66
2	1.89	2.92	4.30	6.96	9.92
3	1.64	2.35	3.18	4.54	5.84
4	1.53	2.13	2.78	3.75	4.60
5	1.48	2.02	2.57	3.36	4.03
...	...	...	...	...	...

The other columns in Table 11.1-1 indicate the critical values corresponding to tail probabilities of $\alpha(2) = 0.20, 0.10, 0.02$, and 0.01 under the curve for the t-distribution. Notice, though, that the column headings in Table 11.1-1 contain $\alpha(1)$ designations, too. This notation, such as $\alpha(1) = 0.025$ in the middle column of the table, indicates areas under the curve at only one tail of the t-distribution (Figure 11.1-3), which is useful when carrying out one-tailed significance tests.

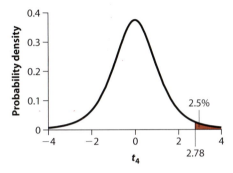

Figure 11.1-3 The critical value corresponding to 2.5% of the area under the curve at only one tail of the t-distribution.

In the remainder of this chapter, we use the t-distribution to compute exact confidence intervals for the population mean and to test hypotheses about the population mean.

11.2   The confidence interval for the mean of a normal distribution

As we discussed in Section 4.3, a confidence interval is a very useful way to express the precision of an estimate of a parameter. It turns out that we can use the t-distribution to calculate an exact confidence interval for the mean of a population having a normal distribution. Example 11.2 illustrates the appropriate method.

| Example 11.2 | **Eye to eye** |

The stalk-eyed fly, *Cyrtodiopsis dalmanni*, is a bizarre-looking insect from the jungles of Malaysia. Its eyes are at the end of long stalks that emerge from its head, like something from the cantina scene in *Star Wars*. These eye stalks are present in both sexes, but they are particularly impressive in males. The span of the eye stalk in males enhances their attractiveness to females and their success in battles against other males.

The span, in millimeters, from one eye to the other was measured in a random sample of nine male stalk-eyed flies.[2] The data are as follows:

8.69 8.15 9.25 9.45 8.96 8.65 8.43 8.79 8.63.

We can use these measurements to estimate the mean eye span in the fly population, and to quantify the uncertainty of our estimate using a 95% confidence interval. Assume that eye span has a normal distribution in the population.

The 95% confidence interval for the mean

The mean and standard deviation of the eye-span sample are

$$\overline{Y} = 8.778 \text{ and } s = 0.398.$$

(Review Section 3.1 if you don't remember how to do these calculations.) How precise is this estimate of the population mean? Let's describe the precision by calculating a 95% confidence interval for the mean. In Section 4.2, we used the "2SE rule of thumb" to calculate this interval, but here in Section 11.2 we will use the t-distribution to give a more exact formula. To do so, though, we must use the fact, learned in Section 11.1, that the standardized difference $(\overline{Y} - \mu)/SE_{\overline{Y}}$ has a t-distribution with $df = n - 1$, assuming that Y is normally distributed. This means that in 95% of random samples from a normal distribution, the standardized difference will lie between $-t_{0.05(2), df}$ and $t_{0.05(2), df}$:

$$- t_{0.05(2), df} < \frac{\overline{Y} - \mu}{SE_{\overline{Y}}} < t_{0.05(2), df}.$$

Rearranging this equation shows that, in 95% of random samples from a normal distribution, $\overline{Y} \pm t_{0.05(2), df} \cdot SE_{\overline{Y}}$ will bracket the population mean:

$$\overline{Y} - t_{0.05(2), df} SE_{\overline{Y}} < \mu < \overline{Y} + t_{0.05(2), df} SE_{\overline{Y}}.$$

This is the exact 95% confidence interval for the mean. It is similar to the 2SE rule of thumb described in Section 4.2, except that we use the critical value from the t-distribution in the formula rather than "2."

> In 95% of random samples from a normal distribution, the interval from $\overline{Y} - [t_{0.05(2), df}]SE_{\overline{Y}}$ to $\overline{Y} + [t_{0.05(2), df}]SE_{\overline{Y}}$ will bracket the population mean. This interval is the 95% confidence interval of the mean.

Let's calculate the 95% confidence interval for mean eye span in male stalk-eyed flies. To begin, we'll need the standard error of the mean:

---

[2]   Data provided by Sam Cotton and Kevin Fowler, University College, London.

$$SE_{\bar{Y}} = \frac{s}{\sqrt{n}} = \frac{0.398}{\sqrt{9}} = 0.133.$$

We also need the degrees of freedom to get the right t-statistic. The sample size is $n = 9$, so the corresponding t has eight degrees of freedom. From Statistical Table C,

$$t_{0.05(2),\,8} = 2.31.$$

Putting all of the numbers from the eye-stalk data into the confidence interval equation we get

$$8.778 - (2.31 \times 0.133) < \mu < 8.778 + (2.31 \times 0.133),$$

which yields

$$8.47 \text{ mm} < \mu < 9.09 \text{ mm}.$$

Thus, the 95% confidence interval for the mean of the eye span in this species, calculated from the random sample, is from 8.47 mm to 9.09 mm.

We do not know whether the population mean eye-stalk span lies between 8.47 and 9.09. All we can say is that the 95% confidence interval for the mean will capture the population mean in 95% of random samples.

The 99% confidence interval for the mean

There's nothing special about 95%; it has just come to be adopted as the conventional level for a confidence interval. We can calculate a confidence interval for any significance level. The more general formula for a $100(1 - \alpha)\%$ confidence interval for the mean is

$$\bar{Y} - t_{\alpha(2),df}\,SE_{\bar{Y}} < \mu < \bar{Y} + t_{\alpha(2),df}\,SE_{\bar{Y}}.$$

After 95%, the 99% confidence interval is the next most popular. Its principal advantage is that it provides better coverage than the 95% interval—that is, it covers the population mean in 99% of random samples.

Let's calculate a 99% confidence interval of the mean for the stalk-eye fly data. All of the numbers are the same as for the 95% interval, except that now we need $t_{0.01(2),8}$, because α is now $1 - 0.99 = 0.01$. To use Statistical Table C, we go to the row for $df = 8$ and then over to the column for $\alpha(2) = 0.01$. Confirm for yourself that this yields $t_{0.01(2),8} = 3.36$. Thus, the formula for the 99% confidence interval is

$$\bar{Y} - t_{0.01(2),df}\,SE_{\bar{Y}} < \mu < \bar{Y} + t_{0.01(2),\,df}\,SE_{\bar{Y}}$$

Applied to the stalk-eye fly data, we get

$$8.778 - (3.36 \times 0.133) < \mu < 8.778 + (3.36 \times 0.133),$$

which yields

$$8.33 \text{ mm} < \mu < 9.22 \text{ mm}.$$

The 99% confidence interval is broader than the 95% interval, because we have to include more possibilities to achieve the higher probability of covering the true mean.

11.3 **The one-sample *t*-test**

A whole series of methods has been developed to test hypotheses about the means of populations. Because these tests are often based on the *t*-distribution, many are called "*t*-tests." In this section we learn about the **one-sample *t*-test**, which is designed to compare the mean of a single sample with a value for the population mean proposed in the null hypothesis. The null hypothesis for a one-sample *t*-test is that the true mean is equal to a specific value, μ_0. The null and alternative hypothesis statements are

H_0: The true mean equals μ_0.
H_A: The true mean does not equal μ_0.

Suppose, for example, that we have a null hypothesis that the mean of a population is $\mu_0 = 0$. We can then use the one-sample *t*-test to determine whether the $\overline{Y}$ that we calculate from a sample is sufficiently different from zero to warrant rejection of the null hypothesis.

> The *one-sample t-test* compares the mean of a random sample from a normal population with the population mean proposed in a null hypothesis.

The test statistic for the one-sample *t*-test is *t* (no surprise), and it is calculated by

$$t = \frac{\overline{Y} - \mu_0}{\mathrm{SE}_{\overline{Y}}},$$

where μ_0 is the population mean proposed by the null hypothesis, $\overline{Y}$ is the sample mean, and $\mathrm{SE}_{\overline{Y}}$ is the sample standard error of the mean. The sampling distribution of this test statistic under H_0 is the *t*-distribution having $n - 1$ degrees of freedom. The *P*-value for the test can be computed by comparing the observed *t* with the Student's *t*-distribution. Example 11.3 shows you how.

Example 11.3   **Human body temperature**

Normal human body temperature, we have all learned, is 98.6 °F. But how well is this supported by data? Researchers obtained body temperature measurements on randomly chosen healthy people (Shoemaker 1996). The data for 25 of those people are as follows:

98.4	98.6	97.8	98.8	97.9
99.0	98.2	98.8	98.8	99.0
98.0	99.2	99.5	99.4	98.4
99.1	98.4	97.6	97.4	97.5
97.5	98.8	98.6	100.0	98.4

The body temperatures are not all identical to 98.6 °F, but are the measurements consistent with a population mean of 98.6 °F?

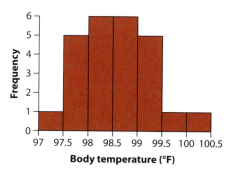

Figure 11.3-1 The frequency distribution of body temperatures in a sample of 25 individuals.

Body temperature is approximately normally distributed, as shown in Figure 11.3-1, so a one-sample *t*-test is appropriate. Our null and alternative hypotheses are

H_0: The mean human body temperature is 98.6 °F.
H_A: The mean human body temperature is different from 98.6 °F.

The null hypothesis is not arbitrary, because we are testing the common wisdom that the mean is $\mu_0 = 98.6$ °F. The test is two-tailed. A true mean higher than 98.6 °F or a mean lower than 98.6 °F would both lead to rejection of the null hypothesis.

The sample mean body temperature is

$$\overline{Y} = 98.524,$$

and the standard deviation is

$$s = 0.678.$$

The standard error is

$$\text{SE}_{\overline{Y}} = 0.678/\sqrt{25} = 0.136.$$

We can now use the test statistic *t* to measure how different the observed mean $\overline{Y}$ is from 98.6. If the sample mean perfectly matches the hypothesized value, then *t* would equal zero. Some difference is expected just by chance, but if the null hypothesis is true, then *t* should have a *t*-distribution with $n - 1$ *df*. If the difference between $\overline{Y}$ and 98.6 is too large, then the observed *t* will lie out in one of the tails of the *t*-distribution. Hence, by comparing this observed *t* to the *t*-distribution, we assess whether 98.6 is a reasonable fit to the data. If the null hypothesis is not a good fit, then we reject H_0 and conclude that the population mean μ does not equal 98.6.

The *t*-statistic is

$$t = \frac{\overline{Y} - \mu_0}{\text{SE}_{\overline{Y}}} = \frac{98.524 - 98.6}{0.136} = -0.56.$$

Under the null hypothesis, the sampling distribution of t is the t-distribution with $n - 1$ degrees of freedom. There are 25 individuals in our sample, so $df = 25 - 1 = 24$.

A perfect match between the null hypothesis and the data would mean that $\bar{Y} = \mu_0$, and t would equal zero. Our question becomes: Is $t = -0.56$ sufficiently different from zero that we should reject the null hypothesis?

The P-value is the probability of obtaining a result as extreme as, or more extreme than, $t = -0.56$, assuming that the null hypothesis is true. This probability is the area under the t-distribution shown in Figure 11.3-2. Both tails of the t-distribution are included in the probability because the test is two-tailed. The P-value is

$$P = \Pr[t < -0.56] + \Pr[t > 0.56].$$

Because the t-distribution is symmetric, this is the same as

$$P = 2\Pr[t > 0.56].$$

Using a computer, we find that

$$P = 0.58.$$

Since P is greater than 0.05, we do *not* reject the null hypothesis. Our results would not surprise someone who believed the null hypothesis.

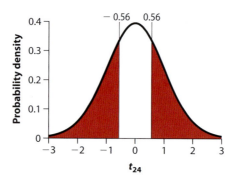

Figure 11.3-2 The t-distribution with 24 degrees of freedom. Shaded areas include all values less than -0.56 and greater than 0.56. These are the values as extreme, or more extreme, as the t-statistic calculated from the data.

The same conclusion is obtained if we use Statistical Table C for the t-distribution. The probability that we need, $\Pr[t > 0.56 \text{ or } t < -0.56]$, is not given directly in Statistical Table C, but we can find the critical values of t for different α values. Remember that the critical value of a test statistic marks off the point or points in the distribution that have a certain probability α in the tails of the distribution. Values of the test statistic that are further in the tails have P-values lower than α. In other words, if the value of t is further from zero than the critical value, then we can reject the null hypothesis. If the calculated value of t is closer to zero than the critical value, then we cannot reject the null hypothesis.

For the present example, we find the row in Statistical Table C corresponding to 24 degrees of freedom and move to the right to find the critical *t*-value corresponding to the column $\alpha(2) = 0.05$. Confirm for yourself that this value is

$$t_{0.05(2),24} = 2.06.$$

This critical value defines 5% in the tails of the *t*-distribution, with 2.5% in each tail for a two-tailed test (Figure 11.3-3). In other words, 5% of the area under the *t*-distribution with $df = 24$ falls outside of either -2.06 and 2.06. The *t*-value of -0.56 that we calculated from the data occurs inside this range. The observed *t*-statistic does not fall into one of the tails. Therefore, $P > 0.05$, and we fail to reject the null hypothesis.

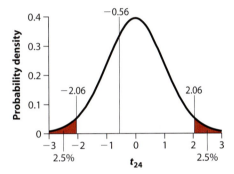

Figure 11.3-3 Five percent of the area under the curve of the *t*-distribution with 24 degrees of freedom lies in the two tails beyond -2.06 and 2.06. The *t*-value calculated from the body temperature data ($t = -0.56$) lies closer to that expected by the null hypothesis. Therefore, the null hypothesis is *not* rejected with these data.

At a significance level of $\alpha = 0.05$, we would not reject the null hypothesis based on this sample of body temperatures. In other words, these data are not very unlikely if the null hypothesis were true. With this sample, we cannot reject the view that mean body temperature in the sampled human population is 98.6 °F.

Does this result imply that the common wisdom about human body temperature is correct? Well, not necessarily: the null hypothesis might still be false, but we lacked sufficient power to detect the difference. What range of values for mean body temperature is most plausible given the data? To answer this question we calculated the 95% confidence interval for the mean using the methods described in Section 11.2. We obtained

$$98.24 \text{ °F} < \mu < 98.80 \text{ °F}.$$

Check the calculations for yourself. This 95% confidence interval for the mean puts fairly narrow bounds on the mean temperature in the population. The value 98.6 is within the 95% confidence interval of the mean for these data, but slightly different values for the mean body temperature between about 98.2 and 98.8 are also consistent with the data.

The effects of larger sample size—body temperature revisited

In the body-temperature study in Example 11.3, we used a sample with only 25 data points to do our calculations. A larger sample is available, with 130 individuals taken from the same population. Using this larger sample, the estimated mean was 98.25 °F with $s = 0.733$ °F. A t-test using this data set finds $t = -5.44$. (Check this value for yourself, for practice.) This t corresponds to $P = 0.000016$, so in reality it is very unlikely that the true value of human body temperature is the canonical 98.6 °F. This value is clearly rejected as the mean body temperature. The confidence interval for the mean, using the larger data set, is

$$98.12 < \mu < 98.38.$$

Why should the larger sample get a different answer than the subset? With a larger sample size, sampling error in the estimate of the mean decreases. Even though the smaller sample by chance had an estimated mean not very different from the larger sample, the larger sample led to an estimate with considerably narrower bounds. Thus, a larger sample is more likely to reject a false null hypothesis.[3] Assuming that the result from the analysis of the larger sample is true, our smaller sample gave us a Type II error.

11.4   The assumptions of the one-sample *t*-test

Methods described for calculating confidence intervals for the mean, and for testing a population mean using the one-sample t-test, make only two assumptions:

► The data are a random sample from the population. (This assumption is shared by every method of statistical inference in this book.)
► The variable is normally distributed in the population.

An excellent way to assess the assumption of normality is to examine the frequency distribution of the data using a histogram or other graphical method and look for skew, bimodality, or other departures from the normal distribution. Chapter 13 discusses methods to investigate this assumption in more detail.

Few variables in biology show an exact match to the normal distribution, but an exact match is not essential. Under certain conditions (discussed in Chapter 13), the t-test and the confidence interval calculations are robust to violations of the assumption of normality. A method is **robust** if the answer it gives is not sensitive to modest

---

[3] How did common wisdom get such a basic value wrong for so long? A hint comes from the fact that 98.6 °F corresponds exactly to 37 °C. The original measures of human body temperature were done on the Celsius scale and rounded to the nearest degree. The estimated mean value of body temperature, 98.25 °F, is 36.8 °C, which when rounded off gives 37 °C. Later, this rounded figure of 37 °C was converted back into Fahrenheit, yielding the erroneous 98.6 °F.

departures from the assumptions. Chapter 13 gives a more thorough account of this issue. Until then, we will consider only data that meet the assumption of normality reasonably well.

11.5 Estimating the standard deviation and variance of a normal population

Up to now, all of our attention has been focused on estimating and testing the population mean. But the standard deviation (or variance) is also interesting in many studies. For example, male stalk-eyed flies (see Example 11.2) often battle each other by pairing head to head and staring at each other for an extended period. The male with the longer eye stalks usually wins these staring matches, thus giving him greater access to females. When paired males have similar eye spans, the outcomes of the staring matches are difficult to predict. It is therefore interesting to know the standard deviation of eye span in the population, because it reflects the typical difference in eye span of any two males.

We must use the sample standard deviation s to estimate the population standard deviation σ. Can we say how precise our estimate s is? We can, provided that the data are from a population having a normal distribution.

Confidence limits for the variance

It is easiest first to work with the *variance*, the square of the standard deviation. The population parameter is σ^2 and the sample estimate is s^2. For the stalk-eyed fly data, the estimate of σ^2 is $s^2 = 0.1587$ mm^2.

To calculate a confidence interval for the sample variance, all we need to know is its sampling distribution. Theory tells us that if a variable Y has a normal distribution, then the sampling distribution of the quantity,

$$\chi^2 = (n - 1)s^2/\sigma^2,$$

is the χ^2 distribution with $n - 1$ degrees of freedom. We encountered the χ^2 distribution in Chapters 8 and 9, where we used it to approximate the sampling distribution of a goodness-of-fit test statistic. Here the same theoretical distribution is useful for calculating the precision of estimates of population variability.

A typical χ^2 distribution is shown in Figure 11.5-1. Notice that χ^2 is always zero or positive, and it extends all the way up to positive infinity. This fits with the properties of a sample variance, which cannot be less than zero but which can be indefinitely large. Notice also that, unlike the normal distribution, the χ^2 distribution is not symmetric around its mean.

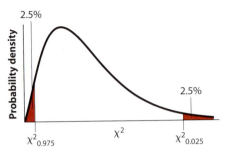

Figure 11.5-1 The χ^2 distribution with critical values for a 95% confidence interval indicated. Ninety-five percent of the area under the χ^2 curve falls between the two critical values, and 5% lies beyond them.

These features make it possible to determine confidence intervals for the variance. The $1 - \alpha$ confidence interval is

$$\frac{df\ s^2}{\chi^2_{\frac{\alpha}{2},df}} < \sigma^2 < \frac{df\ s^2}{\chi^2_{1-\frac{\alpha}{2},df}},$$

where the values $\chi^2_{\frac{\alpha}{2},df}$ and $\chi^2_{1-\frac{\alpha}{2},df}$ represent the critical values of the χ^2 distribution corresponding to the upper and lower tails in Figure 11.5-1. The area under the χ^2 curve to the right of $\chi^2_{\frac{\alpha}{2},df}$ is $\alpha/2$. The other $\alpha/2$ is to the left of $\chi^2_{1-\frac{\alpha}{2},df}$. Because the χ^2 distribution is not symmetric, we are forced to calculate the left and right tails separately. The critical values are available in Statistical Table A in the back of this book. Let's find the 95% confidence interval of the variance for eye span in male stalk-eyed flies. We have already calculated s^2 to be 0.1587. The number of degrees of freedom is just one less than the number of data points, so $df = 9 - 1 = 8$, as before. That only leaves finding the values of $\chi^2_{\frac{\alpha}{2},df}$ and $\chi^2_{1-\frac{\alpha}{2},df}$, which are $\chi^2_{0.025,8}$ and $\chi^2_{0.975,8}$, respectively, for the stalk-eyed fly study. Looking in Statistical Table A under eight degrees of freedom, the χ^2 value that has 2.5% to the right of it is $\chi^2_{0.025,8} = 17.53$, and the χ^2 value that has 2.5% probability to the left (i.e., 97.5% of the area to the right) of it is $\chi^2_{0.975,8} = 2.18$. Thus, the 95% confidence interval for the variance of this distribution is given by

$$\frac{df\ s^2}{\chi^2_{\frac{\alpha}{2},df}} < \sigma^2 < \frac{df\ s^2}{\chi^2_{1-\frac{\alpha}{2},df}}$$

$$\frac{8(0.1587)}{17.53} < \sigma^2 < \frac{8(0.1587)}{2.18}$$

$$0.0724 < \sigma^2 < 0.5824.$$

Any value of the population variance between 0.07 mm and 0.58 mm is reasonably plausible, based on these data. Thus, the true variance would be within the calculated interval 95% of the time that someone calculated a confidence interval for the variance of this population using this much data. This is quite a large range, proportionally, compared to the confidence interval for the mean, but it is more difficult to accurately estimate the variance than the mean.

The estimate of the variance (0.1587) is closer to the lower bound (0.0724) than the upper bound (0.5824) because the χ^2 distribution is asymmetrical. There is an equal chance, though, that a 95% confidence interval for the variance based on a random sample will fall below the true variance as that it will lie above.

Confidence limits for the standard deviation

The 95% confidence interval for the standard deviation can be obtained by taking the square roots of the upper and lower limits of the 95% confidence interval for the variance. Therefore, the 95% confidence interval for the standard deviation of eye span is given by

$$\sqrt{0.0724} < \sigma < \sqrt{0.5824}$$

or

$$0.27 < \sigma < 0.76$$

which surrounds the sample standard deviation of $s = 0.40$.

Assumptions

The assumptions of the method for calculating confidence intervals for the variance and standard deviation are the same as for confidence intervals for the mean—namely, the sample must be a random sample from the population, and the variable must have a normal distribution in the population. Unfortunately, the formulas for the confidence intervals for variance and standard deviation are very sensitive to the assumption of normality. The method is not robust to departures from this assumption.

11.6 Summary

▶ If a variable Y is normally distributed in the population with mean μ and we have a random sample of n individuals, then the sample means $\overline{Y}$ are also normally distributed, with mean equal to μ (the same as the true mean of Y) and standard error $\sigma_{\overline{Y}} = \sigma/\sqrt{n}$, where σ is the true standard deviation in the population.

▶ The estimated standard error of the distribution of sample means is $SE_{\overline{Y}} = s/\sqrt{n}$.

▶ If the population is normally distributed, then the standardized sample mean
$$t = \frac{\bar{Y} - \mu}{SE_{\bar{Y}}}$$ has a Student's t-distribution with $n - 1$ degrees of freedom.

▶ The t-distribution can be used to calculate a confidence interval for the mean.

▶ A one-sample t-test compares the sample mean with μ_0, a specific value for the population mean proposed in a null hypothesis. Under the null hypothesis that the population mean is equal to μ_0, the sampling distribution of the test statistic
$$t = \frac{\bar{Y} - \mu_0}{SE_{\bar{Y}}}$$ is a t-distribution with $n - 1$ degrees of freedom.

▶ The confidence interval for the variance is based on the χ^2 distribution. Take the square root of the limits of the confidence interval for the variance to yield the confidence interval for the standard deviation.

▶ The confidence intervals for mean and variance, as well as the one-sample t-test, assume that the data are randomly sampled from a population with a normal distribution.

11.7 Quick Formula Summary

Confidence interval for a mean

What does it assume? Individuals are chosen as a random sample from a population that is normally distributed.

Estimate: $\bar{Y}$

Parameter: μ

Degrees of freedom: $n - 1$

Formula: $\bar{Y} - t_{\alpha(2),df} SE_{\bar{Y}} \leq \mu \leq \bar{Y} + t_{\alpha(2),df} SE_{\bar{Y}}$

This formula calculates the $1 - \alpha$ confidence interval.

One-sample t-test

What is it for? Compares the sample mean of a numerical variable to a hypothesized value, μ_0.

What does it assume? Individuals are randomly sampled from a population that is normally distributed.

Test statistic: t

Distribution under H$_0$: The t-distribution with $n - 1$ degrees of freedom.

Formula: $t = \dfrac{\bar{Y} - \mu_0}{s/\sqrt{n}}$.

Confidence interval for variance

What does it assume? Individuals are chosen as a random sample from a normally distributed population.

Estimate: s^2

Parameter: σ^2

Degrees of freedom: $df = n - 1$

Formula: $\dfrac{df\, s^2}{\chi^2_{\frac{\alpha}{2},df}} \leq \sigma^2 \leq \dfrac{df\, s^2}{\chi^2_{1-\frac{\alpha}{2},df}}$

PRACTICE PROBLEMS

1. Measurements of the distance between the canine tooth and last molar for 35 wolf upper jaws were made by a researcher, and he found the 95% confidence interval for the mean to be 10.17 cm $< \mu <$ 10.47 cm and the 99% confidence interval to be 10.21 cm $< \mu <$ 10.44 cm. Without seeing the data, explain why he must have made a mistake.

2. Here are the data on wolf teeth from Practice Problem 1, in cm. There are 35 individuals in this data set. Assume that this variable is normally distributed.

    10.2, 10.4, 9.9, 10.7, 10.3, 9.7, 10.3, 10.7, 10.1, 10.6, 10.3, 10.0, 10.2, 10.1, 10.3, 9.9, 9.7, 10.6, 10.4, 10.1, 10.6, 10.3, 10.3, 10.5, 10.2, 10.2, 10.5, 10.1, 11.2, 10.5, 10.3, 10.0, 10.3, 10.7, 11.1

    a. What are the sample mean and the standard error of the mean for this data set? Provide an interpretation of this standard error.
    b. Find the 95% confidence interval for the mean.

    c. Find the 99% confidence interval for the variance.
    d. Find the 99% confidence interval for the standard deviation.

3. In the data set on wolf teeth (Practice Problem 2), each measurement was actually the average of two measurements made on the left and right sides of the jaw of an individual wolf. Thus, a total of 70 measurements were made. Could we use $n = 70$ when calculating confidence intervals for the mean and variance? Why or why not?

4. In a study of mutation and its consequences, researchers propagated a random sample of five lines taken from a clone of the RNA bacteriophage φ6 (a virus) (Burch and Chao 1999). After 100 generations of propagation, the researchers measured the fitness (reproductive rate) of each of the five lines relative to the baseline (starting) population. A value greater than zero represents a fitness greater than the baseline, whereas a negative number represents a fitness lower than the baseline. A value equal

to zero indicates no change. Fitness measurements after 100 generations were as follows:

$$0.063, \; -0.062, 0.064, \; -0.043, 1.34$$

a. With these data, test whether the mean fitness of lines changed over 100 generations.
b. Provide a 95% confidence interval for mean fitness change after 100 generations.
c. What assumptions are you making in parts (a) and (b)?

5. Polyandry is the name given to a mating system in which females mate with more than one male. This mating system leads to competition for fertilization between sperm of different males. The prediction has been made that males in polyandrous populations should evolve larger testes than males in monogamous populations (where females mate with only one male), because larger testes produce more sperm. To test this prediction, researchers carried out an experiment on eight separate lines of yellow dung flies (Hosken and Ward 2001). In four of these lines, each female was allowed to mate with three males (the polyandrous populations). In the other four lines, the females were only allowed to mate once. After 10 generations, the testes size (in cross-sectional area) was measured in each of these lines. The four monogamous lines had testes with areas of 0.83, 0.85, 0.82, 0.89 mm^2, and the polyandrous lines had testes areas of 0.96, 0.94, 0.99, and 0.91 mm^2.

a. Estimate the mean and standard deviation of the testes areas for both monogamous and polyandrous lines.
b. What is the standard error of the mean for each group?
c. What is the 95% confidence interval for the mean testes area in polyandrous lines?
d. What is the 99% confidence interval for the standard deviation of testes area among monogamous lines?

6. A goal of community ecology is to be able to predict who eats what. Community ecologists draw "food webs" that describe the relationships between predators and prey for all organisms in an area. A model has been developed to try to predict the relationships in food webs. This model predicts that a measure of the shape of the food webs called "discontinuity" will be zero. Researchers have collected information about the discontinuity scores for seven food webs, and the values for the discontinuity scores are given below (Cattin et al. 2004). Assume that discontinuity in all natural food webs has a normal distribution.

$$0.35, 0.08, 0.004, 0.08, 0.32, 0.28, 0.17$$

a. What is the 95% confidence interval for this discontinuity score?
b. Is the mean of the discontinuity score consistent with the model's prediction of zero? Carry out an appropriate hypothesis test.

7. Some researchers wanted to test whether a lesion in the hippocampus of rat brains affects their ability to solve a sequential memory test (Fortin et al. 2004). The researchers devised a test that asks the rats to choose between two odors, one of which came first in a series of odors presented earlier. Before they can test the effects of hippocampal lesions, though, they must ensure that normal rats are able to perform this task better than expected by chance. By chance, the rats should score 50% on the test. Seven normal rats were taken through the protocol, and their scores on the memory test were on average 68.4%, with a standard deviation of 7.1%. Do the rats show the ability to perform this task at levels better than that expected by chance? Make all necessary assumptions.

8. About 1/40 of all admissions to Provincial Hospital in Alotau, New Guinea, are injuries due to falling coconuts (Barss 1984). If coconuts weigh on average 3.2 kg, and the upper bound of the 95% confidence interval is 3.5 kg, what is the lower bound of this confidence interval? Assume a normal distribution of coconut weights.

9. The following give the confidence intervals of either the standard deviation or variance from different samples. Provide the confidence interval of the other measure of spread (i.e., either standard deviation or variance).

Standard deviation	Variance
$2.22 < \sigma < 4.78$	
	$425.4 < \sigma^2 < 678.8$
$36.4 < \sigma < 59.6$	
	$185.8 < \sigma^2 < 279.0$

10. Can a human swim faster in water or in syrup? It is unknown whether the increase in the friction of the body moving through the syrup (slowing the swimmer down) is compensated for by the increased power of each stroke. Finally, an experiment was done[4] in which researchers filled one pool with water mixed with syrup (guar gum) and another pool with normal water (Gettelfinger and Cussler 2004). They had 18 swimmers swim laps in both pools. The data are presented as the relative speed of each swimmer (speed in the syrupy pool divided by his or her speed in the water pool). If the syrup has no effect on swimming speed, then these relative measures should be one on average. The data, which are approximately normally distributed, are as follows:

1.08, 0.94, 1.07, 1.04, 1.02, 1.01, 0.95, 1.02, 1.08
1.02, 1.01, 0.96, 1.04, 1.02, 1.02, 0.96, 0.98, 0.99

$\Sigma(Y) = 18.21 \quad \Sigma(Y^2) = 18.4529$

a. Test whether swimming speed in syrup is the same as in water.

b. What is the 99% confidence interval for the relative swim speed in syrup?

ASSIGNMENT PROBLEMS

11. Astronauts increased in height by an average of approximately 40 mm (about an inch and a half) during the Apollo–Soyuz missions, due to the absence of gravity compressing their spines during their time in space. Does something similar happen here on Earth? An experiment supported by NASA measured the heights of six men immediately before going to bed, and then again after three days of bed rest (Styf et al. 1997). On average, they increased in height by 14 mm, with standard deviation 0.66 mm. Find the 95% confidence interval for the change in height after three days of bed rest.

12. Two different researchers measured the weight of two separate samples of ruby-throated hummingbirds from the same population. Each calculated a 95% confidence interval for the mean weight of these birds. Researcher 1 found the 95% confidence interval to be $3.12 \text{ g} < \mu < 3.48 \text{ g}$, while Researcher 2 found the 95% confidence interval to be $3.05 \text{ g} < \mu < 3.62 \text{ g}$.

a. How could the two researchers get different answers?

b. Which researcher most likely had the larger sample?

c. Can you be certain about your answer in part (b)? Why or why not?

13. In the Northern Hemisphere, dolphins swim predominantly in a counterclockwise direction while sleeping. A group of researchers wanted to know whether the same was true for dolphins in the Southern Hemisphere (Stafne and Manger 2004). They watched eight sleeping dolphins and recorded the percentage of time that the dolphins swam clockwise. Assume that this is a random sample and that this variable has a normal distribution. These data are as follows:

77.7, 84.8, 79.4, 84.0, 99.6, 93.6, 89.4, 97.2

a. What is the mean percentage of clockwise swimming for Southern Hemisphere dolphins?

b. What is the 95% confidence interval for the mean time going clockwise in the Southern Hemisphere dolphins?

---

4   After a rather large number of permits were applied for and granted.

c. What is the 99% confidence interval for the mean time going clockwise in the Southern Hemisphere dolphins?

d. What is the best estimate of the standard deviation of the percentage clockwise swimming?

e. What is the median of the percentage of clockwise swimming?

f. Test the null hypothesis that Southern Hemisphere dolphins spend half of their time while asleep swimming clockwise.

14. The mating habits of three-spined stickleback (a fish) have been studied intensively. One experiment examined whether fish are more likely to mate with a member of the opposite sex that was similar in body size, rather than with fish that were different in size (McKinnon et al. 2004). The mating preferences were measured in nine different fish populations, and the preference was measured by an index that is zero if the population shows no preference for mating by size, positive if the population contains fish that prefer to mate with fish of a different size, and negative if the fish mate preferentially with individuals of the same size. Notice that the independent data points here are the indices for each fish population. The nine indices are as follows:

$-32.0, -29.8, -40.6, -90.8, -29.2, -28.8,$
$-78.4, -59.2, -74.3$

Test the hypothesis that, on average, sticklebacks do not mate differently by size. What can you conclude from these data?

15. A new microgram scale is bought by researchers in a lab. Before using it for new results, they want to ensure that the readings on the scale are accurate. They obtain a 10-μg standard weight, and then weigh this standard 30 times. Their measures were approximately normally distributed, with an average of 10.01 μg and standard deviation 0.2 μg. Test whether the scale is accurate, using these data.

16. Pit vipers (including rattlesnakes) have a pit organ located halfway between their eyes and nostrils. These organs have long been known to function in detecting the body heat of unlucky warm-blooded prey (mammals), but it has recently been shown that the snakes also can use the pit organ to detect cooler spots to help in their thermoregulation.

Using the western diamondback rattlesnake, researchers determined that snakes had on average a 73% chance of moving into the right half of a Y-maze that was cooled to 30 °C, instead of the left half which was at 40 °C (Krochmal et al. 2004). This preference was extremely well supported, but is this pattern a preference for the right side of the maze (that just happened to be cooled), or was it a direct response to the difference in heat? To test this, five snakes were put into the same maze several times individually, and the percentage of trials in which they turned right was recorded when both sides were heated to 40 °C. The average of the percentages for the five snakes was 47%, and the standard deviation was 13%. Is there a preference for the right side of the maze when the temperature is equalized? Test the null hypothesis that the mean percentage of snakes that would move right is 50% in this population. Assume that the distribution of scores was normal.

17. In Seychelles warblers, young adult birds known as subordinates sometimes hang around the territories of older birds. Sometimes these subordinates help feed the offspring of the older birds, and sometimes not. In one study, subordinate birds that did *not* help feed the offspring were genetically assessed to discover whether they were related to the offspring (Richardson et al. 2003). A "relatedness" score was assigned to each, with a value of zero meaning no relationship to the offspring. Out of five subordinates examined, the mean relatedness was -0.05 with standard deviation 0.45.

a. What is the 95% confidence interval for relatedness of unhelpful subordinates to the offspring?

b. Is the relatedness of unhelpful subordinates significantly different from zero? Carry out the appropriate hypothesis test. Assume that relatedness is normally distributed.

12

Comparing two means

Biological data are often gathered to compare different group or treatment means. Do female hyenas differ from male hyenas in body size? Do patients treated with a new drug live significantly longer than those treated with the old drug? In Chapter 8, we presented methods to compare proportions of a categorical variable between different groups. In this chapter, we develop procedures for comparing means of a numerical variable between two treatments[1] or groups. We also include methods to compare two variances. All of the methods in the current chapter assume that the measurements are normally distributed in the populations.

We show analyses for two different study designs. In the *paired* design, both treatments have been applied to every sampled unit, such as a subject or a plot, at different times or on different sides. In the *two-sample* design, each group constitutes an independent random sample of individuals. In both cases, we make use of the *t*-distribution to calculate confidence intervals for the mean difference and test hypotheses about means. However, the two different study designs require alternative methods for analysis. We elaborate the reasons in Section 12.1.

---

[1]   We will use the term "treatments" in a broad sense to refer to different states or conditions experienced by subjects, not just to formal experimental treatments.

12.1   Paired sample versus two independent samples

There are two study designs to measure and test differences between the means of two treatments. To describe them, let's use an example: does clear-cutting a forest affect the number of salamanders present? Here we have two treatments (clear-cutting/no clear-cutting), and we want to know if the mean of a numerical variable (the *number of salamanders*) differs between them. "*Clear-cut*" is the treatment of interest, and "*no clear-cut*" is the control. This is the same as asking whether these two variables, *treatment* and *salamander number*, are associated.

There are two ways to design this kind of a study, a two-sample design (see the left panel in Figure 12.1-1) and a paired design (see the right panel in Figure 12.1-1). In the *two-sample* design, we take a random sample of forest plots from the population and then randomly assign each plot to either the clear-cut treatment or the no clear-cut treatment. In this case, we end up with two independent samples, one from each group. The difference in the mean number of salamanders between the clear-cut and no clear-cut areas estimates the effect of clear-cutting on salamander number.[2]

In the *paired* design, we take a random sample of forest plots and clear-cut a randomly chosen half of each plot, leaving the other half untouched. Afterward, we

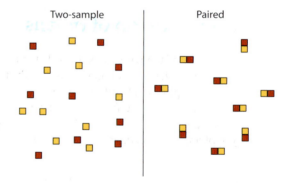

Figure 12.1-1 Alternative designs to compare two treatments. A two-sample design is on the left, whereas the paired design is on the right. Freestanding blocks represent sampling units, such as plots. The red and yellow areas represent different treatments (e.g., clear-cut and not clear-cut). In the paired design (*right*), both treatments are applied to every unit. In the two-sample design (*left*), different treatments are applied to separate units.

---

[2]   We are assuming that you have persuaded an enlightened forest company to go along with this experiment. It is more likely that the forest company has already clear-cut some areas and not others and you must compare the two treatments after the fact. Both the experimental approach and the observational design yield a measure of the association between clear-cutting and salamander number, but only the experimental study can tell us whether clear-cutting is the *cause* of any difference (Chapter 14). The study design issues are the same, however: whether to choose forest plots randomly from each treatment (the two-sample design) or to randomly choose forest plots that straddle both treatments, making it possible to compare cut and uncut sites that are side by side (the paired design).

count the number of salamanders in each half. The mean difference between the two sides estimates the effect of clear-cutting.

Each of these two experimental designs has benefits, and both address the same question. However, they differ in an important way that affects how the data are analyzed. In the paired design, measurements on adjacent plot-halves are not independent. This is because they are likely to be similar in soil, water, sunlight, and other conditions that affect the number of salamanders. As a result, we must analyze paired data differently than when every plot is independent of all the others, as in the case of the two-sample design.

> In the *paired* design, both treatments are applied to every sampled unit. In the *two-sample* design, each treatment group is composed of an independent random sample of units.

Paired designs are usually more powerful than unpaired designs, because they control for a lot of the extraneous variation between plots or sampling units that sometimes obscures the effects we are looking for. Very often, though, it is just not possible to collect data in a paired format.

12.2   Paired comparison of means

The main advantage of the paired design is that it reduces the effects of variation among sampling units that has nothing to do with the treatment itself. For example, forest or agricultural plots likely differ greatly in their local environmental features, and this variation can make it difficult to detect a difference between the effects of two treatments applied to separate plots. The paired design reduces the impact of this variation by applying both treatments to different sides of all plots. Other examples of paired study designs include

- ▶ comparing patient weight before and after hospitalization,
- ▶ comparing fish species diversity in lakes before and after heavy metal contamination,
- ▶ testing effects of sunscreen applied to one arm of each subject compared with a placebo applied to the other arm,
- ▶ testing effects of smoking in a sample of patients, each of which is compared with a nonsmoker closely matched by age, weight, and ethnic background, and
- ▶ testing effects of socioeconomic condition on dietary preferences by comparing identical twins raised in separate adoptive families that differ in their socioeconomic conditions.

The last two examples (the effects of smoking and socioeconomic condition) show that even two unique individuals can constitute a pair if they are similar due to shared physical, environmental, or genetic characteristics.

In a paired study design, the sampling unit is the pair. We must therefore reduce the two measurements made on each pair down to a single number—that is, the *difference* between the two measurements made on each sampling unit (e.g., patient, lake, subject, matched-pair, or twins). This step correctly yields only as many data points as there are randomly sampled units. Thus, if 20 individuals are grouped into 10 pairs, there are 10 measurements of the difference between the two treatments. Ten would be the sample size. We can estimate and test the effect of treatment using the mean of the differences.

> Paired measurements are converted to a single measurement by taking the difference between them.

Estimating mean difference from paired data

We now describe how to estimate mean differences and calculate confidence intervals for those estimates. This method assumes that we have a random sample of pairs and that the differences between members of each pair have a normal distribution. We'll use Example 12.2 to show the concepts and the calculation.

Example 12.2   **So macho it makes you sick?**

In many species, males are more likely to attract females if the males have high testosterone levels. Are males with high testosterone paying for this extra mating success in other ways? One hypothesis is that males with high testosterone might be less able to fight off disease—that is, their high levels of testosterone might reduce their immunocompetence. To test this idea, Hasselquist et al. (1999) experimentally increased the testosterone levels of 13 male red-winged blackbirds by implanting a small permeable tube filled with testosterone. They measured the antibody levels in each bird's blood serum both before and after the implant. The antibody levels were measured optically, in units of log 10^{-3} optical density per minute ($\ln[\text{mOD}/\text{min}]$).

The graph in Figure 12.2-1 shows that the points scatter around the line of equality, and that there is considerable variation among birds in their natural antibody level. What is the mean difference between the two treatments? We can address this question by constructing a confidence interval for the mean change in antibody production.

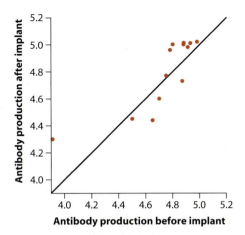

Figure 12.2-1 Antibody production (ln[mOD/min]) of 13 red-winged blackbirds before and after testosterone implants. The black line is the line expected if the antibody production were equal in the two treatments (the line shows the equality of the two measures: *AFTER = BEFORE*).

Table 12.2-1 Antibody production in blackbirds before and after testosterone implants. Each bird is represented by a single row and has a pair of antibody measurements; *d* is the difference ("before" minus "after") between the pair of measurements.

Male identification number	Before implant: Antibody production (ln[mOD/min])	After implant: Antibody production (ln[mOD/min])	*d*
1	4.65	4.44	0.21
4	3.91	4.30	−0.39
5	4.91	4.98	−0.07
6	4.50	4.45	0.05
9	4.80	5.00	−0.20
10	4.88	5.00	−0.12
15	4.88	5.01	−0.13
16	4.78	4.96	−0.18
17	4.98	5.02	−0.04
19	4.87	4.73	0.14
20	4.75	4.77	−0.02
23	4.70	4.60	0.10
24	4.93	5.01	−0.08

The first step is to calculate the difference in antibody production for each male. The data, listed in Table 12.2-1, consist of a pair of measurements for each male: antibody production *before* the testosterone implant and antibody production *after* the implant. We calculated the difference between measurements within each male i as

$$d_i = \text{(antibody production of male } i \text{ before)} - \text{(antibody production of male } i \text{ after)}.$$

For bird 1, for example, $d_1 = 4.65 - 4.44 = 0.21$. It doesn't much matter whether we subtract the "before" measurement from the "after" measurement or vice versa (as in Table 12.2-1), but it is critical that we calculate the differences the same way for each individual. A histogram of the resulting differences is shown in Figure 12.2-2.

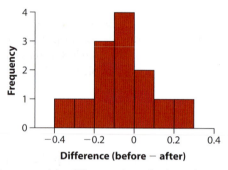

Figure 12.2-2 A histogram of the differences in antibody production in male blackbirds before and after testosterone implants.

We then find the sample mean difference (call it $\bar{d}$), the sample standard deviation of the differences (s_d), and the sample size (n):

$$\bar{d} = -0.056,$$
$$s_d = 0.159, \text{ and}$$
$$n = 13.$$

The confidence interval for the mean of a paired difference is generated in the same way as the confidence interval for any other mean (see Section 11.2). The confidence interval for the mean difference (μ_d) is

$$\bar{d} - t_{\alpha(2),df}\text{SE}_{\bar{d}} < \mu_d < \bar{d} + t_{\alpha(2),df}\text{SE}_{\bar{d}},$$

where

$$\text{SE}_{\bar{d}} = s_d/\sqrt{n}$$

is the standard error of the mean difference. Note that n is the number of *pairs*, not the total number of measurements, because pairs are the independent sampling unit. For this reason, we are carrying out the analysis on the differences.

For the blackbird data, we calculate

$$SE_{\bar{d}} = \frac{0.159}{\sqrt{13}} = 0.044.$$

With $n = 13$, we have $df = 12$, so we look in Statistical Table C to find $t_{0.05(2),12} = 2.18$. Thus, the 95% confidence interval for the mean difference between antibody production before and after testosterone implants is

$$\bar{d} - t_{\alpha(2),df}\, SE_{\bar{d}} \le \mu_d \le \bar{d} + t_{\alpha(2),df}\, SE_{\bar{d}}$$

$$-0.056 - 2.18(0.044) \le \mu_d \le -0.056 + 2.18(0.044)$$

$$-0.152 \le \mu_d \le 0.040.$$

In other words, we can be 95% confident that the true mean difference falls between -0.152 and 0.040 ln[mOD/min]. While this span includes zero, it is also consistent with the possibility of a modest increase in immunocompetence following testosterone implant. A larger sample of individuals would be needed to narrow the interval further.

Paired *t*-test

The **paired *t*-test** is used to test a null hypothesis that the mean difference of paired measurements equals a specified value. The method tests differences when both treatments have been applied to every sampling unit and the data are therefore paired.

The paired *t*-test is straightforward once we reduce the two measurements made on each pair down to a single number: the difference between the two measurements. This difference is then analyzed in the same way as a regular one-sample *t*-test, as described in Chapter 11. We'll continue to analyze the blackbird testosterone data from Example 12.2 to ask whether the antibody production in blackbirds is different after testosterone implants than before.

Because the data are paired, a paired *t*-test is appropriate, with before testosterone and after testosterone representing the two "treatments." The null and alternative hypotheses are

H_0: The mean change in antibody production after testosterone implants was zero.
H_A: The mean change in antibody production after testosterone implants was not zero.

The alternative hypothesis is two-tailed, because either greater or lesser antibody production after the testosterone implants would reject the null hypothesis. These hypothesis statements could also be written as

H_0: $\mu_d = 0$

and

$$H_A: \mu_d \neq 0,$$

where μ_d is the mean difference between treatments.

Again, the first step is to calculate the difference in antibody production before and after the implants, which we have already done in Table 12.2-1. We then need the sample mean ($\bar{d} = -0.056$) and standard error ($SE_{\bar{d}} = 0.044$) of the differences, which we calculated in the previous subsection.

From here on, the paired t-test is identical to a one-sample t-test on the differences. We can calculate the t-statistic as

$$t = \frac{\bar{d} - \mu_{d0}}{SE_{\bar{d}}},$$

where μ_{d0} is the population mean of d proposed by the null hypothesis (0 in this example), and $SE_{\bar{d}}$ is the sample standard error of d. Under the null hypothesis, this t-statistic has a t-distribution with $df = n - 1$ degrees of freedom.

When this formula is applied to the blackbird testosterone data,

$$t = \frac{-0.056 - 0}{0.044} = -1.27.$$

This test statistic has $df = 13 - 1 = 12$. The P-value for the test is

$$P = \Pr[t_{12} < -1.27] + \Pr[t_{12} > 1.27] = 2 \Pr[t_{12} > 1.27].$$

Using a computer, we calculated this probability under a t-distribution with 12 df to be

$$P = 0.23.$$

P is greater than 0.05, so we do *not* reject the null hypothesis that $\mu_d = 0$ with these data.

If we did this test without a computer handy, we could use Statistical Table C instead. In that case, we would find that the critical value for a two-tailed test with $\alpha = 0.05$ and 12 degrees of freedom is

$$t_{0.05(2),12} = 2.18.$$

Because $t = -1.27$ does not fall outside the critical limits of -2.18 and 2.18, we know that $P > 0.05$, and we do not reject the null hypothesis.

The mean difference that we saw (-0.056 ln[mOD/min]) is in the direction of higher immune system function after testosterone implants, but we cannot reject the null hypothesis that the testosterone has no effect on immunocompetence. The confidence interval that we calculated in the previous subsection indicates that a broad range of values is consistent with these data. On the basis of this data set, we do not reject the null hypothesis, but we would want to pursue further study on the problem to resolve the issue more precisely.

Assumptions

The method for calculating a confidence interval for a paired difference and the paired *t*-test make the same assumptions as the single sample methods described in Chapter 11:

▶ The sampling units are randomly sampled from the population.
▶ The paired differences have a normal distribution in the population.

The analysis makes no assumptions about the distribution of either of the single measurements made on each sampling unit. These measurements can have any distribution, as long as the difference between the two measurements is approximately normally distributed.

12.3   Two-sample comparison of means

We now present methods to analyze the difference between the means of two treatments or groups in the case of a *two-sample* design. In a two-sample design, the two treatments are applied to separate, independent sampling units. We can illustrate the process using Example 12.3.

| Example 12.3 | **Spike or be spiked** |

The horned lizard *Phrynosoma mcalli* has many unusual features, including the ability to squirt blood from its eyes. The species is named for the fringe of spikes surrounding the head. Herpetologists recently tested the idea that long spikes help protect horned lizards from being eaten, by taking advantage of the gruesome but convenient behavior of one of their main predators, the loggerhead shrike, *Lanius ludovicianus*. The loggerhead shrike is a small predatory bird that skewers its victims on thorns or barbed wire, to save for later eating.

The researchers identified the remains of 30 horned lizards that had been killed by shrikes and measured the lengths of their horns (Young et al. 2004). As a comparison group, they measured the same trait on 154 horned lizards that were still alive and well. They compared the mean horn lengths of the dead lizards with those of the living lizards. Histograms of the horn lengths of the two groups are shown in Figure 12.3-1. Summary statistics are listed in Table 12.3-1.

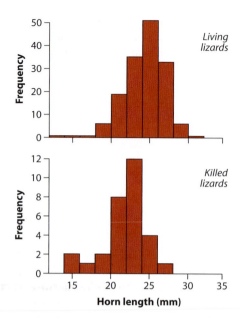

Figure 12.3-1 The frequency distribution of horn lengths for live and killed horned lizards. There are $n_1 = 154$ live lizards and $n_2 = 30$ killed lizards.

Table 12.3-1 Summary statistics for lizard horn lengths.

Lizard group	Sample mean $\bar{Y}$ (mm)	Sample standard deviation s (mm)	Sample size n
Living	24.28	2.63	154
Killed	21.99	2.71	30

The lizards that were killed by shrikes are *different individuals* than the living lizards. They are not paired in any way; instead, each treatment (living or dead) is represented by a separate sample of lizards, belonging to different groups. Therefore, we must analyze the differences using two-sample methods.

Figure 12.3-1 and Table 12.3-1 suggest that the dead lizards have shorter horns than the living lizards on average, as might be expected if shrikes avoid the longest horns. Next, we calculate a confidence interval for the difference, and we test whether the difference is real.

Confidence interval for the difference between two means

How much longer on average are the horns of the surviving lizards? The best estimate of the difference between two population means is the difference between the sample means, $\overline{Y}_1 - \overline{Y}_2$. Here, we will use $\overline{Y}_1$ to refer to the sample mean of the live lizards, and $\overline{Y}_2$ to refer to the sample mean of the dead lizards. Which group we call 1 and which we call 2 is arbitrary, but we have to be consistent with the labels throughout. We use subscripts to indicate the population or sample that a value comes from.

The method for confidence intervals makes use of the fact that, if the variable is normally distributed in both populations, then the sampling distribution for the *difference* between the sample means is also normal. Thus, the Student's *t*-distribution will be very helpful in describing the sampling properties of the standardized difference.

First, we will need the standard error of $\overline{Y}_1 - \overline{Y}_2$, which is given by

$$\mathrm{SE}_{\overline{Y}_1-\overline{Y}_2} = \sqrt{s_p^2\left(\frac{1}{n_1} + \frac{1}{n_2}\right)},$$

where

$$s_p^2 = \frac{df_1 s_1^2 + df_2 s_2^2}{df_1 + df_2}.$$

The quantity s_p^2 is called the **pooled sample variance**. It is a weighted average[3] of the sample variances s_1^2 and s_2^2 (the squared standard deviations) of the two groups. The confidence interval formula assumes that the standard deviations (and variances) of the two populations are the same. The pooled variance s_p^2 uses the information from both samples to get the best estimate of this common population variance. The df_1 and df_2 terms refer to the degrees of freedom for the variances of the two samples:

$$df_1 = n_1 - 1$$

and

$$df_2 = n_2 - 1,$$

where n_1 and n_2 are the sample sizes from the two populations.

> The *pooled sample variance* s_p^2 is the average of the variances of the samples weighted by their degrees of freedom.

Because the sampling distribution of $\overline{Y}_1 - \overline{Y}_2$ is normal, the sampling distribution of the following standardized difference has a Student's *t*-distribution:

$$t = \frac{(\overline{Y}_1 - \overline{Y}_2)}{\mathrm{SE}_{\overline{Y}_1-\overline{Y}_2}}$$

---

[3] A "weighted average" may count each group differently. In this case, each group is weighted by its degrees of freedom, so that the group with the larger sample size contributes proportionately more to the calculated average.

with total degrees of freedom equal to

$$df = df_1 + df_2 = n_1 + n_2 - 2.$$

From these two formulas we can calculate the confidence interval for the difference between two population means:

$$(\bar{Y}_1 - \bar{Y}_2) - t_{\alpha(2),df}\, SE_{\bar{Y}_1 - \bar{Y}_2} < \mu_1 - \mu_2 < (\bar{Y}_1 - \bar{Y}_2) + t_{\alpha(2),df}\, SE_{\bar{Y}_1 - \bar{Y}_2}.$$

Let's use the horned-lizard data to calculate the 95% confidence interval for the difference in horn length between the two groups of lizards. The place to start is with the summary statistics listed in Table 12.3-1. The difference in the means is

$$\bar{Y}_1 - \bar{Y}_2 = 24.28 - 21.99 = 2.29 \text{ mm.}$$

After that, we start from the bottom—we'll need the pooled variance to calculate the standard error of $\bar{Y}_1 - \bar{Y}_2$, and we'll need the standard error to get the confidence interval for $\mu_1 - \mu_2$. The pooled sample variance, which depends on the number of degrees of freedom ($df_1 = 153$ and $df_2 = 29$), is

$$
\begin{aligned}
s_p^2 &= \frac{df_1 s_1^2 + df_2 s_2^2}{df_1 + df_2} \\[6pt]
&= \frac{153(2.63^2) + 29(2.71^2)}{153 + 29} \\[6pt]
&= 6.98.
\end{aligned}
$$

The standard error of the difference between the two means is then

$$SE_{\bar{Y}_1 - \bar{Y}_2} = \sqrt{s_p^2\left(\frac{1}{n_1} + \frac{1}{n_2}\right)} = \sqrt{6.98\left(\frac{1}{154} + \frac{1}{30}\right)} = 0.527.$$

One common mistake is to forget that the standard error equation uses the sample sizes, n_1 and n_2, not the number of degrees of freedom, in the denominators.

There are $154 + 30 - 2 = 182$ degrees of freedom in total, so

$$t_{0.05(2),182} = 1.97.$$

(There is no row in Statistical Table C for $df = 182$, but $t_{0.05(2),df} = 1.97$ for both $df = 180$ and $df = 185$, and so to this order of precision, $t_{0.05(2),182} = 1.97$ as well.)

Plugging these quantities into the formula, the 95% confidence interval for the difference in mean horn length between the living and dead lizards is

$$(\bar{Y}_1 - \bar{Y}_2) - t_{\alpha(2),df}\, SE_{\bar{Y}_1 - \bar{Y}_2} < \mu_1 - \mu_2 < (\bar{Y}_1 - \bar{Y}_2) + t_{\alpha(2),df}\, SE_{\bar{Y}_1 - \bar{Y}_2}$$

$$2.29 - 1.97(0.527) < \mu_1 - \mu_2 < 2.29 + 1.97\,(0.527)$$

$$1.25 < \mu_1 - \mu_2 < 3.33.$$

Thus, the 95% confidence interval for $\mu_1 - \mu_2$ is from 1.25 to 3.33 mm. We can be reasonably confident that the surviving lizards have longer horns than the lizards killed by shrikes, by an amount somewhere between about one and three millimeters.

Two-sample *t*- test

The *two-sample t-test* is the simplest method to compare the means of a numerical variable between two independent groups. Its most common use is to test the null hypothesis that the means of two populations are equal (or, equivalently, that the difference between the means is zero):

$$H_0: \mu_1 = \mu_2$$

and

$$H_A: \mu_1 \neq \mu_2,$$

where μ_1 is the population mean for the first of the two populations and μ_2 is the mean for the second population. It doesn't matter which population we designate as population 1 and which as population 2 as long as we are consistent throughout the analysis. The two-sample *t*-test uses the following test statistic based on the observed difference between the sample means:[4]

$$t = \frac{(\bar{Y}_1 - \bar{Y}_2)}{SE_{\bar{Y}_1 - \bar{Y}_2}}.$$

Provided that the assumptions are met, this *t*-statistic has a *t*-distribution under the null hypothesis with $n_1 + n_2 - 2$ degrees of freedom. The denominator of this formula, $SE_{\bar{Y}_1 - \bar{Y}_2}$, is the standard error of the difference between the two sample means. This standard error is the same as that shown previously when discussing confidence intervals (p. 289):

$$SE_{\bar{Y}_1 - \bar{Y}_2} = \sqrt{s_p^2 \left(\frac{1}{n_1} + \frac{1}{n_2} \right)},$$

The pooled sample variance, s_p^2, was defined in the previous subsection. The null hypothesis is tested by comparing the observed *t*-value to the theoretical *t*-distribution with

$$df = df_1 + df_2 = n_1 + n_2 - 2$$

degrees of freedom.

To apply the two-sample *t*-test to the horned lizard study, begin by writing the null and alternative hypotheses.

> H_0: Lizards killed by shrikes and live lizards do not differ in mean horn length (i.e., $\mu_1 = \mu_2$).

---

[4]   More generally, the null hypothesized value for the difference between the two population means can be any number: $H_0: (\mu_1 - \mu_2) = (\mu_1 - \mu_2)_0$. In this case we would calculate the test statistic as $t = \dfrac{(\bar{Y}_1 - \bar{Y}_2) - (\mu_1 - \mu_2)_0}{SE_{\bar{Y}_1 - \bar{Y}_2}}$ instead.

H$_A$: Lizards killed by shrikes and live lizards differ in mean horn length (i.e., $\mu_1 \neq \mu_2$).

The alternative hypothesis is two-tailed.

Let's apply these equations to the lizard data to perform the two-sample t-test. Once again, the difference between the sample means of the two groups of lizards is

$$\bar{Y}_1 - \bar{Y}_2 = 2.29 \text{ mm}.$$

The standard error of the difference, computed in the previous subsection, is

$$\text{SE}_{\bar{Y}_1 - \bar{Y}_2} = 0.527.$$

Now we can calculate the test statistic, t. According to the null hypothesis, the two means are equal; that is, $\mu_1 - \mu_2 = 0$. With this information,

$$t = \frac{(\bar{Y}_1 - \bar{Y}_2)}{\text{SE}_{\bar{Y}_1 - \bar{Y}_2}} = \frac{2.29}{0.527} = 4.35.$$

We need to compare this test statistic with the distribution of possible values for t if the null hypothesis were true. The appropriate null distribution is the t-distribution, and it will have $df_1 + df_2 = 153 + 29 = 182$ degrees of freedom. The P-value is then

$$P = 2 \Pr[t > 4.35].$$

Using a computer, we find that this t-value corresponds to

$$P = 0.000023.$$

Since $P < 0.05$, we reject the null hypothesis.

We reach the same conclusion using Statistical Table C. The critical value for $\alpha = 0.05$ for a t-distribution with 182 degrees of freedom is

$$t_{0.05(2),182} = 1.97.$$

The $t = 4.35$ calculated from these data is much further in the tails of the distribution than this critical value, so we reject the null hypothesis. In fact, the t calculated for these data is further in the tail of the distribution than all values given in Statistical Table C, including for $\alpha = 0.0002$. From this information we may conclude that $P < 0.0002$. Based on these studies, there is a difference in the horn length of lizards eaten by shrikes, compared with live lizards. It is possible, therefore, that shrikes avoiding the lizards with the longest horns is the reason for the difference.

Assumptions

The two-sample confidence interval for a difference in means and the two-sample t-test are based on the following assumptions:

▶ Each of the two samples is a random sample from its population.
▶ The numerical variable is normally distributed in each population.
▶ The standard deviation (and variance) of the numerical variable is the same in both populations.

We've heard the first two assumptions before—they are required for the one-sample t-test—but the third assumption is new. In Section 12.7 we discuss ways to test the assumption of equal standard deviations.

The two-sample methods we have just discussed are fairly robust to violations of this assumption. With moderate sample sizes (i.e., n_1 and n_2 both greater than 30), the methods work well, even if the standard deviations in the two groups differ by as much as threefold, as long as the sample sizes of the two groups are approximately equal. If there is more than a threefold difference in standard deviations, or if the sample sizes of the two groups are very different, then the two-sample t-test should not be used.

The two-sample t-test is also robust to minor deviations from the normal distribution, especially if the two distributions being compared are similar in shape. For example, in Example 12.3, the distribution of horn size is unlikely to be perfectly normal in either group, but in both cases the measurements do not differ greatly from a normal distribution. The t-test is robust to these kinds of minor deviations from normality. The robustness of the t-test improves as the sample sizes get larger. In Chapter 13 we discuss at greater length the robustness of the t-test to deviations from normality.

A two-sample test when standard deviations are unequal

An important assumption of the two-sample t-test is that the standard deviations of the two populations are the same. If this assumption cannot be met, then the **Welch's approximate t-test** should be used instead of the two-sample t-test. The Welch's t-test is similar to the two-sample t-test except that the standard error and degrees of freedom are computed differently. (See the Quick Formula Summary in Section 12.9 for the equations.) An example using the Welch's t-test is discussed in Section 12.4.

> *Welch's t-test* compares the means of two groups and can be used even when the variances of the two groups are not equal.

In Chapter 13, we describe additional ways to rescue situations in which the assumptions of normality and equal standard deviations are not met.

12.4   Using the correct sampling units

One of the assumptions of the t-test, as with all other tests in this book, is that the samples being analyzed are random samples. Quite often, multiple measurements have been taken on each sampling unit. Because measurements made on the same sampling unit are not independent, they require special handling. One solution is to summarize the data for each sampling unit by a single measurement. One of the biggest challenges when comparing groups is to identify correctly the independent units of replication. This decision determines not only what method is used, but it might also affect the conclusion reached, as Example 12.4 illustrates.

Example 12.4 ## So long; thanks to all the fish

One of the greatest threats to biodiversity is the introduction of alien species from outside their natural range. These introduced species often have fewer predators or parasites in the new area, so they can increase in numbers and out-compete native species. Sometimes these species are introduced accidentally, but often they are introduced intention-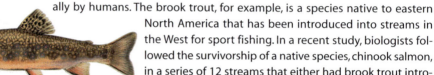ally by humans. The brook trout, for example, is a species native to eastern North America that has been introduced into streams in the West for sport fishing. In a recent study, biologists followed the survivorship of a native species, chinook salmon, in a series of 12 streams that either had brook trout introduced or did not (Levin et al. 2002). Their goal was to determine whether the presence of brook trout affected the survivorship of the salmon. In each stream, they released a number of tagged juvenile chinook and then recorded whether or not each chinook survived over one year. Table 12.4-1 summarizes the data.

Table 12.4-1. The numbers and proportion of chinook released and surviving in streams with and without brook trout. The study included 12 streams in total.

Brook trout	Number of salmon released	Number of salmon surviving	Proportion surviving
Present	820	166	0.202
Present	960	136	0.142
Present	700	153	0.219
Present	545	103	0.189
Present	769	173	0.225
Present	1001	188	0.188
Absent	467	180	0.385
Absent	959	178	0.186
Absent	1029	326	0.317
Absent	27	7	0.259
Absent	998	120	0.120
Absent	936	135	0.144
Total	9211	1865	

In all, 9211 salmon were released, of which 1865 survived and 7346 did not. A quick tally of the fish numbers by treatment yields the 2×2 shown in Table 12.4-2.

Table 12.4-2 Number of salmon surviving and not surviving in each trout treatment.

	Trout absent	Trout present
Survived	946	919
Did not survive	3470	3876

We would like to test whether the proportion of salmon surviving differed between trout treatments. What method shall we use? It is tempting to carry out a χ^2 contingency test of association between treatment and survival.[5]

The problem with the contingency-test approach is that individual salmon are not a random sample. Rather, salmon are grouped by the *streams* in which they were released. If there is any inherent difference between the streams in the probability of survival, over and above any effects of brook trout, then two salmon from the same stream are likely to be more similar in their survival than two salmon picked at random. To lump all the salmon together and analyze with a contingency test is to commit the sin of "pseudoreplication" (Interleaf 4).

The key to solving the problem lies in recognizing that the stream is the independently sampled unit, not the salmon, and there are only 12 streams—six per treatment. As a result, the fates of all the salmon within a stream must be summarized by a single measurement for analysis—namely, the proportion of salmon surviving (given in the last column of Table 12.4-2). This changes the data type, because we are no longer comparing frequencies in categories, but rather differences in the means of a numerical variable. A two-sample test of the difference between the means is therefore required.

Let's label the streams with trout present as group 1 and the streams without trout as group 2. The null and alternative hypotheses are

H_0: The mean proportion of chinook surviving is the same in streams with and without brook trout ($\mu_1 = \mu_2$).

H_A: The mean proportion of chinook surviving is different between streams with and without brook trout ($\mu_1 \neq \mu_2$).

Sample means and standard deviations are listed in Table 12.4-3. The 95% confidence intervals are shown next to the data in Figure 12.4-1.

---

[5]   The results of this folly would be $\chi^2 = 7.2$, $df = 1$, and $P = 0.0071$. Thus, we would reject the null hypothesis of no association.

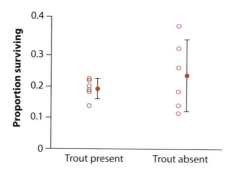

Figure 12.4-1 Proportion of chinook salmon surviving in streams with and without brook trout. Each open circle represents the measurement from a single stream. There are six streams of each type. Means and 95% confidence intervals are indicated to the right of the data.

Table 12.4-3. Sample statistics for the proportion of salmon surviving in streams (Example 12.4), using the stream as the sampling unit.

Group	Sample mean	Sample standard deviation, s_i	Sample size, n_i
Brook trout present	0.194	0.0297	6
Brook trout absent	0.235	0.1036	6

Figure 12.4-1 and the summary statistics in Table 12.4-3 show that streams *without* introduced trout have a sample standard deviation more than three times that of streams *with* trout. This is a case where the Welch's approximate *t*-test is appropriate (see Section 12.3). Using Welch's *t*-test and a computer, we find that the *P*-value for these hypotheses is $P = 0.39$. Hence, $P > 0.05$ and we cannot reject the null hypothesis. In other words, the data do *not* support the claim that the brook trout lower the survivorship of chinook salmon.

The appropriate analysis, in which salmon data within streams are reduced to a single measurement per stream, might seem like a waste of hard-earned data. We started with survival data on 9211 salmon but used only six measurements per treatment. The contingency analysis rejected H_0 but the Welch's two-sample test did not! Have we thrown away data and lost power as a result?

There are two answers to this question. First, if the raw data are not randomly sampled, then it is not legitimate to analyze them as though they were a random sample. You can't lose power that you never had. But the kinder, gentler answer is that the data are not wasted. By pooling together several related data points into a single summary measure such as a proportion, you will have an increasingly reliable measure of the true value of that measure in a given sample unit, such as a stream. As a result, little or no information is "lost."

12.5   The fallacy of indirect comparison

A common error when comparing two groups is to test each group mean separately against the same null hypothesized value, rather than directly comparing the two means with each other. The error might go something like this: "Since group 1 is significantly different from zero, but group 2 is not, group 1 and group 2 must be different from each other." We call this error the "fallacy of indirect comparison." Example 12.5 demonstrates that even papers published in scientific journals with the highest profile can make this mistake.

Example 12.5   **Mommy's baby, Daddy's maybe**

Do babies look more like their fathers or their mothers? The answer matters because, in most cultures, fathers are more likely to contribute to child-rearing if they are convinced that a child is their biological offspring. In this case, it is beneficial for babies to resemble dad, because it would provide evidence of paternity and lead to greater paternal care (mothers do not face the same uncertainty of maternity). Christenfeld and Hill (1995) tested this by obtaining pictures of a series of babies and their mothers and fathers. A photograph of each baby, along with photos of three possible mothers and of three possible fathers, was shown to a large number of volunteers. Each volunteer was asked to pick which woman and which man were the parents of the baby based on facial resemblance. The percentage of volunteers who correctly guessed a parent was used as the measure of a given baby's resemblance to that parent. If there were no facial resemblance of babies to parents, then the mean resemblance should be 33.3, the percentage of correct guesses expected by chance. If babies did resemble a parent, then the mean resemblance should be greater than 33.3%. Figure 12.5-1 shows the means for each parent and the corresponding 95% confidence intervals.

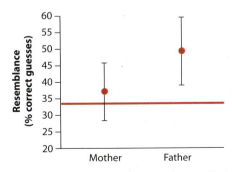

Figure 12.5-1 Resemblance of babies to their biological mothers and fathers, as measured by the percentage of volunteers who correctly guessed the mother and father of each baby from facial photographs. Dots are means, and vertical lines (error bars) are the 95% confidence intervals. The null expectation is 33.3% (shown in red). $n = 24$ babies.

The null hypothesis of no resemblance (i.e., one-third correct guesses) was soundly rejected for fathers, and the null hypothesis of no resemblance was *not* rejected for mothers. So far so good. However, the researchers concluded from these tests that babies therefore resembled their fathers more than they resembled their mothers. This is an "indirect comparison." That is, both groups were tested against the same null expectation (i.e., 33.3%), but mothers and fathers were not directly compared with each other. If they had been, no significant difference would have been found.

The problem with this sort of indirect comparison can be understood by considering a more extreme hypothetical case. In the example shown in Figure 12.5-2, the mean of group 1 is significantly different from the null expectation (indicated by the dashed line), but the mean of group 2 is not. It is false to conclude, therefore, that group 1 has a larger mean than group 2. In this example, it is group 2 that has the larger sample mean!

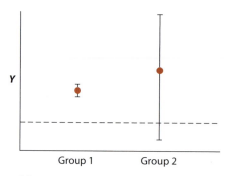

Figure 12.5-2 The 95% confidence intervals for the means of two hypothetical groups. The dashed line represents the null hypothesized value for the means of both groups.

How, then, should mothers and fathers be compared? They should be compared by testing directly whether the mean resemblance of babies to mothers is different from the mean resemblance to fathers. If the null hypothesis of no difference is rejected, then, and only then, can we conclude that babies resemble one parent more than the other.[6]

Indirectly comparing two groups by comparing them separately to the same null hypothesized value will often lead you astray. Groups should always be compared directly to each other.

> Comparisons between two groups should always be made directly, not indirectly by comparing both to the same null hypothesized value.

---

[6]   Larger follow-up studies have failed to find any difference between mothers and fathers in the degree to which their babies resemble them (Brédart and French 1999).

12.6 **Interpreting overlap of confidence intervals**

Scientific papers often report the means of two or more groups, along with their confidence intervals, but they might neglect to test the difference between the means of the two groups. How much information about the difference between group means is contained in the amount of overlap between their separate confidence intervals?

It turns out that reliable conclusions can be drawn only under the two scenarios depicted in panels *a* and *b* of Figure 12.6-1.

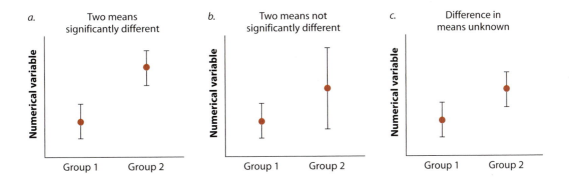

Figure 12.6-1. Each panel shows sample means of two groups with 95% confidence intervals. *a.* The confidence intervals do not overlap; in this case, the null hypothesis of no difference between group means would be rejected. *b.* The confidence interval for one group overlaps the sample mean of the other group; in this case the null hypothesis of no difference would *not* be rejected. *c.* The confidence intervals overlap, but neither includes the sample mean of the other group; in this case, we cannot be sure what the results of a hypothesis test comparing the means would show.

If the 95% confidence intervals[7] of two estimates do not overlap at all (as in Figure 12.6-1, panel *a*), then the null hypothesis of equal means would be rejected at the $\alpha = 0.05$ level. If, on the other hand, the value of one of the sample means is included within the 95% confidence interval of the other mean (as in Figure 12.6-1, panel *b*), then the null hypothesis of equal means would not be rejected at $\alpha = 0.05$. Between these two extremes, where the confidence intervals overlap but neither interval includes the sample mean of the other group (as in Figure 12.6-1, panel *c*), we cannot be sure what the outcome of a hypothesis test would be from a simple inspection of the overlap in confidence intervals.

---

[7] When reading scientific reports, interpret error bars on a graph with care. Sometimes error bars are used to show one standard error above and below the estimate, sometimes they show two standard errors, sometimes they show a confidence interval, and sometimes they just show standard deviations. Figure 12.6-1 shows confidence intervals.

12.7 Comparing variances

Up to now we have focused on comparing group means, but often we want to test the difference between populations in the variability of measurements. Several techniques are available. Here we briefly describe two of these: the *F*-test and Levene's test. In both cases, we describe the tests without much detail. The Quick Formula Summary (Section 12.9) gives the formulas, and most computer statistical programs will perform these tests.

Be warned, though: the *F*-test is highly sensitive to departures from the assumption that the measurements are normally distributed in the populations. It is not recommended for most data, therefore, because real data often show some departure from normality. Nevertheless, we present it here because many researchers continue to use it and you will encounter it in the literature and in statistics packages on the computer. Levene's test is a popular alternative test that is more robust to the assumption of normal populations, but it has its own weaknesses.

The *F*-test of equal variances

The **F-test** evaluates whether two population variances are equal. That is, it tests the null hypothesis that

$$H_0: \sigma_1^2 = \sigma_2^2$$

against the alternative

$$H_A: \sigma_1^2 \neq \sigma_2^2,$$

where σ_1^2 is the variance (the squared standard deviation) of population 1, and σ_2^2 is the variance of population 2. The test statistic is called *F*, and it is calculated from the ratio of the two sample variances:

$$F = s_1^2/s_2^2.$$

If the null hypothesis were true, then *F* should be near one, deviating from it only by chance. Under the null hypothesis, the *F*-statistic has an *F*-distribution with the pair of degrees of freedom $(n_1 - 1, n_2 - 1)$. The first number of the pair refers to the degrees of freedom of the top part (the numerator) of the *F*-ratio, and the second pair is for the bottom part (the denominator) of the *F*-ratio. We present more details in the Quick Formula Summary (Section 12.9). The *F*-distribution is discussed more fully in Chapter 15.

The *F*-test to compare two variances assumes that the variable is normally distributed in both populations. Unfortunately, the test is highly *sensitive* (i.e., not robust) to this assumption. For example, the *F*-test will often falsely reject the null hypothesis of equal variance if the distribution in one of the populations is not normal. For this reason, the *F*-test is not recommended for general use.

Levene's test for homogeneity of variances

Several alternative methods also test the null hypothesis that the variances of two or more groups are equal. One of the best is **Levene's test**, which is available in many statistical packages on the computer. Levene's test assumes that the frequency distribution of measurements is roughly symmetrical within all groups, but it performs much better than the F-test when this assumption is violated. Levene's test has the further advantage that it can be applied to more than two groups; in fact, it can test the null hypothesis that multiple groups all have equal variances.

Levene's test works by first calculating the absolute value of the difference between each data point and the sample mean for its group. These quantities are called "absolute deviations." The method then tests for a difference between groups in the means of these absolute deviations. The test statistic is called W, and it too has an F-distribution under the null hypothesis of equal variances. The calculations are somewhat cumbersome, so we will not detail them here, but we give the formula in the Quick Formula Summary (Section 12.9). Most modern statistical programs on the computer will do a Levene's test, however, and we recommend that you use it when comparing the variances of two or more groups. The online *Engineering Statistics Handbook*[8] is a good place to look for more information on Levene's test.

12.8   Summary

▶ Two study designs are available to compare two treatments. In a paired design, both treatments are applied to every randomly sampled unit. In a two-sample design, treatments are applied to separate randomly sampled units.

▶ Comparing two treatments in a paired design involves analyzing the mean of the differences between the two measurements of each pair. Comparing two treatments in a two-sample design involves analyzing the difference in means of two independent groups of measurements.

▶ A test of the mean difference between two paired treatments uses the paired t-test.

▶ Both the confidence interval for the mean difference and the paired t-test assume that the pairs are randomly chosen from the population and that the differences (d_i) have a normal distribution. These methods are robust to minor deviations from the assumption of normality.

▶ The means of a numerical variable from two groups or populations can be compared with a two-sample t-test.

▶ The two-sample t-test and the confidence intervals for the difference between the means assume that the variable is normally distributed in both populations and

---

[8]   http://www.itl.nist.gov/div898/handbook/eda/section3/eda35a.htm

that the variance is the same in both populations. The methods are robust to minor deviations from these assumptions.

▶ The pooled sample variance is the best estimate of the variance within groups, assuming that the groups have equal variance.

▶ Welch's approximate t-test compares the means of two groups when the variances of the two groups are not equal.

▶ Multiple measurements made on the same sampling unit are not independent and should be summarized for each sampling unit before further analysis.

▶ Indirectly comparing two groups by comparing them separately to the same null hypothesized value will often lead you astray. Groups should always be compared directly to each other.

▶ For variables that are normally distributed, variances of two groups can be compared with an F-test. The F-test, however, is highly sensitive to the departures from the assumption of normal populations.

▶ Levene's test compares the variances of two or more groups. It is more robust than the F-test to departures from the assumption of normality.

12.9 Quick Formula Summary

Confidence interval for the mean difference (paired data)

What does it assume? Pairs are a random sample. The difference between paired measurements is normally distributed.

Parameter: μ_d

Statistic: $\bar{d}$

Degrees of freedom: $n - 1$

Formula: $\bar{d} - t_{\alpha(2),df} \, \mathrm{SE}_{\bar{d}} < \mu_d < \bar{d} + t_{\alpha(2),df} \, \mathrm{SE}_{\bar{d}}$,

where $\bar{d}$ is the mean of the differences between members of each of the pairs, $\mathrm{SE}_{\bar{d}} = s_d/\sqrt{n}$, s_d is the sample standard deviation of the differences, and n is the number of pairs.

Paired t-test

What is it for? To test whether the mean difference in a population equals a null hypothesized value, μ_{d0}.

What does it assume? Pairs are randomly sampled from a population. The differences are normally distributed.

Test statistic: t

Distribution under H$_0$: The t-distribution with $n - 1$ degrees of freedom, where n is the number of pairs.

Formula: $t = \dfrac{\bar{d} - \mu_{d0}}{SE_{\bar{d}}}$,

where the terms are defined as for the confidence interval.

Standard error of difference between two means

Formula: $SE_{\bar{Y}_1 - \bar{Y}_2} = \sqrt{s_p^2 \left(\dfrac{1}{n_1} + \dfrac{1}{n_2} \right)}$,

where s_p^2 is the pooled sample variance: $s_p^2 = \dfrac{df_1 s_1^2 + df_2 s_2^2}{df_1 + df_2}$. The degrees of freedom are $df_1 = n_1 - 1$ and $df_2 = n_2 - 1$.

Confidence interval for the difference between two means (two samples)

What does it assume? Both samples are random samples. The numerical variable is normally distributed within both populations. The standard deviation of the distribution is the same in the two populations.

Degrees of freedom: $n_1 + n_2 - 2$

Statistic: $\bar{Y}_1 - \bar{Y}_2$

Parameter: $\mu_1 - \mu_2$

Formula: $(\bar{Y}_1 - \bar{Y}_2) - t_{\alpha(2),df}\, SE_{\bar{Y}_1 - \bar{Y}_2} < \mu_1 - \mu_2 < (\bar{Y}_1 - \bar{Y}_2) + t_{\alpha(2),df}\, SE_{\bar{Y}_1 - \bar{Y}_2}$,

where $SE_{\bar{Y}_1 - \bar{Y}_2}$ is the standard error of the difference between means, as defined above.

Two-sample *t*-test

What is it for? Tests whether the difference between the means of two groups equals a null hypothesized value for the difference.

What does it assume? Both samples are random samples. The numerical variable is normally distributed within both populations. The standard deviation of the distribution is the same in the two populations.

Test statistic: t

Distribution under H_0: The t-distribution with $n_1 + n_2 - 2$ degrees of freedom.

Formula: $t = \dfrac{(\bar{Y}_1 - \bar{Y}_2) - (\mu_1 - \mu_2)_0}{SE_{\bar{Y}_1 - \bar{Y}_2}}$,

where $(\mu_1 - \mu_2)_0$ is the null hypothesized value for the difference between population means, and $SE_{\bar{Y}_1 - \bar{Y}_2}$ is the standard error of the difference between means, as defined previously.

Welch's approximate *t*-test

What is it for? Tests whether the difference between the means of two groups equals a null hypothesized value when the standard deviations are unequal.

What does it assume? Both samples are random samples. The numerical variable is normally distributed within both populations.

Test statistic: t

Distribution under H_0: t-distribution. The number of degrees of freedom are fewer than in the case of the two-sample t-test: $df = \dfrac{\left(\dfrac{s_1^2}{n_1} + \dfrac{s_2^2}{n_2}\right)^2}{\left[\dfrac{(s_1^2/n_1)^2}{n_1 - 1} + \dfrac{(s_2^2/n_2)^2}{n_2 - 1}\right]}$. Round this number down to the nearest integer.

Formula: $t = \dfrac{(\bar{Y}_1 - \bar{Y}_2) - (\mu_1 - \mu_2)_0}{\sqrt{\dfrac{s_1^2}{n_1} + \dfrac{s_2^2}{n_2}}}$.

F-test

What is it for? Tests whether the variances of two populations are equal.

What does it assume? Both samples are random samples. The numerical variable is normally distributed within both populations.

Test statistic: F

Distribution under H$_0$: F-distribution with $n_1 - 1$, $n_2 - 1$ degrees of freedom. H$_0$ is rejected if $F > F_{\alpha(2),n1-1,n2-1}$. The latter quantity is the critical value of the F-distribution corresponding to the pair of degrees of freedom ($n_1 - 1$ and $n_2 - 1$). Critical values of the F-distribution are provided in Statistical Table D. F is compared only to the upper critical value, because F, as computed below, always puts the larger sample variance in the top (the numerator) of the F-ratio.

Formula: $F = \dfrac{s_1^2}{s_2^2}$,

where s_1^2 is the larger sample variance and s_2^2 is the smaller sample variance.

Levene's test

What is it for? Testing the difference between the variances of two or more populations.

What does it assume? Both samples are random samples, and the distribution of the variable is roughly symmetrical in both populations.

Test statistic: W

Distribution under H$_0$: F-distribution with the pair of degrees of freedom $k - 1$ (for the numerator of W; see the formula below) and $N - k$ (for the denominator of W), where k is the number of groups (two in the case of a two-sample test) and N is the total sample size ($n_1 + n_2$ in the case of two groups). H$_0$ is rejected if $W > F_{\alpha(1),k-1,N-k}$. The latter quantity is the critical value of the F-distribution (see the preceding description of the F-test for an explanation of the critical value).

Formula: $W = \dfrac{(N - k)\sum\limits_{i=1}^{k} n_i(\bar{Z}_i - \bar{Z})^2}{(k - 1)\sum\limits_{i=1}^{k}\sum\limits_{j=1}^{n_i}(Z_{ij} - \bar{Z}_i)^2}$,

where $Z_{ij} = |Y_{ij} - \bar{Y}_i|$ is the absolute value of the deviation between individual observation Y_{ij} (symbolized as the jth data point in the ith group) and the sample mean for its group $\bar{Y}_i$ (symbolized as the sample mean for group i). $\bar{Z}_i$ is the mean of all the Z_{ij} for the ith group, and $\bar{Z}$ is the grand mean of all the Z_{ij}'s, calculated as the average of all the Z_{ij}'s regardless of group. The n_i is the number of data points in the ith group, and k is the number of groups (two in the case of a two-sample test).

PRACTICE PROBLEMS

1. For each of the following scenarios, the researchers are interested in comparing the mean of a numerical variable measured in two circumstances. For each, say whether a paired *t*-test or two-sample *t*-test would be appropriate.

   a. The weight of 14 patients before and after open-heart surgery.

   b. The smoking rates of 14 men measured before and after a stroke.

   c. The number of cigarettes smoked per day by 14 men who have had strokes compared with the number smoked by 14 men who have not had strokes.

   d. The lead concentration upstream from five power plants compared with the levels downstream from the same plants.

   e. The basal metabolic rate (BMR) of seven chimpanzees compared with the BMR of seven gorillas.

   f. The photosynthesis rate of leaves in the crown of 10 Sitka spruce trees compared with the photosynthesis rate of leaves near the bottom of the same trees.

   g. The photosynthetic rates of 10 randomly chosen Douglas-fir trees compared with 10 randomly chosen western red cedar trees.

   h. The photosynthetic rate measured on 10 randomly chosen Sitka spruce trees compared with the rate measured on the Western red cedar growing next to each of the Sitka spruce trees.

2. Every year in Britain there is a No Smoking Day, where many people voluntarily stop smoking for a day. This No Smoking Day occurs on the second Wednesday of March each year. Data are collected about nonfatal injuries on the job, which allows a test of the hypothesis that stopping smoking affects the injury rate (Waters et al. 1998). Many factors affect injury rate, though, such as the year, time of week, etc., so we would like to be able to control some of these factors. One way to do this is to compare the injury rate on the Wednesday of the No Smoking Day to the rate for the previous Wednesday in the same years. Those data for 1987 to 1996 are listed in the following table:

Year	Injuries before No Smoking Day	Injuries on No Smoking Day
1987	516	540
1988	610	620
1989	581	599
1990	586	639
1991	554	607
1992	632	603
1993	479	519
1994	583	560
1995	445	515
1996	522	556

   a. How many more or fewer injuries are there on No Smoking Day, on average, compared with the normal day?

   b. What is the 99% confidence interval for this difference?

   c. In your own words, explain what the 99% confidence interval means.

   d. Test whether the accident rate changes on No Smoking Day.

3. Vertebrates are thought to be unidirectional in growth, with size either increasing or holding steady throughout life. Marine iguanas from the Galápagos (see the photo on the first page of this chapter) are unusual because they might actually shrink during low food periods caused by El Niño events (Wikelski and Thom 2000). During these low food periods, up to 90% of the iguana population can die from starvation. The following histogram plots the changes in body length of 64 surviving iguanas during the 1992–1993 El Niño event:

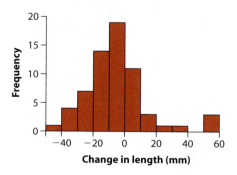

The average change in length was −5.81 mm, with a standard deviation of 19.50 mm.

a. Find the 95% confidence interval for the change in length of marine iguanas during the El Niño event. What assumptions are you making?

b. What is the 95% confidence interval for the standard deviation of the change in length?

c. Using the 95% confidence interval from part (a), say how reasonable it is to think that the iguanas changed in length on average during this time.

d. Test the hypothesis that length did not change on average during the El Niño event.

4. Dung beetles are one of the most common types of prey items for burrowing owls. The owls collect bits of large mammal dung and leave them around the entrance to their burrows, where they spend long hours waiting motionless for something tasty to be lured in. A research team wanted to know whether this dung actually attracted dung beetles or whether it had another use, such as to mask the odor of owl eggs from predators (Levey et al. 2004). They added dung to 10 owls' burrows, randomly chosen, and did not add dung to 10 other owl burrows. The researchers then counted the number of dung beetles consumed by the two types of owls over

Ronald G. Wolff

the next few days. The mean number of beetles consumed in the dung-addition group was 4.8, while the mean number was 0.51 in the control group. The standard deviations for the two groups were 3.26 and 0.89, respectively.

a. Calculate the 95% confidence interval of the mean number of beetles captured for each group. What assumptions are you making?

b. What is an appropriate way to test for a difference in these two groups' beetle-capture rates?

5. Problem 5 in Chapter 11 described an experiment that compared the testes sizes of four experimental populations of monogamous flies to four populations of polyandrous flies. The data are as follows:

Mating system	Testes area (mm²)
Monogamous	0.83
Monogamous	0.85
Monogamous	0.82
Monogamous	0.89
Polyandrous	0.96
Polyandrous	0.94
Polyandrous	0.99
Polyandrous	0.91

a. What is the difference in mean testes size for males from monogamous populations compared to males from polyandrous populations? What is the 95% confidence interval for this difference?

b. Carry out a hypothesis test to compare the means of these two groups. What conclusions can you draw?

6. In many species, some males have body size, shape, and coloration that mimics the look of females. Often this female-mimicry is thought to allow these males to avoid aggression from other males and to allow them to get close enough to guarded females to mate. In garter snakes, some males emerging from over-wintering dens mimic females by producing female pheromones. Males may mimic females in order to warm up. Females are often surrounded by males soon after emergence, and this warms them up faster, because the previously emerged snakes have already warmed in the sun. A prediction based on this idea is that males that

mimic females should be covered by other males more often than males than don't mimic females. Observations on newly emerging garter snakes in Manitoba found that, on average, 58% of a male's body was covered by other males if he emitted female pheromones (with standard deviation 28%, measured on 49 males). In comparison, 32 males that did not emit female pheromones had, on average, 25% of their bodies covered by other males, with standard deviation 24% (Shine et al. 2001).

a. On average, how much more covered by other males are female mimics compared with nonmimics? Give a 95% confidence interval for this parameter.

b. Test the hypothesis that female mimicry has no effect on the proportion of body coverage in these garter snakes. What assumptions are you making?

7. Periodic cicadas have a famously strange life cycle, living for years underground as larvae or pupae, only to emerge as adults every 13 or 17 years. When the adults do emerge, there can be very large numbers of them, and when these adults die, they become a potentially important source of nutrients for the forest. One study measured the phospholipid fatty acids (PLFAs) in fungus, an important part of the breakdown of dead matter in a forest (Yang 2004). The fungal PLFAs were measured before the experiment in a total of 44 one-square-meter plots. For 22 of the plots, the normal levels of cicadas were allowed to die on a given square meter ("controls"), and for the other 22 plots the natural levels of cicadas were supplemented with an extra 120 cicada carcasses ("cicada addition"). Fungal PLFAs were measured again on each of these plots after 28 days had passed from the addition of the cicadas. The following table summarizes, for both treatments, what the PLFA levels were both before and after cicada death:

Treatment	Mean PLFA	SD PLFA	Sample size (n)
Controls—before	7.16	3.39	22
Controls—after	4.50	2.03	22
Cicada addition—before	6.97	3.29	22
Cicada addition—after	6.31	2.12	22

a. Did fungal PLFA differ on average between the control and cicada addition plots before the experiment was done?

b. Did fungal PLFA differ on average between the control and cicada addition plots after the experiment was done?

c. Is it possible to use the results of parts (a) and (b) alone to determine whether the experimental treatment affected the change over time in PLFA level? Explain.

8. **Spot the flaw.** Bluegill sunfish, a species of freshwater fish, prefer to feed in the open water in summer, but, in the presence of predators, they tend to hide in the weeds near shore. A study compared the growth rate of bluegills that fed in the open water with the growth rate of bluegills that fed only in near-shore vegetation. "Open-water" and "near-shore" fish were both measured in eight lakes, and the mean growth rate of open-water fish was compared to the mean growth rate of near-shore fish using a two-sample t-test. What was done wrong in this study?

9. The cichlid fishes of Lake Victoria are amazing in their diversity. This diversity is maintained by the preferences of females for males of their own species. In order to understand how new species arise, it is important to understand the genetic basis of this preference in females. Researchers crossed two species of cichlids, *Pundamilia pundamilia* and *P. nyererei*. They measured an index of preference for *P. pundamilia* males by the female offspring of this cross (the "F_1") (Haeslery and Seehausen 2005). They also crossed these offspring with each other to produce a second generation (the "F_2"). If a small number of genes are important in determining the preference, then these two generations will differ from each other in their variance of the preference index. The researchers measured preference on 20 F_1 individuals and 33 F_2 individuals. The following table summarizes what they found. Assume that the two distributions are normal.

Genotype	Sample mean of preference index	Sample standard deviation of preference index	Sample size (*n*)
F_1	0.0049	0.0642	20
F_2	0.00815	0.1582	33

a. Is the variance of preference index the same for these two groups? Use the appropriate test.

b. Is the mean preference index the same for the two groups? Use the appropriate test.

10. In the early days of genetics, scientists realized that Mendel's laws of segregation could be used to predict that the second generation after a cross between two pure strains (the F_2) ought to have greater variance than the first generation after the cross (the F_1). One early test of this prediction was done measuring the flower length in a cross between two varieties of tobacco (East 1916). The following is a frequency table of the resulting individuals in both the first and second generations:

Flower length (mm)	Number of plants in F_1	Number of plants in F_2
52	0	3
55	4	9
58	10	18
61	41	47
64	75	55
67	40	93
70	3	75
73	0	60
76	0	43
79	0	25
82	0	7
85	0	8
88	0	1
Total	173	444
Mean	63.5	68.8
Variance	8.6	42.4

a. Show how these means and variances were calculated for the F_1 data.

b. Test whether the variances of the two groups of plants differ, making all necessary assumptions. Is one significantly greater than the other? If so, which one?

11. In most years since 1960, a televised debate between the leading candidates for president has been influential in determining the outcome of the U.S. election. One analysis of the transcripts of these debates looked at the number of times that each candidate used the words "will," "shall," or "going to" as an indication of how many promises the candidate made. Also recorded was whether the candidate won the popular vote. (This is not always the same candidate who won the election. In 2000, for example, Al Gore won the popular vote, but George Bush attained the presidency.) The results are shown in the following table. Debates were not held in 1964, 1968, and 1972, and full transcripts were not available from 1984.

Year	Candidate	Won (W) or lost (L) popular vote	Number of "will"s, "shall"s, and "going to"s
1960	Kennedy	W	163
1960	Nixon	L	122
1976	Carter	W	68
1976	Ford	L	32
1980	Reagan	W	19
1980	Carter	L	18
1988	G. Bush	W	111
1988	Dukakis	L	85
1992	Clinton	W	79
1992	G. Bush	L	75
1996	Clinton	W	56
1996	Dole	L	33
2000	Gore	W	68
2000	G.W. Bush	L	48
2004	G.W. Bush	W	176
2004	Kerry	L	149

Was the winner or loser significantly more likely to make promises, as measured by this index? Use an appropriate test.

12. Most bats are very poor at walking, but the vampire bat[9] is an exception. It is not clear why bats are poor walkers from a mechanical perspective, although a leading hypothesis has been that their hind legs are too weak. A test of this hypothesis was performed (Riskin et al. 2005), measuring

[9] Vampire bats can even run. See http://www.nature.com /nature/journal/v434/n7031/extref/434292a-s2.mov.

the strength of the hind legs of an insectivorous bat that walks poorly (*Pteronotus parnellii*) and the hind-leg strength of vampire bats (*Desmodus rotundus*). Six individuals of the *Pteronotus* species were measured, with an average hind-leg strength of 93.5 (in units of percent of body weight), with standard deviation 36.6. The mean hind-leg strength for six vampire bats was 69.3, with standard deviation 8.1.

a. Assuming that these measures of strength were normally distributed within groups, what is the appropriate test for comparing the mean hind-leg strength of these two species?

b. Look at the results closely. Without performing a hypothesis test, comment on the hypothesis that insufficient hind-leg strength is the reason that the insectivorous bat *Pteronotus* cannot walk well.

13. Ostriches live in hot environments, and they are normally exposed to the sun for long periods. Mammals in similar environments have special mechanisms for reducing the temperature of their brain relative to their body temperature. Researchers wanted to know if ostriches could do the same (Fuller et al. 2003). The mean body and brain temperature of six ostriches was recorded at typical hot conditions. The results, in °C, are as follows:

Ostrich	Body temperature	Brain temperature
1	38.51	39.32
2	38.45	39.21
3	38.27	39.2
4	38.52	38.68
5	38.62	39.09
6	38.18	38.94

a. Test for a mean difference in temperature between body and brain for these ostriches.

b. Compare the results to the prediction made from mammals in similar environments.

14. The following graphs are all based on random samples with more than 100 individuals. The red dots represent sample means.

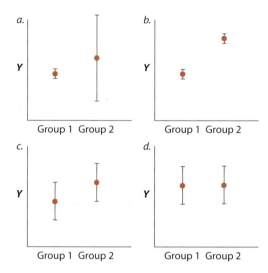

a. Assume that the error bars extend two standard errors above and two standard errors below the sample means. For which graphs can we conclude that group 1 is significantly different from group 2?

b. Assume that the error bars mark 95% confidence intervals. For which graphs can we conclude that group 1 is significantly different from group 2?

c. Assume that the error bars extend one standard error above and one standard error below the sample means. For which graphs can we conclude that group 1 is significantly different from group 2?

d. Assume that the error bars extend two standard deviations above and two standard deviations below the sample means. For which graphs can we conclude that group 1 is significantly different from group 2?

ASSIGNMENT PROBLEMS

15. Sex, with its many benefits, also brings risk. For example, individuals that are more promiscuous are exposed to more sexually transmitted diseases. This is true for other primates as well as for our own species. Different species of primates vary widely in the mean number of sexual partners per individual, and this raises the question, "Are the immune systems of more promiscuous species different from less promiscuous species?" Researchers approached this question by comparing pairs of closely related species, in which one species of the pair was more promiscuous and the other less promiscuous (Nunn et al. 2000). They measured the mean white blood cell (WBC) count for each species. The results are listed in the following table:

WBC count: Less promiscuous species	WBC count: More promiscuous species
5.7	10.4
7.2	10.4
7.4	9.9
8.1	9.1
8.4	9.2
9.2	11.9
9.1	9.3
9.1	8.9
10.6	12.5

a. What is the mean difference in WBC count between less and more promiscuous species? Which type of species (more promiscuous or less promiscuous) has the higher WBC count on average?

b. What is the 99% confidence interval for this difference?

c. Test the null hypothesis that there is no mean difference in WBC count between more and less promiscuous species.

16. The males of stalk-eyed flies (*Cyrtodiopsis dalmanni*) have long eye stalks. The females sometimes use the length of these eye stalks to choose mates. (See Example 11.2 for a similar story in a related species.) Is the male's eye-stalk length affected by the quality of its diet?

An experiment was carried out in which two groups of male "stalkies" were reared on different foods (David et al. 2000). One group was fed corn (considered a high quality food), while the other was fed cotton wool (a food of substantially lower quality). Each male was raised singly and so represents an independent sampling unit. The eye spans (the distance between the eyes) were recorded in millimeters.

The raw data, which are plotted as histograms below, are as follows:

Corn diet: 2.15, 2.14, 2.13, 2.13, 2.12, 2.11, 2.1, 2.08, 2.08, 2.08, 2.04, 2.05, 2.03, 2.02, 2.01, 2, 1.99, 1.96, 1.95, 1.93, 1.89.

Cotton diet: 2.12, 2.07, 2.01, 1.93, 1.77, 1.68, 1.64, 1.61, 1.59, 1.58, 1.56, 1.55, 1.54, 1.49, 1.45, 1.43, 1.39, 1.34, 1.33, 1.29, 1.26, 1.24, 1.11, 1.05.

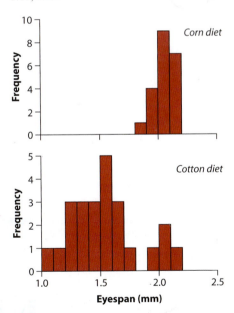

These data can be summarized as follows, where the corn-fed flies represent treatment group 1 and the cotton-fed flies represent treatment group 2.

	Mean (mm)	Variance (mm²)	Sample size, n
Corn diet (group 1)	2.05	0.00558	21
Cotton diet (group 2)	1.54	0.0812	24

a. What is the best test to use to compare the means of the two groups? Why?

b. Carry out the test identified in part (a), using $\alpha = 0.01$.

17. Flies, like almost all other living organisms, have built-in circadian rhythms that keep time even in the absence of external stimuli. Several genes have been shown to be involved in internal time-keeping, including *per* (period) and *tim* (timeless). Mutations in these two genes, and in other genes, disrupt time-keeping abilities. Interestingly, these genes have also been shown to be involved in other time-related behavior, such as the frequency of wing beats in male courtship behaviors. Individuals that carry mutations of *per* and *tim* have been shown to copulate for longer than individuals that have neither mutation. But do these two mutations affect copulation time in similar ways? The following table summarizes some data on the duration of copulation for flies that carry either the *tim* mutation or the *per* mutation (Beaver and Giebultowicz 2004):

Mutation	Mean copulation duration (min)	Standard deviation of copulation duration	Sample size, n
per	17.5	3.37	14
tim	19.9	2.47	17

a. Do these two mutations have different mean copulation durations? Carry out the appropriate test.

b. Do the populations carrying these mutations have different variances in copulation duration?

18. Red-winged blackbird males defend territories and attract females to mate and raise young there. A male protects nests and females on his territory both from other males and from predators. Males also frequently mate with females on adjacent territories. When males are successful in mating with females outside their territory, do they also attempt to protect these females and their nests from predators? An experiment measured the aggressiveness of males towards a stuffed magpie placed on the territory adjacent to the males (Gray 1997). This aggressiveness was measured on a scale where larger scores were more aggressive and lower scores were less aggressive. This aggressiveness score was normally distributed. Later the researchers used DNA techniques to identify whether or not the male had mated with the female on the neighboring territory. They compared the aggressiveness scores of the males who had mated with the adjacent female to those who had not. The results are as follows:

	Mean aggressiveness score	Standard deviation of aggressiveness score	Sample size (n)
Mated with neighbor	0.806	1.135	10
Did not mate with neighbor	−0.168	0.543	36

Test whether there are differences in the mean aggressiveness scores between the two groups of males. Are males aggressive to a different degree depending on whether they had mated with a neighboring female?

19. Researchers studying the number of electric fish species living in various parts of the Amazon basin were interested in whether tributaries affected the number of electric fish species in the rivers (Fernandes et al. 2004). They counted the number of electric fish species above and below the entrance point of a major tributary at 12 different locations. Here's what they found:

Tributary	Upstream number of species	Downstream number of species
Içá	14	19
Jutaí	11	18
Japurá	8	8
Coari	5	7
Purus	10	16
Manacapuru	5	6
Negro	23	24
Madeira	29	30
Trombetas	19	16
Tapajós	16	20
Xingu	25	21
Tocantins	10	12

a. What is the mean difference in the number of species between areas upstream and downstream of a tributary? What is the 95% confidence interval of this mean difference?

b. Test the hypothesis that the tributaries have no effect on the number of species of electric fish.

c. State the assumptions that you had to make to complete parts (a) and (b).

20. Question 17 from Chapter 11 discussed the relatedness of subordinate males to breeding females in the Seychelles warbler. Five subordinates that did not help feed the offspring of the older birds were measured for their relatedness to the offspring of the breeding females, with a mean relatedness of −0.05 and a standard deviation of 0.45. Another eight subordinates that *did* help feed younger offspring were also measured for their relatedness to the younger birds. For these eight, the mean relatedness was 0.27, with a standard deviation of 0.45.

a. Are the helpful and unhelpful subordinates different in their mean relatedness to the younger birds? Carry out an appropriate hypothesis test.

b. Find the 95% confidence interval for the difference in mean relatedness for the two classes of subordinates.

21. In tilapia, an important freshwater food fish from Africa, the males actively court females. They have more incentive to court a female who still has eggs than a female who has

already laid all of her eggs, but can they tell the difference? An experiment was done to measure the male tilapia's response to the smell of female fish (Miranda et al. 2005). Water containing feces from females that were either pre-ovulatory (they still had eggs) or post-ovulatory (they had already laid their eggs) was washed over the gills of males hooked up to an electro-olfactogram machine, which measured when the senses of the males were excited. The amplitude of the electro-olfactogram reading was used as a measure of the excitability of the males in the two different circumstances. Six males were exposed to the scent of pre-ovulatory females; their readings averaged 1.51 with a standard deviation of 0.25. Six different males exposed to post-ovulatory females averaged readings of 0.87 with a standard deviation of 0.31. Assume that the electro-olfactogram readings were approximately normally distributed within groups.

a. Test for a difference in the excitability of the males with exposure to these two types of females.

b. What is the estimated average difference in electro-olfactogram readings between the two groups? What is the standard error of this estimate?

22. A baby dolphin is born into the ocean, which is a fairly cold environment. Water has a high heat conductivity, so the thermal regulation of a newborn dolphin is quite important. It has been known for a long time that baby dolphins' blubber is different in composition and quantity from the blubber of adults. Does this make the babies better protected from the cold compared to adults? One measure of the effectiveness of blubber is its "conductance." This value was calculated on six newborn dolphins and eight adult dolphins (Dunkin et al. 2005). The newborn dolphins had an average conductance of 10.44, with a standard error of the mean equal to 0.69. The adult dolphins' conductance averaged 8.44, with the standard error of this estimate equal to 1.03. All measures are given in watts per square meter per degree Celsius.

a. Calculate the standard deviation of conductance for each group.

b. Test the null hypothesis that adults and newborns do not differ in the conductance of their blubber.

23. Weddell seals live in the Antarctic and feed on fish during long deep dives in freezing water. The seals benefit from these feeding dives, but the food they gain comes at a metabolic cost. The dives are strenuous. A set of researchers wanted to know whether feeding per se was also energetically expensive, over and above the exertion of a regular dive (Williams et al. 2004). They determined the metabolic cost of dives by measuring the oxygen use of seals as they surfaced for air after a dive. They measured the metabolic cost of 10 feeding dives and for each of these also measured a non-feeding dive by the same animal that lasted the same amount of time. The data, in (ml O_2 kg^{-1}), are as follows:

Individual	Oxygen consumption after non-feeding dive	Oxygen consumption after feeding dive
1	42.2	71.0
2	51.7	77.3
3	59.8	82.6
4	66.5	96.1
5	81.9	106.6
6	82.0	112.8
7	81.3	121.2
8	81.3	126.4
9	96.0	127.5
10	104.1	143.1

a. Estimate the mean change in oxygen consumption in feeding dives compared with non-feeding dives.

b. What is the 99% confidence interval for the mean change calculated in part (a)?

c. Test the hypothesis that feeding does not increase the metabolic costs of a dive.

24. Have you ever noticed that, when you tear a fingernail, it tends to tear to the side and not down into the finger? (Actually, the latter doesn't bear too much thinking about.) Why might this be so? One possibility is that fingernails are tougher in one direction than another. A study of the toughness of human fingernails compared the toughness of nails along a transverse dimension (side to side) compared with a longitudinal direction, with 15 measurements of each (Farren et al. 2004). The toughness of fingernails along a transverse direction averaged 3.3 kJ/m^2, with a standard deviation of 0.95, while the mean toughness along the longitudinal direction was 6.2 kJ/m^2, with a standard deviation of 1.48 kJ/m^2.

a. Test for a significant difference in the toughness of these fingernails along two dimensions.

b. As it turns out, all of the fingernails in this study came from the same volunteer. Discuss what the conclusion in part (a) means. What would be required to describe the fingernail toughness of all humans?

25. In a hypothetical study of a new drug called Drug X, researchers found that Drug X reduced chilblains by 10%, rejecting the null hypothesis of no effect ($P = 0.04$). The older drug, Drug Y, showed no statistically significant effect; that is, the null hypothesis of no effect could not be rejected. The researchers concluded that Drug X was better than Drug Y. Was this a valid conclusion? Explain.

Which test should I use?

One of the most challenging parts of statistical analysis is wading through the large number of statistical methods to find the right one for your particular question. In fact, with statistical computer programs so readily available, choosing the right method is usually the main challenge of data analysis left to us humans. The computer does the rest. We make choices to find the right graphical method, the right estimation approach, and the right hypothesis test. How can we choose the right method? Fortunately, the chain of logic involved in choosing the right method is similar in each case. In Section 2.7, we summarized the types of graphs available and when to use them. Here, we focus on choosing the right statistical test. We give four questions that you need to answer to help decide which test to use. The accompanying tables list information about the specific test, depending on your answers to these questions.

Are you testing just one variable, or are you testing the association between two or more variables? Different methods apply, depending on whether we are looking at one or more than one variable at a time. Tests for a single variable may address whether a certain probability distribution fits the data or whether a population parameter (such as a mean or a proportion) equals a specified value. Tests for two variables address whether the variables are associated or whether one variable differs between groups.

Are the variables categorical or numerical? Different tests are suited to different types of data. When testing for association between two variables, it matters whether the variables are categorical, numerical, or a mixture.

Are your data paired? Two treatments can be compared either with two independent samples or with a paired design in which both treatments are applied to every unit of a single sample. Different methods are required for the two approaches.

What are the assumptions of the tests, and do your data likely match those assumptions? For example, many powerful tests assume that the data are drawn from a normal distribution. If this is not true, then another approach must be found. As we discuss each hypothesis test, pay attention to the question it addresses and the assumptions it makes.

In this interleaf, we try to organize the tests that we have already learned and several that are still to come in relation to these questions. Table 1 lists some of the common methods used for hypothesis tests involving a single variable.

Most hypothesis tests are carried out to determine whether two variables are associated or correlated. This kind of problem can be addressed for two variables that are both categorical, both numerical, or one of each. Table 2 lists the most common tests used for each combination when the appropriate assumptions are met.

Many methods allow hypothesis tests of differences in a numerical response variable among different groups (see the bottom left corner of Table 2). Testing differences between groups is a test of association between a categorical explanatory variable (group) and a response variable. Table 3 summarizes these tests and gives the particular circumstances in which each is used, along with alternatives that make fewer assumptions.

If you can organize these tests in your mind according to these classifications, it will be much easier to pick the right one. When you encounter a new test, think about whether it applies to one or more variables, whether the variables are continuous or numerical, and what assumptions it makes.

Table 1 Commonly used statistical tests for comparing a single variable to a constant or to a probability distribution. (Numbers in parentheses refer to the chapter that discusses the test. Some refer to future chapters.)

Data type	Goal	Test
Categorical	Comparing a proportion to a hypothesized value	Binomial test (7)
		χ^2 Goodness-of-fit test with two categories (used if sample size is too large for the binomial test) (8)
	Comparing frequency data to a probability distribution	χ^2 Goodness-of-fit test (8)
Numerical	Comparing mean to a hypothesized value when data are approximately normal (possibly only after a transformation) (13)	One-sample t-test (11)
	Comparing median to a hypothesized value when data are not normal (even after transformation)	Sign test (13)
	Comparing frequency data to a discrete probability distribution	χ^2 Goodness-of-fit test (8)
	Comparing data to the normal distribution	Shapiro–Wilk test (13)

Table 2 Commonly used tests of association between two variables. (Numbers in parentheses refer to the chapter that discusses the test.)

		Type of explanatory variable	
		Categorical	**Numerical**
Type of response variable	**Categorical**	Contingency analysis (9)	Logistic regression (17)
	Numerical	t-tests, ANOVA, Mann–Whitney U-test, etc. [See *Table 3 for more details.*]	Linear and nonlinear regression (17) Linear correlation (16) Spearman's rank correlation (when data are not bivariate normal) (16)

Table 3 A comparison of methods to test differences between group means according to whether the tests assume normal distributions. (Numbers in parentheses refer to the chapter that discusses the test.)

Number of treatments	Tests assuming normal distribution	Tests not assuming normal distributions
Two treatments (independent samples)	Two-sample t-test (12) Welch's t-test (used when variance is unequal in the two groups) (12)	Mann–Whitney U-test (13)
Two treatments (paired data)	Paired t-test (12)	Sign test (13)
More than two treatments	ANOVA (15)	Kruskal–Wallis test (15)

13

Handling violations of assumptions

All of the methods that we have learned about so far to estimate and test population means assume that the numerical variable has an approximately normal distribution. The two-sample *t*-test requires the further assumption that the standard deviations (and variances) are the same in the two corresponding populations. However, frequency distributions often aren't normal and standard deviations aren't always equal. More often than we would like, our study organisms haven't read their stats books carefully enough, and the data they generate do not match the nice neat assumptions of classical statistical methods. What options are available to us when the data do not meet these assumptions?

In the current chapter, we focus on three alternative options:

1. *Ignore the violations of assumptions*. In some situations, we can use a procedure even if its assumptions are not strictly met. Methods for estimating and comparing *means* often work quite well when the assumption of normality is violated, especially if sample sizes are large and the violations are not too drastic.

2. *Transform the data.* For example, taking the logarithm is one way to transform data, with the result that the transformed data may better meet the assumptions. This procedure is often, but not always, effective.

3. *Use a nonparametric method.* A nonparametric method is one of a class of methods that does not require the assumption of normality. These methods can handle even badly behaved data, such as outliers that don't go away even when the data are transformed.

These are not the only alternatives available, but they are the ones most commonly used.[1] We examine each of these three approaches and explain the circumstances under which they should be used. All three assume that each data point in the sample is randomly and independently chosen from the population. We begin by reviewing methods to evaluate the assumption of normality.

13.1 Detecting deviations from normality

A few techniques are available to judge whether numerical data are from a population with a normal distribution.

Graphical methods

The most convenient methods for evaluating whether data fit a normal distribution are graphical. The human eye is a powerful tool for detecting deviations from a pattern. Start by plotting a histogram of the data, separately for each group if there is more than one.

Data can be noisy, especially when the sample size is small, so don't expect your data to follow a perfect bell curve, even when they come from a normal population. For example, Figure 13.1-1 shows two rows of histograms of data all sampled from a perfect normal distribution using a computer. The four in the first row are based on random samples of 10 individuals each and the four in the second row are based on random samples of 20 individuals each. None of the eight histograms resembles a normal distribution precisely, but none is so badly behaved that it would cause us to give up our assumption of a normal population. Of course, the larger the sample size, the more similar a frequency distribution is to that of the population from which it was sampled.

---

[1] Three other options are simulation, randomization, and bootstrapping. They all require a computer to assist with the large number of calculations, and they are discussed together in Chapter 19.

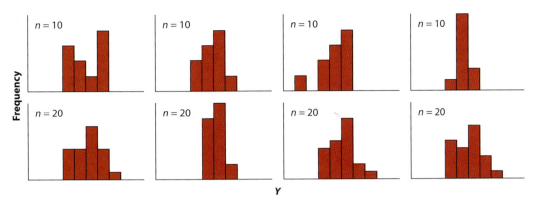

Figure 13.1-1 Top row: Histograms of four random samples of size $n = 10$ from the same normal distribution. Bottom row: Histograms of four random samples of size $n = 20$ from the same normal distribution.

In contrast, if the frequency distribution is strongly skewed or strongly bimodal, or if it has outliers, then it is unlikely to be normal. For example, Figure 13.1-2 shows histograms of four samples drawn from distributions that are not normal. Panels *a* and *c* are from distributions strongly skewed to the right, whereas panel *b* shows a distribution strongly skewed to the left. Panel *d* has an outlier far to the right of the rest of the data. Histograms like these indicate that the data were not from a normal population.

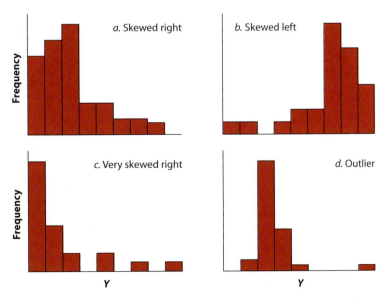

Figure 13.1-2 Frequency distributions of four data sets drawn from non-normal distributions. These distributions are (*a*) skewed right, (*b*) skewed left, (*c*) more extreme skew to the right, and (*d*) with an outlier.

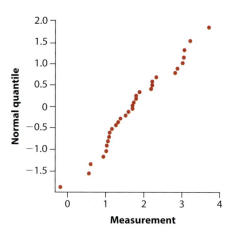

Figure 13.1-3 A normal quantile plot for a random sample of 32 observations from a normal distribution. The points on the plot fall roughly along a straight line.

Besides histograms, the **normal quantile plot** is a second type of graphical technique useful for detecting departures from normality. The normal quantile plot compares each observation in the sample with the corresponding quantile expected from the standard normal distribution. An example is shown in Figure 13.1-3. Each dot on this plot represents one data point, with its measurement indicated by its position on the horizontal axis and its expected normal quantile indicated on the vertical axis.[2] The points in a normal quantile plot should roughly follow a straight line if the data were sampled from a normal distribution. The dots will wiggle around a straight line even in the best of circumstances, as in Figure 13.1-3, because sampling is random. More systematic departures from a straight-line pattern, as judged by eye, indicate that the frequency distribution of the data deviates from the normal distribution. Substantial curvature over a large range of values or substantial jumps in the distribution indicate potential deviations from normality. We recommend using a statistics program on the computer to draw quantile plots.

> The *normal quantile plot* compares each observation in the sample with
> its quantile expected from the standard normal distribution. Points
> should fall roughly along a straight line if the data come from a normal
> distribution.

---

[2]   To find the normal quantile, order the observations from smallest to largest and assign each a rank, called i. The smallest data point has $i = 1$, the next smallest has $i = 2$, and so on up to the highest data point with $i = n$. The estimated proportion of the distribution lying below an observation ranked i is $i/(n + 1)$. The corresponding normal quantile is the standard normal deviate Z having an area under the standard normal curve below it equal to $i/(n + 1)$. For example, if n is 99, the approximate fraction of the probability density lying below $i = 95$ is $95/(99 + 1) = 0.95$. The corresponding normal quantile is 1.64, which has 0.95 of the area under the normal curve below it (and 0.05 above it; see Statistical Table B).

Example 13.1 illustrates the methods used to evaluate normality.

| Example 13.1 | **The benefits of marine reserves** |

Marine organisms do not enjoy anywhere near the same protection from human influence as do terrestrial species. However, marine reserves are becoming increasingly popular for biological conservation and the protection of fisheries. But are reserves effective in preserving marine wildlife?

Halpern (2003) matched each of 32 marine reserves to a control location, which was either the site of the reserve before it became protected or a similar unprotected site nearby. One of the indices of protection evaluated by the study was the "biomass ratio," which is the total mass of all marine plants and animals per unit area of reserve divided by the same quantity in the unprotected control. This biomass ratio would equal one if protection had no effect. The biomass ratio would exceed one if the protection was beneficial, and it would be less than one if protection reduced biomass. The following list gives the biomass ratio for each of the 32 reserves:

1.34,   1.96,   2.49,   1.27,   1.19,   1.15,   1.29,   1.05,   1.10,   1.21,   1.31,

1.26,   1.38,   1.49,   1.84,   1.84,   3.06,   2.65,   4.25,   3.35,   2.55,   1.72,

1.52,   1.49,   1.67,   1.78,   1.71,   1.88,   0.83,   1.16,   1.31,   1.40

Are marine reserves effective? In other words, does the mean biomass ratio differ from one?

The null and alternative hypotheses are

H_0: The mean biomass ratio is unaffected by reserve protection ($\mu = 1$).
H_A: The mean biomass ratio is affected by reserve protection ($\mu \neq 1$).

We might be tempted to use a one-sample t-test, but this test assumes that the biomass-ratio data are drawn from a population having a normal distribution. We must start by asking whether this assumption is valid.

The best starting point is to plot the data, as in Figure 13.1-4. The histogram shows that all is not well. The frequency distribution of the biomass ratio is skewed strongly to the right. The quantile plot, moreover, shows curvature, indicating a poor fit to a normal distribution.

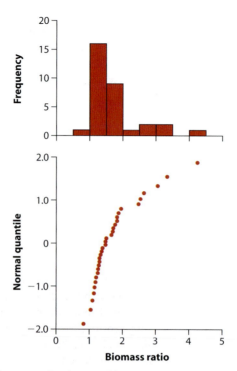

Figure 13.1-4 The frequency distribution of the "biomass ratio" of 32 marine reserves (*top panel*) and the corresponding normal quantile plot (*bottom panel*).

Formal test of normality

The "eyeball test" works quite well to detect departures from normality. However, formal goodness-of-fit tests to the normal distribution are available. These methods test the following hypotheses:

H_0: The data are sampled from a population having a normal distribution.
H_A: The data are sampled from a population not having a normal distribution.

Formal tests of normality should be used with caution. On the one hand, the test can give a false sense of security. A small sample size might not yield enough power to reject the null hypothesis of normality, even when data are drawn from a population without a normal distribution. On the other hand, a sufficiently large sample size will reject the null hypothesis of normality for many data whose departure from the normal distribution is not severe enough to warrant giving up on the methods that assume normality. The assumption of normality becomes less and less important when testing means as sample size increases, for reasons discussed in Section 13.2. As a result, we suggest that graphical methods and common sense are essential when evaluating the assumption of normality.

The **Shapiro–Wilk test** is probably the most powerful formal method for testing departures from normality. This test is carried out by most computer statistics programs, so we don't discuss the calculation details here. The Shapiro–Wilk test first estimates the mean and standard deviation of the population using the sample data. It then tests the goodness of fit to the data of the normal distribution having this same mean and standard deviation.

When we ran the Shapiro–Wilk test on the biomass-ratio data (Example 13.1), we rejected the null hypothesis that the data come from a normal distribution, supporting what we saw from looking at the histogram and quantile plots.

> A *Shapiro–Wilk test* evaluates the goodness of fit of a normal distribution to a set of data randomly sampled from a population.

13.2   When to ignore violations of assumptions

What should be done when the data don't meet the assumptions of normality and (for the two-sample methods) equal standard deviations? Here we consider the option of simply ignoring the violations. The justification is that methods for estimating and testing *means* are not highly sensitive to the assumptions of normality and equality of standard deviations. Under certain conditions, these methods are robust, meaning that the answers they give are not sensitive to modest departures from the assumptions. Here we explain what these conditions are.

> A statistical procedure is *robust* if the answer it gives is not sensitive to violations of the assumptions of the method.

Violations of normality

Even though confidence intervals and *t*-tests require normally distributed data, the methods can sometimes be used to analyze data that are not normally distributed. The reason for this robustness comes from the central limit theorem (Section 10.6), which states that, when a variable does not have a normal distribution, the distribution of sample means is nevertheless approximately normal when sample size is large. With large enough samples, therefore, the sampling distribution of means behaves roughly as assumed by methods based on the *t*-distribution, even when the data are not from a normally distributed population, provided that the violation of normality is not too drastic (Box and Andersen 1955). Robustness applies only to methods for means, not to methods like the *F*-test for testing *variances*, which are not robust to departures from the assumption of normality.

How large must samples be to allow us to ignore the assumption of normality? The answer depends on the shape of the distributions. If two groups are being compared and both differ from normality in different ways, then even subtle deviations from normality can cause errors in the analysis (even with fairly large sample sizes). For example, look back at the skewed distributions in Figure 13.1-2 (i.e., panels *a*, *b*, and *c*). If we compared two groups that both had the same skew as in panel *a* of Figure 13.1-2, we would get reasonably accurate answers from a two-sample *t*-test with sample sizes of about 30 in each group. However, we would require sample sizes of about 500 or more to get sufficient accuracy from a two-sample test if we were to compare one group with right-skewed measurements like those in panel *a* with another group whose data were left-skewed like those in panel *b*. If the distributions are even more skewed than those in panels *a* and *b* of Figure 13.1-2, then the two-sample methods based on the *t*-distribution should be avoided in favor of alternative methods. The frequency distribution in panel *c* of Figure 13.1-2 is so skewed that a *t*-test would not give reliable answers even with extremely large sample sizes. Frequency distributions containing outliers (e.g., panel *d* of Figure 13.1-2) should never be analyzed with a *t*-test or a confidence interval based on the *t*-distribution. Methods that assume normality are very sensitive to outliers.

In the absence of definitive guidelines for every possible case, we recommend a cautious approach to data analysis. If the data show strong departures from normality, such as outliers, or if the frequency distributions in different groups are markedly different, then it is best to adopt one of the other options (i.e., data transformations or nonparametric methods), rather than simply ignoring the violations of assumptions.

Unequal standard deviations

When can we ignore the assumption of equal standard deviations in two-sample methods? With moderate sample sizes (greater than 30 in each group), the two-sample methods for estimation and hypothesis testing using the *t*-distribution will still perform adequately with even a three-fold difference between groups in their standard deviations, as long as the sample sizes of the two groups are approximately equal (Ramsey 1980). If sample sizes are not approximately equal, or if the difference in standard deviations is more than three-fold, then the Welch's *t*-test (Section 12.3) should be used instead of the two-sample *t*-test, assuming that the assumption of normality is met. (Indeed, Welch's *t*-test can be used even if the standard deviations are thought to be equal, but it is not as powerful as the two-sample *t*-test.) If the assumption of normality is also not met, then it is best to try data transformations or one of the methods in Chapter 19).

13.3 Data transformations

One of the best ways to handle data that don't match the assumptions of a statistical method is to try to **transform** the data so that it better meets the assumptions. Transformations can be used to improve the fit of the normal distribution to data and to make the standard deviations more similar in different groups. For example, it is often the case that a sample of data does not follow a normal distribution, but that the logarithms of the data match the normal distribution rather well. The test or estimation procedure could then be carried out on the log-transformed data, and the results can be interpreted in the same way as the analysis of the original variable.

The three most frequently used transformations are the log transformation, the arcsine transformation, and the square-root transformation.[3] In this section, we examine the situations in which these three transformations are most likely to be useful. Keep in mind that, if a transformation is made to one data point, it has to be made to all data points from all samples for that variable if the data are to be compared.

> A *data transformation* changes each measurement by the same mathematical formula.

Throughout this section and beyond, we use a "prime" mark (') to denote transformed data. Thus, if the original variable is Y, we would call the transformed variable Y'.

Log transformation

The most common data transformation in biology is the **log transformation**. Typically, the data are converted by taking the natural log (ln) of each measurement. In mathematical terms,

$$Y' = \ln[Y].$$

Log base-10 is also sometimes used instead of the natural log, but we will use the natural log throughout this text. Again, all observations must be transformed in exactly the same way. (It is not legitimate to compare the natural log of a variable from population 1 to the log base-10 of the variable in population 2.) Finally, the log transformation can only be applied to data when all the values are greater than zero. (The natural log of numbers less than or equal to zero is undefined). If the data include zero, then $Y' = \ln[Y + 1]$ can be tried instead.

In general, the log transformation is most likely to be useful when

- ▶ the measurements are ratios or products of variables,
- ▶ the frequency distribution of the data is skewed to the right (i.e., has a long tail on the right),
- ▶ the group having the larger mean (when comparing two groups) also has the higher standard deviation, or
- ▶ the data span several orders of magnitude.

---

3   Other mathematical operations can also serve as valid transformations as long as there is a one-to-one correspondence between the data on its original scale and on the transformed scale (that is, it must be possible to transform back to get the original value without ambiguity).

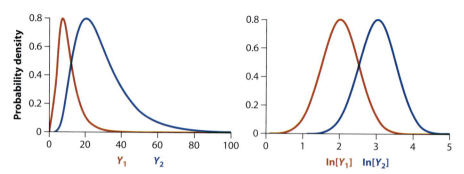

Figure 13.3-1 Left panel: Two right-skewed probability distributions having different standard deviations. In this case, the distribution with the highest standard deviation also has the highest mean. Right panel: The log transformations of the same variables. On the log scale, these two distributions have normal distributions with the same standard deviation.

For example, compare the two probability distributions in the left panel of Figure 13.3-1. Both distributions are right-skewed and the group with the larger mean has the higher standard deviation. The log transformation fixes both problems. That is, both $\ln[Y_1]$ and $\ln[Y_2]$ are no longer skewed (they have normal distributions, instead) and both have the same standard deviation. Don't hesitate to try the log transformation in other situations as well. If it solves the problem, then use it.

> The *log transformation* converts each data point to its logarithm.

In our experience, body measurements such as mass and length often show right-skewed frequency distributions that become more normally distributed after being log-transformed. But log transformation will not always solve the problem, even when frequency distributions are right-skewed or when the group with the larger mean also has the higher standard deviation. In these circumstances, though, the log transformation is worth a try. Just be sure to check the distribution of the transformed data to determine whether it fits the assumptions of the desired method.

Let's look again at the study on biomass ratio in marine reserves (Example 13.1). The raw data are listed in Table 13.3-1 along with the log-transformed data. Try a couple of the log calculations to make sure you get the same numbers as we did. Recall that the frequency distribution of the biomass ratio was right-skewed (Figure 13.1-4).

The effects of the log transformation on the frequency distribution of the data are shown in Figure 13.3-2. The skew has been much reduced in the log-transformed data, which now conforms to a normal distribution much more closely than before (Figure 13.1-4).

Table 13.3-1 Biomass ratios from 32 marine reserves and their log transformations.

Biomass ratio	ln[Biomass ratio]	Biomass ratio	ln[Biomass ratio]
1.34	0.29	3.06	1.12
1.96	0.67	2.65	0.97
2.49	0.91	4.25	1.45
1.27	0.24	3.35	1.21
1.19	0.17	2.55	0.94
1.15	0.14	1.72	0.54
1.29	0.25	1.52	0.42
1.05	0.05	1.49	0.40
1.10	0.10	1.67	0.51
1.21	0.19	1.78	0.58
1.31	0.27	1.71	0.54
1.26	0.23	1.88	0.63
1.38	0.32	0.83	−0.19
1.49	0.40	1.16	0.15
1.84	0.61	1.31	0.27
1.84	0.61	1.40	0.34

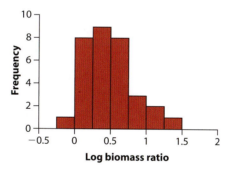

Figure 13.3-2 Frequency distribution of the natural logarithm of the biomass ratio from 32 marine reserves.

We can now apply the one-sample t-test to the transformed data, because they better meet the assumption that the sample came from a population having a normal distribution.

Let's use the data to test the following hypotheses:

H_0: The mean biomass ratio is unaffected by reserve protection ($\mu = 1$).
H_A: The mean biomass ratio is affected by reserve protection ($\mu \neq 1$).

Because we want to test the hypotheses on ln-transformed data, though, we need to modify these hypotheses to reflect that transformation. Thus, a biomass ratio of 1 on the original scale is $\ln[1] = 0$ on the log-transformed scale, so our revised hypotheses are

H_0: The mean of the log biomass ratio of marine reserves is zero ($\mu' = 0$).
H_A: The mean of the log biomass ratio of marine reserves is not zero ($\mu' \neq 0$).

The prime ($'$) denotes the transformed scale. After this, we proceed with the hypothesis test as usual. The sample mean of Y' is

$$\overline{Y}' = 0.479$$

and the sample standard deviation is

$$s' = 0.366.$$

(Note that these are *not* the same as the log of $\overline{Y}$ and the log of the standard deviation of Y.) The corresponding one-sample t-statistic is

$$t = \frac{0.479}{0.366/\sqrt{32}} = 7.40,$$

with $df = n - 1 = 32 - 1 = 31$. Using a computer program, we find that the P-value is

$$P = 2.4 \times 10^{-8}.$$

This P-value is considerably less than 0.05, so we soundly reject the null hypothesis. Marine reserves have more biomass than comparable unprotected areas.

Arcsine transformation

The **arcsine transformation** is used almost exclusively on data that are proportions:

$$p' = \arcsin[\sqrt{p}],$$

where p is a proportion measured on a single sampling unit. The arcsine (abbreviated, with a bizarre lack of brevity, as "arcsin") is the inverse of the sine function from trigonometry.[4] A dedicated transformation for proportions is needed because proportions tend *not* to be normally distributed, especially when the mean is close to zero or to one, and because groups differing in their mean proportion tend *not* to have equal standard deviations. The arcsine transformation often solves both of these problems at the same time.

---

[4]   In other words, if $X = \sin[Y]$ then $Y = \arcsin[X]$. On many calculators, arcsine is given as $\sin^{-1}$.

Don't forget the square-root component of the arcsine transformation. Note that if the original data are given as percentages, they must first be converted to proportions by dividing by 100 before applying the arcsine transformation.

The square-root transformation

The **square-root transformation** is often used when the data are counts, such as number of mates acquired, number of eggs laid, number of bacterial colonies, and so on:

$$Y' = \sqrt{Y + 1/2}.$$

This transformation, like the log transformation, sometimes helps to equalize standard deviations between groups when the group with the higher mean also has the higher standard deviation.

In our experience, the effects of the square-root transformation are usually very similar to those obtained with the log transformation. It makes sense, then, to try both and see which works best. If their effects are the same, then it doesn't matter which transformation you use.

Other transformations

When the frequency distribution of the data is skewed left, try the **square transformation**. This is done by squaring each data point:

$$Y' = Y^2.$$

If the square transformation doesn't work, try the **antilog transformation**,

$$Y' = e^Y,$$

on left-skewed data. When the data are skewed right, try the **reciprocal transformation**,

$$Y' = \frac{1}{Y}.$$

The square transformation and the reciprocal transformation are usable only if all the data points in all samples have the same sign. If all numbers in the samples are negative, then try multiplying each number by -1 before further transforming the new positive numbers.

Confidence intervals with transformations

It may often be necessary to transform the data to meet the assumptions of the methods for confidence intervals. Once the confidence interval is computed, it is usually best to back-transform the lower and upper limits of the confidence interval before presenting.

For example, let's find the confidence interval for the mean of the log biomass ratio of marine reserves (Example 13.1). The original data were not normally distributed, so it is not valid to calculate a confidence interval directly from these data using the method based on the t-distribution. Recall that a log-transformation fixed this problem.

The sample mean of the log-transformed data is $\overline{Y}' = 0.479$, and the sample standard deviation is $s' = 0.366$. There are a total of $n = 32$ data points in the sample. The 95% confidence interval for the mean of log of the biomass ratio is therefore

$$\overline{Y}' - t_{0.05(2),31}\mathrm{SE}_{\overline{Y}'} < \mu' < \overline{Y}' + t_{0.05(2),31}\ \mathrm{SE}_{\overline{Y}'}$$

$$0.479 - (2.04)\frac{0.366}{\sqrt{32}} < \mu' < 0.479 + (2.04)\frac{0.366}{\sqrt{32}}$$

$$0.347 < \mu' < 0.611$$

These limits to the confidence interval are not so easily interpreted because the data were log-transformed. However, we can back-transform the limits to yield a 95% confidence interval on the original scale. In the case of the log transformation, the back-transform is done by taking the antilog (i.e., raise e to the power of the quantity on the transformed scale). Therefore, the 95% confidence interval for the mean on the original scale (called the geometric mean) is

$$e^{0.347} < \text{geometric mean} < e^{0.611}$$

or

$$1.41 < \text{geometric mean} < 1.84.$$

This 95% confidence for the geometric mean biomass ratio indicates that marine reserves have 1.41 to 1.84 times more biomass on average than the control sites do. This is a remarkably narrow interval for the mean effect of marine reserves.

Similar back-transformations can be calculated for all of the other possible transformations. Back-transformations for each transformation are given in the Quick Formula Summary (Section 13.9 at the end of the chapter).

A caveat: Avoid multiple testing with transformations

Trying multiple transformations to find which one works best to meet the assumptions of a test is an excellent plan. It is not legitimate, though, to try multiple transformations to find one that leads to a statistically significant outcome (such as a P-value smaller than 0.05). Each transformation would give a slightly different result to the statistical test, and you might be tempted to choose the result that most agreed with your preconceptions about the data. Repeated testing inflates the Type I error rate, so it should be avoided. Instead, first decide which transformation best meets the assumptions of the method, and then stick with that decision when carrying out the test.

13.4 Nonparametric alternatives to one-sample and paired *t*-tests

If ignoring violations of assumptions and transforming the data fail, try **nonparametric methods** instead. These are methods developed for calculating confidence intervals and hypothesis testing that make fewer assumptions about the probability distribution of the variable of interest in the population from which the data were drawn.[5] This makes them handy when the frequency distribution of the data is not normal—such as when there are outliers. In contrast, the methods that do make assumptions about the distributions are called **parametric methods**. All of the methods we have learned about so far, including those based on the *t*-distribution, are parametric methods.

> A *nonparametric method* makes fewer assumptions than standard *parametric methods* about the distributions of the variable.

Here we focus on nonparametric methods for hypothesis testing. We discuss nonparametric tests that address the same types of questions that we have already considered in Chapters 11 and 12. In this section, we cover alternatives to the one-sample *t*-test and the paired *t*-test. In Section 13.5, we use nonparametric methods to compare two independent groups. In subsequent chapters, we introduce new nonparametric tests in tandem with the parametric tests designed for the same purpose.

Nonparametric tests are usually based on the **ranks** of the data points rather than the actual values of the data. In other words, the data points are ranked from smallest to largest, and the rank (first, second, third, etc.) of each data point is recorded. The actual measurements are not used again for the test. If the null hypothesis were true, these ranks would, on average, be about the same between two groups. However, if one of the groups has most of the smallest measurements and the other group has most of the largest, then we may have evidence to reject a null hypothesis of equality. Using ranks is what frees us from making assumptions about the probability distribution of the measurements, because all distributions make similar predictions about the ranks of the measurements. Nonparametric tests are particularly useful when there are outliers in the data set, because ranks are not unduly affected by outliers.

Sign test

The **sign test** is a nonparametric method that can be used in place of the one-sample *t*-test or the paired *t*-test when the normality assumption of those tests cannot be met.

---

[5] These methods are also sometimes called **distribution-free**.

The sign test assesses whether the *median* of a population equals a null hypothesized value. Measurements lying above the null hypothesized median are designated "+" and the numbers lying below are scored as "−." If the null hypothesis is correct, we expect half of the measurements to lie above the null hypothesized median and half to lie below, except for sampling error. The *P*-value can then be calculated using the binomial distribution (see Section 7.3). The sign test is simply a binomial test in which the number of data points over the null hypothesized median is compared with that expected when $p = 1/2$.

> The *sign test* compares the median of a sample to a constant specified in the null hypothesis. It makes no assumptions about the distribution of the measurement in the population.

Unfortunately, the sign test has very little power compared with the one-sample or paired *t*-test because it discards most of the information in the data. A measurement that is infinitesimally larger than the null hypothesized median and a data point that exceeds the median by several million both count only as a "+." Nonetheless, the sign test is a useful tool to have in your statistical toolbox because sometimes no other test is possible.

| Example 13.4 | **Sexual conflict and the origin of new species** |

The process by which a single species splits into two species is still not well understood. One proposal involves "sexual conflict"—a genetic arms race between males and females that arises from their different reproductive roles.[6] Sexual conflict can cause rapid genetic divergence between isolated populations of the same species, leading to the formation of new species. Sexual conflict is more pronounced in species in which females mate more than once, leading to the prediction that they should form new species at a more rapid rate. To investigate this, Arnqvist et al. (2000) identified 25 insect taxa (groups) in which females mate multiply, and they paired each of these groups to a closely related insect group in which females only mate once. Which type of insect tends to have more species? Table 13.4-1 lists the numbers of insect species in each of the groups.

---

6   For example, in a number of insect species, male seminal fluid contains chemicals that reduce the tendency of the female to mate again with other males. However, the chemicals also reduce female survival, so females have evolved mechanisms to counter these chemicals.

Table 13.4-1 The number of species in 25 pairs of insect groups. Each pair matches a group of insect species in which females mate only once with a related group of insect species in which females mate multiple times.

Taxon pair	Number of species		Difference	Above (+) or below (−) zero
	Multiple-mating group	**Single-mating group**		
A	53	10	43	+
B	73	120	−47	−
C	228	74	154	+
D	353	289	64	+
E	157	30	127	+
F	300	4	296	+
G	34	18	16	+
H	3400	3500	−100	−
I	20	1000	−980	−
J	196	486	−290	−
K	1750	660	1090	+
L	55	63	−8	−
M	37	115	−78	−
N	100	30	70	+
O	21,000	600	20,400	+
P	37	40	−3	−
Q	7	5	2	+
R	15	7	8	+
S	18	6	12	+
T	240	13	227	+
U	15	14	1	+
V	77	16	61	+
W	15	14	1	+
X	85	6	79	+
Y	86	8	78	+

The data are paired. Thus, for each group of insects whose females mate once, there is a corresponding, closely related group of insect species in which females mate more than once. For this reason, the analysis must focus on the paired differences. The differences listed in Table 13.4-1 were calculated by subtracting the number of species in the single-mating group from that of the corresponding multiple-mating group.

First, examine the histogram of the differences in Figure 13.4-1. These data have one outlier at 20,400, and we hardly need a Shapiro–Wilk test (Section 13.1) to tell us that the measurements are not normally distributed. At the same time, there are only

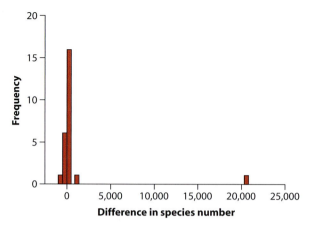

Figure 13.4-1 The distribution of differences in species number between singly mating and multiply mating insect groups. There is an extreme outlier at 20,400.

25 data points, which is too small a sample size to rely on the robustness of the paired *t*-test. There is no obvious transformation that would make these data normal, so we should pursue a nonparametric test instead. We will use the sign test to evaluate whether the median of the difference equals zero.

Our hypotheses are

H_0: The median difference in number of species between insect groups is zero.
H_A: The median difference in number of species between these groups is not zero.

From this point on, the sign test is the same as the binomial test. If the null hypothesis is correct, then we expect half the measurements to fall above zero ("+") and half to fall below ("−"). In fact, 18 out of the 25 measurements fall above zero and only seven fall below (see the last column in Table 13.4-1).

We can use the binomial distribution to calculate the *P*-value for the test. What is the probability of getting seven or fewer "−" observations out of 25 when the probability of a "−" observation is 0.5 under the null hypothesis? The answer is

$$\Pr[X \le 7] = \sum_{i=0}^{7} \binom{25}{i}(0.5)^i(0.5)^{25-i} = 0.02164.$$

The alternative hypothesis requires a two-sided test, so we need to double this calculated probability to account for the other tail. Thus, the *P*-value is

$$P = 2\,(0.02164) = 0.043.$$

Because this *P*-value is less than $\alpha = 0.05$, we reject the null hypothesis. Groups of insects whose females mate multiple times have more species than groups whose females mate singly, consistent with the sexual-conflict hypothesis.

Although it didn't come up when applying the sign test to the data in Example 13.4, what do you do about data points exactly equal to the hypothesized median? The usual approach is to drop all data points exactly equal to the median given by the null hypothesis. The test then proceeds as though those data had never existed. When this occurs, you must reduce the *n* used in the binomial calculations to the number of data points left after culling.

Finally, if the sample size is five or fewer, it is impossible to reject the null hypothesis from a two-sided sign test with $\alpha = 0.05$, no matter how different the true values are from the null hypothesized value. This underscores the fact that sign tests have low power and therefore require large sample sizes.

The Wilcoxon signed-rank test

The **Wilcoxon signed-rank test** is an improvement on the sign test for evaluating whether the median of a population is equal to a null-hypothesized constant. Unlike the sign test, the Wilcoxon signed-rank test retains information about magnitudes— that is, how far above or below the hypothesized median each data point lies. Unfortunately, it assumes that the distribution of measurements in the population is symmetric around the median—in other words, that no skew is present. This assumption is nearly as restrictive as the normality assumption of the one-sample *t*-test, greatly limiting its usefulness. (Skew, after all, is the usual reason the data don't fit the normal distribution, as in Example 13.4.) Because of this limitation, we do not explain the details of its calculation. Most statistics packages on the computer will carry out the Wilcoxon signed-rank test, but without highlighting its limitations.

13.5   Comparing two groups: the Mann–Whitney *U*-test

The **Mann–Whitney *U*-test** can be used in place of the two-sample *t*-test when the normal distribution assumption of the two-sample *t*-test cannot be met.[7] This method uses the ranks of the measurements to test whether the frequency distributions of two groups are the same. If the distributions of the two groups have the same shape, then the Mann–Whitney *U*-test compares the locations (medians or means) of the two groups.

> The *Mann–Whitney U-test* compares the distributions of two groups. It does not require as many assumptions as the two-sample *t*-test.

Example 13.5 shows how the Mann–Whitney *U*-test works.

---

[7]   Some statistical packages on the computer use an equivalent test, the Wilcoxon rank-sum test, instead of the Mann–Whitney *U*-test. The Wilcoxon rank-sum test statistic, *W*, is calculated differently from *U*, but both are based on the rank sums and give equivalent results.

| Example 13.5 | **Sexual cannibalism in sagebrush crickets** |

The sage cricket, *Cyphoderris strepitans*, has an unusual form of mating. During mating, the male offers his fleshy hind wings to the female to eat. The wounds are not fatal,[8] but a male with already nibbled wings is less likely to be chosen by females he meets subsequently. Females get some nutrition from feeding on the wings, which raises the question, "Are females more likely to mate if they are hungry?" Johnson

et al. (1999) answered this question by randomly dividing 24 females into two groups: one group of 11 females was starved for at least two days and another group of 13 females was fed during the same period. Finally, each female was put separately into a cage with a single (new) male, and the waiting time to mating was recorded. The data are listed in Table 13.5-1.

Table 13.5-1 Times to mating (in hours) for female sagebrush crickets that were recently starved or fed. The measurements of fed females are in red to facilitate comparison after ranking (see Table 13.5-2).

Starved	Fed
1.9	1.5
2.1	1.7
3.8	2.4
9.0	3.6
9.6	5.7
13.0	22.6
14.7	22.8
17.9	39.0
21.7	54.4
29.0	72.1
72.3	73.6
	79.5
	88.9

---

[8]   But they are disgusting.

We start by writing down our hypotheses:

H_0: Time to mating is the same for female crickets that were starved and for those that were fed.

H_A: Time to mating differs between these two groups.

This is a two-tailed test. The frequency distributions of the two groups are shown in the histograms in Figure 13.5-1.

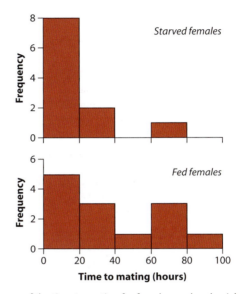

Figure 13.5-1 Histograms of the time to mating for female sagebrush crickets who were starved (*upper panel*) or fed (*lower panel*).

Neither of these groups is normally distributed, and a log transformation did not make them normally distributed either. With the relatively small sample sizes, we cannot count on the central limit theorem to yield a normal sampling distributions of means, so the two-sample *t*-test is a poor choice. Therefore, we will apply a nonparametric alternative test: the Mann–Whitney *U*-test.

There are $n_1 = 11$ crickets that were starved and $n_2 = 13$ that were fed. For convenience, we'll label these group 1 and group 2, respectively.

To use the Mann-Whitney *U*-test, first rank all the data from smallest to largest, combining data from both groups, as shown in Table 13.5-2. The smallest measurement from either data set is from group 2, at 1.5 hours. We assign this data point a rank of 1. The second smallest data point is 1.7 hours, which we assign a rank of 2. We continue to do this, looking at data from both groups together until we have ranked them all.

Table 13.5-2 Times to mating of female crickets from both groups, ordered from smallest to largest and then ranked. Data from group 2 (fed crickets) are highlighted in red to facilitate comparison.

Group	Time to mating	Rank
2	1.5	1
2	1.7	2
1	1.9	3
1	2.1	4
2	2.4	5
2	3.6	6
1	3.8	7
2	5.7	8
1	9.0	9
1	9.6	10
1	13.0	11
1	14.7	12
1	17.9	13
1	21.7	14
2	22.6	15
2	22.8	16
1	29.0	17
2	39.0	18
2	54.4	19
2	72.1	20
1	72.3	21
2	73.6	22
2	79.5	23
2	88.9	24

Second, calculate the *rank-sum* for one of the two groups (it doesn't matter which group, but it's easier to use the group with the fewest data points). R_1 is the sum of all of the ranks in group 1 (the starved group). The ranks for group 1 are 3, 4, 7, 9, 10, 11, 12, 13, 14, 17, and 21, which added together give

$$R_1 = 3 + 4 + 7 + 9 + 10 + 11 + 12 + 13 + 14 + 17 + 21 = 121.$$

Third, use the rank-sum to calculate a new quantity, U_1, as follows:

$$U_1 = n_1 n_2 + \frac{n_1(n_1 + 1)}{2} - R_1 = 11(13) + \frac{11(11 + 1)}{2} - 121 = 88.$$

U_1 is the number of times that a data point from sample 1 is smaller than a data point from sample 2, if we compare all possible pairs of points taken one from each sample.

Also calculate U_2, as follows:

$$U_2 = n_1 n_2 - U_1 = 11(13) - 88 = 55.$$

U_2 is the quantity associated with group 2. If we applied the equation for U_1 to the data from group 2, we would get the same answer (55).

Fourth, choose the larger of U_1 or U_2 as our test statistic, U. In this case, U_1 is larger than U_2, so our test statistic is

$$U = U_1 = 88.$$

Finally, determine the P-value by comparing the observed U with the critical value of the null distribution for U. Critical values for cases in which sample sizes are relatively small are provided in Statistical Table E. The null hypothesis is rejected if U equals or exceeds the critical value for U. For the cricket data, with $\alpha = 0.05$, $n_1 = 11$, and $n_2 = 13$, the critical value of U is

$$U_{0.05(2),\,11,13} = 106.$$

Our U (88) is less than this critical value (106), so $P > 0.05$ and we cannot reject the null hypothesis. The data, therefore, are insufficient evidence that time to mating is different between starved and fed female crickets.

"You slept with her, didn't you?"

Tied ranks

The data from Example 13.5 contain no observations tied for the same value, but often the same value appears more than once in a data set. When measurements are tied, we assign to all instances of the same measurement the average of the ranks that the tied points would have received. Table 13.5-3, for example, contains measurements from two hypothetical groups.

Table 13.5-3 Measurements from two hypothetical groups, illustrating how ties are ranked.

Group	Y	Rank
2	12	1
2	14	2
1	17	3
1	19	4.5
2	19	4.5
1	24	6
2	27	7
1	28	8

The ranks 1, 2, and 3 are assigned, as usual, to the three smallest values: 12, 14, and 17. But there are two measurements with the same value, 19. The two tied measurements, 19, would have been given ranks 4 and 5. Therefore, we assign each of the tied measurements a rank of $(4+5)/2 = 4.5$ (called the "midrank"). Thereafter, we continue to assign ranks to the data by assigning the next value, 24, a rank of 6, because we have already used up ranks 4 and 5 for the two tied measurements (mistakes are often made at this step when ranking by hand).[9]

Data can also be in three-, four-, or even more-way ties, but we apply the same procedure. If we had three values that would otherwise have been given ranks 6, 7 and 8, then all of those individuals would be assigned a midrank of $(6+7+8)/3 = 7$. Thereafter, we would continue by assigning the next higher data point the rank of 9.

Large samples and the normal approximation

Statistical Table E does not include critical values for the U-statistic when sample sizes are large. Fortunately, for medium and large sample sizes (n_1 and $n_2 > 10$), a transformation of the U-statistic,

$$Z = \frac{2U - n_1 n_2}{\sqrt{n_1 n_2 (n_1 + n_2 + 1)/3}},$$

has a sampling distribution that is well-approximated by the standard normal distribution if the null hypothesis is true. For example, the critical value for the Z-statistic at $\alpha = 0.05$ is 1.96 (Statistical Table B).

13.6   Assumptions of nonparametric tests

While nonparametric tests do not rely on the normal distribution, they still make assumptions. Nonparametric tests assume, for example, that both samples are random samples from their populations. Without random samples, nonparametric tests are as likely as parametric tests to give erroneous answers.

As mentioned in Section 13.4, the Wilcoxon signed-rank test assumes that the distribution of measurements in the population is symmetrical. This assumption limits the utility of the method.

The Mann–Whitney U-test compares the distributions of two groups. In practice, the method is used to compare the locations (e.g., the means or medians) of two

---

[9]   The null distribution for the U-statistic is not the same when there are ties as when there are no ties, if the ties are between members of different groups. In this case the test is conservative, meaning that using the critical values in Statistical Table E yields a Type 1 error rate lower than the stated value. Ties also reduce the power of the test, but the effect is not great when there are only a few ties. Corrections for ties exist but are tedious to compute by hand.

groups. To compare locations, however, requires the additional assumption that the distributions of the two groups have the same shape. For example, the two distributions must have the same variance and the same skew. The Mann–Whitney U-test is very sensitive to the assumption of equal shapes. In fact, the Type I error rate can be very different from α if the two distributions being compared have unequal variances or different skews (Zimmerman 2003). In some circumstances, the Mann–Whitney U-test is worse than the two-sample t-test in this respect. This limitation of the Mann–Whitney test is not often appreciated, and you will find many instances of misuse in the scientific literature.

13.7   Type I and Type II error rates of nonparametric methods

When the assumptions of a given test are met, the probability of making a Type I error (i.e., rejecting a true null hypothesis) is constant at α for both parametric and nonparametric tests. When the assumptions of a parametric test are not met, such as by a sharply nonnormal distribution of measurements, then the Type I error rate can become larger than the stated α-value, and sometimes this excess is extremely large (Ramsey 1980). This excess Type I error rate is the main reason why parametric tests should not be used when the assumptions are strongly violated. Under these conditions, a nonparametric test will yield a Type I error rate equal to α, provided its less restrictive assumptions are met.

The story is different for Type II errors (i.e., failing to reject a false null hypothesis). By using only ranks, nonparametric tests use less information from the data than do parametric tests. The actual magnitudes are discarded. This loss of information causes nonparametric tests to have *less power* than the corresponding parametric test, when the assumptions of the parametric test are met. Less power means that the nonparametric test has a lower probability of rejecting a false null hypothesis, and therefore a higher Type II error rate, than does the parametric test. For this reason, most biologists prefer to use parametric tests if the assumptions can be met, and turn to nonparametric methods only after data transformation has failed to meet the assumptions of the parametric methods.

The reduced power of nonparametric tests is immaterial when the assumptions of the parametric test are strongly violated. In that case, the parametric test cannot be used at all.

> Nonparametric tests are typically less powerful than parametric tests.

How much lower is the power of the sign test and the Mann–Whitney U-test compared with their corresponding parametric tests? We can only sensibly make this comparison for cases when the assumptions of the parametric tests are met. In this case, the power of a Mann–Whitney U-test is about 95% as great as the power of a two-sample t-test *when sample sizes are large* (Mood 1954). This is not too bad. The difference in power is greater, however, with smaller sample sizes. In the extreme case, when sample size in each of two groups is only two, the power of the Mann–Whitney U-test is zero.

The sign test has low power, much lower than the one-sample t-test and the paired t-test. Even with large samples, the sign test has only 64% of the power of a t-test (Mood 1954), and its power is even lower with smaller samples. Therefore, the sign test should be used only as a last resort, when the assumptions of the parametric tests simply cannot be met.

13.8   Summary

▶ Statistical methods such as the one-sample, paired, and two-sample t-tests, that make assumptions about the distribution of variables, are called parametric methods. Methods that do not make assumptions about the distribution of variables are called nonparametric methods.

▶ When the assumptions of parametric tests are violated, there are at least three alternative solutions: ignoring the violations, if they are minor; transforming the data to better meet the assumptions; and using nonparametric methods, if the previous two strategies are insufficient. (More options are discussed in Chapter 19.)

▶ A statistical method is robust if violations of its assumptions do not greatly affect its results.

▶ Methods to evaluate the fit of the normal distribution to a data set include visual inspection of histograms, normal quantile plots, and the Shapiro–Wilk test.

▶ Parametric methods for comparing means are robust to minor violations of the assumption of normal populations, especially if the sample size is large. It is acceptable to ignore minor violations with large data sets and to proceed with the parametric tests.

▶ The two-sample t-test and the confidence interval for the difference between two means are robust to minor violations of the assumption of equal standard deviations between populations, if the sample sizes are approximately equal in the two groups.

▶ Many data can be transformed mathematically to a new scale on which the assumptions of parametric methods are met. The parametric method can then be performed on the transformed data.

▶ The most common data transformations are the log transformation, the arcsine transformation, and the square-root transformation. Of these, the log transformation is used most often.

▶ The sign test is a nonparametric alternative to the paired t-test or one-sample t-test. It is much less powerful (it is less likely to reject a false null hypothesis) than the corresponding parametric tests.

> ▶ The Wilcoxon signed-rank test is a nonparametric alternative to the paired *t*-test. However, it assumes a symmetrical distribution, so it should be used with caution.
> ▶ The Mann–Whitney *U*-test is a nonparametric alternative to the two-sample *t*-test. The *U*-test is less powerful than the two-sample *t*-test, but it is almost as powerful when the sample size is large. To use the Mann–Whitney *U*-test to compare the locations of two distributions (medians or means), we must assume that the distributions of the variables in the two populations have the same shape.

13.9 Quick Formula Summary

Transformations

Log: $Y' = \ln[Y]$ or $Y' = \ln[Y + 1]$

Arcsine: $p' = \arcsin[\sqrt{p}]$

Square-root: $Y' = \sqrt{Y + 1/2}$

Back transformations

Log: $Y = e^{Y'}$

Arcsin: $p = (\sin[p'])^2$

Square root: $Y = Y'^2 - 1/2$

Sign test

What is it for? A nonparametric test of whether the median of a population equals a specified constant.

What does it assume? Random samples.

Test statistic: The number of measurements greater than (or less than) the median, according to the null hypothesis.

Formula: Identical to a binomial test with $H_0: p = 0.5$.

Mann–Whitney *U*-test

What is it for? A nonparametric test to compare the locations of two groups.

What does it assume? Random samples. The distributions of the variable in the two populations have the same shape.

Test statistic: U.

Sampling distribution under H_0: The U-distribution, with sample sizes n_1 and n_2.

Formula: $U_1 = n_1 n_2 + \dfrac{n_1(n_1 + 1)}{2} - R_1$ and $U_2 = n_1 n_2 - U_1$, where R_1 is the sum of the ranks for group 1. For a two-tailed test, U is the greater of U_1 or U_2. For $n \le 10$, use Statistical Table E. For large sample sizes, compare to the critical

$$Z = \frac{2U - n_1 n_2}{\sqrt{n_1 n_2 (n_1 + n_2 + 1)/3}}$$ value from the standard normal distribution.

PRACTICE PROBLEMS

1. The following four graphs are normal quantile plots for four different data sets, each sampled randomly from a different population. For each, say whether the distribution is close to a normal distribution. If not, say how the quantile plot differs from what you might expect from a normal distribution.

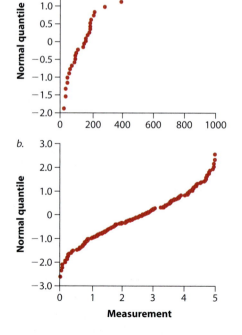

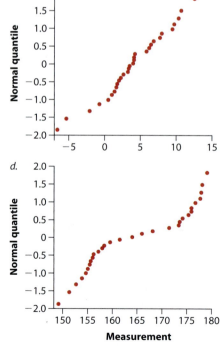

2. Below are frequency distributions of four hypothetical populations.

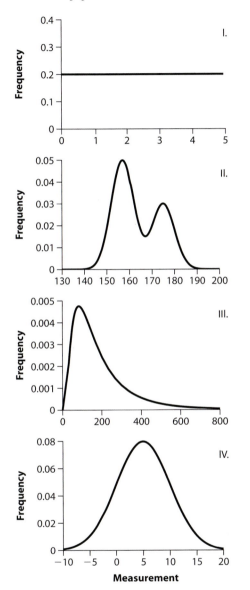

a. For each (I through IV), say whether the distribution appears to fit a normal distribution.

b. For each graph, imagine that you want to test the null hypothesis that the mean or median is zero. What statistical method would you use? If you suggest a transformation of the data, say why you chose that transformation.

c. Match the distributions given in this problem to the quantile plots in Practice Problem 1.

3. For each of the following sets of numbers, calculate the mean and 95% confidence interval of the mean of the log-transformed data. If that is impossible, say why it is impossible.

a. 10.2, 0.105, 67.3, 827
b. 2.1, 8.3, 3.2, 30.1
c. 17, −14, 37, 12
d. 12; 1.2; 125; 12,300
e. 0, 1.2, 4.5, 3.2

4. For the following data, calculate the mean and standard deviation after applying the arcsine transformation:

0.91, 0.99, 0.98, 0.973

5. Can lobsters sense the earth's magnetic field and use this information for navigation? A partial test of this question was carried out by capturing several lobsters from one locality in the Caribbean (Boles and Lohmann 2003). The lobsters were placed in light- and sound-proof boxes and then transported by a circuitous route to another location. Each lobster was then placed individually in a circular box, and, after the lobster had moved to the side of the box, the direction it was facing was recorded. The data are presented as the angle separating the direction the lobster faced relative to the direction to its home. (A lobster with score 0° is facing home. Lobsters with angles between −90° and +90° were in the half of the box towards home, whereas lobsters with angles greater than +90° or less than −90° were in the opposite half. Thus, these data are trickier than they appear, because values of −179 and 179 are almost identical!) The data for 15 lobsters are as follows:

80, 38, 25, 15, 10, 13, −2, −10, −30, −26, −35, −90, 115, −125, −130

Are lobsters able to sense the direction towards home without light, sound, or smell cues?

6. When intruding male lions take over a pride of females, they often kill most or all of the infants in the group. This reduces the time until the females are again sexually receptive. This infanticide has many consequences for the biology of lions. It may be the reason, for example, that female lions band together in groups in the first place (to be better able to repel invading males).[10]

The period after the takeover of a pride by a new group of males is an uncertain time, when the future of the pride is unpredictable. As a result, we might predict that females may delay ovulation until this uncertainty has passed. A long-term project working on the lions of Serengeti, Tanzania, measured the time to reproduction of female lions after losing cubs to infanticide and compared this to the time to reproduction of females that had lost their cubs to accidents (Packer and Pusey 1983). The data are given below in days. Does infanticide lead to a different mean delay to reproduction in females than when cubs die from other causes? The data are not normally distributed within groups, and we have been unable to find a transformation that makes them normal. Perform an appropriate statistical test.

Accidental death: 110, 117, 133, 135, 140, 168, 171, 238, 255

Infanticide: 211, 232, 246, 251, 275

7. Gesturing is common during human speech. Is this completely a learned behavior? A measure was made of the number of gestures produced by each of 12 pairs of sighted individuals while talking to sighted individuals (Iverson and Goldin-Meadow 1998). This was compared with the number of gestures produced while talking by each of 12 pairs of people who had been blind since birth, and therefore presumably unexposed to the gestures of others. The data[11] are as follows:

Sighted: 1, 0, 1, 2, 3, 0, 1, 2, 2, 0, 3, 1

Blind: 0, 1, 1, 2, 1, 1, 1, 3, 1, 0, 1, 1

Test the hypothesis that the number of gestures is related to sightedness, using a nonparametric test.

8. The skin of the rough-skinned newt, *Taricha granulosa*, stores a neurotoxin called tetrodotoxin (TTX), which is extremely poisonous. In some geographical areas, garter snakes, a newt predator, have evolved some resistance to this toxin, and in these areas the newts make up a substantial part of the snakes' diet.

As a first step to understanding the evolution of these traits, researchers wanted to know whether resistance to TTX varied between two specific populations of garter snakes (Geffeney et al. 2002). The data for resistance for two Oregon populations, one near Benton and the other near Warrenton, are listed in the following table:

Locality	Resistance (% reduction in crawl speed after TTX injection)
Benton	0.29
Benton	0.77
Benton	0.96
Benton	0.64
Benton	0.70
Benton	0.99
Benton	0.34
Warrenton	0.17
Warrenton	0.28
Warrenton	0.20
Warrenton	0.20
Warrenton	0.37

a. Calculate summary statistics on these data and prepare an appropriate graph to examine the frequency distributions. Why would a two-sample *t*-test be an inappropriate test?

---

[10] Male infanticide has strong conservation implications, too, because most lion hunting in Africa is targeted towards the trophy males, but killing these males may provoke a takeover of their pride by other males, resulting in infanticide.

[11] Extrapolated from summary statistics in the original paper.

b. List three appropriate methods that could be used to test a difference in the mean or location of these two populations.

c. Use a log transformation to test the hypothesis that the mean resistance does not differ between the two populations. Why might a log transformation be appropriate?

d. Calculate a 95% confidence interval for the mean log-transformed resistance for each population.

e. Back-transform the confidence interval from part (d) to the original scale. What does this back-transformed interval measure?

9. When producing a 95% confidence interval for the difference between the means of two groups, under what circumstances can a violation of the assumption of equal standard deviations be ignored?

10. For each of the sets of samples at the right, state which approach is best if the goal is to compare the mean of group 1 (on the left) to group 2 (on the right). Pay careful attention to differences between samples in the scales of the x-axis. These graphs have not been drawn according to the best practices that we outlined in Chapter 2 (histograms of two samples should be displayed

one above the other and on the same scale), but they are similar to those you might obtain if you plotted them separately with a computer program.

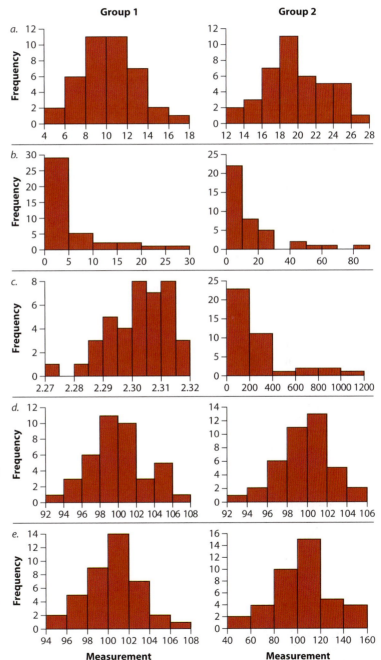

11. The nematode *Caenorhabditis elegans* is often used in studies of development and genetics. This species is somewhat unusual compared to most animals because most *C. elegans* individuals are hermaphrodites. That is, each worm has both ovaries and testes and acts as both a male and a female. Typically, a *C. elegans* individual will produce offspring by mating with itself. Very rarely, worms occur that have testes but not ovaries, and these individuals act as males that must mate with another individual to produce offspring. Therefore, it is important for the males to produce sperm that can compete well for access to eggs. LaMunyon and Ward (1998) measured the size of spermatids in 211 males and 700 hermaphrodites. Histograms of their findings are as follows:

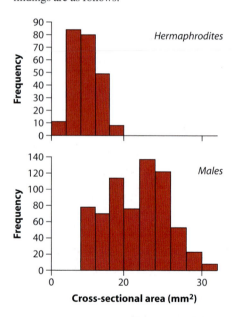

The distribution of spermatid size was significantly different from a normal distribution, so the researchers decided to perform a Mann–Whitney *U*-test. They wanted a two-sided test of the difference in spermatid size between the two types of worms. They got as far as calculating the value for *U*, which was $U = 35,910$.

a. What were their hypotheses?

b. Find the *P*-value for their test.

c. Why might a Mann–Whitney *U*-test be inappropriate for these data?

12. Only about 6% of plant species have separate male and female individuals (a syndrome called *dioecy*). The rest of the species have individuals with both male and female parts (called *monomorphism*). Why are there so many more monomorphic than dioecious species of plants? One possibility is that dioecious plants have low speciation rates or high extinction rates. To test this, Heilbuth (2000) compared the numbers of species in pairs of plant taxa. In each pair, one group was monomorphic and the other group was the most closely related taxon that is dioecious. The data are below. Compare the numbers of species in the two types of taxa. Plot the data to help you decide which test to perform. Do an appropriate test to determine whether there are significantly more or fewer species in the dioecious groups.

Taxon pair	Number of species in dioecious group	Number of species in monomorphic group
1	1	7000
2	1	5000
3	1150	4616
4	9	701
5	1	44
6	4	12
7	11	450
8	2	70
9	650	13
10	15	6
11	6	8
12	17	80
13	405	4
14	3	200
15	450	2770
16	4	11
17	50	53
18	370	2639
19	2	40
20	50	6
21	7	47
22	700	235
23	10	11
24	2	5
25	400	5772
26	400	72
27	1	6143
28	2	31

13. Suppose a researcher wants to test a difference between the means of two groups. Of the tests introduced so far, what is the only option when severe outliers are present in the data?

14. The distribution of body size of mosquitoes (as measured by weight) is known to have a log-normal distribution (that is, it is normally distributed if log-transformed). The (untransformed) weights of 11 female and nine male mosquitoes are given below in milligrams (Lounibos et al. 1995). Do the two sexes weigh the same on average?

Females: 0.291, 0.208, 0.241, 0.437, 0.228, 0.256, 0.208, 0.234, 0.32, 0.34, 0.15

Males: 0.185, 0.222, 0.149, 0.187, 0.191, 0.219, 0.132, 0.144, 0.14

15. One of the founders of modern population genetics, J. B. S. Haldane, was once asked if he could infer anything about God from his study of nature. He replied, "An inordinate fondness for beetles," reflecting the beetles' status as the most species-rich animal group. One hypothesis for why beetles are so common is that a large number of them are plant-eaters, and they may have ridden the coattails of the flowering plants (angiosperms) as the number of flowering plant species increased in numbers over the past 140 million years. A test of this hypothesis (Farrell 1997) compared the numbers of beetles in groups that feed on angiosperms to the number in the most closely related group that feeds on gymnosperms (non-flowering seed plants). Five such pairs of groups were compared in this way. The data are as follows:

Pair	Number of species in angiosperm eaters	Number of species in gymnosperm eaters
1	44,002	85
2	150	30
3	25,000	78
4	400	3
5	33,400	26

a. Are the differences in these groups likely to come from a population with a normal distribution?

b. Do the groups that eat angiosperms and the groups that eat gymnosperms have different numbers of species? Carry out a formal test that does not require the assumption of a normal distribution.

c. Comment on the *P*-value and your conclusion in part (b) in light of the fact that all five groups had more species in the angiosperm group. What does this say about the power of your test?

ASSIGNMENT PROBLEMS

16. Carotenoids are important pigments, responsible for much of the red we see in nature. For example, the bright red beak of the male zebra finch is derived from carotenoids.

Carotenoids are also important as antioxidants for humans and other animals, leading a group of researchers to predict that carotenoids in birds may affect immune system function. To test this, a group of zebra finches was randomly divided into two groups (McGraw and Ardia 2003). Ten individual finches received supplemental carotenoids, and 10 individuals did not. All 20 birds were then measured using an assay that measures cell-mediated immunocompetence (PHA) as well as an assay that measures humoral immunity (SRBC). The data are given below. The researchers had independent reasons

to believe that neither assay score would be normally distributed, so they preferred a non-parametric test.

a. Does PHA differ on average between the birds that received supplemental carotenoids and those that did not?

b. Does SRBC differ on average between the birds that received supplemental carotenoids and those that did not?

Treatment	PHA response	SRBC count
CAROT	1.511	2
CAROT	1.311	2
CAROT	1.460	3
CAROT	1.352	4
CAROT	1.491	5
CAROT	1.599	4
CAROT	1.653	8
CAROT	1.390	5
CAROT	1.779	7
CAROT	1.721	4
NO	1.454	4
NO	1.226	2
NO	1.198	3
NO	1.139	4
NO	1.277	3
NO	1.490	0
NO	0.912	3
NO	1.316	4
NO	1.234	2
NO	1.332	1

c. What assumptions did you require in parts (a) and (b)?

17. The conventional wisdom is that people who do a lot of sports are more successful in finding sexual partners than those who do not. To test this, a group of French researchers asked two groups of students how many sex partners they had had in the previous year (Faurie et al. 2004). One group was composed of 250 physical education majors who regularly participated in sports; the other group was composed of 50 biology majors[12] who did not regularly do sports. The distribution of number of sex partners is shown in the graph

[12]   Sorry, we're not making this up.

below. Biology majors claimed an average of 1.24 sex partners in the previous year, while sports majors claimed an average of 2.4 partners in the previous year.

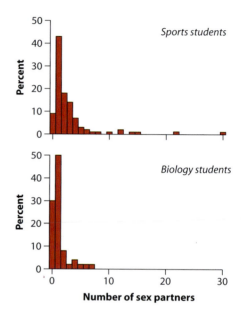

As you can see, the distributions are very different from a normal distribution, and even with a log transformation the distributions are not normal. As a result, the researchers used a Mann–Whitney U-test to compare the median numbers of sex partners for the two groups. They found that the U_1 value corresponding to the biology majors was 8500.5, while the U_2 value corresponding to the sportier group was 3999.5.

a. Finish the Mann–Whitney U-test for them. State the hypotheses and reach a conclusion.

b. Comment on your results in light of the assumptions of the Mann–Whitney U-test.

c. These researchers would like to know the *actual* number of sex partners for the two groups, rather than just the claimed number. What improvements can you suggest to their study design?

18. The males of some cichlid fish species are infertile until a few days after they become the socially dominant male in the presence of females. Males without a territory (and therefore without a hope of mating) have atrophied genitalia, while males who control a territory with females around have well-developed genitalia. They can shift from one state to the other in a matter of days. White et al. (2002) wanted to know the hormonal signal for this shift, and one candidate hormone was gonadotropin-releasing hormone (GnRH). They measured the levels of the messenger RNA (mRNA) of GnRH for five territorial fish (T) and for six non-territorial fish (NT). The data appear in the table at right.

Both distributions are skewed to the right.

a. What procedure or procedures could be used to test for a difference in the average GnRH levels between the T and NT fish?

b. Perform an appropriate hypothesis test for this difference.

c. Calculate a 95% confidence interval for the difference in GnRH mRNA level means between these two groups.

Territorial status	GnRH mRNA level
NT	0.504
NT	0.432
NT	0.744
NT	0.792
NT	0.672
NT	1.344
T	1.152
T	1.272
T	2.328
T	3.288
T	0.888

19. Some genera have far more species than others. Is this just luck, or have some genera hit upon a "key innovation" that gives them a benefit and allows more species to accumulate? One possible key innovation in plants is the ability to climb like a vine, which makes it possible to reach above other plants to compete for light. A study counted the number of species in 48 genera that had all evolved climbing (Gianoli 2004). For each of these genera, the researchers also found the most closely related non-climbing genus and counted the number of species in each. The numbers of species in these pairs of closely related genera are listed in the following table:

Climbing	Non-climbing	Climbing	Non-climbing	Climbing	Non-climbing
14	3	130	5	92	2
20	23	15	1	190	50
13	3	650	25	1041	262
49	42	845	34	13	8
300	9	6	1	1600	1
358	26	30	7	18	2
302	197	307	427	4	3
43	1	2888	710	150	3228
19	62	267	160	350	318
3	1	124	36	142	9
75	60	17	4	525	2
2	3	42	260	24	54
22	11	850	34	180	702
157	39	61	240	320	635
115	7	293	87	625	31
35	12	441	29	306	6

The number of species does not have a normal distribution in either climbing genera or non-climbing genera, and the differences between climbing genera and their most closely related non-climbing genera are also not normally distributed. Plot a histogram of the difference in number of species for each pair. Carry out an appropriate test of the difference in the number of species in climbing and non-climbing genera.

20. Sockeye salmon swim sometimes hundreds of miles from the Pacific Ocean, where they grow up, to rivers for spawning. Kokanee are a type of freshwater sockeye that spend their entire lives in lakes before swimming upriver to mate. In both types of fish, the males are bright red during mating.

This red coloration is caused by carotenoid pigments, which the fish cannot synthesize but get from their food. The ocean environment is much richer in carotenoids than the lake environment, which raises the question of how the kokanee males can get to be as red as the sockeye. One hypothesis is that the kokanee are much more efficient at using available carotenoids than the sockeye. This hypothesis was tested by an

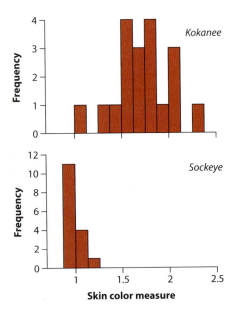

Skin color measure

experiment in which both sockeye and kokanee individuals were raised in the lab with low levels of carotenoids in their diets (Craig and Foote 2001). Their skin color was measured electronically. The data are as follows and are plotted in the accompanying histograms:

Kokanee: 1.11, 1.34, 1.55, 1.53, 1.50, 1.71, 1.87, 1.86, 1.82, 2.01, 1.95, 2.01, 1.66, 1.49, 1.59, 1.69, 1.80, 2.00, 2.30

Sockeye: 0.98, 0.88, 0.97, 0.99, 1.02, 1.03, 0.99, 0.97, 0.98, 1.03, 1.08, 1.15, 0.90, 0.95, 0.94, 0.99

a. List two methods that would be appropriate to test whether there was a difference in average skin color between the two groups.
b. Use a transformation to test the difference in mean between these two groups. Is there a difference in the mean of kokanee and sockeye skin color?

21. One measure of the exposure of a person to tobacco smoke is the cotinine-to-creatinine ratio (CCR). (Cotinine is formed in the body by breaking down nicotine.) Scientists measured this ratio in infants from smoking households (Blackburn et al. 2003). These households were divided into two groups: ones with strict controls to prevent exposure of the infant to smoke (31 babies) and another group with less strict controls (133 babies). The mean (and standard deviation) of the log-transformed CCR was 1.26 (1.58) in the babies from strict households and 2.58 (1.16) from babies from less-strict households. The distribution of the log-transformed cotinine-to-creatinine ratio was approximately normally distributed in both groups.

a. Do babies from strict or less-strict households differ significantly in their exposure to smoke? Perform an appropriate test.
b. On the non-transformed scale, how much higher is the cotinine-to-creatinine ratio in the less-strict group, compared to the strict-household babies?
c. Is this an observational or experimental study? Explain.

22. We are accustomed to thinking that the proportion of males at birth is fixed by genetics in birds and mammals to be close to 50%. Some have suggested, however, that the sex of offspring can be adjusted by females, such as in response to the quality of her mate or the number of helpers she has. West and Sheldon (2002) found a total of 15 studies, all done on birds, that have measured changes in the sex ratio in response to such social factors, and each has expressed its results in terms of a correlation coefficient (see Chapter 16). The correlation coefficient is positive if the change in the sex ratio of offspring is in agreement with evolutionary theory, and zero or negative if the data disagree with the theory. The measures are as follows:

-0.16, -0.037, 0.034, 0.144, 0.137, 0.118, 0.395, 0.363, 0.35, 0.376, 0.253, 0.44, 0.453, 0.46, 0.563

The frequency distribution of the correlation coefficients is shown in the following histogram:

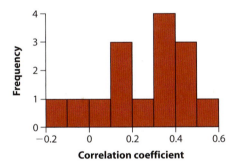

Correlation coefficient

a. What method could be used to test whether these data are consistent with a mean correlation coefficient of zero? Discuss why you would use this method in contrast to specific other methods.

b. Apply the best method to these data to test whether the correlation coefficient is typically positive.

23. Researchers have observed that rainforest areas next to clear-cuts (less than 100 meters away) have a reduced tree biomass. To go further, Laurance et al. (1997) tested whether rainforest areas more distant from the clear-cuts were also affected. They compiled data on the biomass change after clear-cutting (in tons/hectare/year) for 36 rainforest areas between 100 m and several kilometers from clear-cuts. The data are

-10.8, -4.9, -2.6, -1.6, -3, -6.2, -6.5, -9.2, -3.6, -1.8, -1, 0.2, 0.2, 0.1, -0.3, -1.4, -1.5, -0.8, 0.3, 0.6, 1, 1.2, 2.9, 3.5, 4.3, 4.7, 2.9, 2.8, 2.5, 1.7, 2.7, 1.2, 0.1, 1.3, 2.3, 0.5

These measurements are plotted in the following histogram:

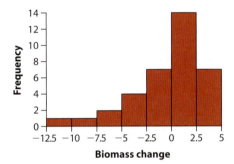

Biomass change

Test whether there is an average change in biomass of these forest areas following clear-cutting.

24. Male zebra finches have bright red beaks (see Assignment Problem 16), and experiments have shown that females prefer males with the reddest beaks. The red coloration in the beak comes from a class of pigments called carotenoids, which must be obtained in the diet of the birds. Pairs of brother finches were randomly assigned to alternative treatments: one was fed extra dietary carotenoids, whereas the other was fed a diet low in carotenoids (Blount et al. 2003). Each of 10 females was given a choice between brothers. Her preference was measured by the percentage of time she sat next to the carotenoid-supplemented male, standardized so that zero means equal time for each brother. The preference data for each of the 10 pairs are

23, 27, 57, 15, 15, 54, 34, 37, -65, 12

Test whether females preferred one type of male over the other type.

14

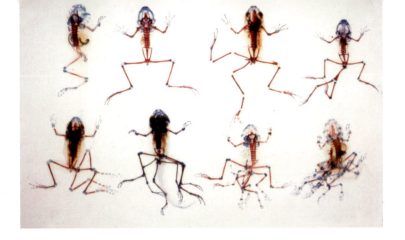

Designing experiments

Two types of investigations are carried out in biology: observational and experimental. In an experimental study, the researcher assigns treatments to units or subjects so that differences in response can be compared. In an observational study, on the other hand, nature does the assigning of treatments to subjects. The researcher has no influence over which subjects receive which treatment.

What's so important about the distinction? Whereas observational studies can identify associations between treatment and response variables, properly designed experimental studies can identify the *causes* of these associations.

How do we best design an experiment to get the most information possible out of it? The short answer is that we must design to eliminate bias and to reduce the influence of sampling error. The present chapter outlines the basics on how to accomplish this feat. We also briefly discuss how to design an observational study: by taking the best features of experimental designs and incorporating as many of them as possible. Finally, we discuss how to plan the sample size needed in an experimental or observational study.

14.1   Why do experiments?

In an experimental study, there must be at least two treatments and the experimenter (rather than nature) must assign them to units or subjects. The crucial advantage of experiments derives from the *random* assignment of treatments to units. Random assignment, or randomization, minimizes the influence of *confounding* variables (Interleaf 4), allowing the experimenter to isolate the effects of the treatment variable.

Confounding variables

Studies in biology are usually carried out with the aim of deciding how an explanatory variable or treatment affects a response variable. How are injury rates in cats with "high-rise syndrome" affected by the number of stories fallen? How does the state of estrus of a cow affect the risk of vampire bat attack? How does the use of supplemental oxygen affect the probability of surviving an ascent of Mount Everest? The easiest way to address these questions is with an observational study—that is, to gather measurements of both variables of interest on a set of subjects and estimate the association between them. If the two variables are correlated or associated, then one may be the cause of the other.

The limitation of the observational approach is that, by itself, it cannot distinguish between two completely different reasons behind an association between an explanatory variable X and a response variable Y. One possibility is that X really does cause a response in Y. For example, taking supplemental oxygen might increase the chance of survival during a climb of Mount Everest. The other possibility is that the explanatory variable X has no effect at all on the response variable Y; they are associated only because other variables affect both X and Y at the same time. For example, the use of supplemental oxygen might just be a benign indicator of a greater overall preparedness of the climbers that use it, and greater preparedness rather than oxygen use is the real cause of the enhanced survival. Variables (like preparedness) that distort the causal relationship between the measured variables of interest (oxygen use and survival) are called *confounding variables*. Recall from Interleaf 4, for example, that ice cream consumption and violent crime are correlated, but neither is the cause of the other. Instead, increases in both ice cream consumption and crime are caused by higher temperatures. Temperature is a confounding variable in this example.

> A *confounding variable* is a variable that masks or distorts the causal relationship between measured variables in a study.

Confounding variables bias the estimate of the causal relationship between measured explanatory and response variables, sometimes even reversing the apparent effect of one on the other. For example, observational studies have indicated that breast-fed babies have lower weight at six and 12 months of age compared with formula-fed infants (Interleaf 4). But an experimental study using randomization found that mean infant weight was actually *higher* in breast-fed babies at six months of age and was not less than that in formula-fed babies at 12 months (Kramer et al. 2000). The observed relationship between breast feeding and infant growth was confounded by unmeasured variables such as the socio-economic status of the parents.

With an experiment, random assignment of treatments to subjects allows researchers to tease apart the effects of the explanatory variable from those of confounding variables. With random assignment, no confounding variables will be associated with treatment except by chance. For example, if women who choose to breast feed their babies have a different average socio-economic background than women who choose to feed their infants formula, randomly assigning the treatments "breast feeding" and "formula feeding" to women in an experiment will break this connection, roughly equalizing the backgrounds of the two treatment groups. In this case, any resulting difference between groups in infant weight (beyond chance) must be caused by treatment.

Experimental artifacts

Unfortunately, experiments themselves might inadvertently create artificial conditions that distort cause and effect. Experiments should be designed to minimize artifacts.

> An *experimental artifact* is a bias in a measurement produced by unintended consequences of experimental procedures.

For example, experiments conducted on aquatic birds have shown that their heart rates drop sharply when they are forcibly submerged in water, compared with individuals remaining above water. The drop in heart rate has been interpreted as an oxygen-saving response. Later studies using improved technology showed that voluntary dives do not produce such a large drop in heart rate (Kanwisher et al. 1981). This suggested that a component of the heart rate response in forced dives was induced by the stress of being forcibly dunked under water, rather than the dive itself. The experimental conditions introduced an artifact that for a time went unrecognized.

To prevent artifacts, experimental studies should be conducted under conditions that are as natural as possible. Observational studies can provide important insight into what is the best setting for an experiment.

14.2 **Lessons from clinical trials**

The gold standard of experimental designs is the **clinical trial**, an experimental study in which two or more treatments are assigned to human subjects. The design of clinical trials has been refined because the cost of making a mistake with human subjects is so high. Experiments on nonhuman subjects are simply called "laboratory experiments" or "field experiments," depending on where they take place. Experimental studies in all areas of biology have been greatly informed by procedures used in clinical trials.

> A *clinical trial* is an experimental study in which two or more treatments are applied to human subjects.

Before we dig into the logic of the main components of experimental design, let's look at the clinical trial in Example 14.2, which incorporates many of these features.

Example 14.2 **Reducing HIV transmission**

Transmission of the HIV-1 virus via sex workers contributes to the rapid spread of AIDS in Africa. How can this be reduced? In laboratory experiments, the spermicide nonoxynol-9 had shown in vitro activity against HIV-1, shown schematically at the right. This finding motivated a clinical trial by van Damme et al. (2002), who tested whether a vaginal gel containing the chemical would reduce the risk of acquiring the disease by female sex workers. Data were gathered on a volunteer sample of 765
HIV-free sex-workers in six clinics in Asia and Africa. Two gel treatments were assigned randomly to women at each clinic. One gel contained nonoxynol-9 and the other contained a placebo (an inactive compound that subjects could not distinguish from the treatment of interest). Neither the subjects nor the researchers making observations at the clinics knew who had received the treatment and who had received the placebo. (A system of numbered codes kept track of who got which treatment.) By the end of the experiment, 59 of 376 women in the nonoxynol-9 group (15.9 %) were HIV-positive (Table 14.2-1), compared with 45 out of 389 women in the placebo group (11.6 %). Thus, the odds of contracting HIV-1 were slightly higher in the nonoxynol-9 group compared with the placebo group, which was the opposite of the expected result.

Design components

A good experiment is designed with two objectives in mind:

▶ to reduce bias in estimating and testing treatment effects, and
▶ to reduce the effects of sampling error.

Table 14.2-1 Results of the clinical trial in Example 14.2 (*n* is the number of subjects).

Clinic	Nonoxynol-9		Placebo	
	n	**Number infected**	*n*	**Number infected**
Abidjan	78	0	84	5
Bangkok	26	0	25	0
Cotonou	100	12	103	10
Durban	94	42	93	30
Hat Yai 2	22	0	25	0
Hat Yai 3	56	5	59	0
Total	376	59	389	45

The most significant elements of the design of the clinical trial in Example 14.2 addressed these two objectives. To reduce bias, the experiment included:

1. A simultaneous control group: the study included both the treatment of interest and a control group (the women receiving the placebo).
2. Randomization: treatments were randomly assigned to women at each clinic.
3. Blinding: neither the subjects nor the clinicians knew which women were assigned which treatment.

To reduce the effects of sampling error, the experiment included:

1. Replication: the study was carried out on multiple independent subjects.
2. Balance: the number of women was nearly equal in the two groups at every clinic.
3. Blocking: subjects were grouped according to the clinic they attended, yielding multiple repetitions of the same experiment in different settings (i.e., "blocks").

> The goal of experimental design is to eliminate bias and to reduce sampling error when estimating and testing the effects of one variable on another.

In Section 14.3, we discuss the virtues of the three main strategies used to reduce bias—namely, simultaneous controls, randomization, and blinding. In Section 14.4, we explain the strategies used to reduce the effects of sampling error—namely, replication, balance, and blocking.

14.3   How to reduce bias

We have seen how confounding variables in observational studies can bias the estimated effects of an explanatory variable on a response variable. The following experimental procedures are meant to eliminate bias.

Simultaneous control group

A **control group** is a group of subjects who are treated like all of the experimental subjects, except that the control group does not receive the treatment of interest.

> A *control group* is a group of subjects who do not receive the treatment of interest but who otherwise experience similar conditions as the treated subjects.

In an uncontrolled experiment, a group of subjects is treated in some way and then measured to see how they have responded. Lacking a control group for comparison, such a study cannot determine whether the treatment of interest is the cause of any of the observed changes. There are several possible reasons for this, including the following:

▶ Sick human subjects selected for a medical treatment may tend to "bounce back" toward their average condition regardless of any effect of the treatment (Interleaf 6).

▶ Stress and other impacts associated with administering the treatment (such as surgery or confinement) might themselves produce a response separate from the effect of the treatment of interest.

▶ The health of human subjects often improves after treatment merely because of their expectation that the treatment will have an effect. This phenomenon is known as the *placebo effect* (Interleaf 6).

The solution to all of these problems is to include a control group of subjects measured for comparison. The treatment and control subjects should be tested simultaneously or in random order, to ensure that any temporal changes in experimental conditions do not affect the outcome.

The appropriate control group will depend on the circumstance. Here are some examples:

▶ In clinical trials, either a placebo or the currently accepted treatment should be provided, such as in Example 14.2. A placebo is an inactive treatment that subjects cannot distinguish from the main treatment of interest.

▶ In experiments requiring intrusive methods to administer treatment, such as injections, surgery, restraint, or confinement, the control subjects should be perturbed in the same way as the other subjects, except for the treatment itself, as far as ethical considerations permit. The "sham operation," in which surgery is carried out without the experimental treatment itself, is an example. Sham operations are very rare in human studies, but they are more common in animal experiments.

▶ In field experiments, applying a treatment of interest may physically disturb the plots receiving it and the surrounding areas, perhaps by trampling the ground by the researchers. Ideally, the same disturbance should be applied to the control plots.

Often it is desirable to have more than one control group. For example, two control groups, where one is a harmless placebo and the other is the best existing treatment, may be used in a study so that the total effect of the treatment and the improvement of the new treatment over the old may both be measured. However, using resources for multiple controls might reduce the power of the study to test its main hypotheses.

Randomization

Once the treatments have been chosen, the researcher should *randomize* their assignment to units or subjects in the sample. **Randomization** means that treatments are assigned to units at random, such as by flipping a coin. Chance rather than conscious or unconscious decision determines which units end up receiving the treatment of interest and which receive the control. A **completely randomized design** is an experimental design in which treatments are assigned to all units by randomization.

> *Randomization* is the random assignment of treatments to units in an experimental study.

The virtue of randomization is that it breaks the association between possible confounding variables and the explanatory variable, allowing the causal relationship between the explanatory and response variables to be assessed. Randomization doesn't eliminate the variation contributed by confounding variables, only their correlation with treatment. It ensures that variation from confounding variables is similar between the different treatment groups. If randomization is done properly, any remaining influence of confounding variables occurs by chance alone, which statistical methods can account for.

Randomization should be carried out using a random process. The following steps describe one way to assign treatments randomly:

1. List all *n* subjects, one per row, in a computer spreadsheet.
2. Use the computer to give each individual a random number.[1]
3. Assign treatment A to those subjects receiving the lowest numbers and treatment B to those with the highest numbers.

This process is demonstrated in Figure 14.3-1, where eight subjects are assigned to two treatments, A and B.

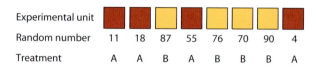

Experimental unit								
Random number	11	18	87	55	76	70	90	4
Treatment	A	A	B	A	B	B	B	A

Figure 14.3-1 A procedure for randomization. Each of eight subjects was assigned a number between 0 and 99 that was drawn at random by a computer. Treatment A (*colored red*) was assigned to the four subjects with the lowest random numbers, whereas treatment B (*gold*) was assigned to the rest.

Other ways of assigning treatments to subjects are almost always inferior because they do not eliminate the effects of confounding variables. For example, the following methods can lead to problems:

▶ Assign treatment A to all patients attending one clinic and treatment B to patients attending a second clinic. (Problem: All of the other differences between the two clinics become confounding variables. If one clinic is better than the other in general, then the difference in clinic quality would show up as a difference in treatments.)

▶ Assign treatments to human subjects alphabetically. (Problem: This might inadvertently group individuals having the same nationality, generating unwanted differences between treatments in health histories and genetic variables.)

It is important to use a computer random-number generator or random-number tables to assign individuals randomly to treatments. "Haphazard" assignment, in which the researcher chooses a treatment while trying to make it random, has repeatedly been shown to be nonrandom and prone to bias.[2]

Blinding

The process of concealing information from participants about which subjects receive which treatments is called **blinding**. Blinding prevents subjects and

---

[1] The Random.org web site at http://random.org/sform.html will also do this.

[2] What do you do if, by chance, the first four of eight units are all assigned treatment A and the last four are assigned treatment B, yielding the arrangement AAAABBBB? Some biologists might randomize again to ensure the interspersion of treatments, but that is not strictly legitimate. If the first four units are different somehow from the last four, apart from treatment, then blocking (Section 14.4) should be considered as a remedy.

researchers from changing their behavior, consciously or unconsciously, which could result from knowing which treatment they were receiving or administering. For example, a researcher who believes that acupuncture helps alleviate back pain might unconsciously interpret a patient's report of pain differently if the researcher knows the patient was assigned the acupuncture treatment instead of a placebo. This might explain why studies that have shown acupuncture has a significant effect on back pain are limited to those without blinding (Ernst and White 1998). Studies implementing blinding, on the other hand, have not found that acupuncture has an ameliorating effect on back pain.

In a **single-blind experiment**, the subjects are unaware of the treatment to which they have been assigned. This requires that the treatments be indistinguishable to the subjects, a particular necessity in experiments involving humans. Single-blinding prevents subjects from responding differently according to their knowledge of their treatment. This is not much of a concern in non-human studies.

In a **double-blind experiment**, the researchers administering the treatments and measuring the response are also unaware of which subjects are receiving which treatments. This prevents researchers interacting with the subjects from behaving differently towards them according to their treatments. Researchers sometimes have pet hypotheses, and they might treat experimental subjects in different ways depending on their hopes for the outcome. Moreover, many response variables are difficult to measure and require some subjective interpretation, which makes the results prone to a bias in favor of the researchers' wishes and expectations. Finally, researchers are naturally more interested in the treated subjects than the control subjects, and this increased attention can itself result in improved response. Reviews of medical studies have revealed that studies carried out without double-blinding exaggerated treatment effects by 16% on average, compared with studies carried out with double-blinding (Jüni et al. 2001).

> *Blinding* is the process of concealing information from participants (sometimes including researchers) about which subjects receive which treatment.

Experiments on nonhuman subjects are also prone to bias from lack of blinding. Bebarta et al. (2003) reviewed 290 two-treatment experiments carried out on animals or on cell lines. They found that the odds of detecting a positive effect of treatment were more than threefold higher in studies without blinding than in studies with blinding.[3] Blinding can be incorporated into experiments on nonhuman subjects using coded tags that identify the subject to a "blind" observer without revealing the treatment (and who measures units from different treatments in random order).

---

[3] This probably overestimates of the effects of a lack of blinding, because the experiments without blinding also tended to have confounding problems, such as a lack of randomization (Bebarta et al. 2003).

14.4   How to reduce the influence of sampling error

Assuming we have designed our experiment to minimize sources of bias, there is still the problem of detecting any treatment effects against the background of variation between individuals ("noise") caused by other variables. Such variability creates sampling error in the estimates, reducing precision and power. How can the effects of sampling error be minimized?

One way to reduce noise is to make the experimental conditions constant. Fix the temperature, humidity, and other environmental conditions, for example, and use only subjects that are of the same age, sex, genotype, and so on. In field experiments, however, highly constant experimental conditions might not be feasible. Constant conditions might not be desirable, either. By limiting the conditions of an experiment, we also limit the generality of the results—that is, the conclusions might apply only under the conditions tested and not more broadly. Until recently, a significant source of bias in medical practice stemmed from the fact that many clinical tests of the effects of medical treatments were carried out only on men, yet the treatments were subsequently applied to women as well (e.g., McCarthy 1994).

In this section, we review replication, balance, and blocking, the three main statistical design procedures used to minimize the effects of sampling error. We also review a strategy to reduce the effect of noise by using extreme treatments.

Replication

Because of variability, **replication**—the repetition of every treatment on multiple experimental units—is essential. Without replication, we would not know whether response differences were due to the treatments or just chance differences between the treatments caused by other factors. Studies that use more units (i.e., that have larger sample sizes) will have smaller standard errors and a higher probability of getting the correct answer from a hypothesis test. Larger samples mean more information, and more information means better estimates and more powerful tests.

> *Replication* is the application of every treatment to multiple, independent experimental units.

Replication is not just about the number of plants or animals used, but true replication depends on the number of independent units in the experiment. An "experimental unit" is the independent unit to which treatments are assigned. Figure 14.4-1 shows three hypothetical experiments designed to compare the effects of two temperature treatments on plant growth. The lack of replication is obvious in the first design (top panel), because there is only one plant per treatment. You won't see many published experiments like it.

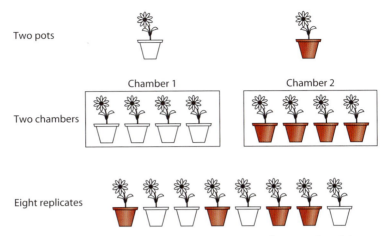

Figure 14.4-1 Three experimental designs used to compare plant growth under two temperature treatments (indicated by the shading of the pots). The upper ("two pots") and middle ("two chambers") designs are unreplicated.

The lack of replication is less obvious in the second design (the middle panel of Figure 14.4-1). Although there are multiple plants per treatment in the second design, all of the plants in one treatment are confined to one temperature-control chamber and all of the plants in the second treatment are confined to another chamber. If there are environmental differences between chambers (e.g., differences in light conditions or humidity) beyond those stemming from the treatment itself, then plants in the same chamber will be more similar in their responses than plants in different chambers, apart from any treatment effects. The plants in the same chamber are not independent. As a result, the chamber, not the plant, is the experimental unit in a test of temperature effects. Because there are only two chambers, one per treatment, the experiment is unreplicated.

Only the third design (the bottom panel) in Figure 14.4-1 is properly replicated, because here treatments have been randomly assigned to individual plants. A giveaway indicator of replication in the third design is *interspersion* of experimental units assigned different treatments, which is an expected (although not inevitable) outcome of randomization. Such interspersion is lacking in the two-chamber design (the middle panel in Figure 14.4-1), which is a clear sign that there is a replication problem.

An experimental unit might be a single animal or plant if individuals are randomly sampled and assigned treatments independently. Or, an experimental unit might be made up of a batch of individual organisms treated as a group, such as a field plot containing multiple individuals, a cage of animals, a household, a Petri dish, or a family. Multiple individual organisms belonging to the same unit (e.g., plants in the same plot, bacteria in the same dish, members of the same family, and so on) should be considered together as a single replicate if they are likely to be more similar on average to each other than to individuals in separate units (apart from the effects of treatment).

Correctly identifying replicates in an experiment is crucial to planning its design and analyzing the results. Erroneously treating the single organism as the independent replicate when the chamber (Figure 14.4-1) or field plot is the experimental unit will lead to calculations of standard errors and P-values that are too small. This is pseudoreplication, as discussed in Interleaf 2.

From the standpoint of reducing sampling error, more replication is always better. As proof, examine the formula for the standard error of the difference between two sample mean responses to two treatments, $\bar{Y}_1 - \bar{Y}_2$:

$$SE_{\bar{Y}_1-\bar{Y}_2} = \sqrt{s_p^2\left(\frac{1}{n_1} + \frac{1}{n_2}\right)}.$$

The symbols n_1 and n_2 refer to the number of experimental units, or replicates, in each of the two treatments. Based on this equation, increasing n_1 and n_2 directly reduces the standard error, increasing precision. Increased precision yields narrower confidence intervals and more powerful tests of the difference between means. On the other hand, increasing sample size also has costs in terms of time, money, and even lives. We discuss how to plan a sufficient sample size in more detail in Section 14.7.

Balance

A study design is **balanced** if all treatments have the same sample size. Conversely, a design is *un*balanced if there are *un*equal sample sizes between treatments.

> In a *balanced* experimental design, all treatments have equal sample size.

Balance is a second way to reduce the influence of sampling error on estimation and hypothesis testing. To appreciate this, look again at the equation for the standard error of the difference between two treatment means (above). For a fixed total number of experimental units, $n_1 + n_2$, the standard error is smallest when the quantity

$$\left(\frac{1}{n_1} + \frac{1}{n_2}\right)$$

is smallest, which occurs when n_1 and n_2 are equal. Convince yourself that this is true by plugging in some numbers. For example, if the total number of units is 20, the quantity $1/n_1 + 1/n_2$ is 0.2 when $n_1 = n_2 = 10$, but it is 1.05 when $n_1 = 19$ and $n_2 = 1$. With better balance, the standard error is much smaller.

To estimate the difference between two groups, we need precise estimates of the means of *both* groups. With an unbalanced design, we may know the mean of one group with great precision, but this does not help us much if we have very little information about the other group that we're comparing it to. Balance allocates the sampling effort in the optimal way.

Balance has other benefits, which we discuss elsewhere in the book. For example, the methods based on the normal distribution for comparing population means are most robust to departures from the assumption of equal variances when designs are balanced or nearly so (see Chapters 12 and 15).

Blocking

Blocking is an experimental design strategy used to account for extraneous variation by dividing the experimental units into groups, called **blocks** or strata, that share common features. Within blocks, treatments are assigned randomly to experimental units. Blocking essentially repeats the same completely randomized experiment multiple times, once for each block, as shown schematically in Figure 14.4-2. Differences between treatments are evaluated only within blocks, and, in this way, much of the variation arising from differences between blocks can be discarded.

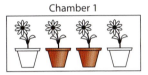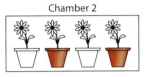

Figure 14.4-2 An experimental design incorporating blocking to test effects of fertilizer on plant growth (see Figure 14.4-1). The shading of the pots indicates which fertilizer treatment each plant received. Chambers might differ in unknown ways and add unwanted noise to the experiment. To remove the effects of such variation, carry out the same completely randomized experiment separately within each chamber. In this design, each chamber represents one block.

The women participating in the nonoxynol-9 HIV study discussed in Example 14.2 were grouped according to the clinic they attended. This made sense because there were age differences between women attending different clinics, as well as differences in condom usage and sexual practices, all of which are likely to affect HIV transmission rates (van Damme et al. 2002). By blocking, the variation in response among clinics is removed, allowing more precise estimates and more powerful tests of the treatment effects.

> *Blocking* is the grouping of experimental units that have similar properties. Within each block, treatments are randomly assigned to experimental units.

The *paired* design for two treatments (Chapter 12) is an example of blocking. In a paired design, both of two treatments are applied to each plot or other experimental unit representing a block. The difference between the two responses made on each block is the measure of the treatment effect. This measure of the treatment effect is not influenced by variation between blocks, heightening precision and power.

The **randomized block design** is analogous to the paired design, but it has more than two treatments, as shown in Example 14.4A.

> The *randomized block design* is like a paired design but for more than two treatments.

Example 14.4A **Holey waters**

The compact size of water-filled tree holes, which can harbor diverse communities of aquatic insect larvae, make them useful microcosms for ecological experiments. Srivastava and Lawton (1998) made artificial tree holes from plastic that mimicked the buttress tree holes of European beech trees (see image on right). They placed the plastic holes next to trees in a forest in southern England to examine how the amount of decaying leaf litter present in the holes affected the number of insect eggs deposited (mainly by mosquitoes and hover flies) and the survival of the larvae emerging from those eggs. Leaf litter is the source of all nutrients in these holes, so increasing the amount of litter might  result in more food for the insect larvae. There were three different treatments. In one treatment (LL), a low amount of leaf litter was provided. In a second treatment (HH), a high level of debris was provided. In the third treatment (LH), leaf litter amounts were initially low but were then made high after eggs had been deposited. A randomized block design was used in which artificial tree holes were laid out in triplets (blocks). Each block consisted of one LL tree hole, one HH tree hole, and one LH tree hole. The location of each treatment within a block was randomized, as shown in Figure 14.4-3.

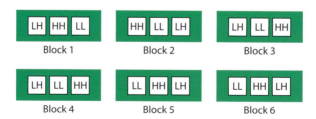

Figure 14.4-3 Schematic of the randomized block design used in the tree-hole study of Example 14.4A. Each block of three tree holes was placed next to its own beech tree in the woods. Within blocks, the three treatments were randomly assigned to tree holes.

As in the paired design, treatment effects in a randomized block design are measured by differences between treatments exclusively within blocks, a strategy that minimizes the influence of variation among blocks.

In the randomized block design, each treatment is applied once to every block. By accounting for some sources of sampling variation, such as the variation among trees, blocking can make differences between treatments stand out. In Chapter 18, we discuss in greater detail how to analyze data from a randomized block design.

Blocking is worthwhile if units within blocks are relatively homogeneous, apart from treatment effects, and units belonging to different blocks vary because of environmental or other differences. For example, blocks can be made up of the following:

- ▶ field plots experiencing similar local environmental conditions,
- ▶ animals from the same litter,
- ▶ aquaria located on the same side of the room,
- ▶ patients attending the same clinic, or
- ▶ runs of an experiment executed on the same day.

One potential drawback to blocking might occur if the effects of one treatment contaminate the effects of the other in the same block. For example, watering one half of a block might raise the soil humidity of the adjacent, unwatered half. Experiments should be designed carefully to minimize these artifacts.

Extreme treatments

Treatment effects are easiest to detect when they are large. Small differences between treatments are difficult to detect and require larger samples, whereas larger treatment differences are more likely to stand out against random variability within treatments. Therefore, one strategy to enhance the probability of detecting differences in an experiment is to include extreme treatments. Example 14.4B illustrates this approach.

Example 14.4B   **Plastic hormones**

Bisphenol-A, or BPA, is an estrogenic compound found in plastics widely used to line food and drink containers and in dental sealants. Human daily exposures are typically in the range of 0.5−1 μg/kg body weight (Gray et al. 2004). Sakaue et al. (2001) measured sperm production of 13-week-old male rats exposed to fixed daily doses of BPA between 0 and 2000 μg/kg  body weight for six days. The results are shown in a "dose–response" curve in Figure 14.4-4.

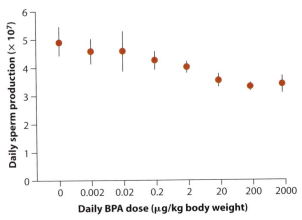

Figure 14.4-4 A dose–response curve showing the results of an experiment measuring the rates of sperm production of male rats exposed to fixed daily doses of bisphenol-A (BPA) (Sakaue et al. 2001). Symbols are the mean $\pm$ 1 SE.

This experiment included doses much higher than the typical doses faced by humans at risk, a strategy that enhanced the ability to detect an effect of BPA. For example, Figure 14.4-4 shows that there was a much larger difference in mean sperm production between the 0 and 2000 μg/kg groups than between the 0 and 0.002 μg/kg treatments.

A larger dose, or stronger treatment, can increase the probability of detecting a response. But be aware that the effects of a treatment do not always scale linearly with the magnitude of a treatment. The effects of a large dose may be qualitatively different from those of a smaller, more realistic dose. Still, as a first step, extreme treatments can be a very good way to detect whether one variable has any effect at all on another variable.

14.5 Experiments with more than one factor

Up to now, we have considered only experiments that focus on measuring and testing the effects of a single "factor." A **factor** is a single treatment variable whose effects are of interest to the researcher. However, many experiments in biology investigate more than one factor, because it is often advantageous (i.e., more efficient) to answer two questions from a single experiment rather than just one.

Another reason to consider experiments with multiple factors is that the factors might *interact*. For example, human activity has driven global increases in atmospheric CO_2 and temperature, as well as greater nitrogen deposition and precipitation. Increases in all of these factors have been shown to stimulate plant growth by experimental studies in which each treatment variable was examined separately. But what are the effects of these factors in combination? The only way to answer this is to design experiments in which more than one factor is manipulated simultaneously. If the climate variables interact when influencing plant growth, then their joint effects can be very different from their separate effects (Shaw et al. 2002).

> A *factor* is a single treatment variable whose effects are of interest to the researcher.

The factorial design is the most common experimental design used to investigate more than one treatment variable, or factor, at the same time. In a **factorial design**, every combination of treatments from two (or more) treatment variables is investigated.

> An experiment having a *factorial design* investigates all treatment combinations of two or more variables. A factorial design can measure interactions between treatment variables.

The main purpose of a factorial design is to evaluate possible interactions between variables. An **interaction** between two explanatory variables means that the effect of one variable on the response depends on the state of a second variable. Example 14.5 illustrates an interaction in a factorial design.

Example 14.5   **Lethal combination**

Frog populations are declining everywhere, spawning research to identify the causes. Relyae (2003) looked at how a moderate dose (1.6 mg/l) of a commonly used pesticide, carbaryl (Sevin), affected bullfrog tadpole survival. In particular, the experiment asked how the effect of carbaryl depended on whether a native predator, the red-spotted newt, was also present. The newt was caged and could cause no direct  harm, but it emitted visual and chemical cues that are known to affect tadpoles. The experiment was carried out in 10-liter tubs, each containing 10 tadpoles. The four combinations of pesticide treatment (carbaryl vs. water only) and predator treatment (present or absent) were randomly assigned to tubs. For each combination of treatments, there were four replicate tubs. The effects on tadpole survival are displayed in the graph in Figure 14.5-1.

The tub, not the individual tadpole, is the experimental unit, because tadpoles sharing the same tub are not independent. The results showed that survival was high, except when pesticide was applied together with the predator—neither treatment alone had

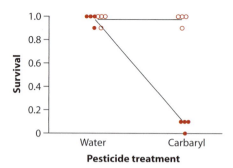

Figure 14.5-1 Interaction between the effects of the pesticide (carbaryl) and predator (red-spotted newt) treatments (○ predator absent; ● predator present) on tadpole survival. Each point gives the fraction of tadpoles in a tub that survived. Lines connect mean survival in the two pesticide treatments, separately for each predator treatment.

much effect (Figure 14.5-1). Thus, the two treatments, predation and pesticide, seem to have interacted—that is, the effect of one variable depends on the state of the other variable. An experiment investigating the effects of the pesticide only would have measured little effect at this dose. Similarly, an experiment investigating the effect of the predator only would not have seen an effect on survival.

> An *interaction* between two (or more) explanatory variables means that the effect of one variable depends upon the state of the other variable.

A factorial design can still be worthwhile even if there is no interaction between explanatory variables. In this case, there are efficiency advantages because the same experimental units can be used to measure the effect of two (or more) variables simultaneously.

14.6 What if you can't do experiments?

Experimental studies are not always feasible, in which case we must fall back upon observational studies. Observational studies can be very important, because they detect patterns and can help generate hypotheses. The best observational studies incorporate all of the features of experimental design used to minimize bias (e.g., simultaneous controls and blinding) and the impact of sampling error (e.g., replication, balance, blocking, and even extreme treatments), except for one: randomization. Randomization is out of the question because, in an observational study, the researcher does not assign treatments to subjects. Instead, the subjects come as they are.

Match and adjust

Without randomization, minimizing bias resulting from confounding variables is the greatest challenge of observational studies. Two types of strategies are used to limit the effects of confounding variables on a difference between treatments in a controlled observational study. One strategy, commonly used in epidemiological studies, is **matching**. With matching, every individual in the target group with a disease or other health condition is paired with a corresponding healthy individual that has the same measurements for known confounding variables, such as age, weight, sex, and ethnic background (Bland and Altman 1994). Matching reduces bias by limiting the contribution of suspected confounding variables to differences between treatments. Unlike randomization in an experiment, matching in an observational study does not account for all confounding variables, only those explicitly measured. Thus, while matching reduces bias, it does not eliminate bias. Matching also reduces sampling error by grouping experimental units into similar pairs, analogous to blocking in experimental studies.

> With *matching*, every individual in the treatment group is paired with a control individual having the same or closely similar values for the suspected confounding variables.

In a weaker version of this approach, a comparison group is chosen that has a frequency distribution of measurements for each confounding variable that is similar to that of the treatment group, but no pairing takes place.

The second strategy used to limit the effects of confounding variables in a controlled observational study is adjustment, in which statistical methods such as analysis of covariance (Chapter 18) are used to correct for differences between treatment and control groups in suspected confounding variables. For example, LaCroix et al. (1998) compared the incidence of cardiovascular disease between two groups of older adults: those who walked more than four hours per week and those who walked less than one hour per week. The ages of the adults were not identical in the two groups, and this could affect the results. To compensate, the authors examined the relationship between cardiovascular disease and age within each exercise group, so that they could compare the predicted disease rates in the two groups for adults of the same age. This approach is discussed in more detail in Chapter 18.

14.7 Choosing a sample size

A key part of planning an experiment or observational study is to decide how many independent units or subjects to include. There is no point in conducting a study whose sample size is too small to detect the expected treatment effect. Equally, there is no point in making an estimate if the confidence interval for the treatment effect is expected to be extremely broad because of a small sample size. Too many subjects is also undesirable, because each replicate costs time and money, and adding one more might put yet another individual in harm's way. If the treatment is unsafe, as the spermicide nonoxynol-9 appears to be (Example 14.2), then we want to injure as few people as possible in coming to this conclusion. Ethics boards and animal-care committees require researchers to justify the sample sizes for proposed experiments. How is the decision made? Here in Section 14.7 we answer this question for two objectives: when the goal is to achieve a predetermined level of *precision* of an estimate of treatment effect or when we want to achieve predetermined *power* in a test of the null hypothesis of no treatment effect. We focus here on techniques for studies that compare the means of two groups. Formulae to help plan experiments for other kinds of data are given in the Quick Formula Summary (Section 14.9).

> An important part of planning an experiment or observational study is choosing a sample size that will give sufficient power or precision.

Plan for precision

A frequent goal of studies in biology is to estimate the magnitude of the treatment effect as precisely as possible. Planning for precision involves choosing a sample size that yields a confidence interval of expected width. Typically, we hope to set the bounds as narrowly as we can afford.

By way of example, let's develop a plan for a two-treatment comparison of means. Let the unknown population mean of the response variable be μ_1 in the treatment group of interest and μ_2 in the control group. When the results are in, we will compute the sample means $\overline{Y}_1$ and $\overline{Y}_2$ and use them to calculate a 95% confidence interval for $\mu_1 - \mu_2$, the difference between the population means of the treatment and control groups. To simplify matters somewhat, we will assume that the sample sizes in both treatments are the same number, n. Let's also assume that the measurement in the two populations is normally distributed and has the same standard deviation, σ.

In this case, a 95% confidence interval for $\mu_1 - \mu_2$ will take the form

$$\overline{Y}_1 - \overline{Y}_2 \pm \text{uncertainty},$$

where "uncertainty" is half the width of the confidence interval. Planning for precision involves deciding the uncertainty we can tolerate in advance. Once we've decided that, then the sample size needed in each group is approximately

$$n = 8 \left(\frac{\sigma}{\text{uncertainty}} \right)^2.$$

This formula is derived from the 2SE rule that was introduced in Section 4.3.[4] According to this formula, a larger sample size is needed if σ, the standard deviation within groups, is large than if it is small. Additionally, a larger sample size is needed to achieve a high precision (a narrow confidence interval) than to achieve a lower precision.

A major challenge in determining sample size is that key factors, like σ, are not known. Typically, a researcher makes an educated guess for these unknown parameters based on pilot studies or previous investigations. (If no information is available, then consider carrying out a small pilot study first, before attempting a large experiment.)

For example, let's plan an experiment to measure the effect of diet on the eye span of male stalk-eyed flies (Example 11.2). The planned experiment will randomly place individual fly larvae into cups containing either corn or spinach. The target parameter is the difference between mean eye spans in the two diet treatments, $\mu_1 - \mu_2$. Assume that we would like to obtain a 95% confidence interval for this difference whose expected uncertainty is 0.1 mm (i.e., the desired full width of the confidence interval is 0.2 mm). How many male flies should be used in each treatment to achieve this goal?

Our sample estimate for σ was about 0.4, based on the sample of nine individuals in Example 11.2. Using these values gives

$$n = 8\left(\frac{\sigma}{\text{uncertainty}}\right)^2 = 8\left(\frac{0.4}{0.1}\right)^2 = 128.$$

This is the sample size in each treatment, so the total number of male flies would be 256. At this point, we would need to decide whether this sample size is feasible in an experiment. If not, then there might be no point in carrying out the experiment. Alternatively, we could revisit the desired width of the 95% confidence interval. That is, could we be satisfied with a higher uncertainty? If so, then we should decide on this new width and then recalculate n.

After all this planning, imagine that the experiment is run and we now have our data. Will the confidence interval we calculate have the precision we planned for? There are two reasons that it probably won't. First, 0.4 was just an educated guess for the value of σ to help our planning, and it was based on only nine individuals. The true value of σ in the population might be larger or smaller. Second, even if we were lucky and the true value of σ really is close to 0.4, the within-treatment standard deviation s from the experiment will not equal 0.4 because of sampling error. The resulting confidence interval will be narrower or wider accordingly. The probability that the resulting confidence interval is less than or equal to the desired precision is only about 0.5. To increase this probability, we would need an even larger sample size.

---

[4] "Uncertainty" is approximately twice the standard error of the difference between sample means (2 SE), or $2\sqrt{\sigma^2\left(\frac{1}{n} + \frac{1}{n}\right)} = 2\sqrt{2\sigma^2/n} = \sqrt{8\sigma^2/n}$. Solving for n gives the rule in the text.

Figure 14.7-1 shows the general relationship between the expected precision of the 95% confidence interval and n, the sample size in each of two groups. The variable on the vertical axis is standardized as uncertainty divided by σ. The effect of sample size from $n = 2$ to $n = 20$ is shown.

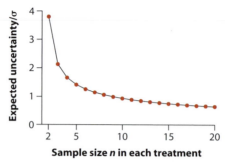

Figure 14.7-1 Expected precision of the 95% confidence interval for the difference between two treatment means depending on sample size n in each treatment. The vertical axis is given in standardized units, uncertainty/σ.

The graph shows that very small sample sizes lead to very wide interval estimates of the difference between treatment means. More data gives better precision. Note also that interval precision initially declines rapidly with increasing sample size (e.g., from $n = 2$ to $n = 10$), but it then declines more slowly (e.g., from $n = 10$ to $n = 20$). Precision is 0.63 at $n = 20$, but it drops to 0.40 by $n = 50$, to 0.28 by $n = 100$, and to 0.20 by $n = 200$. Thus, we get diminishing returns by increasing the sample size past a certain point.

Plan for power

Next we consider choosing a sample size based on a desired probability of rejecting a false null hypothesis—that is, planning a sample based on a desired power. Imagine, for example, that we want to test the following hypotheses on the effect of diet on eye span in stalk-eyed flies:

$$H_0: \mu_1 - \mu_2 = 0$$
$$H_A: \mu_1 - \mu_2 \neq 0.$$

The null hypothesis is that diet has no effect on mean eye span. The power of this test is the probability of rejecting H_0 if it is false. Planning for power involves choosing a sample size that would have a high probability of rejecting H_0 if the absolute value of the difference between the means, $|\mu_1 - \mu_2|$, is at least as great as a specified value D. The value for D won't be the true difference between the means; it is just the minimum we care about. By specifying a value for D in a sample size calculation, we are deciding that we aren't much interested in rejecting the null hypothesis of no difference if $|\mu_1 - \mu_2|$ is smaller than D.

A conventional power to aim for is 0.80. That is, if H_0 is false, we aim to demonstrate that it is false in 80% of the experiments (the other 20% of experiments would fail to reject H_0 even though it is false). If we aim for a power of 0.8 and a conventional significance level of $\alpha = 0.05$, then a quick approximation to the planned sample size n in each of two groups is

$$n \approx 16\left(\frac{\sigma}{D}\right)^2$$

(Lehr 1992). This formula assumes that the two populations are normally distributed and have the same standard deviation (σ), which we are forced to assume is known. A more exact formula is provided in the Quick Formula Summary (Section 14.9), which also allows you to choose other values for power and the significance level.

For a given power and significance level, a larger sample size is needed when the standard deviation σ within groups is large, or if the minimum difference that we wish to detect is small.

Let's return to our experiment to test the effect of diet on the eye span of male stalk-eyed flies. We would like to reject H_0 at $\alpha = 0.05$ with probability 0.80 if the absolute value of the difference between means were truly $D = |\mu_1 - \mu_2| = 0.2$ mm. How many males should be used in each treatment?

Let's assume again that $\sigma = 0.4$. Using this value in the equation for power gives

$$n = 16\left(\frac{0.4}{0.2}\right)^2 = 64.$$

This is the number in each treatment, so the total number of males needed in the experiment would be 128.

These power calculations assume that we know the standard deviation (σ), which is stretching the truth. For this and other reasons, we must always take the results of power calculations with a great deal of caution. The calculations provide useful guidelines, but they do not give infallible answers.

We have only explored sample sizes needed to compare the means of two groups, but similar methods are available for other kinds of statistical comparisons as well. Sample sizes for desired precision and power are available for one- and two-sample means, proportions, and odds ratios in the Quick Formula Summary (Section 14.9). A variety of computer programs are available to calculate sample sizes when planning for power and precision. A good place to start investigating these programs is http://www.cs.uiowa.edu/~rlenth/Power/.

Plan for data loss

The methods given here in Section 14.7 for planning sample sizes refer to sample sizes still available at the *end* of the experiment. But some experimental individuals may die, leave the study, or be lost between the start and the end of the study. The starting sample sizes should be made even larger to compensate.

14.8 **Summary**

▶ In an experimental study, the researcher assigns treatments to subjects.

▶ The purpose of an experimental study is to examine the causal relationship between an explanatory variable, such as treatment, and a response variable. The virtue of experiments is that the effect of treatment can be isolated by randomizing the effects of confounding variables.

▶ A confounding variable masks or distorts the causal relationship between an explanatory variable and a response variable in a study.

▶ A clinical trial is an experimental study involving human subjects.

▶ Experiments should be designed to minimize bias and limit the effects of sampling error.

▶ Bias in experimental studies is reduced by the use of controls, by randomizing the assignment of treatments to experimental units, and by blinding.

▶ In a completely randomized experiment, treatments are assigned to experimental units by randomization. Randomization reduces the bias caused by confounding variables by making non-experimental variables equal (on average) between treatments.

▶ The effect of sampling error in experimental studies is reduced by replication, by blocking, and by balanced designs.

▶ A randomized block design is like a paired design but for more than two treatments.

▶ The use of extreme treatments increases the power of the experiment to detect a treatment effect.

▶ Observational studies should employ as many of the strategies of experimental studies as possible to minimize bias and limit the effect of sampling error.

▶ Although randomization is not possible in observational studies, the effects of confounding variables can be reduced by matching and by adjusting for differences between treatments in known confounding variables.

▶ A factorial design is used to investigate the interaction between two or more treatment variables. The factorial design includes all possible combinations of the treatment variables.

▶ When planning an experiment, the number of experimental units to include can be chosen so as to achieve a desired width of a confidence interval for the difference between treatment means.

▶ Alternatively, the number of experimental units to include when planning an experiment can be chosen so that the probability of rejecting a false H_0 (power) is high for a specified magnitude of the difference between treatment means.

▶ Compensate for possible data loss when planning sample sizes for an experiment.

14.9 **Quick Formula Summary**

Planning for precision

Planned sample size for a 95% confidence interval of a proportion

What is it for? To set the sample size of a planned experiment to achieve approximately a specified half-width ("uncertainty") of a 95% confidence interval for a proportion p.

What does it assume? The population proportion p is not close to zero or one and that n is large.

Formula: $n \approx \dfrac{4p(1 - p)}{\text{uncertainty}^2}$, where p is the proportion being estimated and "uncertainty" is the half-width of the confidence interval for the proportion p. For the most conservative scenario, set $p = 0.50$ when calculating n. The symbol "$\approx$" stands for "is approximately equal to."

Planned sample size for a 95% confidence interval of a log-odds ratio

What is it for? To set the sample size n in each of two groups for a planned experiment to achieve approximately a specified half-width ("uncertainty") of a 95% confidence interval for a log-odds ratio, $\ln(OR)$.

What does it assume? Sample size n is the same in both groups.

Formula: $n \approx \dfrac{4}{\text{uncertainty}^2}\left(\dfrac{1}{p_1} + \dfrac{1}{1 - p_1} + \dfrac{1}{p_2} + \dfrac{1}{1 - p_2}\right)$, where "uncertainty" is the half-width of the confidence interval for $\ln(OR)$, and p_1 and p_2 are the probabilities of "success" in the two treatment groups.

Planned sample size for a 95% confidence interval of the difference between two proportions

What is it for? To set the sample size n in each of two groups for a planned experiment to achieve approximately a specified half-width ("uncertainty") of a 95% confidence interval for a difference between two proportions, $p_1 - p_2$. This is an alternative approach to the one that uses a log-odds ratio to compare the proportion of successes in two treatment groups.

What does it assume? Sample size n is the same in both groups.

Formula: $n \approx \dfrac{8\bar{p}(1 - \bar{p})}{\text{uncertainty}^2}$, where "uncertainty" is the half-width of the confidence interval for $p_1 - p_2$, p_1 and p_2 are the probabilities of "success" in the two treatment groups, and $\bar{p}$ is the average of the two proportions—that is, $\bar{p} = (p_1 + p_2)/2$.

Planned sample size for a 95% confidence interval of the mean

What is it for? To set the sample size of a planned experiment to achieve approximately a specified half-width ("uncertainty") of a 95% confidence interval for a mean μ.

What does it assume? The population is normally distributed with known standard deviation σ.

Formula: $n \approx 4\left(\dfrac{\sigma}{\text{uncertainty}}\right)^2$, where n is the planned sample size and "uncertainty" is the half-width of the confidence interval for the mean μ.

Planned sample size for a 95% confidence interval of the difference between two means

What is it for? To set the sample size of a planned experiment so as to achieve approximately a specified half-width (*uncertainty*) of the 95% confidence interval for $\mu_1 - \mu_2$.

What does it assume? Populations are normally distributed with equal standard deviations σ. The value of σ is known. Sample size n is the same in both groups.

Formula: $n \approx 8\left(\dfrac{\sigma}{\text{uncertainty}}\right)^2$, where n is the planned sample size within each group and "uncertainty" is the half-width of the confidence interval for the difference between means.

Planning for power

Planned sample size for a binomial test of 80% power at $\alpha = 0.05$

What is it for? To set the sample size n of a planned experiment to achieve approximately a power of 0.80 in a binomial test at $\alpha = 0.05$.

What does it assume? The proportion p_0 under the null hypothesis is not close to zero or one, and n is not small. Sample size n is the same in both groups.

Formula: $n \approx \dfrac{8p_0(1 - p_0)}{D^2}$, where p_0 is the proportion under the null hypothesis, and

$D = p - p_0$ is the predetermined difference we wish to be able to detect between the population parameter p and that specified under the null hypothesis.

Planned sample size for 2 × 2 contingency test of 80% power at $\alpha = 0.05$

What is it for? To set the sample size of a planned experiment so as to achieve approximately a power of 0.80 at $\alpha = 0.05$ in a contingency test of the difference between the proportion of successes in two treatment groups (or, equivalently, a test that the odds ratio equals one).

What does it assume? The average probability of "success" in the two treatment groups is known. Sample size n is the same in both treatment groups.

Formula: $n \approx \dfrac{8\bar{p}(1 - \bar{p})}{D^2}$, where $\bar{p}$ is the average of the two probabilities of "success"

[i.e., $\bar{p} = (p_1 + p_2)/2$], and $D = p_1 - p_2$ is the predetermined difference we wish to be able to detect between the two proportions.

Planned sample size for a one-sample or paired *t*-test of 80% power at $\alpha = 0.05$

What is it for? To set the sample size of a planned experiment so as to achieve approximately a power of 0.80 in a one-sample or paired *t*-test at $\alpha = 0.05$.

What does it assume? The population is normally distributed with standard deviation σ. The value of σ is known.

Formula: $n \approx 8\left(\dfrac{\sigma}{D}\right)^2$, where n is the sample size within each group, and $D = \mu$ is

the predetermined value of the mean (or the mean difference in the case of a paired test) that we wish to detect.

Planned sample size for a two-sample *t*-test of 80% power at $\alpha = 0.05$

What is it for? To set the sample size of a planned experiment so as to achieve approximately a power of 0.80 in a two-sample *t*-test at $\alpha = 0.05$.

What does it assume? Populations are normally distributed with equal standard deviation σ. Sample size n is the same in both groups.

Formula: $n \approx 16\left(\dfrac{\sigma}{D}\right)^2$, where n is the sample size within each group, and

$D = |\mu_1 - \mu_2|$ is the predetermined difference between means we wish to detect.

PRACTICE PROBLEMS

1. Identify which goal of experimental design (i.e., reducing bias or limiting sampling error) is aided by the following procedures:
   a. Using a genetically uniform animal stock to test treatment effects.
   b. Using a completely randomized design.
   c. Grouping related experimental units together.
   d. The technician taking the response measurements is unaware of the treatments assigned to experimental units.
   e. Using a computer to randomly assign treatments to experimental units within each block.

2. Using a coin toss for each unit, assign two hypothetical treatments to eight experimental units.
   a. Write down the sequence of eight assignments you ended up with.
   b. Did you end up with an equal number of units in each treatment?
   c. What is the probability of an unbalanced design using this approach?
   d. Recommend a procedure for randomly assigning treatments to units that always results in a balanced design.

3. A series of plots was placed in a large agricultural field in preparation for an experiment to investigate the effects of three fertilizers differing in their chemical composition. Before assigning treatments, it was noticed that plots differed along a moisture gradient. What strategy would you suggest the researchers implement to minimize the impact of this gradient on the ability to measure a treatment effect? Explain with an illustration the experimental design you would recommend.

4. You read the following statement in a journal article: "On the basis of an alpha level of 0.05 and a power of 80 percent, the planned sample size was 129 subjects in each treatment group." State in plain language what this means.

5. Example 12.4 described a study in which salmon were introduced to 12 streams with and without brook trout to investigate the effect of brook trout on salmon survival. Is this an experimental study or an observational study? Explain the basis for your reasoning.

6. Identify the consequences (i.e., increase, decrease, or none) that the following procedures are likely to have on both bias and sampling error in an observational study.
   a. Matching sampling units between treatment and control.
   b. Increasing sample size.
   c. Ensuring that the frequency distribution of subject ages is the same in the two treatments.
   d. Using a balanced design.

7. In 1899, the *British Medical Journal* (page 933) reported the results of a medical procedure involving the subcutaneous infusion of a salt solution for the treatment of extremely severe pneumonia: "Dr. Clement Penrose has tried the effect of subcutaneous salt infusions as a last extremity in severe cases of pneumonia. He con-

tinues this treatment with inhalations of oxygen. He has had experience of three cases, all considered hopeless, and succeeded in saving one. In the other two the prolongation of life and the relief of symptoms were so marked that Dr. Penrose regretted that the treatment had not been employed earlier."

   a. Is this an experimental study? Why or why not?

   b. What design components might Dr. Penrose have included in an experiment to test the effectiveness of his treatment?

8. In a study of the effects of marijuana on the risk of cancer in oral squamous cells, Rosenblatt et al. (2004) examined 407 recent cases of the cancer from western Washington state. They also randomly sampled 615 healthy subjects from the same region having similar frequency distributions of age and sex as the cancer cases. They found that a similar proportion of the cancer cases (25.6%) and healthy subjects (24.4%) reported ever having used marijuana (odds ratio = 0.9; 95% confidence interval, 0.6 − 1.3). In this type of study, the disease itself is the "treatment," and the potential causal factor (marijuana use) is the "response."

   a. Is this an experimental study or an observational study? Explain.

   b. Does this study include a control group? Explain.

   c. What was the purpose of ensuring that the healthy subjects were similar in age and sex to the cancer cases?

   d. Can we conclude that marijuana does not cause cancer in oral squamous cells in this population?

9. After stinging its victim, the honeybee leaves behind the barbed stinger, poison sac, and muscles that continue to pump venom into the wound. Visscher et al. (1996) compared the effects of two methods of removing the stinger left behind: scraping off with a credit card or pinching off with thumb and index finger. A total of 40 stings were induced on volunteers. Twenty were removed with the credit card method, and 20 were removed with the pinching method. The size of the subsequent welt by each sting was measured after 10 minutes. All 40 measurements came from two volunteers (both authors of the study), each of whom received one treatment 10 times on one arm and the other treatment 10 times on the other arm. Pinching led to a slightly smaller average welt, but the difference between methods was not significant.

   a. All 40 measurements were combined to estimate means, standard errors, and the *P*-value for a two-sample *t*-test of the difference between treatment means. What is wrong with this approach?

   b. How should the data be analyzed? Describe how the quantities would be calculated and what type of statistical test would be used.

   c. Suggest two improvements to the experimental design.

10. What is the justification for including extreme doses well outside the range of exposures encountered by people at risk in a dose–response study on animals of the effects of a hazardous substance?

11. A strain of sweet corn has been genetically modified with a gene from the bacterium *Bacillus thuringiensis* (Bt) to express the protein Cry1Ab, which is toxic to caterpillars that eat the leaves. Unfortunately, the pollen of transformed corn plants contains the toxin, too. Corn

pollen dusts the leaves of other plants growing nearby, where it might have negative effects on non-pest caterpillars. You are hired to conduct a study to measure the effects on monarch butterfly caterpillars of ingesting Bt-modified pollen that has landed on the leaves of milkweed, a plant commonly growing in or near corn fields. You decide to use a completely randomized design to compare the effect of two treatments on monarch pupal weight. In one treatment, you place potted milkweed plants in plots of Bt-modified corn, where their leaves receive pollen carrying the toxin. In the other treatment, you place milkweed plants in plots with ordinary corn that has not been transformed with the Bt gene. You place a monarch larva on every milkweed plant. Previous studies have estimated that the standard deviation of pupal weight in monarch butterflies is about 0.25 g.

a. Suppose your goal at the end of the experiment is to calculate a 95% confidence interval for the difference between treatments in mean monarch pupal weights. How many plots would you plan in each treatment if your goal was to produce a confidence interval for the difference in mean pupal weights between treatments having a total width of 0.4 g?

b. What sample size would you need if you decided that 0.4 was not precise enough, and that you wished to halve this interval to 0.2?

c. Imagine that your permits allow you to plant only five plots of Bt-transformed corn, so that the only way you can increase the total sample size for the whole experiment is to increase the number of plots in the ordinary corn treatment. To achieve the same width of confidence interval as in part (a), would the total sample size needed (both treatments combined) likely be greater, smaller, or no different from that calculated in part (a)? Explain.

d. In designing the experiment, why would you not simply place all the milkweed plants for one treatment at random locations in a single large Bt-transformed corn field, and all the milkweed plants for the other treatment at random locations in a single large normal corn field?

12. In the Bt and monarch study described in the previous problem, how many plots would you plan per treatment if your goal were to carry out a test having 80% power to reject the null hypothesis of no treatment effects when the difference between treatments means is at least 0.25 g?

ASSIGNMENT PROBLEMS

13. Consider the results of a six-year observational study that documented health changes related to homeopathic care (Spence and Thompson 2005). Homeopathic treatment was defined as "stimulating the body's autoregulatory mechanisms using microdoses of toxins." Every one of the 6544 patients in the study was assigned to a hospital outpatient unit for homeopathic treatment. Of these, 4627 patients (70.7%) reported positive health changes following treatment. Suggest a major improvement to the design of this study.

14. The fish species *Astyanax mexicanus* includes blind, cave-inhabiting populations whose eyes degenerate during embryonic development. To understand how eye degeneration worked, Yamomoto and Jeffery (2000) replaced the lens of the degenerate eye on one side (randomly chosen) of a blind cave fish embryo with a lens from the embryo of a "normal," sighted fish. This procedure was repeated on all individuals in a sample of blind cave fish. Final eye size was measured on both sides of each experimental fish, after embryonic development was complete. Remarkably, a normal-sized eye was restored on the transplant eyes of blind cave fish

but not on the unmanipulated side. Based on the above description of a laboratory experiment, identify which of the six main strategies of experimental design (listed in Section 14.2) were incorporated.

15. Identify the consequences (i.e., increase, decrease, or none) that the following procedures are likely to have on both bias and sampling error in an experimental study.
    a. Assigning treatments to subjects alphabetically, not randomly
    b. Increasing sample size
    c. Calculating power
    d. Applying every treatment to every experimental unit in random order
    e. Using a sample of convenience instead of a random sample
    f. Testing only one treatment group, without a control group
    g. Using a balanced design
    h. Informing the human subjects which treatment they will receive

16. Blaustein et al. (1997) used a field experiment to investigate whether UV-B radiation was a cause of amphibian deformities (see the photo at the beginning of this chapter). They measured long-toed salamanders either exposed to natural UV-B radiation or under UV-B shields. It was not possible to carry out all replicates simultaneously, so the researchers carried them out over several days. They made sure that both treatments were included on each day. In their analysis, they grouped replicates together that were carried out on the same day.
    a. By grouping experiments carried out on the same day, what experimental procedure were they using?
    b. What is the main reason for adopting this procedure in an experimental study?

17. The experiment described in Example 12.2 compared antibody production in 13 male red-winged blackbirds before and after testosterone implants. The units of antibody levels were log 10^{-3} optical density per minute (ln[mOD/min]). The mean change in antibody production was $\bar{d} = 0.056$ and the standard deviation was

$s_d = 0.159$. If you were assigned the task of repeating this experiment to test the hypothesis that testosterone changed antibody levels, what sample size (i.e., number of blackbirds) would you plan to ensure that a mean change of 0.05 units could be detected with probability 0.8? Explain the steps you took to determine this value.

18. Two clinical trials were designed to test the effectiveness of laser treatment for the treatment of acne. Seaton et al. (2003) randomly divided subjects into two groups. One group received the laser treatment whereas the other group received a sham treatment. Orringer et al. (2004) used an alternate design in which laser treatment was applied to one side of the face, randomly chosen, and the sham treatment was applied to the other side. The number of facial lesions was the response variable.
    a. Identify the main component of experimental design that differs between the two studies. Give the statistical term identifying this component in experimental design.
    b. Under what circumstances would there be an advantage to using the "divided-face" design over the completely randomized (two-sample) design?
    c. Assuming that the advantage identified in part (b) is met, can you think of a disadvantage of the divided-face design?[5]

19. What is the main advantage of experimental studies over observational studies?

20. Michalsen et al. (2003) conducted a study to examine the effects of "leech therapy" for pain resulting from osteoarthritis of the knee. Two treatments were randomly assigned to 51 patients with osteoarthritis of the knee. Patients in the leech treatment received 4–6 medicinal leeches applied to the soft tissue of the affected knee in a single session. The animals were left to feed *ad libitum* until they detached themselves, on average 70 minutes later. Patients in the control treatment were given diclofenac

---

[5]   Other than a possible social dilemma.

gel and were told to apply it twice daily to the affected area. Pain was assessed by questionnaire by personnel unaware of the treatments applied to each subject. The results showed that seven days after the start of treatment pain was significantly lower in the leech group.

a. Does this study include a control group? Explain.

b. Is this an experimental study or an observational study? Explain.

c. Is this a completely randomized design or a randomized block design? Explain.

d. Which strategy for reducing bias was not adopted in this study? How might its absence have affected the results?

21. Design a study to compare the reaction times of the left and right hands of right-handed subjects using a computer mouse. Two design choices are available to you. In the first, a sample of right-handed subjects is randomly divided into two groups. Reaction time with the left hand is measured in one group and reaction time with the right hand is tested in the other group.

a. What is the second design choice available to you?

b. Under what circumstances would the second choice be the preferred choice?

c. Assume that you decide to go with the completely randomized design and that, at the end of the experiment ,you aim to calculate a 95% confidence interval for the difference between the mean reaction times of left and right hands. To achieve a confidence interval of specified width, what information would you require to plan an appropriate sample size?

22. Identify the single most significant flaw in each of the following experimental designs. Use statistical language to identify what's missing.

a. In a test of the effectiveness of acupuncture in treating migraine headaches, a random sample of patients at a migraine clinic was provided with a novel acupuncture treatment daily for six months. The patients were interviewed at the start of the study and at the end to determine whether there had been any change in the severity of their migraines.

b. In a modified study, a second sample of patients was chosen after the acupuncture treatment was completed on the first set of patients. This second sample of patients received a placebo in pill form for six months. At the end of the study, perceived pain levels in the two groups were compared.

c. In a modified study, the sample of patients was divided into two groups according to gender. The women received the acupuncture treatment and the men received the placebo medication in pill form. At the end of the experiment, perceived pain levels in the two groups were compared.

d. In a modified study, patients were randomly divided into two groups. One group received the acupuncture treatment and the other received a fixed dose of placebo medication in pill form. At the end of the experiment, perceived pain levels in the two groups were compared.

23. Young et al. (2006) took measurements of subordinate female meerkats to determine the changes in reproductive physiology experienced by females that are evicted from their social

groups. They compared evicted females and those not evicted in their level of plasma luteinizing hormone following a GnRH hormone challenge. They found that nine evicted females had a sample average of 6.2 miu/ml of plasma luteinizing hormone (milli-International Units per milliliter) compared with 12.1 miu/ml in 18 females that had not been evicted. The pooled sample variance was 28.4.

a. Is this an experimental study or an observational study? Explain.

b. The sample size was unequal between the two groups of females compared. How would this affect the power of a hypothesis test of the difference between group means compared with a more balanced design? Explain.

c. How would the imbalance of the sample sizes affect the width of the confidence interval for the difference between group means compared with a more balanced design? Explain.

d. If you were planning to repeat the comparison of plasma luteinizing hormone between these two groups of females, what sample size would you plan to achieve an expected half-width of 3 miu/ml for a 95% confidence interval of the difference between means? Explain the steps you took to determine this value.

24. Diet restriction is known to extend life and reduce the occurrence of age-related diseases. To understand the mechanism better, you propose to carry out a study to look at the separate effects of age and diet restriction, and the interaction between age and diet restriction, on the activity of liver cells in rats. What experimental design should you consider employing? Why?

Data dredging

In the spoof journal *Annals of Improbable Research*, a satirical article reported on a study of the so-called "butterfly effect" (Inaudi et al. 1995). This effect, a mainstay of the popular representation of chaos theory, says that small initial causes, like the flapping of a butterfly's wings, can ultimately have large effects, like a hurricane, on the other side of the world. The fearless researchers set out to measure this effect by capturing several dozen butterflies and holding them in captivity in Switzerland. Each day, they checked the butterflies and recorded whether or not they flapped their wings. Then, using the lab's phone, they called their girlfriends in Paris each day to ask whether or not it was raining.[1] At the end of the study, the students tested each butterfly for an association between its daily flapping behavior and the daily weather in Paris. They found that the flapping days of one of the butterflies closely matched the rainfall days in Paris ($P < 0.05$). They exulted, "Not only have we proven that the butterfly effect exists, we have found the butterfly."

These guys were clearly joking, but statistically speaking, where did they go wrong? The answer is that they went "*data dredging*." They performed many statistical tests and eventually one of them was significant. Data dredging (also called "data snooping" or "data

fishing") is the carrying out of many statistical tests in hope of finding at least one statistically significant result.

The problem with data dredging is that the probability of making *at least one* Type I error (i.e., of obtaining a "false positive") is greater than the significance level α when many tests are conducted, if the null hypothesis is true (as it surely is in the butterfly example). Each hypothesis test has some chance of error, and these errors compound over multiple tests. There is a much larger probability of getting an error out of several tries than in any one try. By analogy, we might get away with playing Russian roulette once, but we would be unlikely to survive a month of playing once a night.

It's useful to do a few calculations to see how big the problem might be. The probability of making no Type I errors in N independent tests is $(1 - \alpha)^N$. Thus, the chance of making at least one Type I error from N independent tests is $1 - (1 - \alpha)^N$. This means that, if we use $\alpha = 0.05$ and carry out 20 independent tests of true null hypotheses, the probability that at least one of these tests will falsely reject the null hypothesis is about 65%. If we carry out 100 tests, then the chance of rejecting at least one of the null hypotheses becomes 99.4%, even if all the null hypotheses are true. With data dredging, a false positive result is almost inevitable.

Nevertheless, multiple testing is common in biology, and for good reasons. A dedicated

---

[1] They continued the experiment "until the first phone bill reached our Office of Financial Services."

experimentalist on human subjects might measure many conceivable responses (e.g., blood pressure, body temperature, red blood cell count, white blood cell count, speed of recovery, appetite, and weight change) and perhaps even a few extra variables that might be long shots. The result is that the clinician might end up carrying out 10 or 20 tests of treatment effects, raising the probability of a false positive result. This level of multiple testing pales next to that seen in gene mapping. Locating a gene for a single trait, such as a genetic disease, typically involves 100 or more statistical tests (one for each section of the genome). What should be done about the soaring Type I error rates resulting from so much testing?

The answer to this question depends on your goals. If your goal is simply to *explore* the data, to discover the possibilities but not to provide rigorous tests, then you need do nothing special about multiple testing except report the number of tests that you carried out and which ones yielded a significant result. If you admit that you dredged the data, your results can still be useful. New hypotheses and unexpected discoveries can emerge from a thorough fishing expedition. However, the individual significant results that pop up from data dredging cannot yet be taken seriously, because of the high probability of one or more Type I errors. Some of the significant results might indeed be real, but it will be difficult to establish which ones. Rather, a new study must be carried out with new data to test any promising results that emerged from the exploratory approach. Another strategy sometimes used when exploring data is to divide the data randomly into two parts. One part is used for data dredging, and the other part is used to confirm any positive results suggested by the dredging.

If your goals from multiple testing are more rigorous (e.g., you want to determine which

variable really did respond to treatment in a clinical trial, or which location in the genome really does contain a gene for a heritable disease), then steps must be taken to *correct* for the inflation of Type I error rates that occur with multiple testing. The simplest way to accomplish this is to use a more stringent significance level—that is, one smaller than the usual $\alpha = 0.05$.

The most common correction for multiple comparisons is the **Bonferroni correction**. In the simplest version of this method, each test uses a significance level α^* rather than α, where

$$\alpha^* = \frac{\alpha}{number\ of\ tests}$$

For example, if we typically adopt the significance level $\alpha = 0.05$ when carrying out a single test, then to carry out 12 separate tests we should use the significance level $\alpha^* = 0.05/12 = 0.00417$, instead. In this case, we would only reject H_0 in each test if P were less than or equal to 0.00417. With the Bonferroni correction, the probability of getting at least one Type I error during the course of carrying out all 12 tests is approximately equal to the initial α-value (i.e., 0.05 in this case).

Keep in mind, though, that applying the Bonferroni correction greatly reduces the power of single tests. This is the price paid for asking many questions of the data. More than ever, we should be mindful not to "accept the null hypothesis." It is okay to be skeptical when a null hypothesis is not rejected and power is so limited, but there is little to do about it except to repeat the study and look again.

Another, increasingly popular approach to correct for multiple comparisons is called the **false discovery rate** (FDR). To use this approach, carry out all of the multiple tests at a fixed significance level α (e.g., the usual 0.05). Gather all of the tests that yield a statistically significant result (i.e., all of the tests for which $P \leq \alpha$). We can call this subset of tests the "discoveries." The FDR estimates the proportion of discoveries that are false positives. In other words, the FDR is the proportion of tests for which the null hypothesis was rejected yet the null hypothesis was true. For example, Brem et al. (2005) carried out hundreds of statistical hypothesis tests of interactions between pairs of yeast genes. Of these tests, 225 yielded a statistically significant result (the "discoveries"). Using the false discovery rate method, they estimated that 12 of these 225 tests were false positives, leaving 213 "true" discoveries.

An extension of the FDR calculates a quantity called the q-value for each discovery. The q-value is analogous to a P-value, providing a measure of the strength of support from the data that the null hypothesis is false in a specific test. The smaller the q-value, the stronger is the evidence that H_0 is false and should be rejected. Unlike the P-value, the q-value takes into account other tests carried out at the same time. The idea is that, by choosing to reject H_0 only if the q-value is 0.05 or less, we reject the null hypothesis falsely in only 5% of tests. FDR and q-values are a more powerful approach to dealing with multiple comparisons, and we expect their use to increase in biological applications over the next decade. Consult Benjamani and Hochberg (1995) or Storey and Tibshirani (2003) for more details.

15

Comparing means of more than two groups

How would we analyze the results of a clinical trial that randomly assigns not two but *three* treatments to a sample of patients? Two of the treatments might be different medications and the third a placebo control. Such a design can answer more questions than a two-treatment experiment because more comparisons can be made in the same experiment. For example, are both medications better than the placebo? If they are, then by how much? Is one medication superior to the other? If so, how much better is it?

How do we compare the means of the three groups? At first glance, it might seem reasonable to compare them two at a time: first compare the means of groups 1 and 2, then compare groups 2 and 3, and finally, compare groups 1 and 3. This analysis-by-twos quickly runs into problems, because testing multiple pairs of means inflates the probability of committing at least one Type I error (recall the data dredging discussed in Interleaf 8). The danger is modest when comparing only three groups, but it escalates rapidly with an increasing number of groups. Comparing five groups would require 10 tests, which would give as much as a 40% chance of falsely rejecting at least one of those null hypotheses if they were all true.

The best solution is the **analysis of variance**, or **ANOVA**, which compares means of multiple groups simultaneously in a single analysis. Analysis of *variance* might seem like a misnomer, given our intention to compare *means*, but testing for variation among groups is equivalent to asking whether the means differ. ANOVA was developed by the biologist and statistician, R. A. Fisher, who was first mentioned in Interleaf 1.

This chapter discusses "one-way" or "single-factor" analysis of variance, which investigates the means of several groups differing by one explanatory variable or factor. "Two-way" or "two-factor" ANOVA is discussed in Chapter 18.

15.1   The analysis of variance

Analysis of variance is the most powerful approach known for simultaneously testing whether the means of k groups are equal. It works by assessing whether individuals chosen from different groups are, on average, more different than individuals chosen from the same group. Example 15.1 introduces the method.

Example 15.1   **The knees who say night**

Traveling to a different time zone can cause jet lag, but people adjust as the schedule of light to their eyes in the new time zone gradually resets their internal, circadian clock. This change in their internal clock is called a "phase shift." Campbell and Murphy (1998) reported that the human circadian clock can also be reset by exposing the back of the *knee* to light, a finding that was met with skepticism by some, but was hailed as a major discovery by others. Aspects of the experimental design were subsequently challenged.[1] The data in Table 15.1-1 are from a later experiment by Wright and Czeisler (2002) that re-examined the phenomenon. The new experiment measured circadian rhythm by the daily cycle of melatonin production in 22 subjects randomly assigned to one of three light treatments. Subjects were awakened from sleep and subjected to a single three-hour episode of bright lights applied to the eyes only, to the knees only, or to neither (control group). Effects of treatment on the circadian rhythm were measured two days later by the magnitude of phase shift in each subject's daily cycle of melatonin production.

---

[1]   In the 1998 experiment, subjects' eyes had been exposed to low levels of light while their knees were being illuminated.

Results are plotted in Figure 15.1-1. A negative measurement indicates a delay in melatonin production, which is the predicted effect of light treatment; a positive number indicates an advance. Does light treatment affect phase shift?

Table 15.1-1 Raw data and descriptive statistics of phase shift, in hours, for the circadian rhythm experiment.

Treatment	Data (h)	$\overline{Y}$	s	n
Control	0.53, 0.36, 0.20, −0.37, −0.60, −0.64, −0.68, −1.27	−0.3088	0.6176	8
Knees	0.73, 0.31, 0.03, −0.29, −0.56, −0.96, −1.61	−0.3357	0.7908	7
Eyes	−0.78, −0.86, −1.35, −1.48, −1.52, −2.04, −2.83	−1.5514	0.7063	7

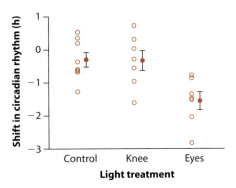

Figure 15.1-1 "Dot plot" showing the phase shift in the circadian rhythm of melatonin production in 22 experimental subjects given alternative light treatments (*open circles*). Filled dots and vertical lines ("error bars") are group means ±1 standard error.

Hypotheses

The null hypothesis of ANOVA is that the population means μ_i are the same for all treatments. (Throughout this chapter, we'll use subscripts on various quantities to indicate the group that each refers to. Generically, we'll refer to population i, which will have mean μ_i; for example, the mean of group 3 will be written as μ_3.)

Under the null hypothesis, the sample means $\overline{Y}_i$ differ from each other solely because of random sampling error. The alternative hypothesis is that the mean phase shift is not the same in all three light-treatment populations:

H_0: $\mu_1 = \mu_2 = \mu_{3.}$

H_A: At least one μ_i is different from the others.

The alternative hypothesis does not state that every mean is different from all the others, but only that at least one mean stands apart.

> Rejecting H_0 in ANOVA indicates that the mean of at least one group is different from the others.

ANOVA in a nutshell

Even if the null hypothesis of an ANOVA is true, that all groups have equal population means, the sample means calculated from data will differ from one another by chance. The job of ANOVA is to determine whether the variation among the group sample means is *greater* than expected by chance. This would indicate that at least one of the population means is different from the others and that the null hypothesis should be rejected.

ANOVA achieves this result by comparing two sources of variation in the data. These sources are illustrated in Figure 15.1-2 for the circadian rhythm data. We've spread out the data points horizontally so you can see them better (their vertical positions are unchanged). The left panel of the figure shows all the variation in the data. A vertical line shows the deviation between each data point and the grand mean (the mean of all the data). If we squared these deviations and took their average, we would get the ordinary sample variance for all the data combined.

ANOVA breaks the total deviation for every observation into two parts, representing the two sources of variation in the data. One part is the deviation between every observation and the mean of its group (the right panel of Figure 15.1-2). This deviation is called the "error" or the residual. The other part is the deviation between an observation's group mean and the grand mean (the middle panel of Figure 15.1-2).

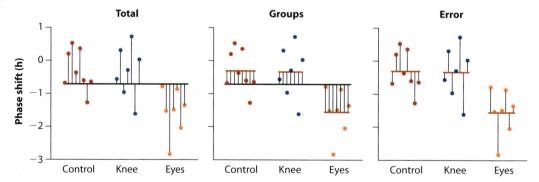

Figure 15.1-2 Depiction of the portions of variation in the circadian rhythm data (Example 15.1). Each dot is the measurement of phase shift in a single subject. The long horizontal line in black is the grand mean, $\bar{Y}$. Short horizontal lines in red are the sample means of the three light-treatment groups. Vertical lines represent deviations.

If we square the "error" deviations in the right panel of Figure 15.1-2 and take their average, we get the pooled sample variance. In ANOVA jargon[2] this is called the **error mean square**, and it represents the variation among subjects belonging to the same group.

> The *error mean square* of ANOVA is the pooled sample variance, a measure of the variability within groups.

The other portion of variation is the **group mean square**, which takes an average over groups of the squared deviations shown in the middle panel of Figure 15.1-2. The group mean square represents variation among subjects that belong to different groups.

> The *group mean square* of ANOVA represents variation among individuals belonging to different groups. It should be similar to the error mean square if population means are equal.

If the null hypothesis is true, and the group means are the same, then the two mean squares should be roughly equal—that is, variation among subjects belonging to different groups should be no different than that among subjects belonging to the same group (except by chance). If the group mean square is significantly greater than the error mean square, however, then there are real differences among the population means. The final outcome of ANOVA is a test that compares these two mean squares.[3]

Calculating the mean squares

The error mean square, symbolized as MS_{error}, measures variance within groups. ANOVA assumes that σ^2 (i.e., the variance of Y) is the same in every population. In

---

[2] Unfortunately, analysis of variance has its own special jargon. You need to learn it so that the terms are familiar when you see them used in published articles and in the output of statistical packages on the computer.

[3] The concept of standard error of the mean is another way to gain intuition for why these two mean squares should be the same under H_0. In Chapter 4, we learned that, if we took repeated samples from a population (or from multiple populations having the same mean and variance, as assumed by H_0 in ANOVA), then the separate sample means would differ by chance. We learned that the true variance among these sample means is just the variance within the populations divided by the sample size (equivalently, the variance within populations equals the variance among means multiplied by the sample size). The mean squares in ANOVA are based on the same principle. The error mean square (MS_{error}) estimates the variance within the populations, whereas the group mean square (MS_{groups}) represents the variance among the sample means multiplied by the sample size. These two quantities are expected to be the same under H_0.

other words, we assume that $\sigma^2 = \sigma_1^2 = \sigma_2^2 = ... = \sigma_k^2$ for all k groups. The best estimate of this variance within groups is the pooled sample variance, just as in the two-sample t-test (Chapter 12). In ANOVA, the error mean square is calculated as

$$MS_{error} = \frac{\sum s_i^2 (n_i - 1)}{N - k}.$$

The s_i values in the numerator of this formula are the sample standard deviations from each group, and the n_i values are the group sample sizes; $n_i - 1$ is the number of degrees of freedom of group i. The denominator of this formula is the number of degrees of freedom for error, and it is just the sum of the degrees of freedom for the different groups: $\sum (n_i - 1) = N - k$. N is the total number of data points in all groups combined: $N = \sum n_i$.

For the circadian rhythm data in Table 15.1-1,

$$MS_{error} = \frac{(0.6176)^2(8 - 1) + (0.7908)^2(7 - 1) + (0.7063)^2(7 - 1)}{22 - 3}$$

$$= \frac{9.415}{19}$$

$$= 0.4955.$$

The group mean square, symbolized by MS_{groups}, is calculated from the deviations of group sample means ($\overline{Y}_i$) around the grand mean of all of the measurements ($\overline{\overline{Y}}$):

$$MS_{groups} = \frac{\sum n_i (\overline{Y}_i - \overline{\overline{Y}})^2}{k - 1},$$

where k is the total number of groups. The denominator of this formula (i.e., $k - 1$) is the number of degrees of freedom for groups. The grand mean $\overline{\overline{Y}}$ is the mean of all the data from all groups combined. The grand mean is not the same as the average of the group sample means if the sample size is not the same in every group. It is calculated by

$$\overline{\overline{Y}} = \frac{\sum n_i \overline{Y}_i}{N}.$$

This equation is equivalent to adding up all of the individual measures from all of the groups and dividing the sum by the total number.

For the circadian rhythm data, the grand mean is

$$\overline{\overline{Y}} = \frac{8(-0.3087) + 7(-0.3357) + 7(-1.5514)}{22} = -0.7127,$$

and so

$$\text{MS}_{\text{groups}} = \frac{8[-0.3087 - (-0.7127)]^2 + 7[-0.3357 - (-0.7127)]^2 + 7[-1.5514 - (-0.7127)]^2}{3 - 1}$$

$$= 3.6122.$$

The variance ratio, F

Under the null hypothesis that the population means of all groups are the same, the variation among individuals belonging to different groups (represented by $\text{MS}_{\text{groups}}$) should be the same as the variation among individuals belonging to the same group (estimated by MS_{error}). In ANOVA, therefore, we test for a difference by calculating the "variance ratio" of $\text{MS}_{\text{groups}}$ over MS_{error}:

$$F = \frac{\text{MS}_{\text{groups}}}{\text{MS}_{\text{error}}}.$$

This F-ratio is the test statistic in analysis of variance. Under the null hypothesis, F should lie close to one. If the null hypothesis is false, however, and the alternative hypothesis is correct, then $\text{MS}_{\text{groups}}$ should *exceed* MS_{error} and we expect F to be greater than one.

For the circadian-rhythm data,

$$F = \frac{3.6122}{0.4955} = 7.29.$$

To calculate the P-value, we need the sampling distribution for the F-statistic under H_0. This null distribution for the F-statistic is called the F-distribution. The F-distribution has a pair of degrees of freedom, one for the numerator (top) of the F-ratio and a second for the denominator (bottom). The numerator ($\text{MS}_{\text{groups}}$) has $k - 1$ degrees of freedom, and the denominator (MS_{error}) has $N - k$ degrees of freedom.

For the circadian rhythm data, $N = 22$ and $k = 3$, so there are $k - 1 = 3 - 1 = 2$ degrees of freedom for the numerator and $N - k = 22 - 3 = 19$ degrees of freedom for the denominator. The F-distribution with 2 and 19 *df* is therefore the appropriate null distribution for F. The number of degrees of freedom for the numerator is always presented first when specifying the F-distribution. This is important because the F-distribution with 2 and 19 degrees of freedom is *not* the same as the F-distribution having 19 and 2 degrees of freedom.

Figure 15.1-3 illustrates the F-distribution with 2,19 degrees of freedom. The distribution ranges from zero to positive infinity. The right tail of the curve is the part we are interested in. This is because if H_0 is false, then $\text{MS}_{\text{groups}}$ should *exceed* MS_{error} and lead to an F-ratio in the right tail of the F-distribution. F-ratios less than one might occur, but only by chance.

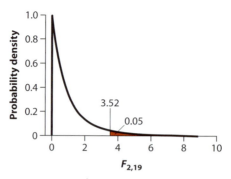

Figure 15.1-3 The F-distribution with 2,19 degrees of freedom. The value of F ranges from zero to positive infinity. The area under the curve to the right of the critical value $F = 3.52$ (*shaded*) is 0.05. (See Statistical Table D.)

The critical value corresponding to the area of 0.05 in the right tail of the F-distribution is found in Statistical Table D. An excerpt is shown in Table 15.1-2.

Table 15.1-2 An excerpt from Statistical Table D, with critical values of the F-distribution corresponding to the significance level $\alpha(1) = 0.05$.

Denominator df	Numerator df									
	1	**2**	**3**	**4**	**5**	**6**	**7**	**8**	**9**	**10**
10	4.96	4.10	3.71	3.48	3.33	3.22	3.14	3.07	3.02	2.98
11	4.84	3.98	3.59	3.36	3.20	3.09	3.01	2.95	2.90	2.85
12	4.75	3.89	3.49	3.26	3.11	3.00	2.91	2.85	2.80	2.75
13	4.67	3.81	3.41	3.18	3.03	2.92	2.83	2.77	2.71	2.67
14	4.60	3.74	3.34	3.11	2.96	2.85	2.76	2.70	2.65	2.60
15	4.54	3.68	3.29	3.06	2.90	2.79	2.71	2.64	2.59	2.54
16	4.49	3.63	3.24	3.01	2.85	2.74	2.66	2.59	2.54	2.49
17	4.45	3.59	3.20	2.96	2.81	2.70	2.61	2.55	2.49	2.45
18	4.41	3.55	3.16	2.93	2.77	2.66	2.58	2.51	2.46	2.41
19	4.38	**3.52**	3.13	2.90	2.74	2.63	2.54	2.48	2.42	2.38
20	4.35	3.49	3.10	2.87	2.71	2.60	2.51	2.45	2.39	2.35

To find the critical value of the F-distribution having 2,19 degrees of freedom, locate the cell in the table corresponding to 2 *df* in the numerator and 19 *df* in the denominator. This value, 3.52, is highlighted in Table 15.1-2. We write it as

$$F_{0.05(1),2,19} = 3.52.$$

The "(1)" indicates that we are looking only at the right tail of the F-distribution. In other words, the area under the curve in Figure 15.1-3 to the right of 3.52 is 0.05. Because our observed value of F (i.e., 7.29) is larger than 3.52, it lies farther out in the right tail of the F-distribution, so P must be less than 0.05. Therefore, we reject the null hypothesis.

The exact P-value is the area under the curve of the F-distribution to the *right* of the observed F-value:

$$P = \Pr[F > 7.29].$$

If you analyzed the data with a statistics package on the computer, you would obtain this probability directly: $P = 0.004$. Again, $P < 0.05$, so we reject H_0. The mean phase shift differs significantly among light treatments.

Rejecting H_0 indicates only that at least one of the population means μ_i is different from the others, not that every μ_i is different from all of the others. We show in Section 15.4 how to take this further to decide which means are different.

ANOVA tables

Typically, all of the results of an analysis of variance are presented in an **ANOVA table**. Computer programs, for example, present the output of an analysis of variance in this way. Table 15.1-3 is the ANOVA table for the circadian-rhythm experiment discussed in Example 15.1.

An ANOVA table lists the group and error mean squares as well as their ratio, F. It also lists the group sum of squares and the error sum of squares, two other quantities that were used in the calculation of the mean squares. The group sum of squares is the numerator from the formula that we used to calculate the group mean square; the formula is given again in the Quick Formula Summary in Section 15.8 at the end of this chapter. The error sum of squares is the numerator from the formula used to calculate the error mean square. The mean squares are obtained by dividing the sums of squares by the corresponding degrees of freedom, which are also listed in the ANOVA table (Table 15.1-3).

Table 15.1-3 ANOVA table for the results of the circadian rhythm experiment (Example 15.1).

Source of variation	Sum of squares	df	Mean squares	F-ratio	P
Groups (treatment)	7.224	2	3.6122	7.29	0.004
Error	9.415	19	0.4955		
Total	16.639	21			

Conveniently, the group and error sums of squares add up to the total sum of squares, which is the sum of the squared differences between every measurement and the grand mean (the left panel of Figure 15.1-2). The number of degrees of freedom for error and groups add up to the total number of degrees of freedom, which is $N - 1$. The mean squares don't sum, so there is no entry for "Total" under "Mean squares" in the ANOVA table.

Variability explained: R^2

The **R^2 value** ("R-squared") is used in ANOVA to summarize the contribution of group differences to the variability in the data. The quantity is based on the fact that the total sum of squares can be split into its two parts, the error sum of squares and the group sum of squares:

$$SS_{total} = SS_{groups} + SS_{error}.$$

R^2 is the group portion of variation expressed as a fraction of the total:

$$R^2 = \frac{SS_{groups}}{SS_{total}}.$$

It can be thought of, loosely, as the "fraction of the variability in Y that is explained by groups." It is a reflection of how much narrower the scatter of measurements is around the group means compared with that around the grand mean (see Figure 15.1-2).

R^2 takes on values between zero and one. When R^2 is close to zero, the group means are all very similar and most of the variability is within groups—that is, the explanatory variable defining the groups explains *very little* of the variability in Y. Conversely, an R^2 close to one indicates that there is little variation in Y left over after the different group means are taken into account—that is, the explanatory variable explains *most* of the variability in Y.

For the circadian rhythm data,

$$R^2 = \frac{7.224}{16.639} = 0.43.$$

In other words, 43% of the total sum of squares among subjects in the magnitude of phase shift is explained by differences among them in light treatment. The remaining 57% of the variability among subjects is not accounted for by light treatment; it is simply "error."

> R^2 measures the fraction of the variation in Y that is explained by group differences.

ANOVA with two groups

The analysis of variance works even when $k = 2$. ANOVA and the two-sample t-test give identical results[4] when testing the null hypothesis—that is, H_0: $\mu_1 = \mu_2$. An advantage of the two-sample t-test is that it generalizes more easily to other hypothesized differences between means, such as H_0: $\mu_1 - \mu_2 = 10$. Welch's t-test (Section 12.3) is additionally useful when the variances are very different between groups, whereas ANOVA requires more similar variances.

15.2 Assumptions and alternatives

The assumptions of analysis of variance are the same as those of the two-sample t-test, but they must hold for all k groups. To review,

▶ The measurements in every group represent a random sample from the corresponding population.
▶ The variable is normally distributed in each of the k populations.
▶ The variance is the same in all k populations.

Methods to evaluate these assumptions were discussed in Chapter 13.

The robustness of ANOVA

The ANOVA is surprisingly robust to deviations from the assumption of normality, particularly when the sample sizes are large. This robustness stems from a property of sample means described by the central limit theorem (Section 10.6)—that is, within each group, the sampling distribution of means is approximately normal when sample size is large, even if the variable itself does not have a normal distribution (Chapter 13).

ANOVA is also robust to departures from the assumption of equal variance in the k populations, but only if the samples are all about the same size.

Data transformations

If the data do not conform to the assumptions of ANOVA, they can be transformed as described in Chapter 13. Any of the transformations discussed there (e.g., log transformations and arcsine transformations) can be applied to ANOVA. If you get lucky,

---

[4] The test statistics and null distributions are basically the same. The F-ratio for two means (having one degree of freedom for groups and "df" degrees of freedom for error) is the same as the square of the two-sample t-statistic having "df" degrees of freedom.

transforming the data will simultaneously make the data more normal and make the variances more equal, but this does not always happen.

Nonparametric alternatives to ANOVA

If the normality assumption of ANOVA is not met and transformations are unsuccessful, then there is a nonparametric alternative to single-factor ANOVA. The **Kruskal–Wallis test**, a nonparametric method based on ranks, is the equivalent of the Mann–Whitney U-test (Section 13.5) when there are more than two groups. It is sometimes referred to as "analysis of variance based on ranks." It makes all the same assumptions as the Mann–Whitney U-test, but it is applied to more than two groups:

▶ All group samples are random samples from the corresponding populations.
▶ To use Kruskal–Wallis as a test of differences among populations in means or medians, the distribution of the variable must have the same shape in every population.

As in the Mann–Whitney U-test, the Kruskal–Wallis test begins by ranking the data from all groups together (employing the strategy for tied observations first described in Section 13.5). The sum of the ranks for each group, R_i, is then used to calculate the Kruskal–Wallis test statistic, H. The formula is given in the Quick Formula Summary (Section 15.8). In general, we recommend using a computer program for this procedure. Under the null hypothesis of no difference among populations, the sampling distribution of H is approximately χ^2 with $k - 1$ degrees of freedom, where k is the number of groups. The null hypothesis is rejected if H is greater than or equal to the critical value from the appropriate χ^2 distribution, $\chi^2_{k-1,\alpha(1)}$.

As with the Mann–Whitney U-test, the Kruskal–Wallis test has little power when sample sizes are very small. Remember that power is the probability of rejecting a false null hypothesis, so more power is better. Therefore, ANOVA is preferred if its assumptions can be met. The Kruskal–Wallis test has nearly the same power as ANOVA when sample sizes are large.

15.3 Planned comparisons

Analysis of variance is often the start, but not the end, of efforts to compare the means of more than two groups. Researchers might want to answer two additional questions, such as "Which means are different?" and "What is the magnitude of the difference between means?" ANOVA by itself answers neither of these questions.

There are two approaches to figuring out which means are different and by how much—namely, planned and unplanned comparisons of means. A **planned comparison** is a comparison between means identified as being of crucial interest during the design of the study, *prior* to obtaining the data. A planned comparison must have a strong prior justification, such as an expectation from theory or a prior study.[5] Only one or a small number of planned comparisons is allowed, to minimize inflating the Type I error rate. An **unplanned comparison** is one of multiple comparisons, such as between all pairs of means, carried out to determine where differences between means lie. Unplanned comparisons represent a kind of data dredging (Interleaf 8), so it is necessary to protect against rising Type I errors. Here we briefly describe an example of a planned comparison. Unplanned comparisons are covered in Section 15.4.

> A *planned comparison* is a comparison between means planned during the design of the study, before the data are examined.

Planned comparison between two means

A good example of a planned comparison between two means comes from the circadian rhythm experiment (Example 15.1). The main point of that experiment was to contrast the mean phase shift of melatonin production between the knee-treatment group and the control group, which received no extra light. Because the experiment was built around this contrast, we are justified in using methods for planned comparisons to ask, "How big is the difference between the knee-treatment and control groups?" and "Is the difference between these two groups statistically significant?"

The method for a planned comparison between two means is almost the same as the two-sample comparison based on the t-distribution that we learned about in Chapter 12. Only the standard error is calculated differently: the planned comparison uses the pooled sample variance (the error mean square) based on all k groups (and the corresponding error degrees of freedom), rather than that based only on the two groups being compared. This step increases precision and power. The planned comparison method assumes, just as in ANOVA, that the variance is the same within all groups. Details of the modified formulas for a planned comparison are provided in the Quick Formula Summary (Section 15.8).

For example, let's examine the confidence interval for the difference between the means of the "knee" and "control" treatment groups. The difference between the sample means (knee treatment minus control) is small:

$$\bar{Y}_2 - \bar{Y}_1 = (-0.336) - (-0.309) = -0.027 \text{ h}.$$

The standard error for this difference is

$$SE = \sqrt{MS_{error}\left(\frac{1}{n_1} + \frac{1}{n_2}\right)}.$$

---

[5] For this reason they are also called a priori comparisons.

For the knee versus control comparison, the standard error of the difference between these two means is

$$SE = 0.364,$$

which has $N - k = 22 - 3 = 19$ degrees of freedom, the same as that for the error mean square (Table 15.1-3). The critical value from the t-distribution is $t_{0.05(2),19} = 2.09$ (Statistical Table C), which leads to the planned 95% confidence interval for the difference between the population means:

$$(-0.027) - 0.364(2.09) < \mu_2 - \mu_1 < (-0.027) + 0.364(2.09)$$

or

$$-0.788 < \mu_2 - \mu_1 < 0.734,$$

where the units are in hours. If we avoided using the planned-comparison method and simply used the two-sample method for a confidence interval introduced in Chapter 12, we would obtain the following 95% confidence interval instead:

$$-0.813 < \mu_2 - \mu_1 < 0.759.$$

Thus, the planned-comparison method has a slightly higher precision. Similarly, the planned-comparison method for testing the null hypothesis of no difference between these two means has slightly higher power. These are the main advantages of using the planned-comparison methods, provided that the assumption of equal variance is met.

Planned comparisons make all of the same assumptions as ANOVA—namely, random samples from populations, a normal distribution of the variable in every population, and equal variances in all populations. Because each comparison typically involves only one pair of means, planned comparisons are not as robust as ANOVA to violations of the assumptions.

15.4 Unplanned comparisons

The formulas for planned comparisons are not valid for unplanned comparisons, because unplanned comparisons need to make adjustments for the inflated false positive rate (Type I errors) that accompanies multiple testing (Interleaf 8). The number of possible comparisons involving k groups is potentially large. Unplanned comparisons basically represent data dredging or "snooping" because we are poring through data to find differences. This is not inherently bad, provided that the method used corrects for the number of comparisons.

Testing all pairs of means to find out which groups stand apart from the others is the most common type of unplanned comparison, and the **Tukey–Kramer test** is the most commonly used procedure for accomplishing this.[6] The method assumes that we have already carried out an ANOVA and that the null hypothesis of no differences among means has been rejected. Example 15.4 illustrates unplanned comparisons with the Tukey–Kramer method.

| Example 15.4 | **Wood wide web** |

Most plants have underground associations with fungi, called mycorrhizae, which provide minerals and antibiotics to the plant in exchange for sugars. The fungi's hyphae (branching filaments) extend through the soil and may connect to other plants, even to other plant species, creating an underground network that might cause a trickle of nutrients to flow from plant to plant. Simard et al. (1997) measured the flow of carbon between seedlings of birch and Douglas fir and tested whether the carbon flow rate depended on shading. Shaded trees may draw more carbon via the mycorrhizae than trees in full sun. In each of three shade treatments, five pairs of birch and Douglas-fir

seedlings were planted and allowed to grow for one year. Then, each birch was covered in a sealed bag for two hours and supplied with carbon dioxide (CO_2) whose carbon consisted entirely of the carbon-13 isotope (atmospheric CO_2 is made up almost entirely of carbon-12). The same was done to its partnered fir seedling except that CO_2 with carbon-14 was used. The amounts of carbon-13 and carbon-14 present in the tissues of both plants of each pair were measured nine days later. Because different isotopes of carbon were used on birch and Douglas-fir, it was possible to calculate the amount of carbon transferred from each plant to the other. Most transfer occurred from birch to Douglas-fir. Descriptive statistics for the average net carbon gain by Douglas-fir, in milligrams, are given in Table 15.4-1.

Table 15.4-1 Summary of the net amount of carbon transferred from birch to Douglas-fir (Example 15.4).

Shade treatment	Sample mean $\overline{Y}_i$ (mg)	Sample standard deviation, s_i	n_i
Deep shade	18.33	6.98	5
Partial shade	8.29	4.76	5
No shade	5.21	3.00	5

---

[6]   Testing all pairs of means is not the only kind of unplanned comparison. For example, the Scheffé method tests any linear combination of means, but it is too conservative when applied only to pairs of means. Three other methods to test all pairs of means—Duncan's multiple-range test, Fisher's least-significant-difference test, and the Newman–Keuls test—provide less protection than the Tukey–Kramer test against inflated Type I error rates. The Tukey–Kramer method is also called "Tukey's honestly significant difference (HSD)" test. Unplanned comparisons are also called "post hoc tests," "a posteriori tests," or simply "multiple comparison tests."

The ANOVA results for these data, testing the null hypothesis for no differences among treatment means, are shown in Table 15.4-2. The different shade treatments indeed led to differences in the mean net carbon gain by Douglas-fir seedlings, as shown by the high F-ratio and low P-value. It remains to be seen, though, which means are different. That is, are all of the means detectably different from each other? If not, which of the means are different from the others?

Table 15.4-2 ANOVA table summarizing results of the Douglas-fir carbon-transfer data (Example 15.4).

Source of variation	Sum of squares	df	Mean squares	F-ratio	P
Groups (treatments)	470.704	2	235.352	8.784	0.004
Error	321.512	12	26.793		
Total	792.216	14			

Testing all pairs of means using the Tukey–Kramer method

The Tukey–Kramer method works like a series of two-sample t-tests, but it uses a larger critical value to limit the Type I error rate. With the Tukey–Kramer test, the probability of making at least one Type I error throughout the course of testing *all pairs* of means is no greater than our significance level α.

To carry out the Tukey–Kramer procedure, the means must first be ordered from smallest to largest:

No shade	Partial shade	Deep shade
$\bar{Y}_3$	$\bar{Y}_2$	$\bar{Y}_1$
5.21	8.29	18.33

Then, compare every pair of means in turn. For example, the hypotheses for the comparison between the means for "deep shade" (call it μ_1) and "partial shade" (call it μ_2) are

H_0: $\mu_1 - \mu_2 = 0$.
H_A: $\mu_1 - \mu_2 \neq 0$.

The test statistic (q) is calculated using a standard error based on the MS_{error}. This test statistic is then compared with the q-distribution having k and $N - k$ degrees of freedom. Critical values are provided in Statistical Table F. Calculation details are given in the Quick Formula Summary (Section 15.8).

The results for all the pairwise tests in the carbon-transfer data are given in Table 15.4-3.

Table 15.4-3 Summary of Tukey–Kramer tests carried out on the results of Example 15.4.

Group i	Group j	$\bar{Y}_i - \bar{Y}_j$	SE	Test statistic q	Critical value $q_{0.05,3,12}$	Conclusion
Deep	No	13.12	3.2737	4.008	2.67	Reject H_0
Deep	Partial	10.04	3.2737	3.067	2.67	Reject H_0
Partial	No	3.08	3.2737	0.941	2.67	Do not reject H_0

The results are unambiguous. The mean of the "deep shade" group is different from that of both the "partial shade" and "no shade" groups, whereas the means of the "partial shade" and "no shade" groups are not significantly different from each other.

The results of the Tukey–Kramer procedure are often indicated using symbols, such as those shown in Figure 15.4-1 for displaying the means. Groups in the figure are assigned the same symbol if their means are not significantly different, based on the unplanned comparisons (e.g., the partial shade and no shade treatments are both assigned the symbol "b" in Figure 15.4-1). Sample means in the figure are assigned a unique symbol if they are different from all other means (e.g., the deep shade treatment is given the symbol "a" in Figure 15.4-1).[7]

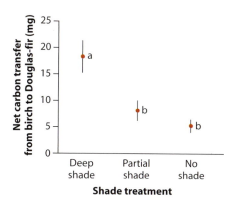

Figure 15.4-1 Using symbols to indicate the outcome of Tukey–Kramer tests of all pairs of means. Dots and error bars indicate means $\pm$ 1 SE of the three treatment groups in Example 15.4. Two means are assigned a different symbol if they are significantly different ("a" vs. "b"), whereas means are assigned the same symbol ("b" in this case) if they are not significantly different.

---

[7] One or more groups with intermediate means might be assigned two symbols if the Tukey-Kramer results were ambiguous. For example, if it had turned out that the partial-shade mean was not significantly different from the means of either the deep shade or no shade groups, yet the deep shade and no shade means were different from each other, then partial shade would have been assigned both symbols "a" and "b" to indicate this ambiguity.

The Tukey–Kramer results do not imply that the "partial shade" and "no shade" treatments have the same mean carbon transfer rates. Assuming that they do would be making the mistake of accepting a null hypothesis. The results merely indicate that a test of their differences did not reject H_0.

> With the Tukey–Kramer method, the probability of making at least one Type I error throughout the course of testing all pairs of means is no greater than the significance level α.

Assumptions

The Tukey–Kramer method makes all of the same assumptions as ANOVA—namely, random samples, a normal distribution of the variable in every population, and equal variances in all groups. Because each comparison tests only one pair of means at a time, rather than all means simultaneously, the method is not as robust as ANOVA to violations of these assumptions.

The P-value for the Tukey–Kramer test is exact when the experimental design is balanced—that is, when the sample size is the same in every group ($n_1 = n_2 = \ldots = n_k$). If the sample sizes are different, then the Tukey–Kramer test is conservative, which means that the real probability of making at least one Type I error, when testing all pairs of means, is smaller than the stated α. This makes it harder to reject H_0, which is why the test is deemed "conservative."

15.5 Fixed and random effects

Up to now, we have been analyzing *fixed* groups—namely, studies in which the different categories of the explanatory variable are predetermined, of direct interest, and repeatable. ANOVA on fixed groups is called **fixed-effects** analysis of variance. Other examples of fixed effects include

▶ alternative medical treatments in a clinical trial,
▶ fixed doses of a toxin,
▶ different heights above low tide in the intertidal zone, and
▶ different sexes or age categories of individuals.

Any conclusions reached about differences among fixed groups apply only to those fixed groups. If a difference among drug treatments was found in some response variable, for example, we could not generalize the results to other drugs not included in the study.

By contrast, there is a second type of ANOVA in which the groups are not fixed, but instead are *randomly chosen*. Randomly chosen groups are not predetermined, but instead are randomly sampled from a much larger "population" of groups. ANOVA applied to random groups is called **random-effects** ANOVA.[8] Examples of random effects include

▶ family, in a study of resemblance among relatives in IQ scores, and
▶ individual, in a study of measurement error involving repeated measurements of individuals.

Because random effects are randomly sampled from a population, conclusions reached about differences among groups *can* be generalized to the whole population.

With random effects, the specific groups included in a study are ephemeral and would not typically be reused. For example, consider a study to investigate whether families differ from one another in the mean IQ scores of their children. This study would begin with a random sample of families from a population of families. "Family" would be used as the group variable in an ANOVA, and the replicates would be the different children making up each family. A later study attempting to address the same question in the same population would not attempt to relocate the same families used in the previous study, because the population, not the families themselves, is the target of study. Rather, a new study would begin with a new random sample of families and would again be able to generalize the results to the whole population. In contrast, a new study of the effects of shade treatment on net carbon transfer—a fixed effect—could quite easily use the same shade treatments as previous studies.

> An explanatory variable is called a *fixed effect* if the groups are predefined and are of direct interest. An explanatory variable is called a *random effect* if the groups are randomly sampled from a population of possible groups.

For a single-factor ANOVA, *F*-tests of differences among group means are the same whether the effects are fixed or random. However, the difference between fixed and random effects becomes important when ANOVA is expanded to examine the effects of more than one explanatory variable simultaneously. We discuss some of these issues in Chapter 18.

15.6   ANOVA with randomly chosen groups

Because the groups are not of specific interest in a random-effects ANOVA, planned and unplanned comparisons aren't particularly useful. Instead, random-effects

---

[8] Some people refer to fixed-effects ANOVA as "Model 1" and random-effects ANOVA as "Model 2."

ANOVA can be used to estimate *variance components*—that is, the magnitude of the variances within and among groups. Among other uses, variance components are employed in animal and plant breeding to identify the contributions of genes and the environment to variance in traits. Variance components are also useful for quantifying measurement error in data, as shown in Example 15.6.

Example 15.6   ## Walking stick limbs

The walking stick *Timema cristinae* is a wingless herbivorous insect that lives on plants in chaparral habitats of California. In a study of the insect's adaptations to different plant species, Nosil and Crespi (2006) measured a variety of traits using digital photographs of specimens collected from

a study site in California. They measured various traits on the photographs using the computer. Because the researchers were concerned about measurement error, they took two separate photographs of each specimen. After measuring traits on one set of photographs, they repeated the measurements on the second set. Very often the result was different the second time around, indicating measurement error. How large was the measurement error compared with real variation among the traits of individuals? Figure 15.6-1 illustrates the two measurements of femur length made on 25 specimens. The raw data are listed in Table 15.6-1.

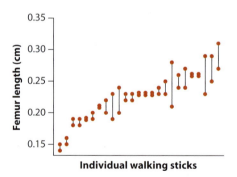

Figure 15.6-1 "Dot plot" showing the pair of femur length measurements (connected by a line segment) obtained from separate photographs of each of 25 walking sticks.

Table 15.6-1 Femur length, in centimeters, measured from separate photographs of 25 walking sticks. Two measurements were made per specimen to evaluate measurement error.

Specimen	Femur length (cm)	Specimen	Femur length (cm)
1	0.26, 0.26	14	0.27, 0.24
2	0.23, 0.19	15	0.23, 0.29
3	0.25, 0.23	16	0.23, 0.23
4	0.26, 0.26	17	0.14, 0.15
5	0.23, 0.22	18	0.19, 0.19
6	0.23, 0.23	19	0.31, 0.27
7	0.22, 0.23	20	0.23, 0.24
8	0.21, 0.28	21	0.16, 0.15
9	0.24, 0.26	22	0.22, 0.20
10	0.24, 0.20	23	0.19, 0.18
11	0.29, 0.25	24	0.21, 0.21
12	0.23, 0.23	25	0.19, 0.20
13	0.18, 0.19		

ANOVA table

The ANOVA results, which are needed to calculate the variance components of the walking-stick data, are presented in Table 15.6-2. With only one explanatory variable, the calculations for random-effects ANOVA are the same as for fixed effects. The "groups" are the individual insects, and the replicates are the repeat measurements made of each insect.

The results of the ANOVA indicate that significant differences are present among groups. In other words, despite measurement error, there are still detectable differences in the femur lengths among insect specimens. This result implies that there is significant variation in femur length among insect individuals *in the population*, which is the real aim of the study. We can generalize the conclusions of our ANOVA to the whole insect population if the 25 specimens are a random sample.

Table 15.6-2 ANOVA table with results of repeat femur-length measurements of 25 walking sticks.

Source of variation	Sum of squares	df	Mean squares	F-ratio	P
Groups (individual insects)	0.059132	24	0.002464	6.902	< 0.0001
Error	0.008900	25	0.000356		
Total	0.068032	49			

Variance components

Single-factor ANOVA with random effects differs from fixed-effects ANOVA by having *two* levels of random variation[9] in the response variable (Y). The first level is variation within groups and the second level is variation between groups. In our stick insect study (Example 15.6), the first level (variation within groups) is the variance among repeat measurements made on the same individual, which is exclusively due to measurement error in the study. ANOVA assumes that the "true" variance between measurements is the same in every group. In Example 15.6, this means that measurement error is the same for every individual insect, except by chance. We use the symbol σ^2 to indicate the value of the variance within groups in the population. The single best estimate of σ^2 is MS_{error}, the error mean square. In the walking sticks, $MS_{error} = 0.000356$ (Table 15.6-2).

To evaluate the second level of random variation (i.e., variation between groups), each group is assumed to have its own mean. For the walking sticks, the mean femur length of an individual insect is its "true" femur length—that is, the length we would obtain if we measured its femur a great many times and took the average measurement. Random-effects ANOVA assumes that group means have a normal probability distribution in the population with a grand mean μ_A (the mean of all the group means) and a variance σ_A^2 (the variance among the group means in the population of groups). In our study of the walking sticks, we assume that "true" femur length varies between individual insects in the population according to a normal distribution. The parameter σ_A^2 is the variance among insects in the population in their average femur length.

In random-effects ANOVA, the parameters σ^2 and σ_A^2 are called **variance components**. Together they describe all the variance in the response variable Y—that is, σ^2 describes the variance within groups (e.g., the measurement error in the walking-stick study) and σ_A^2 describes the variance among groups (e.g., the differences between the true femur lengths of individual insects).

> The *variance components* in a random-effects ANOVA make up all the variance in the response variable: variance within groups (σ^2) and variance among groups (σ_A^2).

The group mean square from the ANOVA results (Table 15.6-2) can be used to estimate σ_A^2. Using the symbol s_A^2 to indicate the estimate of σ_A^2,

$$s_A^2 = \frac{MS_{groups} - MS_{error}}{n},$$

---

[9] Fixed-effects ANOVA also has two levels of variation, but only variation within groups is random. Differences among fixed groups are not random.

where n is the number of measurements taken within each group.[10]

There were $n = 2$ measurements for each insect specimen in the data for Example 15.6, so

$$s_A^2 = \frac{0.002464 - 0.000356}{2} = 0.00105.$$

Thus, the best estimate of the variance in femur length among individuals in this population is 0.00105.

Repeatability

Repeatability is the fraction of the summed variance that is among groups:

$$\text{Repeatability} = \frac{s_A^2}{s_A^2 + MS_{error}}.$$

The denominator of the repeatability equation (i.e., $s_A^2 + MS_{error}$) estimates the total amount of variance in the population, summing the variance among groups and the variance within groups.

Repeatability measures the overall similarity of repeat measurements made on the same group.[11] A repeatability near zero, the lowest possible value, indicates that nearly all of the variance in the response variable results from differences between separate measurements made on the same group. In our example, this would mean that measurement error greatly dominates the variation found in the data. In contrast, a repeatability near one, the maximum value, indicates that repeated measurements on the same group give nearly the same answer every time.

Calculated on the walking-stick data,

$$\text{Repeatability} = \frac{0.00105}{0.00105 + 0.000356} = 0.75.$$

In other words, an estimated 75% of the total variance in femur length measurements in the population is the result of differences in true femur length between individual insects. The remaining 25% is the result of measurement error.

The repeatability of femur length is not higher, because the walking sticks are small and their femurs are even smaller. Slight variation in the position of the insect when a photograph is taken can lead to different length measurements. It is important, therefore, to report repeatability estimates in your research papers. If measure-

---

[10] This formula is modified if the number of measurements is not identical in every group. Also, s_A^2 can sometimes be negative by chance, even though σ_A^2 can only be zero or positive. When this happens, s_A^2 is set to zero.

[11] The repeatability is also called the "intraclass correlation."

ment error is present, then it is best to take several measurements of each specimen and take their average.

Don't confuse repeatability with the R^2 value (Section 15.1). Repeatability reflects the magnitudes of variance components, which estimate specific population parameters. It applies only to random effects. R^2, on the other hand, isn't an estimate of any population parameter. R^2 just measures the reduction in the amount of scatter around the group means compared with that around the grand mean. R^2 is based on the sums of squares rather than variances, and it can be applied to fixed or random effects.

Assumptions

Random-effects ANOVA makes all of the same assumptions as fixed-effects ANOVA, but adds two more. It assumes that groups are randomly sampled. Also, random-effects ANOVA assumes that the group means have a normal distribution in the population.

15.7 Summary

▶ The analysis of variance (ANOVA) tests differences among the means of multiple groups.

▶ ANOVA works by comparing the variance among subjects within groups (the error mean square, MS_{error}) with the variance among individuals belonging to different groups (represented by the group mean square, MS_{groups}).

▶ The test statistic in ANOVA is the F-ratio, where $F = MS_{groups}/MS_{error}$.

▶ Under H_0, the F-ratio should be close to one except by chance. Under H_A, the F-ratio is expected to exceed one. H_0 is rejected if F is much larger than one. When F is significantly greater than one, it implies that there is more variance among group means than expected by chance.

▶ The quantities needed to test hypotheses or estimate variance components in single-factor ANOVA are summarized in an ANOVA table.

▶ R^2 measures the fraction of the variability explained by the explanatory (group) variable in analysis of variance. It measures the reduction in the amount of scatter of the measurements around their group means compared with that around the grand mean.

▶ ANOVA assumes that the Y-variable has a normal distribution in each population, with equal variance in all populations.

▶ ANOVA is robust to departures from the assumption of normal populations, especially if the sample size is large. ANOVA is also robust to departures from the assumption of equal standard deviations, but only if the study design is balanced.

▶ The Kruskal–Wallis test is an alternative, nonparametric method used to test the null hypothesis of equal means or medians in different groups if the normality assumption of ANOVA cannot be met.

▶ The Kruskal–Wallis test assumes that the frequency distributions of measurements in different groups have the same shape (i.e., the same assumption as the Mann–Whitney U-test).

▶ Planned comparisons between means are few in number and represent only comparisons identified as crucial before the data are collected and analyzed. Unplanned comparisons are a more comprehensive set of comparisons done in search of interesting patterns. Unplanned comparisons require special methods to protect against high Type I error rates.

▶ The Tukey–Kramer method to compare all pairs of means is the most commonly used method for unplanned comparisons.

▶ In fixed-effects ANOVA, the groups are predetermined, repeatable categories of direct interest. The results of ANOVA apply only to those groups included in the study.

▶ In random-effects ANOVA, groups are a random sample from a population of groups. The results of random-effects ANOVA can be generalized to the population of groups.

▶ The repeatability estimates the fraction of the total variance that is among groups in random-effects ANOVA. Repeatability is frequently used to evaluate the importance of measurement error.

15.8   Quick Formula Summary

Analysis of variance (ANOVA)

What is it for? Testing the difference among means of k groups simultaneously.

What does it assume? The variable is normally distributed with equal standard deviations (and therefore equal variances) in all k populations. Each sample is a random sample.

Test statistic: F

Sampling distribution under H₀: F-distribution with $k - 1$ and $N - k$ degrees of freedom. Use the right tail of the F-distribution in ANOVA.

Formulae:

Source of variation	Sum of squares	df	Mean squares	F-ratio
Groups	$SS_{groups} = \sum n_i(\bar{Y}_i - \bar{Y})^2$	$k-1$	$\dfrac{SS_{groups}}{df_{groups}}$	$\dfrac{MS_{groups}}{MS_{error}}$
Error	$SS_{error} = \sum s_i^2(n_i - 1)$	$N-k$	$\dfrac{SS_{error}}{df_{error}}$	
Total	$SS_{groups} + SS_{error}$	$N-1$		

In these formulae, n_i is the sample size in group i, k is the number of groups, $\bar{Y} = \dfrac{\sum n_i(\bar{Y}_i)}{N}$ is the grand mean, and $N = \sum n_i$ is the total sample size.

R squared (R^2)

What is it for? Measuring the fraction of the variation in Y that is "explained" by differences among groups.

Formula: $R^2 = \dfrac{SS_{groups}}{SS_{total}}$,

where SS_{groups} is the sum of squares for groups and SS_{total} is the total sum of squares.

Kruskal–Wallis test

What is it for? Testing differences among k populations in the means or medians of their distributions.

What does it assume? Random samples. The frequency distributions of measurements in the different groups have the same shape.

Test statistic: H

Sampling distribution under H₀: Approximately χ^2 with $k-1$ degrees of freedom.

Formula: $H = \dfrac{12}{N(N+1)} \sum \dfrac{R_i^2}{n_i} - 3(N+1),$

where N is the total sample size and R_i is the sum of the ranks for group i.

Observations are ranked from small to large, as described in Section 13.5 for the Mann–Whitney U-test.

Planned confidence interval for the difference between two of *k* means

What is it for? Estimating the difference between means of two out of k populations when the comparison is *planned*.

What does it assume? Random samples. The variable is normally distributed with equal variances in all k populations.

Estimate: $\bar{Y}_i - \bar{Y}_j$, where i and j are the two populations, with $i \neq j$.

Parameter: $\mu_i - \mu_j$

Formula: $(\bar{Y}_i - \bar{Y}_j) - \text{SE } t_{0.05(2),N-k} < \mu_i - \mu_j < (\bar{Y}_i - \bar{Y}_j) + \text{SE } t_{0.05(2),N-k}$,

where $\text{SE} = \sqrt{\text{MS}_{\text{error}}\left(\dfrac{1}{n_i} + \dfrac{1}{n_j}\right)}$.

Planned test of the difference between two of *k* means

What is it for? Testing the difference between two out of k population means when the comparison is *planned*.

What does it assume? Random samples. The variable is normally distributed with equal variances in all k populations.

Test statistic: t

Distribution under H₀: The t-distribution with $N - k$ degrees of freedom.

Formulae: $t = \dfrac{(\bar{Y}_i - \bar{Y}_j)}{\text{SE}}$,

where $\bar{Y}_i - \bar{Y}_j$ is the observed difference between the means of two groups i and j (with $i \neq j$), and $\text{SE} = \sqrt{\text{MS}_{\text{error}}\left(\dfrac{1}{n_i} + \dfrac{1}{n_j}\right)}$ is the standard error of the difference between the two means.

Tukey–Kramer test of all pairs of means

What is it for? Testing the differences of all pairs of k means. These tests are *unplanned*.

What does it assume? Random samples. The variable is normally distributed with equal variances in all k populations. An ANOVA has already rejected the null hypothesis of equal means for all k groups.

Test statistic: q

Sampling distribution under H$_0$: The q-distribution with k means and $N - k$ degrees of freedom.

Formula: $q = \dfrac{\overline{Y}_i - \overline{Y}_j}{\text{SE}}$,

where $\overline{Y}_i - \overline{Y}_j$ is the observed difference between the means of two groups i and j (with $i \neq j$), $\text{SE} = \sqrt{\text{MS}_{\text{error}}\left(\dfrac{1}{n_i} + \dfrac{1}{n_j}\right)}$, and n_i and n_j are the sample sizes for the two groups i and j, respectively.

Repeatability and variance components

What is it for? Repeatability is the fraction of the total variance that is among groups in a random-effects ANOVA.

What does it assume? Groups are a random sample from a population of groups. Repeat measurements within groups are randomly sampled and normally distributed, with an equal standard deviation (and variance) in all groups. Group means have a normal distribution in the population.

Formulae: $\text{Repeatability} = \dfrac{s_A^2}{s_A^2 + \text{MS}_{\text{error}}}$,

where $s_A^2 = \dfrac{\text{MS}_{\text{groups}} - \text{MS}_{\text{error}}}{n}$, and n is the sample size within each group (assumed to be equal for our formulas).

PRACTICE PROBLEMS

1. An important issue in conservation biology is how dispersal among populations influences the persistence of species in a fragmented landscape. Molofsky and Ferdy (2005) measured this in the annual plant *Cardamine pensylvanica*, a weed that produces explosively dispersed seeds. Four treatments were used to manipulate seed dispersal by changing the distance among experimental plant populations. These treatments were adjacent (continuous treatment), separated by 23.2 cm (medium), separated by 49.5 cm (long), or separated by partitions that blocked all seed dispersal among populations (isolated). Treatments were randomly assigned to plant populations. The data below are the number of generations that the populations persisted in four replicates of each treatment.

Treatment	Generations persisted
Isolated	13, 8, 8, 8
Medium	14, 12, 16, 16
Long	13, 9, 10, 11
Continuous	9, 13, 13, 16

a. What is the explanatory variable in this analysis? What is the response variable?
b. Is this an experimental study or an observational study? Explain.
c. Calculate the sample means for each group, and then calculate a confidence interval for the mean of each group. What assumptions have you made?
d. Display the data along with the means and confidence intervals in a graph.

2. Analysis of variance carried out on the *Cardamine* data of the previous problem yielded the following results:

Source of variation	Sum of squares	df	Mean squares	F-ratio	P
Groups (treatments)	63.188				0.035
Error	63.250				
Total					

a. What are the null and alternative hypotheses being tested?
b. Fill in the rest of the table.
c. What is the sampling distribution of the *F*-statistic under the null hypothesis?
d. What does *P* measure?
e. In words, explain what each "sum of squares" measures.
f. If you wanted to determine the fraction of the variation in the number of generations that populations persisted that is "explained" by treatment, what quantity would you use?
g. Calculate the quantity identified in part (f).

3. The Tukey–Kramer procedure was carried out on the results of Practice Problems 1 and 2, yielding the data at the bottom of this page:

a. What are the null and alternative hypotheses being tested?
b. Are these comparisons considered planned or unplanned? Why?
c. Only the largest pairwise difference between means, that between the "medium" and "isolated" treatments, is statistically significant. How is this possible, given that neither of these two means is significantly different from the means of the other two groups?
d. Using symbols, summarize the results of these tests on the figure you created in part (d) of Practice Problem 1.
e. Explain in words why the critical value for each test (2.97) is larger than the critical value of the *t*-distribution having 12 degrees of freedom (2.18).

4. Imagine a hypothetical experiment with multiple treatments but relatively small sample sizes within treatments. The goal is to test whether the treatment means are equal. Calculations and graphical analysis indicate that the data differ markedly from a normal distribution.
a. What other two options are available to carry out a test?
b. Which of the other two options should be attempted first? Why?

5. In a field experiment designed to investigate the role of genetic diversity in ecosystems, Reusch et al. (2005) planted eelgrass in plots in a shallow estuary in the Baltic Sea. Eighteen eelgrass shoots were planted in every plot. Some plots (randomly chosen) were planted with only one eelgrass genotype, others were planted with three genotypes, and others were planted with six different eelgrass genotypes. At the end of

Data Table, Problem 3.

Group *i*	Group *j*	$\bar{Y}_i - \bar{Y}_j$	SE	q	$q_{0.05,4,12}$	Conclusion
Medium	Isolated	5.25	1.623	3.234	2.97	Reject H_0
Medium	Long	3.75	1.623	2.310	2.97	Do not reject H_0
Medium	Continuous	1.75	1.623	1.078	2.97	Do not reject H_0
Continuous	Isolated	3.50	1.623	2.156	2.97	Do not reject H_0
Continuous	Long	2.00	1.623	1.232	2.97	Do not reject H_0
Long	Isolated	1.50	1.623	0.924	2.97	Do not reject H_0

the experiment, the total number of shoots was counted in each plot. The results from 32 plots are given in the following table:

Treatment (number of genotypes)	Number of shoots at end of experiment
1	11, 14, 21, 27, 28, 30, 32, 36, 38, 49, 61, 71
3	20, 35, 36, 41, 46, 47, 52, 53, 58, 67
6	31, 45, 45, 47, 48, 62, 64, 69, 84, 86

a. What are the hypotheses for the test?
b. Carry out the test using ANOVA. Summarize your results in an ANOVA table.
c. What assumptions have you made?
d. Is this a fixed-effects or random-effects ANOVA? Explain your reasoning.

6. Finding the bases of the major mental illnesses, schizophrenia and bipolar disorder, is a major activity in medical research. In an amazing illustration of the "new anatomy," Tkachev et al. (2003) compared expression levels of several genes involved in the production and activity of myelin, a tissue important in nerve function, in the brains of 15 persons with schizophrenia, 15 with bipolar illness, and 15 control individuals. The results presented in the table at the bottom of this page summarize the expression levels of the proteolipid protein 1 gene (PLP1) in the 45 brains.
   a. The main objective of the study was to compare PLP1 gene expression in persons having schizophrenia with that of control individuals. Using a planned comparison approach, compute a 95% confidence inter-

val for the difference between the means of these two groups.
   b. Is a planned comparison appropriate in part (a)? Explain.
   c. What are your assumptions in part (a)?

7. Use the data from Practice Problem 6 to solve this problem.
   a. Test whether mean PLP1 gene expression differs among the schizophrenia, bipolar, and control groups.
   b. What are your assumptions in part (a)?
   c. Is the analysis in part (a) a random-effects or fixed-effects ANOVA? Explain.
   d. What quantity would you use to describe the fraction of the variation in expression levels explained by group differences? Calculate this quantity for the data from Practice Problem 6.
   e. What method would you use after ANOVA to determine which group means were different from each other?

8. The bright yellow head of the adult Egyptian vulture (see the photo at the beginning of the chapter) requires carotenoid pigments. These pigments cannot be synthesized by the vultures, though, so they must be obtained through their diet. Unfortunately, carotenoids are scarce in rotten flesh and bones, but they are readily available in the dung of ungulates. Perhaps for this reason, Egyptian vultures are frequently seen eating the droppings of cows, goats, and sheep in Spain, where they have been studied.[12]

---

[12] Spaniards have nicknamed the Egyptian vulture "moñiguero," which politely translates as "dung-eater."

Group	Raw data (normalized units)	Mean	Standard deviation
Control	−0.02, −0.27, −0.11, 0.09, 0.25, −0.02, 0.48, −0.24, 0.06, 0.07, −0.3, −0.18, 0.04, −0.16, 0.25	−0.004	0.218
Schizophrenia	−0.1, −0.31, −0.05, 0.11, −0.38, 0.23, −0.23, −0.28, −0.36, −0.22, −0.4, −0.19, −0.34, −0.29, −0.12	−0.195	0.182
Bipolar	−0.34, −0.39, −0.22, −0.32, −0.32, −0.05, −0.43, −0.33, −0.41, −0.36, −0.25, −0.29, 0.06, −0.3, 0.01	−0.263	0.151

Ungulates are common in some areas but not in others. Negro et al. (2002) measured plasma carotenoids in wild-caught vultures at four randomly chosen locations in Spain as part of a study to determine the causes of variation among sites in Spain in carotenoid availability.

Site	Mean concentration (μg/ml)	Standard deviation	n
1	1.86	1.22	22
2	5.75	2.46	72
3	6.44	3.42	77
4	11.37	1.96	11

a. Is the explanatory variable a random effect or a fixed effect? Explain.
b. Use the data provided in the table to test whether the mean plasma concentration of carotenoids in wild Egyptian vultures differs among sites.
c. What are the assumptions of your analysis in part (b)?

9. Dalterio et al. (1982) carried out an experiment to examine the effects of cannabinoids on fertility measures of male mice. Four treatment groups were established with 18 male mice per treatment. Treatments included three forms of cannabinoids and a sesame oil control. Cannabinoids were administered orally three times a week for five weeks at a dose roughly equivalent in humans to three marijuana cigarettes. In the last three weeks of the study each male was placed with a virgin female. At the end of the study, the state of pregnancy in the companion female was recorded, as was the weight of each male's testes. The results are in the table at the bottom of the page.

a. Test whether the cannabinoid treatment affected mean testes weight.
b. What are your assumptions in part (a)?
c. Was this an experimental study in the statistical sense? Explain your answer.
d. What fraction of the variation in testes weight is "explained" by the treatment?

10. For the data in Practice Problem 9, test whether the cannabinoid treatment of males affected the proportion of females impregnated.

11. One way to assess whether a trait in males has a genetic basis is to determine how similar the measurements of that trait are among his offspring born to different, randomly chosen females. In a lab experiment, Kotiaho et al. (2001) randomly sampled 12 male dung beetles, *Onthophagus taurus*, and mated each of them to three different virgin females. The average body-condition score of offspring born to each of the three females is listed for each male in the following table.

Data Table, Problem 9.

Treatment	Females impregnated	Females not impregnated	Mean testes weight (mg)	Standard error of mean
Oil	14	4	332	10
THC	13	5	310	10
Cannabinol	11	7	299	12
Cannabidiol	13	5	288	9

Male	Offspring body-condition scores
1	0.82, 0.44, 0.92
2	0.35, 0.19, 1.39
3	0.12, 0.84, 0.16
4	0.49, 0.59, −0.23
5	0.44, 0.33, 0.07
6	0.00, 0.29, 0.30
7	0.69, −0.49, −0.60
8	0.13, −0.43, 0.66
9	0.21, −0.34, −0.09
10	−0.35, −1.40, −0.45
11	−1.04, −0.34, −0.82
12	−1.27, −0.75, −0.86

ANOVA was used to test whether males differed in the mean condition of their offspring using the three measurements for each male. The results were $SS_{groups} = 9.940$ and $SS_{error} = 4.682$.

a. Is this a random-effects or a fixed-effects ANOVA? Explain the reason behind your answer.

b. With these data, test whether males differed in the mean condition of their offspring.

c. What is the repeatability of the offspring condition of males mated to different females?[13]

12. Imagine that you are a statistical consultant and a researcher comes to you with the following problem. The researcher has a random sample of data from each of six groups and she wants to test whether the means of the six groups differ. In each of the following situations, recommend the best procedure.

a. The data are approximately normally distributed with equal standard deviations in all groups. Sample size is small.

b. The data are *not* normally distributed and do not have equal standard deviations in all groups. Sample size is small.

c. The data are *not* normally distributed. Groups have nearly equal standard deviations. Sample size is large and equal among groups.

d. The null hypothesis of no difference among group means was rejected. Now the researcher wants to find out which means are different. The data are approximately normally distributed with equal standard deviations in all groups, and the sample size is small.

ASSIGNMENT PROBLEMS

13. Dormant eggs of the zooplankton *Daphnia* survive in lake sediments for decades, making it possible to measure their physiological traits in past years. Hairston et al. (1999) extracted *Daphnia* eggs from sediment cores of Lake Constance in Europe to examine trends in resistance to dietary cyanobacteria, a toxic food type

that has increased in density since 1960 in response to increased nutrients in the lake. The table and histogram on the next page give the resistance of 32 *Daphnia* clones, each initiated from single eggs extracted from deposits laid down during years of low, medium, and high cyanobacteria density between 1962 and 1997. Resistance is the average growth rate of individuals fed cyanobacteria divided by the growth rate when individuals from the same clone are fed a high-quality algal food instead. We wish to test whether mean resistance differs among the three groups of *Daphnia* clones.

[13] The repeatability of a trait among the offspring of males born from different females helps estimate the "heritability" of that trait, the fraction of variation in the trait in the population that is genetic rather than environmental. Differences among males have a genetic basis, because they were randomly mated and their offspring were raised in a common (lab) environment.

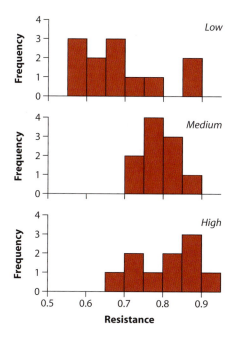

Cyanobacterium density	Measurements of resistance
Low	0.56, 0.57, 0.58, 0.62, 0.64, 0.65, 0.67, 0.68, 0.74, 0.78, 0.85, 0.86
Medium	0.70, 0.74, 0.75, 0.76, 0.78, 0.79, 0.80, 0.82, 0.83, 0.86
High	0.65, 0.73, 0.74, 0.76, 0.81, 0.82, 0.85, 0.86, 0.88, 0.90

a. Examine the histograms of the data. Give the two main reasons why caution is warranted before using ANOVA to test for differences among group means.

b. Give the main reason why caution is warranted before using the Kruskal–Wallis method to test for differences between group means.

14. Tsetse flies are the vectors of human sleeping sickness and animal trypanosomiasis in Africa. The tsetse fly species *Glossina palpalis* feeds on the blood of a variety of animals, including humans, and an important question is whether the feeding preferences of individuals can be affected by learning. To investigate this, Bouyer et al. (2007) provided cohorts of male tsetse flies

with a first blood meal of either cows or lizards. After two days, the flies were offered a second blood meal of cows only. The data below measure the proportion of flies in each cohort that took a meal from the cows (the remaining individuals chose not to feed).

Treatment: first blood meal	Proportion of flies taking second meal from cow
Lizard	0.66, 0.58, 0.52, 0.37, 0.35, 0.34, 0.29
Cow	1.00, 0.98, 0.97, 0.96, 0.87, 0.83

a. Display the results of the study in a graph.

b. What assumptions must be met before using ANOVA to test for differences between the two treatment groups in the mean proportion of flies taking the second blood meal from cows? In view of your results in part (a), do you see a problem meeting these assumptions?

c. Consider using a transformation to fix the problems identified in part (b). Given the data, what transformation should be attempted first? Try this transformation on the data. Did it fix the problems?

d. Using the transformed data, test whether the means of the two blood-meal groups are different.

15. When using analysis of variance, what are the main advantages of
a. a large sample size?
b. a balanced design?

16. Examine Figure 15.4-1 (p. 409) which shows means and standard errors for three treatment groups. If this graph was to be the only graph to display the results in a scientific report, what additional feature(s) would you recommend?

17. Tukey–Kramer tests carried out on the results in Practice Problem 5 yielded the following table of results.

Group i	Group j	$\bar{Y}_i - \bar{Y}_j$	SE	q	$q_{0.05,k,N-k}$	Conclusion
6	1	23.26	7.13	3.26	2.47	
6	3	12.60	7.45	1.69	2.47	
3	1	10.67	7.13	1.50	2.47	

a. Fill in the conclusions in the preceding table.
b. Are these planned or unplanned comparisons? Explain.
c. Why not use a series of two-sample *t*-tests instead of the Tukey–Kramer method?
d. In the preceding analysis, what is the probability of making at least one Type I error during the course of carrying out all of the pairwise tests of differences between means?

18. A study by Hovatta et al. (2005) examined the genetic basis of anxiety-like behavior using four strains of inbred mice. The data, summarized in the following table, give the mean number of minutes (out of 5 min) that mice spent in an "open field" chamber; more time in the open field indicates less anxiety. $SE_{\bar{Y}}$ is the standard error of the mean. Assume that the measurement is normally distributed within each strain, and that the four inbred strains were randomly sampled from a large population of inbred strains available.

Strain	$\bar{Y}$ (min)	$SE_{\bar{Y}}$	n
A	0.61	0.123	7
B	0.34	0.164	7
C	1.07	0.140	7
D	1.49	0.093	7

a. What does $SE_{\bar{Y}}$ measure? Explain.
b. Use ANOVA to test whether there is significant variation among strains in the mean time spent in the open field chamber.
c. Is this a random-effects or a fixed-effects ANOVA? Explain.
d. Estimate the variance components within and among strains in the measurement of time spent in the open field.
e. What is the estimated repeatability of the measurement?
f. What does the quantity in part (e) measure?

19. As the "baby boom" generation ages, interest in finding treatments that extend lifespan has surged. Experimental research mainly uses nonhuman animals. In a recent experiment, Evason et al. (2005) tested the influence of the anticonvulsant medication trimethadione on the lifespan of the nematode worm, *Caenorhabitis*

elegans. The study compared the effects of three trimethadione treatments (provided at the larval stage, provided at the adult stage, and provided at both stages) and a water treatment (the control). The resulting lifespans are shown in the following histograms for 50 worms in each treatment. Assume that each worm was treated independently.

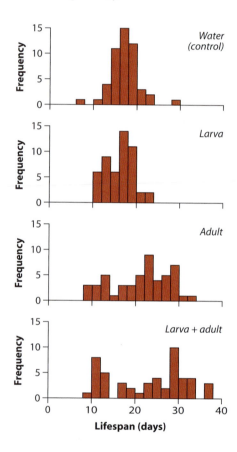

a. ANOVA would be the preferred method to compare the means of each group. What problem or problems do you foresee in applying this method to the data shown in the preceding histograms?
b. Applying a Kruskal–Wallis test to the data yielded the following results.

	Trimethadione treatment			
	None (water)	Larval stage	Adult stage	Both stages
Rank sums	4201	3672	6003.5	6223.5

$H = 29.27$; $\chi^2_{3,0.05} = 7.81$.

Can we conclude from this result that the means of the treatment groups are unequal? Explain.

20. An observational study gathered data on the rate of progression of multiple sclerosis in patients diagnosed with the disease at different ages. Differences in the mean rate of progression were tested among several groups that differed by age-of-diagnosis using ANOVA. The results gave $P = 0.12$. From the following list, choose all of the correct conclusions that follow from this result (Borenstein 1997). Explain the basis of your answer.
    a. The mean rate of progression does not differ among age groups.
    b. The study failed to show a difference among means of age groups, but the existence of a difference cannot be ruled out.
    c. If a difference among age groups exists, then it is probably small.
    d. If the study had included a larger sample size it probably would have detected a significant difference among age groups.

21. Head width was measured twice, in cm, on the random sample of 25 walking stick insects described in Example 15.6. The data are as follows:

Specimen	Head width (cm)	Specimen	Head width (cm)
1	0.15, 0.15	14	0.15, 0.18
2	0.18, 0.18	15	0.19, 0.21
3	0.17, 0.18	16	0.18, 0.19
4	0.21, 0.21	17	0.15, 0.15
5	0.15, 0.16	18	0.17, 0.17
6	0.19, 0.19	19	0.20, 0.21
7	0.18, 0.17	20	0.18, 0.18
8	0.18, 0.23	21	0.18, 0.16
9	0.18, 0.17	22	0.17, 0.18
10	0.19, 0.21	23	0.16, 0.13
11	0.17, 0.20	24	0.18, 0.16
12	0.19, 0.21	25	0.17, 0.17
13	0.16, 0.18		

The following table presents some of the results of an ANOVA of the head-width measurements in which the individual specimen was the group.

Source of variation	Sum of squares	df	Mean squares	F-ratio	P
Groups (specimens)	0.015788				
Error	0.004150				
Total	0.019938				

    a. Complete the ANOVA table.
    b. What is the estimate of the variance within groups for head width?
    c. What is the estimate of the variance among groups?
    d. What is the repeatability of the head-width measurements?
    e. Compare your result in part (d) with that for femur length analyzed in the text. Which trait has higher repeatability? Which trait is more affected by measurement error?

22. The accompanying table presents mean cone size (mass) of lodgepole pine in 16 study sites in three types of environments in western North America (Edelaar and Benkman 2006). The three environments were islands of lodgepole pines in which pine squirrels were absent (an "island" here refers to a patch of lodgepole pine surrounded by other habitat and separated from the large tracts of contiguous lodgepole pine forests), islands with squirrels present, and sites within the large areas of extensive lodgepole pines ("mainland") that all have squirrels.

Habitat type	Raw data (g)	Mean	SD
Island, squirrels absent	9.6, 9.4, 8.9, 8.8, 8.5, 8.2	8.90	0.53
Island, squirrels present	6.8, 6.6, 6.0, 5.7, 5.3	6.08	0.62
Mainland, squirrels present	6.7, 6.4, 6.2, 5.7, 5.6	6.12	0.47

    a. The main comparison of interest in this study, identified before the data were gathered, was that between islands with and without squirrels, because this comparison controls for any effects of forest isolation on the mass of lodgepole pine cones. What do we label this type of comparison?

b. Taking into account the type of comparison identified in part (a), calculate a 95% confidence interval for the difference in cone mass between islands with and without squirrels. Assume that sites were randomly sampled.

c. Using these data, carry out a test of the differences among the means of all three groups.

23. People with an autoimmune disease like lupus produce antibodies that react to their own tissues. Research has shown that lupus-prone strains of mice have B cells with reduced expression levels of the receptor gene $Fc\gamma RIIB$, suggesting that low expression of this gene might contribute to the autoimmune reaction. To test this, McGaha et al. (2005) experimentally enhanced expression of the $Fc\gamma RIIB$ gene in bone marrow taken from a lupus-prone mouse strain and transplanted it back into irradiated mice of the same strain. Other mice of the same strain were subjected to the same procedures but received bone marrow that was not enhanced for $Fc\gamma RIIB$ expression (i.e., the sham treatment). Mice in a third group were left untreated. Autoimmune reactivity was measured six months later. The following table is a frequency table indicating the highest dilution of blood serum, in a fixed series of dilutions, at which reactivity could be detected (a high dilution reflects a high autoimmune reactivity).

Number of mice

Dilution measured	Enhanced	Sham-treated	Untreated
100	3	0	0
200	4	0	0
400	2	3	2
800	0	2	4
Total	9	5	6

a. Is this a balanced design? Explain.

b. What distinct purposes do the sham-treated and untreated groups serve in this experiment?

c. Calculate the mean and standard deviation of dilution measurements in each group.

d. We would like to test whether the mean dilutions of the three groups are different. Based on your answers to parts (a) and (c), why should we be cautious about employing ANOVA?

e. Choose a transformation that overcomes the main difficulty in part (d). Display your resulting sample means and standard deviations.

f. Using the transformed data, test whether there is a difference among treatment groups in the mean of dilution measurements.

g. What method would we use next to help decide which group means differed from the others?

24. Huey and Dunham (1987) measured the running speed of fence lizards, *Sceloporus merriami*, in Big Bend National Park in Texas. Individual lizards were captured and placed in a 2.3-meter raceway, where their running speeds were measured. Lizards were then tagged and released. The researchers returned to the park the following year, captured many of the same lizards again, and measured their sprint speed in the same way as previously. The pair of measurements for 34 individual lizards is as follows:

Lizard	Sprint speed (m/s)	Lizard	Sprint speed (m/s)
1	1.43, 1.37	18	2.28, 2.05
2	1.56, 1.30	19	2.44, 1.92
3	1.64, 1.36	20	2.23, 2.12
4	2.13, 1.54	21	2.53, 2.11
5	1.96, 1.82	22	2.20, 2.22
6	1.89, 1.79	23	2.16, 2.27
7	1.72, 1.72	24	2.25, 2.39
8	1.80, 1.80	25	2.42, 2.33
9	1.87, 1.87	26	2.61, 2.33
10	1.61, 1.88	27	2.62, 2.39
11	1.60, 1.98	28	3.09, 2.17
12	1.71, 2.08	29	2.13, 2.54
13	1.83, 2.16	30	2.44, 2.63
14	1.92, 2.08	31	2.76, 2.69
15	1.90, 2.01	32	2.96, 2.64
16	2.06, 2.03	33	3.13, 2.81
17	2.06, 1.97	34	3.27, 2.88

a. With these data, calculate the repeatability of running speed.

b. What does repeatability measure?

Experimental and statistical mistakes

Science is mostly done by intelligent people who want to get it right. There are cases of outright fraud, but these are rare (Panel on Scientific Responsibility and the Conduct of Research, 1992). On average, scientists are hardworking and careful, and they pay attention to details. But, sometimes things don't go as planned. Sometimes replicates get lost, vials get swapped, labels get blurred, tired hands write down the wrong numbers, the wrong button is pushed, or the wrong statistical test is applied. Many of these mistakes are caught by the researchers or by the peer review process, but not all. Sometimes, as President Richard M. Nixon once said, "Mistakes were made."

"Experimental mistakes" are errors made during the process of an experiment, in which the protocol actually followed was not the same as the one that was intended. No one knows how common important mistakes are in research, but we do know that they happen. For example, a study of the dopamine neurotoxicity of MDMA ("ecstasy") found significant toxic effects in the brains of 15 monkeys (Ricaurte et al. 2002). Before this study, this sort of brain damage was thought to be unlikely from MDMA, but it was well known from methamphetamine. This paper was published in *Science*, a high-profile scientific weekly. It turned out, though, that the lab that did the study had received its shipment of MDMA at the same time that it had received

Life is like a sewer. What you get out of it depends on what you put into it.

—Tom Lehrer

some methamphetamine, and the labels on the two bottles were swapped! Unbeknownst to them, the entire experiment had been done with methamphetamine rather than MDMA (Ricaurte 2003). The mistake was caught by careful follow-up work by this lab, but most mistakes of this type would not have been caught. Fundamental experimental errors sometimes make their way into the literature, even in the most prominent journals.

It is difficult, if not impossible, to know how often experimental mistakes are made or how important they are. It is a little easier to investigate how often mistakes are made when analyzing the data using statistics. A few studies have looked at the rates of statistical errors in published papers. The number of mistakes that are found is sobering. Surveys of papers in medical journals routinely find that one-third to one-half of papers that use statistics make at least one minor mistake (Gore et al. 1977, Kantoer and Taylor 1994, McGuigan 1995). This is likely to be an underestimate, since not all statistical mistakes made are detectable in the actual published paper. More importantly, though, about 8% of medical papers make statistical mistakes important enough to alter the conclusions of the paper (Gore et al. 1977). These statistical errors are not limited to the medical literature. Even in a field like ecology, which prides itself on its sophisticated use of statistics, a survey found statistical mistakes in about half of the papers (Hurlbert and White 1993). These mistakes are sometimes jokingly referred to as "Type III errors."

It is interesting to compare the 8% rate at which conclusions are changed by statistical mistakes to the 5% rate of Type I error that we normally tolerate. If we demand this level of confidence against errors of chance, we should also keep in mind the relatively high rates of other errors that might affect the result. It is one more reason why all studies should be repeated.

The moral is, be careful when doing science and when reading about science. Trust the authors to have done a good job, but don't expect their work to be perfect. Watch for mistakes. Very often, great scientific advances come from spotting, and fixing, the mistakes of our predecessors.

16

Correlation between numerical variables

When two numerical variables are associated we say that they are **corre-lated**. For example, brain size and body size are positively correlated across mammal species, as the accompanying graph[1] demonstrates. Large-bodied species tend to have large brains, and small-bodied species tend to have small brains.

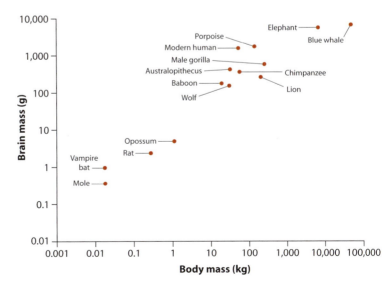

[1] Modified from Jerison (2006).

The correlation coefficient is a quantity that describes the strength and direction of a linear association between two numerical variables. In this chapter, we show how to measure a correlation, put confidence bounds on estimates of a correlation coefficient, and test hypotheses about correlation.

16.1   Estimating a linear correlation coefficient

The **correlation coefficient** is one of the most useful numbers in all of biology.[2] It measures the tendency of two numerical variables (call them X and Y) to "co-vary"—that is, to change together. We use the lower-case Greek letter ρ (rho, pronounced "row"[3]) to represent the correlation between X and Y in the population. We use r to represent the correlation between X and Y in a sample taken from the population.

> The *correlation coefficient* measures the strength and direction of the linear association between two numerical variables.

The correlation coefficient

The formula for the sample correlation coefficient r has three parts, two of which may look familiar and one of which is new:

$$r = \frac{\sum (X - \bar{X})(Y - \bar{Y})}{\sqrt{\sum (X - \bar{X})^2}\sqrt{\sum (Y - \bar{Y})^2}}.$$

The term in the numerator of the formula is called the sum of products,[4] and it measures how deviations in X and Y vary together. A deviation is the difference between an observation and its mean (Figure 16.1-1). If an observation i is above the mean for both X and Y (upper right corner of Figure 16.6-1), then its deviations are both positive and so the product of its deviations $(X_i - \bar{X})(Y_i - \bar{Y})$ is a positive number. If an observation is below the mean in both X and Y (lower left corner of Figure 16.6-1), then both its deviations are negative and so the product of the deviations $(X_i - \bar{X})(Y_i - \bar{Y})$ is again *positive*. Observations lying in the other two corners of

---

[2] The linear correlation coefficient is usually called simply the "correlation coefficient." It is also called the Pearson correlation coefficient after Karl Pearson, who first defined it. Pearson was one of the founders of the modern field of statistics (see Interleaf 1).

[3] Be careful: Don't mix up the Greek letter ρ with the roman letter p, which we use to represent proportion (Chapter 7).

[4] We provide a shortcut formula for this quantity in the Quick Formula Summary (Section 16.8 at the end of the chapter) to help when you are doing calculations by hand.

the plane have a positive deviation for one variable and a negative deviation for the other, so they have a negative product of deviations. The sum of products adds all the products of deviations. This sum will be positive if most of the observations are in the lower left and upper right corners of the plane, like that shown in Figure 16.1-1. The sum will be negative if most observations lie in the upper left and lower right corners. If the scatter of observations fills all four corners of the plane, then the positive and negative values cancel in the sum, yielding a sum of products close to zero.

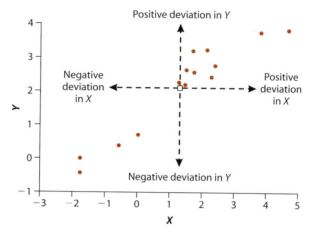

Figure 16.1-1 The position of X and Y observations in relation to the means, $\overline{X}$ and $\overline{Y}$ (indicated by an open square at the intersection of the two dashed lines). Observations lying in the upper right corner are above both $\overline{X}$ and $\overline{Y}$, and so have positive deviations in X and Y. Those lying in the lower left corner fall below both $\overline{X}$ and $\overline{Y}$, and so have negative deviations in X and Y. In the other two corners, observations have a positive deviation for one trait and a negative deviation for the other.

The denominator in the formula for r includes the sums of squares for X and for Y (under square root signs). You will have calculated quantities like these in Section 3.1 as part of calculating a standard deviation.[5]

The population correlation coefficient, ρ, is calculated using the same formula as r except it is measured on all individuals in the population. The correlation coefficient ρ and its sample estimate r lie between -1 and 1.

The correlation coefficient has no units, which means it is readily interpretable whatever the variables (provided they are numerical). A negative correlation means that one variable decreases as the other increases, whereas a positive correlation means that both variables increase and decrease together (Figure 16.1-2).

The maximum correlation (i.e., $r = 1.0$) occurs when all points lie along a straight line (the left panel in Figure 16.1-3), which is why we refer to the correlation as "linear."

---

[5]   Another way to write the equation for the correlation coefficient is $r = \dfrac{\text{Covariance}(X,Y)}{s_X s_Y}$, where the covariance of X and Y is in the numerator, and the standard deviations of X and Y are in the denominator. The covariance of X and Y is a measure of their relationship, analogous to the variance of a single variable. Its formula is provided in the Quick Formula Summary (Section 16.8).

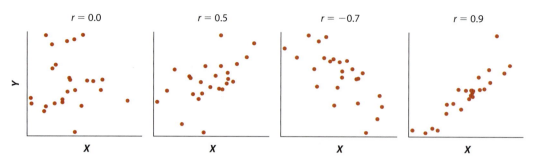

Figure 16.1-2 Scatter plots illustrating correlations between two numerical variables. In each case, $n = 25$.

Two variables might be strongly associated yet have no correlation (i.e., $r = 0.0$) if the relationship between them is nonlinear (the right panel in Figure 16.1-3). Example 16.1 illustrates the use of the correlation coefficient.

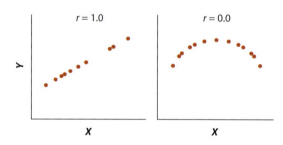

Figure 16.1-3 Correlation between two variables that are strongly associated. On the left, measurements of X and Y lie along a straight line, producing the maximum correlation possible ($r = 1.0$). On the right, the relationship is non-linear and exhibits no correlation ($r = 0.0$).

| Example 16.1 | **Manly digits** |

The ratio of the lengths of the second and fourth digits (the 2D:4D ratio) is smaller in men than in women, an effect of the greater exposure to testosterone by males when still in the womb. (Because testosterone also influences sexual development, it has been suggested that differences among men in the 2D:4D ratio might predict aspects of their adult sexual behavior.) Manning et al. (2003) investigated a possible connection between the 2D:4D ratio in male subjects and the structure of their androgen receptor gene. The number of CAG repeats[6] in a specific region of the gene is correlated with sensitivity to testosterone. The question was whether the number of CAG repeats also correlates with the 2D:4D ratio. Measurements on 46 subjects are listed in Table 16.1-1.

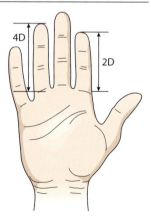

---

[6]   The number of CAG repeats counts the number of times this triplet of nucleotides occurs in succession. For example, the sequence . . .CAGCAGCAG. . . represents three repeats.

Table 16.1-1 The ratio of lengths of the second and fourth digits on the right hands of a sample of men (*n* = 46) and the number of CAG repeats in a region of their androgen receptor gene.

Number of CAG repeats	2D:4D ratio	Number of CAG repeats	2D:4D ratio	Number of CAG repeats	2D:4D ratio
18	1.00	20	0.96	22	0.99
19	1.03	20	0.95	22	0.94
19	0.98	20	0.94	22	0.93
19	0.98	20	0.94	23	0.96
19	0.98	20	0.92	24	1.02
19	0.95	20	0.91	24	1.01
19	0.95	21	1.06	24	0.97
19	0.91	21	0.99	24	0.97
19	0.90	21	0.99	25	1.06
20	1.03	21	0.98	25	1.02
20	0.99	21	0.97	25	0.99
20	0.98	21	0.97	26	0.99
20	0.98	21	0.97	27	0.95
20	0.98	21	0.96	28	1.00
20	0.98	21	0.94		
20	0.96	22	1.06		

A scatter plot of the data, shown in Figure 16.1-4, suggests that the 2D:4D ratio is positively associated with the number of CAG repeats in the gene, although the association is not strong. The larger the number of repeats (reflecting lower sensitivity to testosterone), the greater (and more female-like) the ratio of second and fourth digit lengths.

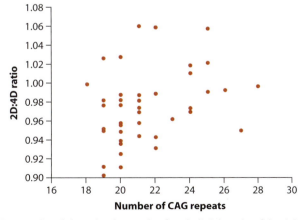

Figure 16.1-4 Scatter plot of the ratio of second to fourth digit lengths of the right hand of a sample of men and the number of CAG repeats in a region of their androgen receptor gene. Each dot in the graph represents the pair of measurements for a single male subject. *n* = 46.

The correlation coefficient r quantifies the strength of this association. To calculate r, we first obtained the following three quantities:

$$\sum (X - \bar{X})(Y - \bar{Y}) = 1.186$$
$$\sum (X - \bar{X})^2 = 250.435$$
$$\sum (Y - \bar{Y})^2 = 0.0633,$$

yielding

$$r = \frac{1.186}{\sqrt{250.435}\sqrt{0.0633}} = 0.298.$$

The sample correlation coefficient r between the two variables is 0.298.

Standard error

The data are merely a sample taken to estimate the correlation between the same two variables in a population, ρ. The standard error of r—that is, the standard deviation of its sampling distribution—is one way to assess how close our estimate is likely to be to the population parameter ρ. The standard error is

$$SE_r = \sqrt{\frac{1 - r^2}{n - 2}}.$$

For our example data set, the standard error is

$$SE_r = \sqrt{\frac{1 - (0.298)^2}{46 - 2}} = 0.144.$$

It is not ideal to use this quantity to calculate a confidence interval, because the sampling distribution of r is not normal. However, we will use SE_r later in hypothesis testing. Calculating a confidence interval requires the modified method shown next.

Approximate confidence interval

The 95% confidence interval for ρ puts bounds on our estimate, identifying the range of values that are compatible with the data. Fisher discovered an approximate confidence interval for ρ. The approximation is best when sample size is large. With this method, we convert r to a new quantity called z that approximately follows a normal sampling distribution. The Fisher's z-transformation is

$$z = 0.5 \ln\left(\frac{1 + r}{1 - r}\right),$$

where "ln" is the natural logarithm. The z-transform of ρ is symbolized as ζ (the lower-case Greek letter zeta), so z represents the value in a sample and ζ is the true value in the population. The standard error of the sampling distribution for z is approximately

$$\sigma_z = \sqrt{\frac{1}{n-3}}.$$

For the 2D:4D ratio data (Example 16.1),

$$z = 0.5 \ln\left(\frac{1+0.298}{1-0.298}\right) = 0.307,$$

and

$$\sigma_z = \sqrt{\frac{1}{46-3}} = 0.1525.$$

The sampling distribution of the statistic z is approximately normal, so we can use the standard normal distribution to generate the 95% confidence interval for ζ:

$$z - 1.96\,\sigma_z < \zeta < z + 1.96\,\sigma_z.$$

The quantity 1.96 is the value of Z_{crit}, the two-tailed critical value of the standard normal distribution corresponding to $\alpha = 0.05$ (Statistical Table B; $\Pr[Z > 1.96] = 0.025$). More generally, to obtain a $1 - \alpha$ confidence interval, find the value of Z_{crit} such that $\Pr[Z > Z_{\mathrm{crit}}] = \alpha/2$.

For the 2D:4D ratio data, the 95% confidence interval for ζ is

$$0.307 - 1.96\,(0.1525) < \zeta < 0.307 + 1.96\,(0.1525)$$
$$0.0081 < \zeta < 0.6059.$$

To complete the analysis, we convert the lower and upper bounds of this confidence interval back to the original correlation scale using the inverse of Fisher's transformation,

$$r = \frac{e^{2z} - 1}{e^{2z} + 1},$$

where e is the base of the natural logarithm. For our example, this yields

$$\frac{e^{2(0.0081)} - 1}{e^{2(0.0081)} + 1} < \rho < \frac{e^{2(0.6059)} - 1}{e^{2(0.6059)} + 1}$$

or

$$0.0081 < \rho < 0.5412,$$

which, when rounded to two digits, is

$$0.01 < \rho < 0.54.$$

This calculation shows that the data are consistent with a fairly broad range of values for the population correlation between the number of CAG repeats and the 2D:4D ratio. The confidence interval ranges from values near zero to the moderately strong value of 0.54. We can be reasonably confident that ρ is greater than zero and is well below one.

The formula for σ_z gives only an approximation to the standard deviation of the sampling distribution for z, so the 95% confidence interval for ζ (and therefore ρ) is also an approximation.

16.2   Testing the null hypothesis of zero correlation

The most common use of hypothesis testing in correlation analysis is to test the null hypothesis that the population correlation ρ is exactly zero:[7]

$$H_0: \rho = 0.$$
$$H_A: \rho \neq 0.$$

Example 16.2 shows how the method works.

Example 16.2   **What big inbreeding coefficients you have**

By 1970, the wolf (*Canis lupus*) had been wiped out in Norway and Sweden, but around 1980 two wolves immigrated from farther east and founded a new population. By 2002 the new population reached approximately 100 wolves. Liberg et al. (2005) compiled observations on reproduction made between 1983 and 2002 and constructed the pedigree of the wolves in the small population. The data listed in Table 16.2-1 show the inbreeding coefficients of litters produced by mated pairs and the number of pups of each litter surviving their first winter. An inbreeding coefficient is zero if parents of the litter were unrelated, 0.25 if parents were brother and sister whose own parents were unrelated, and greater than 0.25 if inbreeding had continued for more generations.

Does the number of pups of a litter surviving their first winter correlate with the inbreeding coefficient of the litter? A scatter plot of the data, shown in Figure 16.2-1, indicates a negative association between inbreeding coefficient and the number of surviving pups.

The correlation coefficient between inbreeding coefficient and the number of pups can be calculated from the following quantities:

---

[7] Testing the more general null hypothesis $H_0: \rho = \rho_0$, where ρ_0 is a number other than zero, requires a different method that makes use of the Fisher's z-transformation. We do not present it here.

Table 16.2-1 Inbreeding coefficients of pups of mated wolf pairs and the number surviving their first winter. $n = 24$ **litters.**

Inbreeding coefficient	Number of pups	Inbreeding coefficient	Number of pups
0.00	6	0.24	3
0.00	6	0.24	2
0.13	7	0.24	2
0.13	5	0.25	6
0.13	4	0.27	3
0.19	8	0.30	5
0.19	7	0.30	3
0.19	4	0.30	2
0.22	4	0.30	1
0.24	3	0.36	3
0.24	3	0.37	2
0.24	3	0.40	3

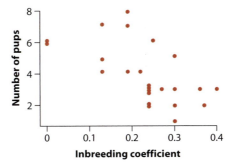

Figure 16.2-1 The number of surviving wolf pups in a litter and their inbreeding coefficient. Overlapping points have been offset slightly to render them visible. The total number of litters, $n = 24$.

$$\sum (X - \bar{X})(Y - \bar{Y}) = -2.612$$
$$\sum (X - \bar{X})^2 = 0.228$$
$$\sum (Y - \bar{Y})^2 = 80.958.$$

Putting these into the formula for r gives

$$r = \frac{-2.612}{\sqrt{0.228}\sqrt{80.958}} = -0.608.$$

The observed correlation coefficient is less than zero, but we want to test whether this correlation is sufficiently strong to warrant rejection of the null hypothesis that the population correlation ρ is zero. Our two hypotheses are as follows:

H_0: There is no relationship between the inbreeding coefficient and the number of pups ($\rho = 0$).

H_A: Inbreeding depression and the number of pups are correlated ($\rho \neq 0$).

To test the hypotheses, we calculate the t-statistic,

$$t = \frac{r}{SE_r},$$

where the standard error (SE_r) is calculated as

$$SE_r = \sqrt{\frac{1 - r^2}{n - 2}}.$$

Under the null hypothesis of zero correlation, the sampling distribution of the t-statistic is a Student's t-distribution with $n - 2$ degrees of freedom.[8]

For the wolf data,

$$SE_r = \sqrt{\frac{1 - (-0.608)^2}{24 - 2}} = 0.169$$

and so

$$t = \frac{-0.608}{0.169} = -3.60.$$

The P-value for this t-statistic is $P = 0.002$, obtained using a computer. Using Statistical Table C, instead, the critical value for the t-distribution having 22 degrees of freedom is $t_{0.05(2),22} = 2.075$ with $\alpha = 0.05$. Since $t = -3.60$ is less than -2.075 and farther from the value expected by the null hypothesis, P must be less than 0.05. Therefore, we reject the null hypothesis of zero correlation. Inbreeding is indeed correlated with the number of surviving offspring in this isolated wolf population.

16.3 Assumptions

The methods used to estimate and test a population correlation assume that the sample of individuals is a random sample from the population. In addition, correlation analysis assumes that the measurements have a **bivariate normal distribution** in the population. A bivariate normal distribution is a bell-shaped probability distribution in three dimensions rather than two (Figure 16.3-1).

---

[8] There are n independent data points when calculating a correlation coefficient, but there are two fewer degrees of freedom because we have to use two summaries of the data, $\overline{X}$ and $\overline{Y}$, when we calculate r.

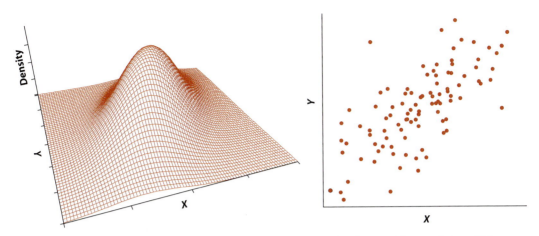

Figure 16.3-1 The left panel shows a bivariate normal distribution with a correlation of 0.7 between X and Y. Height above the plane represents the probability density of each pair of values of X and Y. The right panel shows a random sample of 100 observations from the bivariate normal distribution shown on the left.

A bivariate normal distribution has the following features:

▶ The relationship between X and Y is linear.
▶ The cloud of points in a scatter plot of X and Y has a circular or elliptical shape.
▶ The frequency distributions of X and Y separately are normal.

This is not a complete list of the features of the bivariate normal distribution, but it is the set that matters most and is easiest to evaluate with data. Inspecting the scatter plot of the data is probably the best way to check the assumption of bivariate normality. The right panel of Figure 16.3-1, a random sample from a bivariate normal distribution having a correlation of $\rho = 0.70$, shows what a scatter of points should look like when the assumption of bivariate normality is met. All of the examples shown in the scatter plots of Figure 16.1-1 are also random samples from bivariate normal distributions.

The most noticeable departures from bivariate normality in a scatter plot, shown in Figure 16.3-2, are also those that most seriously affect correlation analysis—namely,

▶ a cloud of points that is funnel shaped (i.e., wider at one end than at the other),
▶ the presence of outliers, and
▶ a relationship between X and Y that is not linear.

Histograms depicting the frequency distributions of X and Y separately are also helpful. If either X or Y has a decidedly skewed distribution, then the frequency distribution of the two variables is not bivariate normal.

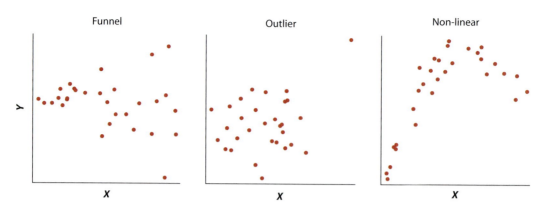

Figure 16.3-2 Data from three distributions that differ from bivariate normality. Scatter plots show a funnel shape (*left*), an outlier (*middle*), and a non-linear relationship between X and Y (*right*).

What do we do if the assumption of bivariate normality is not met? Two strategies are available—namely, transforming the data and nonparametric methods. It is best to try first to transform X, Y, or both variables to see if the assumptions are better met on a new scale. The usual transformations, first described in Chapter 13, are

▶ the log transformation [an all-purpose transformation, as long as the data are not negative; use $\log(X + 1)$ or $\log(Y + 1)$ if there are zeros],
▶ the square-root transformation (which often works for data that are counts), and
▶ the arcsine transformation (for data that are proportions).

Log transformations are good to try if the relationship between the two variables is non-linear or if the variance in one variable seems to increase with the value of the other variable.

If transforming the data is unsuccessful, use a nonparametric method instead, such as Spearman's rank correlation. This method is explained in Section 16.5.

16.4 The correlation coefficient depends on the range

The correlation between two variables X and Y depends on the range of values included. For example, the top panel of Figure 16.4-1 shows that the population density of different species of stream invertebrates (Y) is strongly correlated with their body mass (X) when both variables are log-transformed (Schmid et al. 2000). Points falling between the two dashed lines in the top panel are replotted in the lower panel. The correlation is weaker when a smaller range of X-values is present, because the other factors that affect Y become relatively more important.

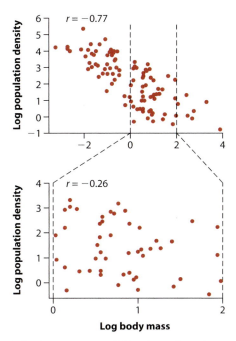

Figure 16.4-1 The correlation between two variables depends on the range of X-values included. The bottom graph plots those data points lying between the two dashed lines in the top graph. The data drawn from a smaller range of X have a smaller correlation coefficient. The data are log (base 10) of population density (individuals/m²) and body mass (μg) of different species of stream invertebrates (Schmid et al. 2000).

This effect means that the correlation coefficient between the same two variables is not comparable between separate studies unless the studies use the same range of values for the X-variable.

16.5 Spearman's rank correlation

Many situations require a test of zero correlation between variables that do not meet the assumption of bivariate normality, even after data transformation. For these cases, the nonparametric **Spearman's rank correlation** is used. Spearman's rank correlation uses the ranks of both the X- and Y-variables to calculate a measure of correlation. It does not make assumptions about the distribution of the variables, but it still assumes that the individuals are randomly chosen from the population. Example 16.5 illustrates the method.

> The *Spearman's rank correlation* measures the strength and direction of the linear association between the ranks of two variables.

Example 16.5   ## The miracles of memory

Of the many illusions performed by magicians, none is more renowned than the Indian rope trick. In the most sensational version of the trick, a magician tosses one end of a rope in the air, which forms into a rigid pole. A boy climbs up the rope and disappears at the top. The magician scolds the boy to return but gets no reply, whereupon he grabs a knife, climbs the rope, and also disappears. The boy's body then falls in pieces from the sky into a basket on the ground. The magician descends the rope and retrieves the boy from the basket, revealing him to be unharmed and in one piece.

Wiseman and Lamont (1996) tracked down 21 first-hand, written accounts of the Indian rope trick. They gave a score to each description according to how impressive it was. For example, a score of 1 was given if the observer saw only that "boy climbs up rope, then climbs down again." The most impressive accounts in the sample, "boy climbs rope, vanishes at the top, reappears in basket in full view of audience," were given a score of 5, the highest possible. For each account, the researchers also recorded the number of years that had elapsed between the date that the trick was witnessed and the date the memory of it was written down. The measurements of impressiveness score and number of years elapsed are shown in the scatter plot in Figure 16.5-1. Is there an association between the impressiveness of eyewitness accounts and the time elapsed until the writing of the description? If so, then it might indicate the fallibility of human memory.

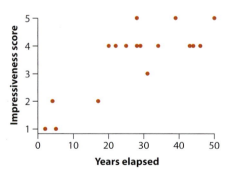

Figure 16.5-1 Impressiveness of written accounts of the Indian rope trick by first-hand observers and the number of years elapsed between witnessing the event and writing of the account. $n = 21$.

A test of the null hypothesis of zero correlation is what we would like to carry out, but the assumption of bivariate normality is clearly violated because the impressiveness score is not even a numerical variable. Rather, it is an ordered categorical variable, ruling out the use of the parametric correlation coefficient that we have learned before. A test of zero correlation is nevertheless possible using a nonparametric method, because the different categories for the impressiveness score can be ranked, as shown in Table 16.5-1. The Spearman's rank correlation measures the association between the ranks of the two variables. Spearman's rank correlation is measured by the parameter ρ_S, which is estimated by r_S.

Table 16.5-1 Raw data from Example 16.5-1, and their ranks. Each variable is ranked separately. Midranks are assigned when there are ties. $n = 21$.

Years elapsed	Rank years	Impressiveness score	Rank impressiveness
2	1	1	2
5	3.5	1	2
5	3.5	1	2
4	2	2	5
17	5.5	2	5
17	5.5	2	5
31	13	3	7
20	7	4	12.5
22	8	4	12.5
25	9	4	12.5
28	10.5	4	12.5
29	12	4	12.5
34	14.5	4	12.5
43	17	4	12.5
44	18	4	12.5
46	19	4	12.5
34	14.5	4	12.5
28	10.5	5	19.5
39	16	5	19.5
50	20.5	5	19.5
50	20.5	5	19.5

The Spearman's correlation coefficient is the linear correlation coefficient computed on the *ranks* of the data. The two variables must be ranked separately, from low to high.

The data from Figure 16.5-1 are listed in Table 16.5-1. The table also includes the separate rankings of each variable. As first discussed in Section 13.5, we assign midranks when there are ties. The midrank is the average of the ranks associated with a set of tied observations. For example, the three measurements with the lowest impressiveness score (1) were all assigned the midrank 2, which is the average of the three ranks associated with the three lowest values: 1, 2, and 3. The 10 values having impressiveness score 4 are associated with the ranks 8 through 17, and so were assigned the midrank 12.5. In the following calculations, R refers to the rank of years elapsed and S refers to the rank of the impressiveness score:

$$\sum (R - \bar{R})(S - \bar{S}) = 566$$
$$\sum (R - \bar{R})^2 = 767.5$$
$$\sum (S - \bar{S})^2 = 678.5,$$

yielding

$$r_S = \frac{566}{\sqrt{767.5}\sqrt{678.5}} = 0.784.$$

The hypotheses for the test are

$$H_0: \rho_S = 0$$

and

$$H_A: \rho_S \neq 0,$$

where ρ_S refers to the Spearman's correlation in the population. To determine the P-value for the test, compare r_S to the critical value[9] given in Statistical Table G corresponding to a sample size of 21:

$$r_{S(0.05,21)} = 0.435.$$

Since $r_S = 0.784$ is greater than 0.435, $P < 0.05$, and so we reject the null hypothesis (a computer program gave $P = 0.0003$). We conclude that there is a positive correlation between the impressiveness score of the eyewitness accounts of the Indian rope trick and the number of years elapsed between viewing the trick and the retelling of it in writing. The likely explanation for these findings is that eyewitness accounts of the Indian rope trick are exaggerated,[10] becoming more so with time.

Procedure for large *n*

Statistical Table G provides critical values for the Spearman's rank correlation for sample sizes up to $n = 100$. For larger n, use the procedure for the linear correlation coefficient, but applied to the ranks. Calculate the t-statistic,

---

[9] Different computer programs might yield slightly different P-values for the same data, depending on the approximation used to compute it. Some use a normal approximation, which is not accurate for small n. Most calculations assume no ties in the data. Corrections for ties exist, but they are tedious to compute and usually make little difference.

[10] According to the authors of the report, one eyewitness claimed to have seen the trick performed and had taken a photograph. Examination of the photograph showed only a boy balancing on the end of a long pole.

$$t = \frac{r_S}{\text{SE}[r_S]},$$

where

$$\text{SE}[r_S] = \sqrt{\frac{1 - r_S^2}{n - 2}}.$$

Under the null hypothesis of no Spearman rank correlation in the population ($\rho_S = 0$), t is approximately t-distributed with $n - 2$ degrees of freedom. Reject the null hypothesis of zero rank correlation if $t \geq t_{0.05(2),n-2}$ or $t \leq -t_{0.05(2),n-2}$.

Assumptions of Spearman's correlation

The Spearman's rank correlation assumes that observations are a random sample from the population. It also assumes that the relationship between the ranks of the two numerical variables is linear.

16.6   The effects of measurement error on correlation

When a variable is not measured perfectly, we say that there is **measurement error**. Measurement error is difficult to avoid. Some biological traits are extremely challenging to measure, and measurement error can sometimes be an important component of variation (Chapter 15). For example, behavioral traits are notorious for having low repeatability: a behavior measured on one individual might be quite different the next time it is measured on the same individual.

Measurement error in either X or Y tends to weaken the observed correlation between the variables. The same thing happens if there is measurement error in both X and Y, if the errors in X and Y are uncorrelated. With measurement error, r will tend to underestimate the magnitude of ρ (it will tend to be closer to zero on average than the true correlation), a bias called **attenuation** (Figure 16.6-1).

In the Quick Formula Summary (Section 16.8 at the end of the chapter), we include an equation that corrects the estimate of correlation for the effects of measurement error. The method requires that repeated measurements have been made on the same individuals. This corrected correlation, r^*, can't be used in place of r in confidence intervals for ρ or in hypothesis testing. However, it might be useful for comparison with the uncorrected correlation to evaluate how measurement error is affecting the observed correlation between two variables. In general, measurement error can be reduced by taking precise measurements. If this is not possible, then it is best to measure each individual multiple times and use the average measurement in subsequent analysis.

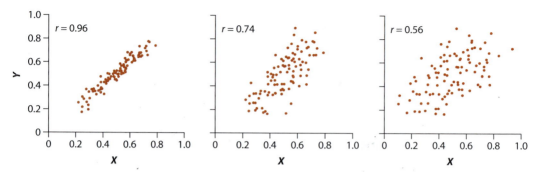

Figure 16.6-1   Attenuation. In the left panel, X and Y are measured without error and are highly correlated. In the middle panel, Y is measured with error. In the right panel, both X and Y are measured with error, and these errors are uncorrelated. In all cases, the true correlation is very strong, but the correlation *appears* weaker when the variables are measured with error.

16.7 Summary

▶ The correlation coefficient (r) measures the strength and direction of the linear association between two numerical variables.

▶ The correlation coefficient ranges from -1 (the maximum negative correlation) to 0 (no correlation) to 1 (the maximum positive correlation).

▶ Analysis using the linear correlation assumes that the two numerical variables have a bivariate normal distribution and that the individuals are randomly sampled.

▶ With a bivariate normal distribution, the relationship between X and Y is linear, the cloud of points in a scatter plot of X and Y is circular or elliptical in shape, and the frequency distributions of X and Y separately are normal. (This is just a partial list of its features.)

▶ The scatter plot is a useful tool for examining the assumption of bivariate normality. Histograms of X and Y should both appear normal.

▶ The Spearman's rank correlation measures the linear correlation between the ranks of two variables, where each variable is ranked separately from low to high.

▶ The correlation between two variables is expected to be weaker when only a narrow range of X values is represented.

▶ Measurement error biases the estimate of a correlation coefficient towards zero.

16.8 **Quick Formula Summary**

Shortcuts

Sum of Products: $\sum (X - \bar{X})(Y - \bar{Y}) = \sum (XY) - \dfrac{(\sum X)(\sum Y)}{n}$

Sum of Squares: $\sum (X - \bar{X})^2 = \sum (X^2) - \dfrac{(\sum X)^2}{n}$

$$\sum (Y - \bar{Y})^2 = \sum (Y^2) - \dfrac{(\sum Y)^2}{n}$$

Covariance

What is it for? Measuring the strength of an association between two numerical variables.

Estimate: Covariance(X, Y)

Formula: $\text{Covariance}(X,Y) = \dfrac{\sum (X - \bar{X})(Y - \bar{Y})}{n - 1}$

Correlation coefficient

What is it for? Measuring the strength of a linear association between two numerical variables.

What does it assume? Bivariate normality and random sampling.

Parameter: ρ

Estimate: r

Formula: $r = \dfrac{\sum (X - \bar{X})(Y - \bar{Y})}{\sqrt{\sum (X - \bar{X})^2}\sqrt{\sum (Y - \bar{Y})^2}}$

Standard error: $\text{SE}_r = \sqrt{\dfrac{1 - r^2}{n - 2}}$

Degrees of freedom: $n - 2$

Alternate formula: $r = \dfrac{\text{Covariance}(X,Y)}{s_X s_Y}$, where Covariance($X$, Y) is the covariance between X and Y and where s_X and s_Y are the sample standard deviations of X and Y, respectively.

Confidence interval (approximate) for a population correlation

What does it assume? The sample is a random sample. The numerical variables X and Y have a bivariate normal distribution in the population. The approximation improves with increasing sample size.

Parameter: ρ

Estimate: r

Formula: $z - Z_{\text{crit}}\, \sigma_z < \zeta < z + Z_{\text{crit}}\, \sigma_z$, where $z = 0.5 \ln\left(\dfrac{1 + r}{1 - r}\right)$ is Fisher's z-transformation of r, $\sigma_z = \sqrt{\dfrac{1}{n - 3}}$ is the approximate standard error of z, and Z_{crit} is the critical value of the standard normal distribution for which $\Pr[Z > Z_{\text{crit}}] = \alpha/2$. To obtain the confidence interval for ρ, back-transform the limits of the resulting confidence interval using the inverse of Fisher's transformation, $r = \dfrac{e^{2z} - 1}{e^{2z} + 1}$.

The t-test of zero linear correlation

What is it for? To test the null hypothesis that the population parameter (ρ) is zero.

What does it assume? Bivariate normality and random sampling.

Test statistic: t

Distribution under H$_0$: t-distributed with $n - 2$ degrees of freedom.

Formula: $t = \dfrac{r}{\text{SE}_r}$,

where $\text{SE}_r = \sqrt{\dfrac{1 - r^2}{n - 2}}$ is the standard error of r.

Spearman's rank correlation

What is it for? To measure correlation between the two variables, when the variables do not meet the assumptions of correlation.

What does it assume? A linear relation between the ranks of X and Y, and random sampling.

Parameter: ρ_S

Estimate: r_S

Formula: Same as for linear correlation but calculated on ranks.

Spearman's rank correlation test

What is it for? To test the null hypothesis that the rank correlation in the population (ρ_S) is zero.

What does it assume? A linear relation between the ranks of X and Y, and random sampling.

Test statistic when n ≤ 100: r_S

Distribution under H$_0$: Distribution of the Spearman's rank correlation (Statistical Table G).

Test statistic when n > 100: t

Distribution under H$_0$: t-distributed with $n - 2$ degrees of freedom.

Formula: $t = \dfrac{r_S}{\text{SE}[r_S]}$,

where $\text{SE}[r_S] = \sqrt{\dfrac{1 - r_S^2}{n - 2}}$ is the standard error of r_S.

Correlation corrected for measurement error

What does it assume? That X and Y have been measured two or more times independently on all individuals, and that measurement error in X is uncorrelated with measurement error in Y. The formula below assumes that the correlation r between X and Y is measured using the average of the repeat measurements for every individual.

Parameter: ρ

Estimate: r^*

Formula: $r^* = \dfrac{r}{\sqrt{R_X R_Y}}$,

where r is calculated from the average of the repeat measurements for every individual (Adolph and Hardin 2007). R_X and R_Y are repeatabilities of X and Y, respectively. Repeatability here is similar to that described in Chapter 15, except that here the repeatability is for the average values of repeat measurements made on each individual rather than single measurements. The formula for R_X is $R_X = \dfrac{s_A^2}{s_A^2 + (\text{MS}_{\text{error}}/m)}$, where m is the number of repeat measurements made on each individual and $s_A^2 = \dfrac{\text{MS}_{\text{groups}} - \text{MS}_{\text{error}}}{m}$. $\text{MS}_{\text{groups}}$ and MS_{error} are calculated from a random effects ANOVA on the repeat measurements for X, as explained in Section 15.6. The formula for R_Y is calculated in the same way using measurements of the Y-variable.

PRACTICE PROBLEMS

1. Visually estimate the value of the correlation coefficient in each of the four following scatter plots.

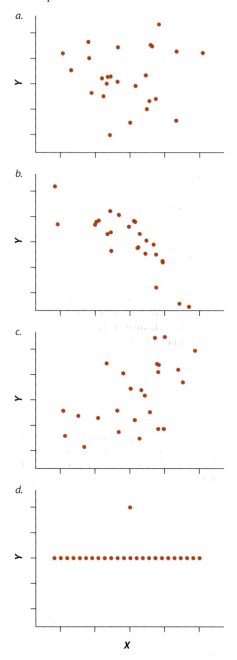

a.

b.

c.

d.

X

2. Calculate a 95% and a 99% confidence interval for ρ in each of the following cases.
   a. $r = 0.45, n = 60$
   b. $r = -0.89, n = 24$
   c. $r = 0.04, n = 88$

3. Birds of many species retain the same breeding partner year after year. In some of these species, male and female partners migrate separately and spend the winter in different places, often thousands of kilometers apart. How do partners find one another again each spring? In a field study of individually banded pairs of black-tailed godwits, Gunnarsson et al. (2004) recorded spring arrival dates of males and females on the breeding grounds in the year after they were observed breeding together. The data for 10 pairs are provided in the accompanying table. Arrival date is measured as the number of days since March 31.

   a. Display the relationship between arrival dates of males and females in a graph. What type of graph did you use?
   b. Describe the pattern in part (a) briefly. Is there a relationship? Is it positive or negative? Is it linear or non-linear? Is it weak or strong?

Arrival date of female partner (days)	Arrival date of male partner (days)
24	22
36	35
35	35
35	44
38	46
50	50
55	55
56	56
57	56
69	59

   c. Calculate the correlation coefficient between arrival dates of male and female godwits. Include a standard error for your estimate.

   d. What does the standard error in part (c) refer to?

   e. Calculate an approximate 95% confidence interval for ρ.

4. Answer the following questions using the data for the godwits in Practice Problem 3.

   a. Adding 30 to each of the observations for males converts arrival dates to "days since March 1" rather than March 31. How is the correlation coefficient between arrival dates of males and females affected? What can you conclude about the effects on the correlation coefficient of adding a constant to one or both of the variables?

   b. Dividing female arrival dates by seven converts their arrival dates to "weeks since March 31" rather than days. How does this affect the correlation between male and female arrival dates? What can you conclude about the effects on the correlation coefficient of multiplying one or both of the variables by a constant?

5. Use the godwit data in Practice Problem 3 to test whether the mean arrival dates of male and female partners differ significantly. What assumptions are required?

6. When measuring a correlation between two variables, under what circumstances would it be best to make and average several repeat measurements of each subject for a given variable rather than measure the variable only once on each subject?

7. In large wolf populations, most inbreeding coefficients of litters are close to zero, and very few are as high or higher than 0.25. What effect is this narrower range of inbreeding coefficients expected to have on the correlation between the number of pups surviving and inbreeding coefficient, compared with that measured in the Scandinavian population of Example 16.2? Explain.

8. Large males of the European earwig, *Forficula auricularia*, develop abdominal forceps, which are used in fighting and courtship. Smaller males do not develop the forceps. Tomkins and Brown (2004) compared the proportion of males having forceps on islands in the North Sea with the population density of earwigs. A subset of their data, for seven islands only, is listed in the following table.

Earwig density (number per trap)	Proportion of males with forceps
0.3	0.04
5.2	0.02
12.7	0.66
20.0	0.19
25.6	0.06
32.7	0.55
33.8	0.62

   a. The distribution of the two variables is not bivariate normal, and transforming the data does not improve matters. Choosing the most appropriate method, test whether the two variables are correlated.

   b. What are your assumptions in part (a)?

9. Earwig density on an island and the proportion of males with forceps are estimates, so the measurements of both variables include sampling error. In light of this fact, would the true correlation between the two variables tend to be larger, smaller, or the same as the measured correlation?

10. According to the "immunocompetence handicap hypothesis," males of a species evolve high reproductive effort to the point that they divert resources away from immune function. To test this, Simmons and Roberts (2005) measured sperm viability and immune function in lab-raised male crickets to test whether male reproductive effort affects immune function. The data displayed in the accompanying graph shows male sperm viability and lysozyme activity, an important defense against bacterial infection. Each point is the average of the males in a single family of crickets. Lysozyme activity is measured as the area of clear region around an inoculation of $2 \, \mu l$ of hemolymph onto an agar plate containing bacteria. The total sample size (n) is 41.

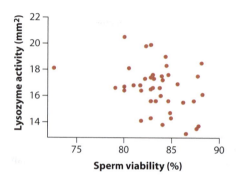

a. What assumption of linear correlation analysis is violated by these data? Explain.
b. Assuming that transforming the data doesn't help, what is the most appropriate method to test the null hypothesis of no correlation between male sperm viability and male lysozyme activity?
c. When we applied the most appropriate method to these data, we obtained the following numbers based on the ranks of sperm viability and lysozyme activity:

  Sum of products: -1744.5
  Sum of squares (sperm viability): 5726.0
  Sum of squares (lysozyme activity): 5729.5

Using these figures, test the hypothesis that sperm viability and male lysozyme activity are correlated.
d. What assumptions have you made in part (c)?

11. The filefish, *Paraluteres prionurus,* looks similar to *Canthigaster valentini,* a toxic pufferfish common on the same coral reefs. To test whether resemblance to the pufferfish gives an advantage, Caley and Schluter (2003) painted plastic fish with color patterns that differed in

their degree of resemblance to this pufferfish. Plastic dummies were scored as 1 (strongest resemblance), 2, 3, or 4 (weakest resemblance) according to how similar they were in appearance to the pufferfish. The researchers then placed the dummies at randomly chosen locations on the reef and recorded the number of predatory fish that approached them during five-minute observation periods. The number of predators recorded for each dummy is given in the accompanying table. Does resemblance to the pufferfish give protection from predators?

Resemblance to pufferfish	Number of predators/5-min
1	2
1	6
1	5
1	5
1	0
1	3
1	11
2	6
2	8
2	11
2	7
3	11
3	9
3	15
3	15
4	11
4	11
4	11
4	18
4	14

a. Display the data in a graph. Describe the pattern briefly.
b. Test the hypothesis that the number of predators approaching the dummy is correlated with its degree of resemblance to the pufferfish.

ASSIGNMENT PROBLEMS

12. Does stress age you? As part of an investigation, Epel et al. (2004) measured telomere length in blood mononuclear cells of healthy pre-menopausal women, each of whom was the biological mother and caregiver of a chronically ill child. Telomeres are complexes of DNA and protein that cap chromosomal ends. They tend to shorten with cell divisions and with advancing age. A scatter plot of the data is shown below. Telomere length is measured as a ratio compared to a standard. Chronicity is the number of years since the child's diagnosis.

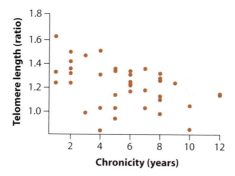

The data in the scatter plot can be summarized as follows:

Sum of products: -8.636
Sum of squares for chronicity: 327.436
Sum of squares for telomere length: 1.228
Total sample size (n): 38

a. Describe the pattern in the scatter plot briefly in words. Is there a relationship? Is it positive or negative? Is it linear or non-linear? Is it weak or strong?

b. Calculate the linear correlation between telomere length and the chronicity of care giving.

c. Calculate a 95% confidence interval for the population correlation.

d. Provide an interpretation for the interval in part (c). What does the interval represent?

e. What are your assumptions in part (c)? Does the scatter plot support these assumptions? Explain.

13. Does learning a second language change brain structure? Mechelli et al. (2004) tested 22 native Italian speakers who had learned English as a second language. Proficiencies in reading, writing, and speech were assessed using a number of tests whose results were summarized by a proficiency score. Gray-matter density was measured in the left inferior parietal region of the brain using a neuroimaging technique, as mm^3 of gray matter per voxel. (A voxel is a picture element, or "pixel" in three dimensions.) The data are listed in the accompanying table.

Proficiency score for second language	Gray-matter density (mm³/voxel)	Proficiency score for second language	Gray-matter density (mm³/voxel)
0.26	−0.070	2.75	−0.008
0.44	−0.080	3.25	−0.006
0.89	−0.008	3.85	0.022
1.26	−0.009	3.04	0.018
1.69	−0.023	2.55	0.023
1.97	−0.009	2.50	0.022
1.98	−0.036	3.11	0.036
2.24	−0.029	3.18	0.059
2.24	−0.008	3.52	0.062
2.58	−0.023	3.59	0.049
2.50	−0.006	3.40	0.033

a. Display the association between the two variables in a scatter plot.

b. Calculate the correlation between second language proficiency and gray-matter density.

c. Test the null hypothesis of zero correlation.

d. What are your assumptions in part (c)?

e. Does the scatter plot support these assumptions? Explain.

f. Do the results demonstrate that second language proficiency affects gray-matter density in the brain? Why or why not?

Site	Attachment	Area (ha)	Number of butterfly species	Number of bird species	Ln number of plant species
A	4.4	23.8	6	12	5.1
B	4.5	16.0	14	18	5.5
C	4.7	6.9	8	8	6.4
D	4.5	2.3	10	17	4.7
E	4.3	5.7	6	7	5.3
F	3.8	1.2	5	4	4.6
G	4.4	1.4	5	8	4.5
H	4.6	15.0	7	22	5.5
I	4.1	3.1	9	7	5.2
J	4.2	3.8	5	4	4.6
K	4.6	7.6	10	11	4.5
L	4.2	12.9	9	11	5.0
M	4.3	4.0	12	13	5.0
N	4.4	5.6	11	16	5.6
O	4.2	4.9	7	7	5.4

14. In an increasingly urban world, are there psychological benefits to biodiversity? Fuller et al. (2007) measured the number of plant, bird, and butterfly species in 15 urban green spaces of varying size in Sheffield, England, a city of more than a half-million people. They also interviewed 312 green-space users and asked a series of questions related to the degree of psychological well-being obtained from green-space use. From the answers, the researchers obtained a measure of user "attachment" to green spaces (strength of emotional ties). Their results are in the table at the top of this page.

    a. Which of the three measures of green-space biodiversity (number of butterfly species, number of bird species, and ln number of plant species) is most strongly correlated with the "attachment" variable? Provide a standard error with each of your correlations.

    b. Provide an approximate 95% confidence interval for each of your correlations in part (a).

15. Use the data in Assignment Problem 14 to calculate a 95% confidence interval for the correlation between attachment and green-space area.

16. The following data are from a laboratory experiment by Smallwood et al. (1998) in which liver preparations from five rats were used to measure the relationship between the administered concentration of taurocholate (a salt normally occurring in liver bile) and the unbound fraction of taurocholate in the liver.

Rat	Concentration (μM)	Unbound fraction
1	3	0.63
2	6	0.44
3	12	0.31
4	24	0.19
5	48	0.13

    a. Calculate the correlation coefficient between the taurocholate unbound fraction and the concentration.

    b. Plot the relationship between the two variables in a graph.

    c. Examine the plot in part (b). The relationship appears to be maximally strong, yet the correlation coefficient you calculated in part (a) is not near the maximum possible value. Why not?

    d. What steps would you take with these data to meet the assumptions of correlation analysis?

17. If you are having trouble solving homework problems, should you sleep on it and try again in the morning? Huber et al. (2004) asked 10 human subjects to perform a complex spatial learning task on the computer just before going to sleep. EEG recordings were then taken of the electrical activity of brain cells during their sleep. The magnitude of the increase in their "slow-wave" sleep after learning the complex task, compared to baseline amounts, is listed in the accompanying table for all 10 subjects. Also provided is the increase in performance recorded when the subjects were challenged with the same task upon waking.

Increase in slow-wave activity during sleep (percent)	Improvement in task performance after waking (percent)
8	8
14	3
13	0
15	0
17	8
18	15
31	14
32	10
44	27
54	26

a. Calculate the correlation coefficient between the magnitude of the increase in slow-wave sleep and the magnitude of the improvement in performance upon waking.
b. What is the standard error for your estimate in part (a)?
c. Provide an interpretation of the quantity you calculated in part (b). What does it measure?
d. Test the hypothesis that the two variables are correlated in the population.
e. Is this an observational or an experimental study? Explain.

18. Both of the variables in Assignment Problem 17 are measurements that include some measurement error.
a. How would this measurement error effect the correlation between the two variables?
b. What steps could be taken in the design of the study to minimize the effect of measurement error?

c. For a given variable, what quantity is used to estimate the proportion of the variance among subjects not attributable to measurement error?

19. Left-handed people have an advantage in many sports, and it has been suggested that left-handedness might have been advantageous in hand-to-hand fights in early societies. (Left-handed people can get a lot of practice against right-handed opponents, whereas right-handers are less experienced against lefties.) To explore this potential advantage, Faurie and Raymond (2005) compared the frequency of left-handed individuals in traditional societies with measures of the level of violence in those societies. The following table lists the data for one index of violence, the rate of homicide.

Society	Percent left-handed	Homicide rate (number/1000 people/year)
Dioula	3.5	0.01
Ntumu	8.2	0.02
Kreyol	6.6	0.03
Inuit	6.4	0.17
Baka	10.2	0.50
Jimi	13.0	5.37
Eipo	20.4	3.02
Yanomamo	22.7	3.98

a. Use a graph to illustrate the association between the two variables.
b. What assumption of linear correlation analysis is violated by these data? Explain.
c. Before resorting to a nonparametric method, what strategy is available to test for a correlation between percent left-handedness and homicide rate?
d. Carry out this strategy, using a scatter plot to assess your success in meeting assumptions.
e. Using your results from part (d), test whether the two variables are correlated.

20. Logging in western North America impacts populations of western trillium, a long-lived perennial that inhabits conifer forests (*Trillium ovatum*; see the photo at the beginning of the chapter). Jules and Rathcke (1999) measured attributes of eight local populations of western trillium confined to forest patches of varying size created by logging in southwestern Oregon. Their data, which are presented in the following table, compare estimates of recruitment (the density of new plants produced in each population per year) at each site with the distance from the site to the edge of the forest fragment.

Local population	Distance to clearcut edge (m)	Recruitment
1	67	0.0053
2	65	0.0021
3	61	0.0069
4	30	0.0006
5	84	0.0124
6	97	0.0045
7	16	0.0028
8	332	0.0182

a. Display these data in an appropriate graph. Examine the graph and describe the shape of the distribution. What departures from the assumption of correlation analysis do you detect?

b. Choose a transformation and transform one or both of the two variables. Plot the results. Did the transformation solve the problem? If not, try a different transformation.

c. Using the transformed data, estimate the correlation coefficient between the two variables. Provide a standard error with your estimate.

d. Calculate an approximate 95% confidence interval for the correlation coefficient.

21. Cocaine is thought to affect the brain by blocking the dopamine receptor. To investigate this, Volkow et al. (1997) administered intravenous doses of 0.3 to 0.6 mg/kg of cocaine to volunteers. They used PET scans to compare the magnitude of the perceived "high" of regular cocaine users with the percentage of dopamine receptors blocked. The results for 34 subjects are illustrated below.

a. Using the following quantities, calculated from these data, estimate the correlation between the percentage of dopamine receptors blocked and subjects' ratings of the

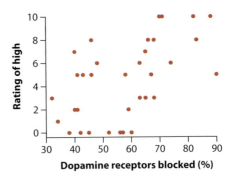

cocaine "high." Provide a standard error with your estimate.

> Sum of products: 957.5
> Sum of squares (receptors blocked): 8145.441
> Sum of squares (rating of high): 372.5

b. Calculate a 99% confidence interval for the correlation in the population.

c. What are your assumptions in part (b)?

d. Imagine the following scenario: A second team of researchers carried out a similar study using the same population and sample size. They used a narrower range of intravenous doses of cocaine in their experiment, which led to a smaller range of values than in the above study for the percentage of dopamine receptors blocked. When they analyzed their results, they found only a low correlation between percentage dopamine receptors blocked and perceived "high." In their published report, they concluded that the true correlation between these variables is much lower than estimated in the Volkow et al. study. Who is right? Explain.

22. An experiment by Collins and Bell (2004) investigated the impacts of elevated carbon dioxide (CO_2) concentrations on plant evolution. They raised separate lines of the unicellular algae *Chlamydomonas* under normal or high CO_2 levels. After 1000 generations they measured the growth rate of all of the experimental lines in a high CO_2 environment. The results for 14 experimental lines are presented in the

accompanying table. Growth rate is measured relative to the starting strain and has no units. Use these data to test whether the mean growth rate is associated with the CO_2 treatment.

CO_2 treatment	Growth rate
Normal	2.31
Normal	1.95
Normal	1.86
Normal	1.59
Normal	1.55
Normal	1.30
Normal	1.07
High	2.37
High	1.89
High	1.55
High	1.49
High	1.26
High	1.20
High	0.98

Publication bias

When we read an article in a journal we respect, we tend to believe what we read. The paper has been carefully vetted by expert referees, and the decision to publish it was made by an editor who is likely among the best scientists in the field. We might look at the authors of the papers and at the methods section, and we often find that both are above reproach. Very often the authors will have used statistical analysis, and usually it seems to be correctly done. We see that P is less than 0.05, or even less than 0.01, and we interpret that to mean that there was a low probability of getting results as extreme as those observed if the null hypothesis were true. When we've finished reading, we've just learned something. Haven't we?

It turns out that the papers that are actually published, especially those published in the "better quality," more widely read journals, are not a random sample of all studies done. This is true in at least two ways. First, the papers that get published are, on average, reporting on science that is done better than those submitted to journals but rejected by the editors. This is as it should be, for far too much science is done poorly and journal space is limited.

Second, papers that are more "interesting" get published more often than boring papers. Again, there is nothing intrinsically wrong with this, but it can raise a very serious problem: papers that do *not* reject the null hypothesis of no effect are usually thought, by most editors and even most authors, to be less interesting than those that do reject the null hypothesis. Thus, papers that reject the null hypothesis are more likely to be published than those that do not. Moreover, papers that

The odds of publishing a study whose main outcome was a *P*-value less than 0.05 are about 2.5 times higher than those of studies obtaining *P* > 0.10

describe large effects of an experiment are more interesting than those showing smaller effects, so published papers tend to show larger effects than unpublished studies.

As a result, the science that gets published is a biased selection of all science done. The difference between the true effect and the average effect published in journals is called **publication bias**. Publication bias can seriously skew our perception of nature. It is difficult to quantify, though, because we don't know much about the research that isn't published, such as how much there is and what results were obtained. After all, publication is how we normally find out about research.

One way to detect publication bias takes advantage of the fact that scientists must file for permits from an Ethics Review Board to do research on humans in research hospitals. Records of these reviews allow people studying publication bias to know how many studies were carried out, so that they can follow up the results and publication outcome of each study. Such reviews consistently find that the odds of publishing a study whose main outcome was a P-value less than 0.05 are about 2.5 times

higher than those of studies obtaining $P > 0.10$
(Easterbrook et al. 1991,
Dickersin et al. 1992).[1]
Moreover, researchers are
slower to publish non-
significant results ($P >
0.05$) when they do pub-
lish them (Stern and
Simes 1997). Studies
with $P < 0.05$ are more
likely to get into more
widely read journals
(Ioannidis et al. 1997),
and subsequently, they
are more likely to be
cited by other papers
(Gøtzsche 1987). These
findings are disturbing—
the papers that we read
are not necessarily representative of the truth.

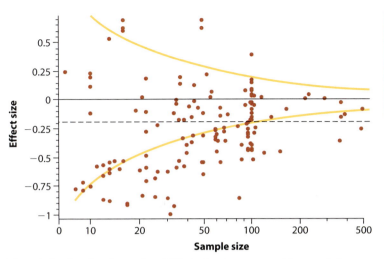

A funnel plot showing the results of 140 studies that measured the association between left–right asymmetry and male mating success. The effect size is the correlation coefficient between asymmetry and mating success. Adapted from Palmer (1999).

Another, perhaps more troubling, source of
evidence about the possibility of publication bias
comes from statistical analyses of drug trials
according to their funding source. Most analyses
of this sort have found that studies funded by
drug manufacturing companies are about 3.5
times more likely to yield a result favorable to the
company than are publicly funded studies
(Melander et al. 2003, Leopold et al. 2003,
Bekelman et al. 2003). The implication is that
research is unlikely to be published if it reflects
poorly on the interests of the company funding it.

Small studies finding minor or non-
significant effects are more likely to be left
unpublished than are large studies with similarly weak results. Perhaps scientists who have
done a large study are more determined to
publish whatever the result, to get some pay-

back for all their work. On the other hand, a
researcher who has carried out a small study
and gets inconclusive results is quite likely to
assume that the study had insufficient power
and leave it unpublished. If publication bias
is present, therefore, then there should be
a relationship between publication, the size of
the effect, and the sample size, as depicted in
a **funnel plot**. A funnel plot is a scatter plot of
the magnitude of the effect detected in
published studies and their corresponding
sample sizes.

The figure in this interleaf is a funnel plot
showing the results of 140 published studies,
each of which examined the relationship
between the mating success of individual males
in a study species and the degree of left–right
asymmetry in a male trait (Palmer 1999). In
studies of the mating success in flies, for example, asymmetry might measure the absolute
value of the difference between the lengths of
the left and right wings. In human studies,
asymmetry might measure the difference in
proportions of the two sides of the face. The
140 studies devoted solely to one male feature

---

[1] These studies also show that the main source of the bias
is the authors themselves, not the journal editors. People
are less likely to go to all the trouble of writing a paper if
they are not as interested in the results, or if they fear that
it will not be published anyway.

might seem excessive, but the causes of romantic success in nature are of great interest to biologists. Most of the 140 studies followed on the heels of claims that the symmetry of traits may be even more important than the traits themselves in explaining why, in nature and in human societies, some males get more than their fair share of mates and others get less. But does asymmetry really matter?

Each point in the figure is from a different study. The X-axis gives the sample size of each study, whereas the Y-axis gives the "effect size," which in this case is the estimated correlation coefficient between asymmetry and mating success. A negative effect size means that males with greater asymmetry had lower mating success than males with less asymmetry. The plot combines studies of many types of animals carried out by many researchers in many jungles, shopping malls, and laboratories.

The solid horizontal line in the funnel plot represents the null hypothesis tested in every study that the true correlation coefficient between male asymmetry and mating success is zero. The yellow curves mark the critical values for tests of the null hypothesis. Points falling outside these bounds are statistically significant at $\alpha = 0.05$. The dashed horizontal line marks the average of the observed effect sizes of all 140 published studies.

This funnel plot is highly revealing. For example, note that the range of published correlation coefficients is broad when the sample size is small and narrow when the sample size is large. This is expected, though, because larger sample sizes should yield more precise estimates (see Chapter 4). This expectation gives the funnel plot its name.

Other features of the funnel plot are unexpected and are cause for concern. In the first place, very few small studies yielded estimates close to the average effect size. Instead, most small studies found very large effects, in contrast to the larger studies, which tended to find smaller effects. Even more disturbing is the

fact that many results for the smaller studies are statistically significant, clumping outside the lower critical value for significance (indicated in the figure by the lower yellow line). Again, this contrasts with the largest studies, most of which found no statistically significant effect. What is behind these unexpected patterns?

The probable answer is that most small studies—those finding weak and non-significant effects—were never published. As a result, the published papers are not representative of all studies done. There must be many studies with weaker, non-significant effects still sitting in file drawers and on hard drives in universities around the world, never to see the light of day. Another implication is that the average effect size of all of the published studies is overestimated, so reading just the published papers gives a biased view of the strength of the relationship between asymmetry and male mating success. Symmetry apparently gives the hopeful male at most a slight edge in romance.

As mentioned previously, the more interesting a result is, the more likely it is to be published in a high-profile journal. One of the things that makes a result interesting is an effect that is stronger than previously believed. This can occur because our previous assumptions were wrong, but it can also occur because the effect was overestimated. As a consequence, dramatic claims in published papers often turn out later to be exaggerated or even false. This is not always or even usually the result of bad science, but rather it is due to publication bias.

The lesson to take home about publication bias is that flashy new results of published studies should always be repeated, preferably by different researchers working with larger sample sizes and who are bent upon publishing no matter what the outcome. In this way, the excesses of publication bias can be detected and corrected, yielding more accurate views of the patterns in nature.

17

Regression

Regression is the method used to predict the value of one numerical variable from that of another. For example, the following scatter plot shows how genetic diversity in a local contemporary human population is predicted by its dispersal distance from East Africa.[1] Modern humans emerged from Africa around 50,000

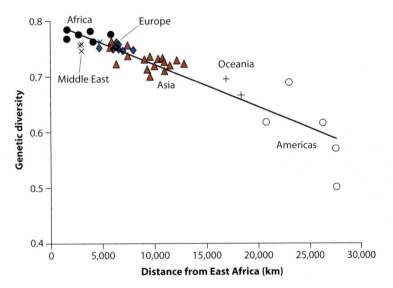

---

[1]   Modified from Prugnolle et al. (2005).

years ago, and our ancestors lost some genetic variation at each step as they spread to new lands.

The line embedded within the plot is the regression line. It describes the linear relationship between the two variables, genetic diversity and distance from East Africa. The regression line can be used to *predict* the genetic diversity of a local human population (the response variable), even for a locale not included in this study, based on its dispersal distance from East Africa (the explanatory variable). The regression line also indicates the *rate of change* of genetic diversity with distance. Both features of the relationship are captured in the equation for the line.

In this chapter, we show how to estimate the regression line, how to put bounds on its predictions, and how to test hypotheses about the coefficients describing it. Our focus is linear regression, but we also introduce some general principles of non-linear regression.

> *Regression* is a method that predicts the value of one numerical variable from that of another.

Regression and correlation both describe features of a scatter plot, and the two methods are often used interchangeably in the scientific literature. Both techniques measure a relationship between two numerical variables, and some of their calculations are shared. But correlation treats both variables equally, whereas regression predicts the value of one variable (the response variable) from the other (the explanatory variable). Correlation measures the strength of association between the two variables, whereas regression measures how steeply the response variable changes, on average, with change in the explanatory variable.

17.1   **Linear regression**

The most common type of regression is **linear regression**, which draws a straight line through a scatter plot to predict the response variable (Y, shown on the vertical axis) from the explanatory variable (X, shown on the horizontal axis). One important assumption of the linear regression method is that the relationship between the two variables really is linear. Example 17.1 shows how to use linear regression to predict the value of a response variable.

| Example 17.1 | **The lion's nose** |

Managing the trophy-hunting of African lions is an important part of maintaining viable lion populations. Knowing the ages of the male lions helps, because removing males older than six years has little impact on lion social structure, whereas taking younger males is more disruptive. Whitman et al. (2004) showed that the amount of black pigmentation on the nose of male lions increases as they get older and so might be used to calculate the age of unknown lions. The relationship between age and the proportion of black pigmentation on the noses of 32 male

lions of known age in Tanzania is shown in the scatter plot in Figure 17.1-1. The raw data are listed in Table 17.1-1. We can use these data to predict a lion's age from the proportion of black in his nose.

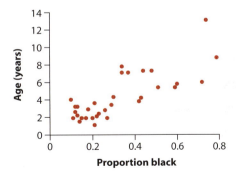

Figure 17.1-1 Scatter plot of the known ages of 32 male lions (Y, vertical axis) and the proportion of black on their noses (X, horizontal axis).

Table 17.1-1 The proportion of black on the noses of 32 male lions of known age.

Proportion black	Age (years)	Proportion black	Age (years)
0.21	1.1	0.30	4.3
0.14	1.5	0.42	3.8
0.11	1.9	0.43	4.2
0.13	2.2	0.59	5.4
0.12	2.6	0.60	5.8
0.13	3.2	0.72	6.0
0.12	3.2	0.29	3.4
0.18	2.9	0.10	4.0
0.23	2.4	0.48	7.3
0.22	2.1	0.44	7.3
0.20	1.9	0.34	7.8
0.17	1.9	0.37	7.1
0.15	1.9	0.34	7.1
0.27	1.9	0.74	13.1
0.26	2.8	0.79	8.8
0.21	3.6	0.51	5.4

The scatter plot in Figure 17.1-1 puts age as the response variable (the vertical axis) and the proportion of black on the nose as the explanatory variable (the horizontal axis), rather than the reverse, because we want to predict age from the proportion of black, not the other way around.

The method of least squares

There are many straight lines that can be drawn through a scatter of points, so how do we find the "best" one? Ideally we would find a line that leads to the smallest possible deviations in Y (the vertical axis) between the data points and the regression line (Figure 17.1-2).

More specifically, we want to find the line for which the sum of all the *squared* deviations is smallest. We square the deviations from the regression line for the same reason that we square deviations from the mean when calculating an ordinary variance—that is, to overcome the fact that some deviations are positive (the points above the regression line) and others are negative (the points below the regression line), thus canceling each other out. The method is called **least squares regression.** To find the least squares regression line for a set of data, we need to describe the line in mathematical terms, as we show next.

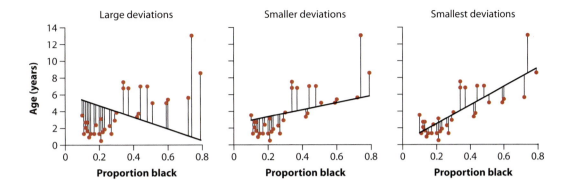

Figure 17.1-2 Illustration of the deviations between the data and several possible regression lines (the heavy black lines) drawn through the scatter of points originally plotted in Figure 17.1-1. Vertical lines are the deviations in *Y* between each point and the regression line. The line in the right panel is the least squares regression line.

Formula for the line

The regression line through a scatter plot is described mathematically by the following equation:

$$Y = a + bX.$$

The symbol *Y* is the response variable (displayed on the vertical axis), whereas *X* is the explanatory variable (the horizontal axis). The formula has two coefficients, *a* and *b*. The coefficient *a* is the *Y*-intercept, or just the **intercept**. Mathematically, *a* is the value of *Y* when *X* is zero. Its units are the same as for the *Y*-variable.

The coefficient *b* is the **slope** of the regression line. It measures how much *Y* changes per unit change in *X*. Its units are the ratio of the units of *Y* and *X*. If *b* is positive, then larger values of *X* predict larger values of *Y*. If *b* is negative, then larger values of *X* predict smaller values of *Y*. Figure 17.1-3 shows the slope of the line

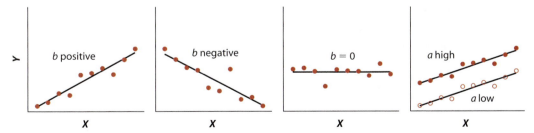

Figure 17.1-3 Comparing the slope of a line when *b* is positive (*far left*), negative (*left*), and zero (*right*); comparing a line with a high intercept *a* and one with a low intercept *a* (*far right*).

when b is positive, negative, and equal to zero, as well as the difference between a line with a high value of a and a low value of a.

> The *slope* of a linear regression is the rate of change in Y per unit of X.

Calculating the slope and intercept

Typically, you would use a computer to calculate the regression line, but we provide the formulas here for use with a calculator. The slope of the least squares regression line is computed as[2]

$$b = \frac{\sum (X_i - \overline{X})(Y_i - \overline{Y})}{\sum (X_i - \overline{X})^2},$$

where $\overline{X}$ and $\overline{Y}$ are the sample means of the two variables, and X_i and Y_i refer to the X and Y measurements of individual i. The top of this formula is the sum of products, something we saw originally in Section 16.1. The bottom is the sum of squares for X. Shortcut formulas for these sums are given in the Quick Formula Summary (Section 17.10) at the end of the chapter.

Once we have the slope b, getting the intercept is relatively straightforward, because the least squares regression line always goes through the point $(\overline{X}, \overline{Y})$. As a result,

$$\overline{Y} = a + b\overline{X}.$$

So, we find a by simple algebra:

$$a = \overline{Y} - b\overline{X}.$$

We can now use these formulas to calculate the coefficients of the least squares regression line for the lion data in Example 17.1. First, though, we need the following quantities that can be calculated from the data in Table 17.1-1:

$$\overline{X} = 0.3222 \qquad\qquad \overline{Y} = 4.3094$$
$$\sum (X - \overline{X})^2 = 1.2221 \qquad\qquad \sum (Y - \overline{Y})^2 = 222.0872$$
$$\sum (X - \overline{X})(Y - \overline{Y}) = 13.0123.$$

---

[2] Another way to write the formula for the slope is $b = \dfrac{\text{Covariance}(X,Y)}{s_X^2}$. The term in the numerator is the covariance between X and Y and the term in the denominator is the variance in X. The formula for the covariance can be found in the Quick Formula Summary of the correlation chapter (Section 16.8).

The slope is then

$$b = \frac{\sum (X_i - \bar{X})(Y_i - \bar{Y})}{\sum (X_i - \bar{X})^2} = \frac{13.0123}{1.2221} = 10.647.$$

The slope b measures the change in age of male lions per unit increase in the proportion of black in the nose. Its units are years per unit proportion black.

The intercept, in years, is

$$a = \bar{Y} - b\bar{X} = 4.3094 - 10.647(0.3222) = 0.879.$$

The formula for the line that predicts age from the proportion of black pigmentation on the nose in these lions can be written by putting all of this together, with appropriate rounding:

$$Y = 0.88 + 10.65X.$$

This equation could also be written as

$$\text{Age} = 0.88 + 10.65(\text{proportion black})$$

Figure 17.1-4 shows what this line looks like when it is plotted on the scatter plot[3] shown originally in Figure 17.1-1.

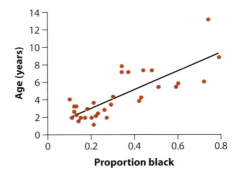

Figure 17.1-4 The regression line for the lion data from Example 17.1.

Populations and samples

The regression line is not just calculated for the sake of the data. It is typically used to estimate the "true" regression of Y on X in the population from which the data are a sample. The regression equation for the population is

$$Y = \alpha + \beta X,$$

---

[3]   To draw the line by hand, pick two points and use a ruler to connect them. Two convenient points are the intercept (the predicted Y at $X = 0$) and the mean, because the regression line always passes through the point $(\bar{X}, \bar{Y})$. Don't extend the actual line to pass through the Y-intercept at $X = 0$ if this point lies beyond the range of X in the data (as in Figure 17.1-4).

where β is the slope in the population and α is the intercept. The quantities α and β are population parameters, whereas a and b are their sample estimates.

The concept of population in regression is different from that in correlation. In correlation (Chapter 16), we assume that we have a random sample of (X, Y) pairs of measurements from a population. In regression, we assume instead that there is a population of *possible* Y-values for every value of X. The mean Y-value for every value of X lies on the true regression line. In other words, the true regression line connects the mean Y-values for every X-value.

For example, one of the lions in the data for Example 17.1 has a value of 0.6 for the proportion of black on its nose. According to Table 17.1-1, this lion is 5.8 years old. We assume that there is a population of lions having the value $X = 0.6$ for the proportion of black on their noses. The ages of the lions in this population vary, but their mean age lies on the true regression line.

Predicted values

Now that we have the regression line, we can use it to determine points on the line that correspond to specified values of X. These points on the regression line are called **predictions**, and we will symbolize them as $\hat{Y}$ ("Y-hat") to distinguish them from values of Y (i.e., actual data points), which lie above or below the line but not usually on it. The predicted value of Y for a given value of X estimates the mean of Y for the whole population of individuals that have that value of X. For example, to predict the age of a male lion corresponding to a proportion of 0.50 black on the nose, plug the value $X = 0.50$ into the regression formula:

$$\hat{Y} = a + b(0.50) = 0.88 + 10.65(0.50) = 6.2.$$

In other words, the regression line predicts that lions with a proportion of black $X = 0.50$ will be 6.2 years old on average.

According to Table 17.1-1, the value $X = 0.50$ was not represented in the sample, although it falls within the range of observed X-values (i.e., 0.10–0.79). For reasons that are explained in Section 17.2, we can only reliably make predictions using values of X that lie within the range of values in the sample.

> The predicted values of Y from a regression line estimate the mean value of Y for all individuals that have a given value of X.

Residuals

Residuals measure the scatter of points above and below the least squares regression line. They are crucial for evaluating the fit of the line to the data. Each observation in a scatter plot has a corresponding residual, measuring the vertical deviation from the least squares regression line (see the right panel in Figure 17.1-2). The point on the regression line used to calculate the residual for individual i is $\hat{Y}_i$, the value predicted when its corresponding value for X_i is plugged into the regression formula:

$$\hat{Y}_i = a + bX_i.$$

For example, the 31st lion in the sample ($i = 31$) has a proportion $X_{31} = 0.79$ of black on its nose (see Table 17.1-1). The corresponding age $\hat{Y}_{31}$ predicted for a lion with this much black on the nose is

$$\hat{Y}_{31} = 0.88 + 10.65(0.79) = 9.3$$

(rounded to one decimal place). The actual age of the lion was 8.8 years, which is below the predicted value. The residual is the observed value minus the predicted value:

$$\text{residual}_{31} = (Y_{31} - \hat{Y}_{31}) = (8.8 - 9.3) = -0.5 \text{ years.}$$

The variance of the residuals, symbolized as MS_{residual}, quantifies the spread of the scatter of points above and below the line. In regression jargon, this variance is called the "residual mean square":

$$MS_{\text{residual}} = \frac{\sum (Y_i - \hat{Y}_i)^2}{n - 2}.$$

The MS_{residual} is like an ordinary variance, but it has $n - 2$ degrees of freedom[4] rather than $n - 1$. It is analogous to the error mean square in the analysis of variance (Section 15.1). The following alternate formula is easier to use, though, because you don't need to calculate each $\hat{Y}_i$:

$$MS_{\text{residual}} = \frac{\sum (Y_i - \bar{Y})^2 - b \sum (X_i - \bar{X})(Y_i - \bar{Y})}{n - 2}.$$

Shortcuts for calculating the sum of squares and products are provided in the Quick Formula Summary (Section 17.10).

All of the quantities needed to determine MS_{residual} for the lion data have been calculated previously on page 468. Inserting these values into the equation for MS_{residual} yields

$$MS_{\text{residual}} = \frac{222.0872 - 10.647(13.0123)}{32 - 2} = 2.785.$$

---

[4] The degrees of freedom are $n - 2$ rather than $n - 1$ because we couldn't calculate the predicted values $\hat{Y}_i$ without first calculating *two* other quantities using the data—namely, the slope and the intercept of the regression line.

Standard error of slope

Like any other estimate, there is uncertainty associated with the sample estimate b of the population slope β. Uncertainty is measured by the standard error, the standard deviation of the sampling distribution of b. The smaller the standard error, the higher the precision, and the lower the uncertainty of the estimate of the slope. If the assumptions of linear regression are met (Section 17.5), then the sampling distribution of b is a normal distribution having a mean equal to β and a standard error estimated from data as

$$SE_b = \sqrt{\frac{MS_{residual}}{\sum (X_i - \bar{X})^2}}.$$

The quantity on top of the fraction under the square root sign is the residual mean square, and the quantity on the bottom is the sum of squares for X.

The standard error of b for the lion data is

$$SE_b = \sqrt{\frac{MS_{residual}}{\sum (X_i - \bar{X})^2}} = \sqrt{\frac{2.785}{1.2221}} = 1.510.$$

The standard error of the slope has the same units as the slope itself (i.e., years per unit of proportion black for the lion data in Example 17.1).

Confidence interval for the slope

A confidence interval for the parameter β is given by

$$b - t_{\alpha(2),df} SE_b < \beta < b + t_{\alpha(2),df} SE_b,$$

where $t_{\alpha(2),df}$ is the two-tailed critical value of the t-distribution having $df = n - 2$ degrees of freedom. For a 95% confidence interval, $\alpha = 0.05$, and for a 99% confidence interval, $\alpha = 0.01$. For the lion data, $t_{0.05(2),30} = 2.042$ (Statistical Table C), so the 95% confidence interval for the slope is

$$10.647 - 2.042(1.510) < \beta < 10.647 + 2.042(1.510)$$

$$7.56 < \beta < 13.73.$$

This is a modest range of most-plausible values for the slope. The mean age of lions increases by as little as 7.6 years per unit proportion of black on the nose, or by as much as 13.7 years.

17.2   Confidence in predictions

The purpose of regression is to predict. Thus, the regression line calculated from data predicts the mean value of Y for any specified value of X lying between the smallest and largest X in the data. This line is calculated with error, however, so we must know how precise the predictions are. Here in Section 17.2 we quantify the precision of predictions. We also discuss the hazards of *extrapolating*, making predictions for values of X beyond the range of X-values in the data.

Confidence intervals for predictions

There are two subtly different types of predictions that can be made using the regression line. The first predicts the *mean Y* for a given X. What, for example, is the mean age of all male lions in the population whose noses are 60% black (i.e., $X = 0.60$)? The second type predicts a *single Y* for a given X. (That is, how old is that lion over there, given that 60% of its nose is black?) Usually we just want to predict the mean Y for each X (i.e., the first prediction) because we are interested in the overall trend. In special situations, though, we also want to predict an individual Y-value (i.e., the second prediction). This is especially true in the lion study (Example 17.1). A hunter who encounters a male lion would want to know the age of that specific lion if he wishes to avoid shooting a young male.

Both types of predictions generate the same value for $\hat{Y}$. In the case of lions with 60% black on their noses,

$$\hat{Y} = a + bX = 0.88 + 10.65(0.60) = 7.27 \text{ years.}$$

Regardless of goals, this is the best prediction of age. The precision of the prediction is lower, however, if the goal is to predict the age of an individual lion rather than the mean age of lions having the specified proportion of black on their noses. This is because the prediction for a single Y-value includes uncertainty stemming from differences in Y between individuals having the same value of X (i.e., not all male lions having 60% black noses are the same age). The two graphs in Figure 17.2-1 illustrate these differences in precision.

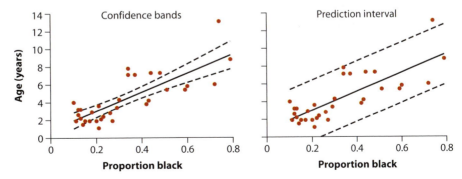

Figure 17.2-1 *Left:* 95% confidence bands for the predicted mean age of male lions at every value of proportion of black on their noses. *Right:* 95% prediction intervals for the predicted age of single lions. $n = 32$.

The 95% confidence intervals for the predicted mean lion age at every X are displayed in the left panel of Figure 17.2-1. The upper curve connects the upper bounds of all of the 95% confidence intervals for the predicted mean Y-values, one for every X between the smallest and largest X in the data. The lower curve connects the lower bounds of these same confidence intervals. Together the upper and lower curves are called the 95% **confidence bands**. These bands are narrowest in the vicinity of $\overline{X}$, the mean value for proportion of black on the nose, and they flare outward toward the extremes of the range of data. The uncertainty of predictions always increases the farther the X-value is from the mean X in the data. In 95% of samples, the confidence bands will bracket the true regression line in the population.

The 95% **prediction intervals** are displayed in the right-hand panel of Figure 17.2-1. The upper and lower curves connect the upper and lower limits of the 95% prediction intervals for a *single Y* over the range of X-values in the data. These are much wider than the confidence bands because predicting an individual's age from the color of its nose is more uncertain than predicting the mean age of all lions having the same proportion of black on their noses. Prediction intervals bracket the majority of individual data points in the sample, because they incorporate the variability in Y from individual to individual at a given X.

> *Confidence bands* measure the precision of the predicted mean Y for each value of X. *Prediction intervals* measure the precision of the predicted single Y-values for each X.

Most statistical packages on the computer will calculate and display confidence bands and prediction intervals. We haven't given calculation details, but we provide the formulas in the Quick Formula Summary (Section 17.10).

Extrapolation

We've stressed that regression can be used to predict Y for any value of X lying between the smallest and largest values of X in the data set. Regression cannot be used to predict the value of the response variable when an X-value lies well outside the range of the data, though, because there is no way to ensure that the relationship between X and Y continues to be linear beyond the range of the data. Predicting Y for X-values beyond the range of the data is called **extrapolation**. The graph in Figure 17.2-2 illustrates the problem.

The data are measurements of ear length[5] taken on a sample of adults at least 30 years old (Heathcote 1995). The linear regression equation calculated from these data (in millimeters) is

---

[5]   The measurements were made with a ruler on the left external ear from the top of the ear to the lowest part.

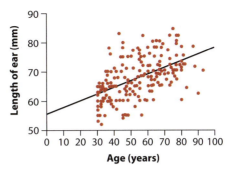

Figure 17.2-2 Ear lengths of 206 adults 30 years old or more as a function of their ages. Modified from Heathcote (1995).

$$\text{ear length} = 55.9 + 0.22 \,(\text{age}).$$

The results suggest that our ears grow longer by about 0.22 mm per year on average as we age. The intercept of this equation, which predicts the ear length at birth (i.e., when age is zero), is 56 mm. This makes no sense, though. To quote the authors of the study (Altman and Bland 1998), "A baby with ears 5.6 cm long would look like Dumbo." The rela-

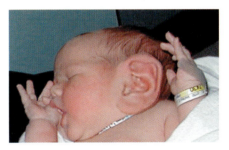

tionship between ear length and age is not linear from birth, but we wouldn't know this unless we took measurements over the complete range of ages.

> *Extrapolation* is the prediction of the value of a response variable outside the range of *X*-values in the data.

17.3   Testing hypotheses about a slope

Hypothesis testing in regression is used to evaluate whether the population slope equals a null hypothesized value, β_0, which is typically (but not always) zero. The test statistic t is

$$t = \frac{b - \beta_0}{\text{SE}_b},$$

where b is the estimate of the slope in the sample and SE_b is the standard error of b. Under the null hypothesis, this test statistic has a t-distribution with $n - 2$ degrees of freedom. Example 17.3 shows how to use this test.

Example 17.3 ## Chickadee alarms

Animals vary their calls depending on context, suggesting that the calls convey information. Black-capped chickadees make a "chick-a-dee-dee-dee" alarm call when they encounter a nonflying predator (they produce a completely different "seet" call when the predator is flying). Templeton et al. (2005) presented live, perched individuals of 13 differently sized avian predators to flocks of chickadees and recorded the

alarm calls. The 13 predator species ranged in size from small maneuverable hawks and owls, whose diets frequently include small birds, to large and less maneuverable predators that eat few small birds (Table 17.3-1). The average number of "dee" notes produced per call by chickadees, broken down by predator, is listed in Table 17.3-1. A scatter plot of these data is shown in Figure 17.3-1. Assuming that the measurements made using different predators are independent, does predator body size predict the average number of "dees" in the alarm calls?

Table 17.3-1 The average number of "dee" notes per alarm call by black-capped chickadees presented with a live, perched predator.

Predator species	Predator body mass (kg)	Number of "dee" notes per call
Northern pygmy-owl	0.07	3.95
Saw-whet owl	0.08	4.08
American kestrel	0.12	2.75
Merlin	0.19	3.03
Short-eared owl	0.35	2.27
Cooper's hawk	0.45	3.16
Prairie falcon	0.72	2.19
Peregrine falcon	0.72	2.80
Great horned owl	1.40	2.45
Rough-legged hawk	0.99	1.33
Gyrfalcon	1.40	2.24
Red-tailed hawk	1.08	2.56
Great gray owl	1.08	2.06

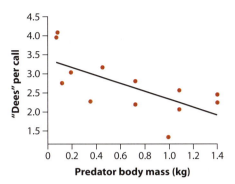

Figure 17.3-1 Body masses of 13 avian predator species and the average number of "dees" per call by chickadees encountering them.

The *t*-test of regression slope

The null hypothesis is that the average number of "dees" per alarm call cannot be predicted from the predator body mass—that is, the slope of the linear regression of "dee" rate on predator size is zero. The alternative hypothesis is that the "dee" rate either increases or decreases with increasing predator body mass. This is a two-tailed test:

H_0: The slope of the regression of the number of "dees" per alarm call by chickadees on predator body size is zero ($\beta = 0$).

H_A: The slope of the regression of the number of "dees" per alarm call by chickadees on predator body size is not zero ($\beta \neq 0$).

The following quantities calculated from the data are needed for the *t*-test of zero slope:

$$\overline{X} = 0.6654 \qquad\qquad \overline{Y} = 2.6823$$

$$\sum (X - \overline{X})^2 = 2.9009 \qquad \sum (Y - \overline{Y})^2 = 6.8210$$

$$\sum (X - \overline{X})(Y - \overline{Y}) = -3.0118.$$

The best estimate of the slope is

$$b = \frac{\sum (X - \overline{X})(Y - \overline{Y})}{\sum (X - \overline{X})^2} = \frac{-3.0118}{2.9009} = -1.0382.$$

Calculating the intercept as before, and rounding to two decimal points, we get the least squares regression line as

$$Y = 3.37 - 1.04X,$$

which can also be written as

$$\text{dee rate} = 3.37 - 1.04(\text{predator mass}).$$

This line has a negative slope, as shown in Figure 17.3-1. Notice that the number of "dees" per alarm call drops by about one for every 1 kg increase in predator body mass.

To calculate the standard error of the slope, we need the mean square residual:

$$MS_{residual} = \frac{\sum (Y_i - \bar{Y})^2 - b\sum (X_i - \bar{X})(Y_i - \bar{Y})}{n - 2}$$

$$= \frac{6.8210 - (-1.0382)(-3.0118)}{13 - 2}$$

$$= 0.3358.$$

Thus, the standard error of b is

$$SE_b = \sqrt{\frac{MS_{residual}}{\sum (X_i - \bar{X})^2}} = \sqrt{\frac{0.3358}{2.9009}} = 0.3402$$

We now have all of the elements to calculate the t-statistic:

$$t = \frac{b - \beta_0}{SE_b} = \frac{-1.038 - 0}{0.3402} = -3.051.$$

We must compare this t-statistic to the t-distribution having $df = n - 2 = 13 - 2 = 11$ degrees of freedom. Using a computer, we find that $t = -3.051$ corresponds to $P = 0.011$, so we reject the null hypothesis. We reach the same conclusion if we use the critical value for the t-distribution with $df = 11$ (Statistical Table C),

$$t_{0.05(2),11} = 2.201.$$

Since $t = -3.051$ is less than -2.201, $P < 0.05$ and we reject H_0. In other words, increasing predator body mass predicts a lower number of "dees" per call by chickadees. A 95% confidence interval for the population slope, calculated from the formula in Section 17.1, is

$$-1.79 < \beta < -0.29.$$

The ANOVA approach

In the literature, and in the output of regression analyses conducted on the computer, you may encounter tests of regression slopes that use an F-test rather than the t-test. Just as ANOVA can be used to compare two population means in place of the two-sample t-test, ANOVA can be used to test for a significant slope in place of the t-test of slope. The resulting P-values are equivalent. Table 17.3-2 shows the ANOVA table for the chickadee data. Formulas for the quantities are given in the Quick Formula Summary (Section 17.10).

Table 17.3-2 ANOVA table testing the effect of predator body mass on the number of "dees" per call by chickadees.

Source of variation	Sum of Squares	df	Mean Squares	F-ratio	P
Regression	3.1268	1	3.1268	9.3106	0.011
Residual	3.6942	11	0.3358		
Total	6.8210	12			

The analysis of variance approach takes the deviation between each Y-measurement and $\overline{Y}$ and breaks it into two parts, the residual and a "regression component," which represent the two sources of variation in Y in the data. The residual is the deviation in Y between a data point and its predicted value on the regression line (i.e., $Y_i - \hat{Y}_i$ analogous to the "error" component in ANOVA). This component measures variation among observations left over after the regression line is drawn. The "regression component," on the other hand, is the difference between the predicted value for each point and $\overline{Y}$ (i.e., $\hat{Y}_i - \overline{Y}$ is analogous to the "groups" component in ANOVA). If the null hypothesis is true, and the slope of the population regression β is zero, then the mean squares corresponding to these two components should be equal (except by chance). If the slope of the regression is not zero, however, then the regression mean square should be significantly greater than the residual mean square. The test statistic is an F-ratio of the two mean squares (the mean square regression divided by the mean square residuals).

The ANOVA approach can be used when the test is two-tailed and the null hypothesized slope is zero.

Using R^2 to measure the fit of the line to data

We can measure the fraction of variation in Y that is "explained" by X with the quantity R^2:

$$R^2 = \frac{SS_{regression}}{SS_{total}}.$$

R^2 is calculated from the sums of squares in the ANOVA table, and it is analogous to the R^2 in analysis of variance (Section 15.1), which measures the fraction of variation in the sample of Y accounted for by differences between groups.[6] If R^2 is close to one (i.e., its maximum possible value), then X predicts most of the variation in the values of Y. In this case, the Y-observations will be clustered tightly around the regression line with little scatter. If R^2 is close to zero (i.e., its minimum value), then X does not

---

[6] R^2 is sometimes written as r^2. In regressions that use only one explanatory variable (such as those we are discussing here), the R^2 value is the same as r^2, the square of the correlation coefficient.

predict much of the variation in Y, and the data points will be widely scattered above and below the regression line.

R^2 for the chickadee predator study can be calculated from the quantities in the ANOVA table, Table 17.3-2:

$$R^2 = \frac{3.1268}{6.8210} = 0.46.$$

Thus, predator body mass explained about 46% of the variation in the "dee" rate per alarm call by chickadees, a moderately high percentage.

17.4   Regression toward the mean

Suppose a study measured the cholesterol levels of 100 men randomly chosen from a population. After their initial measurement, the men were put on a new drug therapy, designed to reduce cholesterol levels. After one year, the cholesterol level of each man was measured again and compared with the first measurement. Figure 17.4-1 shows a scatter plot of the results. The researchers were delighted to find that cholesterol levels had dropped on average in the men who had previously had the highest levels. Their excitement dimmed, though, when they realized that cholesterol levels had *increased* on average in the men who had previously had low levels of cholesterol. What happened? Had they discovered a drug with complex effects?

In this hypothetical example, there was no effect at all due to the drug—the trend resulted entirely from a general phenomenon known as **regression toward the mean**. If two variables, such as consecutive measures of cholesterol, have a correlation less than one, then individuals that are far from the mean on the first measure will

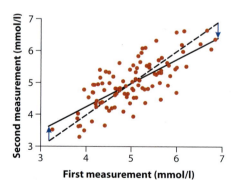

Figure 17.4-1 Regression toward the mean. These hypothetical data are two cholesterol measurements taken on the same 100 men. The dashed line is the one-to-one line with a slope of one. The solid line is the regression line predicting the second measure from the first. It has a slope less than one, as indicated by the blue arrows.

on average lie closer to the mean for the second measure. Even without an effect of the drug treatment, average cholesterol levels of the men with the highest levels on the first measure were expected to drop by the second measurement, and average levels of the men who originally had low levels were expected rise.[7]

> *Regression toward the mean* results when two variables have a correlation less than one. Individuals that are far from the mean for one of the measurements will, on average, lie closer to the mean for the other measurement.

Regression toward the mean is a tricky concept, which is perhaps why it is often overlooked. Think of it this way: each of the men in the study has a "true," underlying cholesterol value, but his "measured" cholesterol value varies randomly with time and circumstance around the true value. The subset of men who scored highest on the first measurement therefore likely included a disproportionate number of men whose cholesterol measurement was higher than its "true" value the first time. The second measurement made on each of these men is expected to be closer to its "true" value on average, bringing down the average for the subset of men as a whole. Similarly, the subset of men who initially scored lowest likely included a disproportionate number of men whose measured values were lower than their "true" values, so on the second measurement they would seem to improve.[8]

Regression toward the mean is a potentially large problem in any study that tends to focus on individuals in one tail of the distribution. In many medical studies, for example, only sick people are included in the research, as indicated by their initial assessment before the study. Because of regression toward the mean, many of these people will appear to improve even if the treatment has no effect. Interpreting this improvement as if it were a response to the treatment, instead of a mathematical fact of regression, is called the **regression fallacy**. It is one of the reasons why experiments should always include a control group for comparison.

Regression toward the mean is an issue only in observational studies, not in randomized experiments, where the value of the explanatory variable (X) is set by the experimenter. Kelly and Price (2005) discuss ways to disentangle biologically meaningful trends from the effect of regression toward the mean.

---

[7] Regression toward the mean happens even when the overall mean changes from the first to the second measurement.

[8] Regression was named for this property by Galton (1886), who studied heights of parents and their grown children. He noticed that tall fathers tended to have sons shorter than their fathers on average, while the reverse was true for short fathers. Galton was concerned at first that this tendency would eventually eliminate variation in height in the population. He later realized that, in each generation, very tall (and very short) individuals are still present, but they are not necessarily the offspring of the tallest (or shortest) parents.

17.5 Assumptions of regression

When using linear regression, the following assumptions must be met for confidence intervals and hypothesis tests to be accurate:

▶ At each value of X, there is a population of possible Y-values whose mean lies on the "true" regression line (this is the assumption that the relationship must be linear).
▶ At each value of X, the distribution of possible Y-values is normal.
▶ The variance of Y-values is the same at all values of X.
▶ At each value of X, the Y-measurements represent a random sample from the population of possible Y-values.

Figure 17.5-1 illustrates the first three of these assumptions.

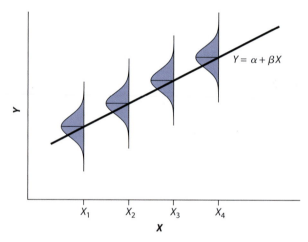

Figure 17.5-1 Illustration of the assumptions of linear regression. At each value of X there is a normally distributed population of Y-values with the mean on the true regression line. The variance of the Y-values is assumed to be the same for every value of X.

In the next few sections, we explore some of the ways to examine deviations from these assumptions and methods to try when the assumptions are not supported by the data.

No assumptions are made about X when using regression. Many assumptions about the distribution of X are made when using *correlation*, but *not* when using regression. In regression, for example, it is not necessary that the distribution of X-values is normal. It is not even necessary that X-values are randomly sampled—they might be fixed by the experimenter, instead.

Outliers

Besides creating a non-normal distribution of Y-values at the corresponding value of X, and violating the assumption of equal variance in Y, outliers disproportionately affect estimates of the regression slope and intercept. If an outlier is present, biologists usually examine its influence on the results by comparing the regression line produced with and without the outlier (Figure 17.5-2). Outliers are especially likely to be influential if they occur at or beyond the range of X-values in the rest of the data. If the outlier has a large effect, then alternative approaches should be sought. One is to transform X or Y, such as by taking logarithms, to see if this brings the outlier closer to the rest of the distribution. Another approach is to abandon regression and use rank correlation (Section 16.4), instead, to investigate the association only.

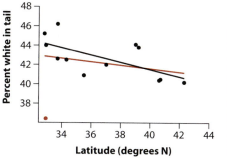

Figure 17.5-2 Graph showing the effect of an outlier on an estimate of the regression line. The data are the percentage of white in the tail feathers of the dark-eyed junco at sites at different latitudes in California (Yeh 2004). The black regression line was calculated after excluding the point on the lower left, whereas the red regression line included it.

Detecting non-linearity

Visually inspecting the scatter plot is a useful method for detecting departures from the assumption of a linear relationship between Y and X. Often, this approach is enough to conclude that the relationship between X and Y is not linear. Forcing a linear regression through the scatter plot can sometimes make the non-linearity even more obvious (see the left panel of Figure 17.5-3).

Scatter-plot "smoothing," a method discussed in greater detail in Section 17.8, can also aid the eye in detecting a non-linear relationship (see the right panel of Figure 17.5-3). More and more statistics packages on the computer are able to carry out scatter-plot smoothing.

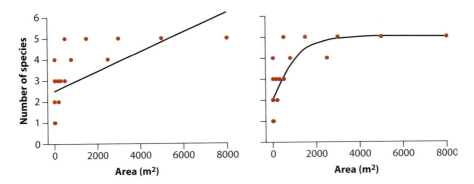

Figure 17.5-3 A scatter plot showing the relationship between the number of fish species and the surface area of 20 desert pools (Kodric-Brown and Brown 1993). The left panel fits a linear regression to the data to highlight how poorly a straight line matches the data. The right panel adds a "smoothed" fit to the same data (see Section 17.8).

Detecting non-normality and unequal variance

It is often difficult to decide whether the assumptions of normally distributed residuals and equal variance of residuals are met. Visual inspection of a **residual plot** can help. In a residual plot, the residual for every data point $(Y_i - \hat{Y}_i)$ is plotted against X_i, the corresponding value of the explanatory variable. This plot is best made with the aid of a computer. Two examples of residual plots are shown in Figure 17.5-4.

If the assumptions of normality and equal variance of residuals are met, then the residual plot should have the following features:

▶ a roughly symmetric cloud of points above and below the horizontal line at zero, with a higher density of points close to the line than away from the line,
▶ little noticeable curvature as we move from left to right along the X-axis, and
▶ approximately equal variance of points above and below the line at all values of X.

The blue tit data in the left panel of Figure 17.5-4 fit these requirements reasonably well. The density of observations peaks near the horizontal line and spreads outward above and below in a fairly symmetrical fashion. The spread of points above and below the line is similar across the range of X-values. (The spread may seem low at the extreme left end, but we can't tell because there are only two data points). The cockroach data in the right panel of Figure 17.5-4 do not fit these requirements as well. The spread of points above and below the horizontal line is considerably higher at low values of X than at high values of X.

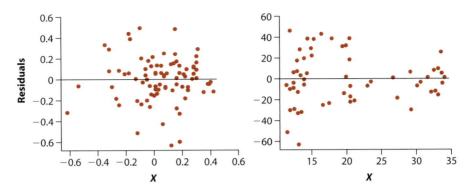

Figure 17.5-4 Two examples of residual plots. Data on the left are from a linear regression of the cap color of offspring on that of their parents in blue tits, a British bird (Hadfield et al. 2006). Those on the right are from a linear regression of firing rates of cockroach neurons on temperature (Murphy and Heath 1983).

> A *residual plot* is a scatter plot of the residuals $(Y_i - \hat{Y_i})$ against the X_i, the values of the explanatory variable.

Normal quantile plots (Section 13.1) and histograms of the residuals are yet another way to evaluate the assumption that the residuals are normally distributed.

17.6 Transformations

Some (but not all) non-linear relationships can be made linear with a suitable transformation. The most versatile transformation in biology is the log transformation (Section 13.3). The power and exponential relationships, which are commonly encountered in biology, are two non-linear relationships that can be made linear with a log transformation, as shown in Figure 17.6-1.

To illustrate the transformation of data, review the scatter plot in Figure 17.5-3, which compared the number of fish species in 20 desert pools of different surface area. The relationship between the number of species and pool area might fit a power curve or an exponential curve—we won't know until we try. When we log-transform both the number of species and the surface area, we get the graph shown in Figure 17.6-2, where the resulting relationship fits a straight line much better. We can now proceed to estimate parameters and test hypotheses about this relationship using the methods of linear regression, setting Y to be the log of the number of fish species and X to be the log of the surface area of the pools.

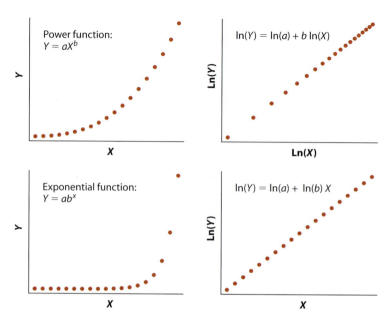

Figure 17.6-1 The power function (*upper left*) and the exponential function (*lower left*) are two types of non-linear relationships that can be made linear using the log transformation. Plot log of Y against log of X if the two variables are described by a power function (*upper right*). Plot the log of Y against X if the relationship between these two variables is described by an exponential function (*lower right*).

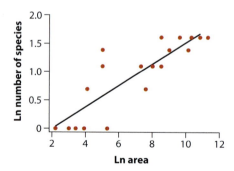

Figure 17.6-2 A scatter plot of the log-transformed number of fish species and surface area of desert pools. This relationship is more linear than the one in Figure 17.5-3, which is based on the untransformed data.

Transformation can also be used to help meet other assumptions of linear regression. For example, if a residual plot reveals that the variance of Y increases with increasing X, then log-transforming Y can often improve matters (it may be necessary, though, to transform X as well to keep the relationship linear). For example, testes mass in bat species increases with increasing body mass (Pitnick et al. 2005). In a residual plot from a regression using the untransformed data (see

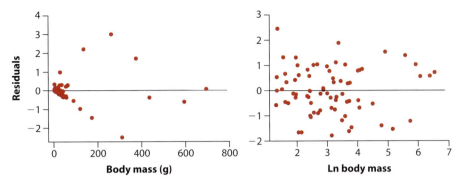

Figure 17.6-3 The effect of a log-transformation on the residuals from a linear regression of testes mass on body mass of bat species (Pitnick et al. 2005). Residuals from a linear regression calculated on the original data (*left panel*) do not fit the equal-variance assumptions of linear regression, but residuals from a regression using the log-transformed data (*right panel*) have more equal variances.

the left panel of Figure 17.6-3), the variance of residuals increases from the smallest X-values to the middle of the range of X-values. This problem goes away when X and Y are log-transformed (see the right panel of Figure 17.6-3).

Two other useful transformations originally described in Section 13.3 are the arcsine transformation (often effective when Y is a proportion) and the square-root transformation (when Y is a count). When analyzing data that violate the assumptions of regression, try simple transformations of X and/or Y to see if they help meet the assumptions of linear regression.

17.7 The effects of measurement error on regression

Recall from Section 16.6 that measurement error occurs when a variable is not measured with complete accuracy. Many biological traits, such as behavior or aspects of physiology, can be difficult to measure accurately, so measurement error can be an important component of variation (Section 15.6).

The effects of measurement error on regression are different than the effects on correlation (Section 16.6). The effect depends on the variable. Measurement error in Y increases the variance of the residuals, as shown in Figure 17.7-1 when you compare the scatter in the middle panel (measurement error in Y) with that in the left panel (no measurement error). This increases the standard error of the estimate of the slope and of the predictions.

Measurement error in X (the right panel in Figure 17.7-1) also increases the variance of the residuals, but it also causes bias in the expected estimate of the slope. With measurement error in X, b will tend to lie closer to zero on average than the population quantity β.

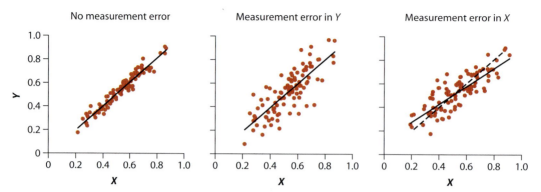

Figure 17.7-1 The effects of measurement error on the estimate of regression slope. X and Y are measured without error in the left panel. Y is measured with error in the middle panel, which has little effect on expected slope but increases the variability in the residuals. X is measured with error in the right panel, which causes the expected estimate of the slope to decline (*solid line*) compared with the slope in the absence of measurement error (*dashed line*). The variability of the residuals also increases.

17.8 Non-linear regression

Transformations won't always successfully convert a non-linear relationship into one that can be analyzed using linear regression. Non-linear regression methods, however, are readily available in most statistics packages on the computer. Here in Section 17.8 we outline some basic principles for non-linear regression.

The assumptions of non-linear regression are almost the same as those of linear regression (Section 17.5), except that here we usually assume that the true relationship between X and Y has a specific non-linear form.

The immediate problem when turning to non-linear regression is that there are a nearly unlimited number of options. There are so many mathematical functions to choose from that it can be difficult to know where to begin. The appropriate choice depends on the data, but a few guidelines can help you make a sensible choice.

A curve with an asymptote

The best advice we can offer is to keep things simple, unless the data suggest otherwise. Figure 17.8-1 illustrates this principle. The left panel shows a non-linear regression model fitted to data on the population growth rate of a species of phytoplankton and the concentration of iron, a limiting nutrient. The mathematical function fit to the data passes through every data point. The resulting $MS_{residual}$ is zero.

This is not the best possible outcome, however, even though each data point fits precisely. The problem with the curve in the left panel of Figure 17.8-1 is that it would probably do a terrible job of predicting any *new* observations obtained from the same population because the curve does not describe the general trend. Such a complicated curve is also difficult to justify biologically—that is, is there good

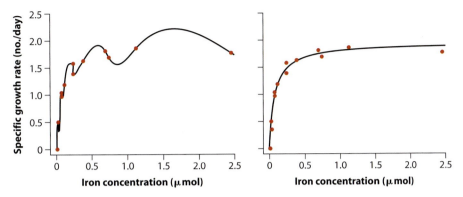

Figure 17.8-1 Population growth rate of a species of phytoplankton in culture in relation to the concentration of iron in the medium (the data are from Sunda and Huntsman 1997). The curve in the left panel is an arbitrarily complex function that passes through all of the data points. The curve in the right panel is a Michaelis–Menten curve that fits the data more simply.

reason to think that all the peaks and dips in this curve truly reflect the effects of iron on the growth of phytoplankton?

Greater simplicity, as demonstrated by the fitted curve in the right panel of Figure 17.8-1, solves both of these problems. The data are the same as in the left panel but this time we've fit the much simpler function,

$$Y = \frac{aX}{b + X}.$$

This is the *Michaelis–Menten* equation used frequently in biochemistry. The curve rises from a Y-intercept at zero and increases at a declining rate with increasing X, eventually reaching a saturation point, or asymptote. The asymptote is represented in the formula by the constant a, whereas b determines how fast the curve rises to the asymptote. Reminiscent of linear regression, the Michaelis–Menten equation has only two parameters to estimate (the true asymptote α and the true rate parameter β). These parameters are very different, however, from those of linear regression.

The virtue of linear regression is simplicity. We should strive to retain this property as we look at the wide range of non-linear functions available.

Quadratic curves

The *quadratic* curve is often used in biology to fit a humped-shape curve to data, such as the relationship shown in Figure 17.8-2 between the number of plant species present in ponds (Y) and pond productivity (X).

The curve is a symmetric parabola described by the quadratic (second degree polynomial) equation,

$$Y = a + bX + cX^2.$$

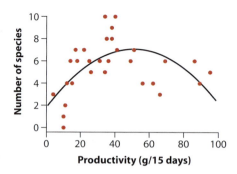

Figure 17.8-2 A quadratic curve fit to the relationship between the number of plant species present in ponds and pond productivity (Chase and Leibold 2002).

This equation is similar to the formula for a straight line except that one more term has been added for the square of X, and another regression coefficient c must be computed. When c is negative, the curve is humped, as in Figure 17.8-2. When c is positive, the parabola curves upward in a U-shape.

Asymptotic and quadratic curves are just two of several non-linear functions commonly used in biology. The choice between them must depend on the data: Is the relationship asymptotic or humped? If humped, is the hump symmetric or does it fall more steeply on one side than the other? If it falls more steeply on one side, then we must search for another function altogether. A good guide to a variety of curves used in biology can be found in Motulsky (1999).

Formula-free curve fitting

With the aid of a computer, it is possible to fit curves to data without specifying a formula. Often called **smoothing**, this approach gathers information from groups of nearby observations to estimate how the mean of Y changes with increasing values of X. There are several methods, including "kernel," "spline," and "loess" smoothing. We bypass the technical details here, but Example 17.8 illustrates the utility of the approach.

Example 17.8 | **The incredible shrinking seal**

Trites (1996) amassed a large set of measurements of the ages and sizes of northern fur seals (*Callorhinus ursinus*) in the Pacific Ocean, gathered over decades by many researchers. Most measurements were taken between spring and fall. In summer, females spend a lot of time on land giving birth and nursing young. The graph in Figure 17.8-3 shows the measurements for nearly 10,000 adult females. In a prelimi-

nary analysis, a line was fitted to the data, but the relationship appeared non-linear. Average length increased with age but appeared to taper off by about 4500 days (i.e., about 12 years old).

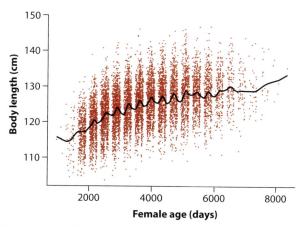

Figure 17.8-3 Measurements of body length as a function of age for female fur seals, with the "spline" fit in black.

To fit the data, we used a "spline" technique to calculate a smoothed fit of body length on female age. The result is plotted in Figure 17.8-3. Astonishingly, the curve indicates that average female body length does not rise steadily with age, but oscillates up and down each year. Female fur seals become longer each summer, and then shorter again by winter (keep in mind that these changes are in length, not weight). Elongation results in part because the seals are heavier on land than in water, and the added weight stretches the skeleton during summer breeding. The skeleton shrinks back after the seals return to the water in the winter. It would have been difficult to come up with a formula that captured this complex relationship.

The seal example illustrates that it is not always easy to anticipate or see the type of relationship present in the data by visual inspection alone. Smoothing techniques can help. Example 17.8 also demonstrates that we don't always need a mathematical formula if all we want to do is fit the data and use it to improve our understanding of a biological system.

The fit obtained by smoothing is controlled by a "smoothing coefficient" that determines how bumpy the curve is. This coefficient can be altered in statistical computer programs so that you can explore its effects. A low value for the coefficient results in a bumpy curve that, in the extreme, would pass through all the data points. A larger value of the coefficient gives a smoother fit. Computer programs usually use rules of thumb to choose the best value for the smoothing parameter, but it is wise to try alternatives to see what effect varying the smoothing coefficient has on the curve.

Fitting a binary response variable

Logistic regression is a special type of non-linear regression developed for a *binary* response variable—that is, when Y is either zero or one. A common use for logistic regression is the dose–response curve, where mortality (or survival) of individuals (the "response") is plotted against the concentration of a drug, toxin, or other chemical (the "dose"). Here we provide a quick description with a minimum of calculations.

The graph in Figure 17.8-4 shows the results of an experiment in which 72 rhesus macaques were exposed to different fixed concentrations of aerosolized anthrax spores. We would like to use these data to predict the probability of death based on anthrax concentration. Such a curve must take account of the fact that the Y-values are binary (zero or one) instead of normally distributed, and that the predicted probability must remain between zero and one.

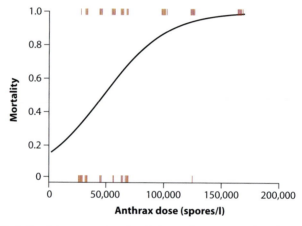

Figure 17.8-4 Mortality of rhesus monkeys in relation to the concentration of aerosolized anthrax spores in an exposure chamber (data from Hans 2002). Each vertical tick indicates a different individual (offset to minimize the overlap of points). $Y = 0$ if the individual survived, whereas $Y = 1$ if the individual died. The line is a logistic regression predicting the probability of death. $n = 72$.

The method of logistic regression deals with both of these issues by fitting the following equation to the data:

$$\text{Log-odds}(Y) = a + bX.$$

The log-odds is also known as the logit function.[9] Note that the expression on the right side of the equation (i.e., $a + bX$) is linear. In other words, the log-odds of the probability of mortality is fitted with an ordinary straight line. Applied to the anthrax data, we estimated

$$\text{Log-odds(mortality)} = -1.74 + 0.0000364 \, (\text{dose}).$$

---

9   Log-odds$(Y) = \ln[Y/(1 - Y)]$.

This is the curve shown in Figure 17.8-4. Using this approach, we can calculate standard errors and test hypotheses about the coefficients in the population from which the data were obtained. Routines for these calculations are provided in most computer statistical packages.

17.9 **Summary**

▶ Regression is a method used to predict the value of a numerical response variable Y from the value of a numerical explanatory variable X.

▶ Linear regression fits a straight line through a scatter of points. The equation for the regression line is $Y = a + bX$, where b is the slope of the line and a is the intercept.

▶ The least squares regression line is found by minimizing the sum of the squared differences between the observed Y-values and the values predicted by the line.

▶ The residuals are the differences between the observed values of Y and the values predicted by the least squares regression line, $\hat{Y}$. The variance of the residuals, $MS_{residual}$, measures the spread of the points above and below the regression line.

▶ Linear regression calculated on a sample of points estimates the straight line relationship between the two variables in the population. The formula for the population regression line is $Y = \alpha + \beta X$, where β is the slope of the line and α is the intercept.

▶ If the assumptions of regression are met, then the sampling distribution of b is normal with mean β and standard deviation estimated by the standard error of the estimate of the slope, SE_b.

▶ The confidence interval for the slope β is based on the t-distribution.

▶ There are two types of prediction in regression: the mean Y at a given X, and a single Y observation at X. Both predictions generate the same value $\hat{Y}$, but they have very different precision. Precision is lower when predicting an individual Y because it includes the variability between individuals having the same X-value.

▶ Confidence intervals for predicted mean Y-values at each X are represented by confidence bands. Analogous intervals for the predicted Y-values of a single individual are called prediction intervals.

▶ Extrapolation is the prediction of Y at values of X beyond the range of X-values in the sample. Extrapolation is problematic, though, because there is no way to ensure that the relationship between X and Y continues to be linear beyond the data.

▶ If the null hypothesis is correct, that the slope β of a population regression line is zero, then the test statistic $t = b/SE_b$ has a t-distribution with $n - 2$ degrees of freedom.

▶ An ANOVA table and F-test can also be used to test the null hypothesis that the population slope $\beta = 0$.

▶ R^2 is a measure of the fit of a regression line to the data. It measures the "fraction of the variation in Y that is explained by X."

▶ Regression toward the mean results when two imperfectly correlated variables are compared by regression. Individuals that are far from the mean for one of the measurements will on average lie closer to the mean for the other measurement.

▶ Methods for regression assume that the relationship between X and Y falls along a straight line, that the Y-measurements at each value of X are a random sample from a population of Y-values, that the distribution of Y-values at each value of X is normal, and that the variance of Y-values is the same at all values of X.

▶ The scatter plot and the residual plot are graphical devices for detecting departures from the assumptions of linear regression.

▶ Transformations of X and/or Y can be used to render a non-linear relationship linear and to correct violations of the assumption of equal variance of residuals at every X.

▶ The log-transformation is the most versatile transformation. It is useful when Y is related to X by a power function or by an exponential function.

▶ If transformations do not work, non-linear regression is an option.

▶ Non-linear regression curves should be kept as simple as the data warrant. Overly complex curves may be biologically unjustified and have low predictive power.

▶ Smoothing methods make it possible to fit non-linear curves to data without specifying a formula.

17.10 Quick Formula Summary

Shortcuts

Sum of products: $\sum (X - \bar{X})(Y - \bar{Y}) = \sum (XY) - \dfrac{(\sum X)(\sum Y)}{n}$

Sum of squares for X: $\sum (X - \bar{X})^2 = \sum (X^2) - \dfrac{(\sum X)^2}{n}$

Regression slope

What is it for? Estimating the slope of the linear equation $Y = \alpha + \beta X$ between an explanatory variable X and a response variable Y.

What does it assume? The relationship between X and Y is linear; each Y-measurement at a given X is a random sample from a population of Y-measurements; the distribution of Y-values at each value of X is normal; and the variance of Y-values is the same at all values of X.

Parameter: β

Estimate: b

Formula: $b = \dfrac{\sum (X_i - \bar{X})(Y_i - \bar{Y})}{\sum (X_i - \bar{X})^2}$

Standard error: $\mathrm{SE}_b = \sqrt{\dfrac{\mathrm{MS}_{\mathrm{residual}}}{\sum (X_i - \bar{X})^2}}$,

where $\mathrm{MS}_{\mathrm{residual}}$ is the mean squared residual (the estimated variance of the residuals); $\mathrm{MS}_{\mathrm{residual}} = \dfrac{\sum (Y_i - \hat{Y}_i)^2}{n - 2}$. A quicker formula for $\mathrm{MS}_{\mathrm{residual}}$, not requiring you to calculate the $\hat{Y}$ first, is $\mathrm{MS}_{\mathrm{residual}} = \dfrac{\sum (Y_i - \bar{Y})^2 - b\sum (X_i - \bar{X})(Y_i - \bar{Y})}{n - 2}$.

Regression intercept

What is it for? Estimating the intercept of the linear equation $Y = \alpha + \beta X$.

What does it assume? Same as the assumptions for the regression slope.

Parameter: α

Estimate: a

Formula: $a = \bar{Y} - b\bar{X}$

Standard error: Set $X = 0$ in the formula for the standard error of the predicted mean Y at a given X [see "Confidence interval for the predicted mean Y at a given X (confidence bands)" on p. 496]. This is only valid if the value $X = 0$ falls within the range of X-values in the data.

Confidence interval for the regression slope

What is it for? An interval estimate of the population slope.

What does it assume? Same as the assumptions for the regression slope.

Statistic: b

Parameter: β

Formula: $b - t_{\alpha(2),df}\,SE_b < \beta < b + t_{\alpha(2),df}\,SE_b$, where SE_b is the standard error of the slope (see the formula given previously with the regression slope), and $t_{\alpha(2),n-2}$ is the two-tailed critical value of the t-distribution having $df = n - 2$.

Confidence interval for the predicted mean Y at a given X (confidence bands)

What is it for? An interval estimate of the predicted *mean Y* of all individuals in the population having the given value of X.

What does it assume? Same as the assumptions for the regression slope.

Statistic: $\hat{Y}$

Formula: $\hat{Y} - t_{\alpha(2),n-2}\,SE[\hat{Y}] < $ predicted $Y < \hat{Y} + t_{\alpha(2),n-2}\,SE[\hat{Y}]$,

where $SE[\hat{Y}] = \sqrt{MS_{residual}\left(\dfrac{1}{n} + \dfrac{(X - \bar{X})^2}{\sum (X - \bar{X})^2}\right)}$ is the standard error of the

prediction, $MS_{residual}$ is the mean square residual (see the formula given previously under "Regression slope"), and $t_{\alpha(2),n-2}$ is the two-tailed critical value of the t-distribution having $df = n - 2$.

Interval for the predicted individual Y at a given X (prediction intervals)

What is it for? An interval estimate of the predicted Y for a *single individual* having the given value of X.

What does it assume? Same as the assumptions for the regression slope.

Statistic: $\hat{Y}$

Formula: $\hat{Y} - t_{\alpha(2),n-2}SE_1[\hat{Y}] < $ predicted $Y < \hat{Y} + t_{\alpha(2),n-2}SE_1[\hat{Y}]$,

where $SE_1[\hat{Y}] = \sqrt{MS_{residual}\left(1 + \dfrac{1}{n} + \dfrac{(X - \bar{X})^2}{\sum (X - \bar{X})^2}\right)}$ is the standard error of the

prediction, $MS_{residual}$ is the mean square residual (see the formula given previously under "Regression slope"), and $t_{\alpha(2),n-2}$ is the two-tailed critical value of the t-distribution having $df = n - 2$.

The *t*-test of a regression slope

What is it for? To test the null hypothesis that the population parameter β equals a null hypothesized value β_0.

What does it assume? Same as the assumptions for the regression slope.

Test statistic: t

Distribution under H₀: t-distributed with $n - 2$ degrees of freedom.

Formula: $t = \dfrac{b - \beta_0}{\text{SE}_b}$, where SE_b is the standard error of b (see the formula given previously under "Regression slope").

The ANOVA method for testing zero slope

What is it for? To test the null hypothesis that the slope β equals zero, and to partition sources of variation.

What does it assume? Same as the assumptions for the regression slope.

Test statistic: F

Distribution under H₀: F distribution. F is compared with $F_{\alpha(1),1,n-2}$.

Source of variation	Sum of squares	df	Mean squares	F-ratio
Regression	$\text{SS}_{\text{regression}} = \sum (\hat{Y}_i - \bar{Y})^2$	1	$\dfrac{\text{SS}_{\text{regression}}}{df_{\text{regression}}}$	$\dfrac{\text{MS}_{\text{regression}}}{\text{MS}_{\text{residual}}}$
Residual	$\text{SS}_{\text{residual}} = \sum (Y_i - \hat{Y}_i)^2$	$n - 2$	$\dfrac{\text{SS}_{\text{residual}}}{df_{\text{residual}}}$	
Total	$\text{SS}_{\text{total}} = \sum (Y_i - \bar{Y})^2$	$n - 1$		

R squared (*R*²)

What is it for? Measuring the fraction of the variation in Y that is "explained" by X.

Formula:: $R^2 = \dfrac{\text{SS}_{\text{regression}}}{\text{SS}_{\text{total}}}$,

where $\text{SS}_{\text{regression}}$ is the sum of squares for regression and SS_{total} is the total sum of squares.

PRACTICE PROBLEMS

1. What is the formula for each of the following four regression lines?

a.
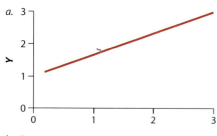

b.
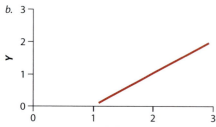

c.
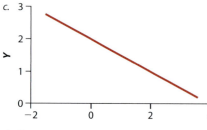

d.
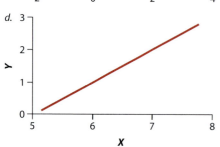

2. Some species seem to thrive in captivity, whereas others are prone to health and behavior difficulties when caged. Maternal care problems in some captive species, for example, lead to high infant mortality. Can these differences be predicted? The following data are measure-

ments of the infant mortality (percentage of births) of 20 carnivore species in captivity along with the log (base-10) of the minimal home range sizes (in km^2) of the same species in the wild (Clubb and Mason 2003).

Log$_{10}$ home range size	Captive infant mortality (%)	Log$_{10}$ home range size	Captive infant mortality (%)
−1.3	4	0.1	29
−0.5	22	0.2	33
−0.3	0	0.4	33
0.2	0	1.3	42
0.1	11	1.2	33
0.5	13	1.4	20
1.0	17	1.6	19
0.3	25	1.6	25
0.4	24	1.8	25
0.5	27	3.1	65

a. Draw a scatter plot of these data, with the log of home range size as the explanatory variable. Describe the association between the two variables in words.

b. Estimate the slope and intercept of the least squares regression line, with the log of home range size as the explanatory variable. Add this line to your plot.

c. Does home range size in the wild predict the mortality of captive carnivores? Carry out a formal test. Assume that the species data are independent.

d. Outliers should be investigated because they might have a substantial effect on the estimates of the slope and intercept. Recalculate the slope and intercept of the regression line from part (c) after excluding the outlier at large home range size (which corresponds to the polar bear). Add the new line to your plot. By how much did it change the slope?

3. The following two graphs display data gathered to test whether the exercise performance of women at high elevations depends on the stage of their menstrual cycle (Brutsaert et al. 2002). In the upper panel, the explanatory variable is the progesterone level and the response variable

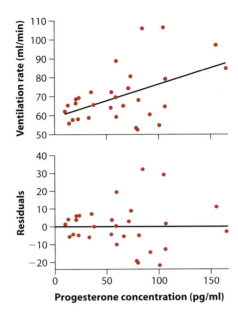

Body leanness (m²/kg)	Heat loss rate (°C/min)
7.0	0.103
7.0	0.097
6.2	0.090
5.0	0.091
4.4	0.071
3.3	0.024
3.6	0.014
2.8	0.041
2.4	0.031
2.1	0.010
2.1	0.006
1.7	0.002

b. Does rate of heat loss change with body leanness? Using the following intermediate calculations, carry out a formal test.

$$\overline{X} = 3.9667 \qquad \overline{Y} = 0.04833$$

$$\sum (X - \overline{X})^2 = 41.1467 \quad \sum (Y - \overline{Y})^2 = 0.01696$$

$$\sum (X - \overline{X})(Y - \overline{Y}) = 0.7805.$$

c. Calculate a 95% confidence interval for the regression slope.

d. What are your assumptions in parts (b) and (c)?

is the ventilation rate at submaximal exercise levels. The line is the least squares regression. The lower panel is the corresponding residual plot.

a. What is a "least squares" regression line?

b. What are residuals?

c. Assume that the random sampling assumption is met. By viewing these plots, assess whether each of the three other main assumptions of linear regression is likely to be met in this study.

4. To investigate whether subcutaneous fat provides insulation in humans, Sloan and Keatinge (1973) measured the rate of heat loss by boys swimming for up to 40 min in water at 20.3 °C and expending energy at about 4.8 kcal/min. Heat loss was measured by the change in body temperature, recorded using a thermometer under the tongue, divided by time spent swimming, in minutes. The authors measured an index of body "leanness" on each boy as the reciprocal of the skin-fold thickness adjusted for total skin surface area (in meters squared) and body mass (in kg). Their data are listed in the following table.

a. Draw a scatter plot of these data, showing how the rate of heat loss is related to body leanness.

5. The slopes of the regression lines on the accompanying graph show that the winning Olympic 100-m sprint times for men and women have been getting shorter and shorter over the years, with a steeper trend in women than in men (the graph is modified from Tatem et al. 2004). If trends continue, women are predicted to have a shorter winning time than men by the year 2156. What cautions should be applied to this conclusion? Explain.

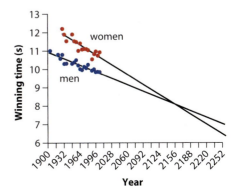

6. In an analysis of the performance of Major League Baseball players, Schaal and Smith (2000) found that the batting scores of the top 10 players in the 1997 baseball season dropped on average in 1998. What is the best interpretation of this finding?
   a. Players who did well in 1997 reduced their effort the following year, realizing that they didn't need to work as hard to get an above-average result.
   b. Players performing above average in 1997 were older and more worn out by 1998.
   c. Regression toward the mean.
   d. Possibly (a) and (b), but (c) is likely and cannot be ruled out.

7. Hybrid offspring of parents of different species are often sterile. How different must the parent species be, genetically, to produce this effect? The accompanying table lists the proportion of pollen grains that are sterile in hybrid offspring of crosses between pairs of species of *Silene* (bladder campions—see the photo on the first page of this chapter) (Moyle et al. 2004). Also listed is the genetic difference between the pair of species, as measured by DNA sequence divergence. Assume that different species pairs are independent.
   a. We would like to predict the proportion of hybrid pollen that is sterile (*Y*) from the genetic distance between the species (*X*). Since the response variable is a proportion, what transformation would be your first choice to help meet the assumptions of linear regression?
   b. Transform the proportions and then produce a scatter plot of the data. Estimate and draw the regression line.
   c. Calculate the 95% confidence interval for the slope of the line.

Silene species pair	Genetic distance	Proportion of pollen that is sterile
1	0.00	0.02
2	0.00	0.06
3	0.00	0.14
4	0.00	0.24
5	0.00	0.30
6	0.03	0.62
7	0.02	0.28
8	0.03	0.23
9	0.04	0.15
10	0.04	0.45
11	0.05	0.84
12	0.11	0.65
13	0.12	0.77
14	0.12	1.00
15	0.13	1.00
16	0.13	0.93
17	0.13	0.91
18	0.14	0.93
19	0.13	0.96
20	0.13	1.00
21	0.15	1.00
22	0.16	0.97
23	0.18	1.00

8. Rattlesnakes often eat large meals that require significant increases in metabolism for efficient digestion. Snakes are known to adjust their thermoregulatory behavior after feeding, seeking out warmer spots to increase their metabolic rates. Can snakes increase body temperature, though, even without this behavior? Tattersall et al. (2004) measured the change in body temperature of snakes after meals of various sizes, and we have used their data in an inappropriate way in the following graph, fitting a non-linear mathematical function indicated by the curve.

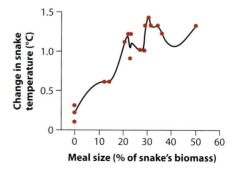

a. Why is the non-linear fit shown inappropriate?
b. What alternative procedure would you recommend to achieve the goal of predicting snake body temperature change from meal size?

9. Male lizards in the species *Crotaphytus collaris* use their jaws as weapons during territorial interactions. Lappin and Husak (2005) tested whether weapon performance (bite force) predicted territory size in this species. Their measurements for both variables are listed in the following table for 11 males.

Bite force (N)	Territory area (m²)
4.3	14.0
4.6	16.0
4.3	18.8
5.0	19.8
5.0	22.5
5.2	28.0
5.3	29.8
5.4	34.8
4.8	30.8
4.7	27.4
4.6	32.4

a. Estimate the slope of the regression line to predict territory area from bite force. Provide a standard error for your estimate.
b. Provide a 99% confidence interval for β.
c. Provide an interpretation for the 99% confidence interval in part (b). What does it measure?
d. Bite force is difficult to measure accurately, and so the values shown probably include some measurement error. Is the slope of the true regression line most likely to be under-

estimated, overestimated, or unaffected as a result?
e. Territory area is difficult to measure accurately, so the values shown probably include some measurement error. Is the slope of the true regression line most likely to be underestimated, overestimated, or unaffected as a result?

10. An ANOVA carried out to test the null hypothesis of zero slope for the regression of lizard territory area on bite force (see Practice Problem 9) yielded the following results:

Source of variation	Sum of squares	df	Mean squares	F-ratio
Regression	194.3739	1		
Residual	301.5516	9		
Total				

a. Complete the ANOVA table.
b. Using the F-statistic, test the null hypothesis of zero slope at the significance level $\alpha = 0.05$.
c. What are your assumptions in part (b)?
d. What does the $MS_{residual}$ measure?
e. Calculate the R^2 statistic. What does it measure?

11. James et al. (1997) demonstrated that the chemical hypoxanthine in the vitreous humour (the colorless jelly filling the eye) shows a postmortem linear increase in concentration with time since death. This suggests that hypoxanthine concentration might be useful in predicting time of death when this is unknown. The graph below shows measurements collected by the researchers on 48 subjects whose time of

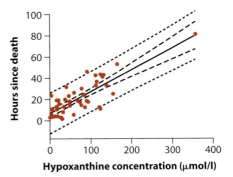

death was known. The regression line, the 95% confidence bands, and the 95% prediction interval are included on the graph.

a. The data set depicted in the graph includes one conspicuous outlier on the far right. If you were advising the forensic scientists who gathered these data, how would you suggest they handle the outlier?

b. What do the confidence bands measure?

c. Are the inner dashed lines the confidence bands or the prediction interval?

d. If the regression depicted in the graph was to be used to predict the time of death in a murder case, which bands would provide the most relevant measure of uncertainty, the confidence bands or the prediction interval? Why?

12. Social spiders live together in kin groups where they build communal webs and cooperate in gathering prey. The following web measurements were gathered on 17 colonies of the social spider *Cyrtophora citricola* in Gabon (Rypstra 1979),

a. Use these data to draw a scatter plot of the relationship between the colony height above ground and the number of spiders in the colony.

b. Examine the scatter plot and determine any impediments that might make it difficult to use linear regression to predict one variable from another.

c. In view of what you discerned in part (b), what method would you recommend to test whether an association is present between the colony height and the number of spiders?

Colony	Height of web above ground (cm)	Number of spiders
1	90	17
2	150	32
3	270	96
4	320	195
5	180	372
6	380	135
7	200	83
8	120	36
9	240	85
10	120	20
11	210	82
12	250	95
13	140	59
14	300	89
15	290	152
16	180	62
17	280	64

ASSIGNMENT PROBLEMS

13. Identify the assumption(s) of linear regression that is violated in each of the following residual plots.

a.

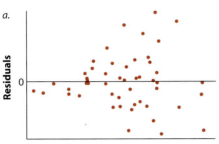

b.

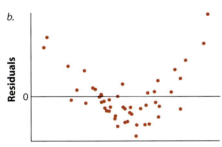

c.

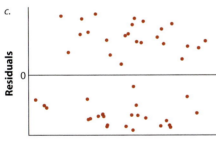

d.

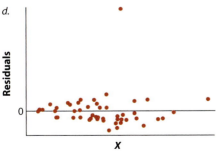

14. You might think that increasing the resources available would elevate the number of plant species that an area could support, but the evidence suggests otherwise. The data in the accompanying table are from the Park Grass Experiment at Rothamsted Experimental Station in the UK, where grassland field plots have been fertilized annually for the past 150 years (collated by Harpole and Tilman 2007). The number of plant species recorded in 10 plots is given in response to the number of different nutrient types added in the fertilizer treatment (nutrients types include nitrogen, phosphorus, potassium, and so on).

a. Draw a scatter plot of these data. Which variable should be the explanatory variable (X) and which should be the response variable (Y)?

Plot	Number of nutrients added	Number of plant species
1	0	36
2	0	36
3	0	32
4	1	34
5	2	33
6	3	30
7	1	20
8	3	23
9	4	21
10	4	16

b. What is the rate of change in the number of plant species supported per nutrient type added? Provide a standard error for your estimate.

c. Add the least squares regression line to your scatter plot. What fraction of the variation in the number of plant species is "explained" by the number of nutrients added?

d. Test the null hypothesis of no treatment effect on the number of plant species.

15. The forests of the northern US and Canada have no native terrestrial earthworms, but many exotic species (including those used as bait when fishing) have been introduced. These immigrant species are dramatically changing the soil. The following data were gathered to predict the nitrogen content of mineral soils of 39 hardwood forest plots in Michigan's Upper Peninsula from the number of earthworm species found in those plots (Gundale et al. 2005).

Number of earthworm species	Total nitrogen content (%)
0	0.22, 0.19, 0.16, 0.08, 0.05
1	0.33, 0.30, 0.26, 0.24, 0.20, 0.18, 0.14, 0.13, 0.11, 0.09, 0.08
2	0.27, 0.24, 0.23, 0.18, 0.16, 0.13
3	0.32, 0.32, 0.29, 0.24, 0.22, 0.12, 0.40
4	0.34, 0.33, 0.23, 0.21, 0.18, 0.17, 0.15, 0.14
5	0.20, 0.54

a. Draw a scatter plot of these data, with the number of earthworm species as the explanatory variable.

b. Using the following intermediate calculations, calculate the regression line to predict the total nitrogen content of the soil from the number of earthworm species present. Add the line to your plot.

$$\bar{X} = 2.205 \qquad \bar{Y} = 0.215$$

$$\sum (X - \bar{X})^2 = 86.359$$

$$\sum (Y - \bar{Y})^2 = 0.366$$

$$\sum (X - \bar{X})(Y - \bar{Y}) = 2.453.$$

c. What are the units of your estimate of slope, b?

d. What is the predicted nitrogen content of soil having five earthworm species?

e. Calculate a standard error of the slope.

f. Produce a 95% confidence interval for the slope.

16. Huesner (1991) assembled the following data on the mass and basal metabolic rate of 17 species of primates.

Species	Mass (g)	Basal metabolic rate (watts)
Alouatta palliata	4670.0	11.6
Aotus trivirgatus	1020.0	2.6
Arctocebus calabarensis	206.0	0.7
Callithrix jachus	190.0	0.9
Cebuella pygmeaa	105.0	0.6
Cheirogaleus medius	300.0	1.1
Euoticus elegantilus	261.5	1.2
Galago crassicaudatus	1039.0	2.9
Galago demidovii	61.0	0.4
Galago elegantulus	261.5	1.2
Homo sapiens	70,000.0	82.8
Lemur fulvus	2330.0	4.2
Nycticebus coucang	1300.0	1.7
Papio anubis	9500.0	16.0
Perodicticus potto	1011.0	2.1
Saguinus geoffroyi	225.0	1.3
Saimiri sciureus	800.0	4.4

Previous research has indicated that basal metabolic rate (R) of mammal species depends on body mass (M) in the following way: $R = \alpha M^\beta$, where α and β are constants.

a. Use linear regression to estimate β for primates. Call your estimate "b."

b. Plot your line and the data in a scatter plot.

c. Provide a standard error for b and a 95% confidence interval for β. Assume that the species data are independent.

17. Previous evidence and some theory predict that the exponent β describing the relationship between metabolic rate and mass should equal 3/4. Using the data from Assignment Problem 16, test whether the exponent differs from the expected value of 3/4.

18. The white forehead patch of the male collared flycatcher is important in mate attraction. Griffith and Sheldon (2001) found that the length of the patch varied from year to year. They measured the forehead patch on a sample of 30 males in two consecutive years, 1998 and 1999, on the Swedish island of Gotland. The scatter plot provided gives the pair of measurements for each male. The solid regression line predicts the 1999 measurement from the 1998 measurement. The dashed line is drawn through the means for 1998 and 1999, but it has a slope of 1. The difference between the two lines indicates that males with

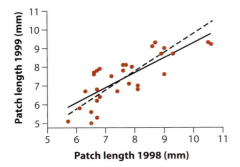

Tree	Mean fleck duration (min)	Relative growth rate (mm/mm/week)
1	3.4	0.013
2	3.2	0.008
3	3.0	0.007
4	2.7	0.005
5	2.8	0.003
6	3.2	0.003
7	2.2	0.005
8	2.2	0.003
9	2.4	0.000
10	4.4	0.009
11	5.1	0.010
12	6.3	0.009
13	7.3	0.009
14	6.0	0.016
15	5.9	0.025
16	7.1	0.021
17	8.8	0.024
18	7.4	0.019
19	7.5	0.016
20	7.5	0.014
21	7.9	0.014

the longest patches in 1998 had smaller patches in 1999, relative to the other birds. Similarly, the males with the smallest patches in 1998 had larger patches, on average, in 1999, relative to other birds.

a. The following table summarizes the data. Use these numbers to calculate the regression slope.

	Mean	Sum of squares	Sum of products
Patch length 1998	7.62	45.43	36.26
Patch length 1999	7.40	47.03	

b. Now let the patch length in 1998 be the response variable (Y). Use the patch length in 1999 to predict patch length in 1998. What is the slope of this new regression?

c. What is the most likely reason why the slope is less than one in both regressions?

19. Seedlings of understory trees in mature tropical rainforests must survive and grow using intermittent flecks of sunlight. How does the length of exposure to these flecks of sunlight (fleck duration) affect growth? Leakey et al. (2005) experimentally irradiated seedlings of the Southeast Asian rainforest tree *Shorea leprosula* with flecks of light of varying duration while maintaining the same total irradiance to all the seedlings. Their data for 21 seedlings are listed in the following table.

$\bar{X} = 5.062 \qquad \bar{Y} = 0.0111$

$$\sum (X - \bar{X})^2 = 100.210$$

$$\sum (Y - \bar{Y})^2 = 0.001024$$

$$\sum (X - \bar{X})(Y - \bar{Y}) = 0.2535.$$

a. What is the rate of change in relative growth rate per minute of fleck duration? Provide a standard error for your estimate.

b. Using these data, test the hypothesis that fleck duration affects seedling growth rate.

c. Calculate a 99% confidence interval for the slope of the regression.

d. What are your assumptions in parts (a)–(c)?

e. What is the main procedure you would employ to evaluate those assumptions?

20. How do we estimate a regression relationship when each subject is measured multiple times over a series of *X*-values? The easiest approach is to use a *summary* slope for each individual and then calculate the average slope. Green et al. (2001) dealt with exactly this type of problem in their study of macaroni penguins exercised on treadmills. Each penguin was exercised at a range of speeds, and its oxygen consumption was measured in relation to its heart rate (a proxy for metabolic rate). The graph provided shows the relationship for just two individual penguins.

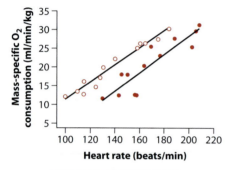

The following table lists the estimated regression slopes for each of 24 penguins in three categories.

Group	Regression slope
Breeding males	0.31, 0.34, 0.3, 0.38, 0.35, 0.33, 0.32, 0.32, 0.37
Breeding females	0.30, 0.32, 0.23, 0.38, 0.31, 0.26, 0.42, 0.28, 0.35
Molting females	0.25, 0.41, 0.32, 0.34, 0.27, 0.23

a. Calculate the mean, standard deviation, and sample size of the slope for penguins in each of the three groups. Display your results in a table.

b. Test whether the means of the slopes are equal between the three groups.

21. Many species of beetle produce large horns that are used as weapons or shields. The resources required to build these horns, though, might be diverted from other useful structures. To test this, Emlen (2001) measured the sizes of wings and horns in 19 females of the beetle species *Onthophagus sagittarius*. Both traits were scaled for body-size differences and hence are referred to as relative horn and wing sizes. Emlen's data are shown in the following scatter plot along with the least squares regression line ($Y = -0.13 - 132.6\,X$).

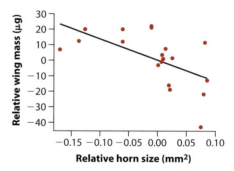

We used this regression line to predict the horn lengths at each of the 19 observed horn sizes. These are given in the following table along with the raw data.

Relative horn size (mm²)	Relative wing mass (µg)	Predicted relative wing mass (µg)
0.074	−42.8	−9.9
0.079	−21.7	−10.6
0.019	−18.8	−2.6
0.017	−16.0	−2.4
0.085	−12.8	−11.4
0.081	11.6	−10.9
0.011	7.6	−1.6
0.023	1.6	−3.2
0.005	3.7	−0.8
0.007	1.1	−1.1
0.004	−0.8	−0.7
−0.002	−2.9	0.1
−0.065	12.1	8.5
−0.065	20.1	8.5
−0.014	21.2	1.7
−0.014	22.2	1.7
−0.132	20.1	17.4
−0.143	12.5	18.8
−0.177	7.0	23.3

a. Use these results to calculate the residuals.
b. Use your results from part (a) to produce a residual plot.
c. Use the graph provided and your residual plot to evaluate the main assumptions of linear regression.
d. In light of your conclusions in part (c), what steps should be taken?

22. The parasitic bacterium *Pasteuria ramose* castrates and later kills its host, the crustacean *Daphnia magna*. The length of time between infection and host death affects the number of spores eventually produced and released by the parasite, as the following scatter plot reveals. The X-axis measures age at death for 32 infected host individuals, and the response variable is the square-root transformed number of spores produced by the infecting parasite (Jensen et al. 2006).

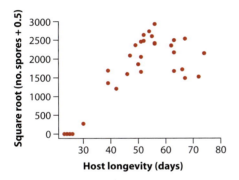

a. Describe the shape of the relationship between the number of spores and host longevity.
b. What equation would be best to try first if you wanted to carry out a nonlinear regression of *Y* on *X*?

23. Human brains have a large frontal cortex with excessive metabolic demands compared with the brains of other primates. However, the human brain is also three or more times the size of the brains of other primates. Is it possible that the metabolic demands of the human frontal cortex are just an expected consequence of greater brain size? Sherwood et al. (2006) investigated this question in a number of ways. Their data in the accompanying scatter plot shows the relationship between the glia–neuron ratio (an indirect measure of the metabolic requirements of brain neurons) and the log-transformed brain mass in non-human primates. A linear regression is drawn through these data. The black dot at the upper right indicates the human measurements. Compared with other primates, does the human brain have an excessive glia–neuron ratio for its mass? Explain the reasoning behind your answer.

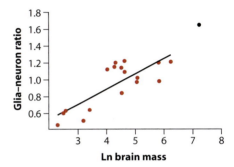

24. Golenda et al. (1999) carried out a human clinical trial to investigate the effectiveness of a formulation of DEET (*N,N*-diethyl-*m*-toluamide) in preventing mosquito bites. The study applied DEET to the underside of the left forearm of volunteers. Cages containing 15 fresh mosquitoes were then placed over the skin for five minutes, and the number of bites was recorded. This was repeated four times at intervals of three hours. The scatter plot provided displays the total number of bites (square-root transformed) received by 52 women in the study in relation to the dose of DEET they received.

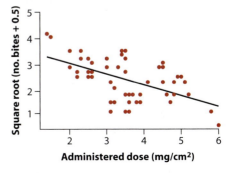

a. What are the uses of the square-root transformation in linear regression?

b. What feature of this study justifies our calling it an *experimental* study rather than just an observational study?

c. Complete the ANOVA table for these data:

Source of variation	Sum of squares	df	Mean squares	F-ratio
Regression	9.97315			
Residual				
Total	32.0569			

d. Use the *F*-statistic to test the null hypothesis of zero slope.

e. Calculate the R^2 statistic. What does it measure?

25. Calculating the year of birth of cadavers is a tricky enterprise. Recently, a method has been proposed based on the radioactivity of the enamel of the body's teeth. The proportion of the radioisotope ^{14}C in the atmosphere increased dramatically during the era of above-ground nuclear bomb testing between 1955 and 1963. Given that the enamel of a tooth is non-regenerating, measuring the ^{14}C content of a tooth tells when the tooth developed, and therefore the year of birth of its owner. Predictions based on this method seem quite accurate (Spalding et al. 2005), as shown in the accompanying graph.

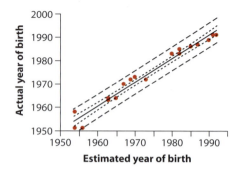

There are three sets of lines on this graph. The solid line represents the least squares regression line, predicting the actual year of birth from the estimate based on tooth radioactivity. One pair of dashed lines shows the 95% confidence bands and the other shows the 95% prediction interval.

a. What is the approximate slope of the regression line?

b. Which pair of lines shows the confidence bands? What do these confidence bands tell us?

c. Which pair of lines shows the prediction interval? What does this prediction interval tell us?

Using species as data points

Many types of studies in biology use species measurements as data points. We've encountered several cases in this book. In Chapter 2, for instance, we looked at the frequency distribution of bird abundances using the abundance measurements of different bird species as data points (Example 2.1B). In Chapter 16, we illustrated the association between brain and body mass in mammals using measurements of different mammal species as data points. What we haven't told you yet is how tricky it can be to analyze such data. The trouble is that species data are not usually *independent*. The reason is that species share an evolutionary history. Here we explain the situation and what can be done about it.

The following study of lilies illustrates the problem created by shared evolutionary history. Patterson and Givnish (2002) found that lily species flowering in the low-light environment of the forest understory, such as the blue bead lily (*Clintonia borealis*; below left), tend to have small and inconspicuous flowers

whitish or greenish in color. Lilies that live in sunny, open habitats, or that live in deciduous woods but flower before the tree leaves come out, such as the Turk's-cap lily (*Lilium superbum*; below right), tend to have large, showy flowers. Data from 17 lily species, shown in the branching figure on the next page, indicate an almost perfect association between habitat and flower type. All 10 species flowering in open habitats had large and showy flowers. Six of the seven species flowering in shaded habitats had relatively small and inconspicuous flowers. A χ^2 contingency test with these data soundly rejects the null hypothesis of no association ($\chi^2 = 13.24$, $df = 1$, $P = 0.0003$).

However, this contingency test assumes that the data from the 17 species of lilies are independent. The figure indicates that this assumption is likely false. The branching tree in the figure is a *phylogeny*, indicating the ancestor–descendant relationships among the 17 lily species in the data set, which are at the tips of the tree. Branching points, or nodes,

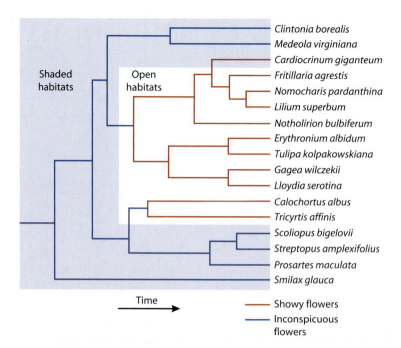

Shaded habitats

Open habitats

Clintonia borealis
Medeola virginiana
Cardiocrinum giganteum
Fritillaria agrestis
Nomocharis pardanthina
Lilium superbum
Notholirion bulbiferum
Erythronium albidum
Tulipa kolpakowskiana
Gagea wilczekii
Lloydia serotina
Calochortus albus
Tricyrtis affinis
Scoliopus bigelovii
Streptopus amplexifolius
Prosartes maculata
Smilax glauca

Time

—— Showy flowers
—— Inconspicuous flowers

in the tree represent points in history when a single ancestor split into two descendant species. Two lily species at the tips of the tree are relatively closely related if they have a recent ancestor in common, such as *Nomocharis pardanthina* and *Lilium superbum*. Two species are more distantly related if their common ancestor is deeper in the tree, such as *Lilium superbum* and *Prosartes maculata*. The color of the branches in the figure indicates flower type, and the background shading indicates habitat type. (We can only guess as to what the habitat types and flower types were of the ancestors because they are not alive any more. The colors and shading used in the figure represents just one of the more likely possible scenarios for the transitions in habitat and flower type through history.)

The crucial insight from the tree is that closely related lily species tend to have the same flower type. They also tend to have the same habitat type. Both attributes were likely inherited from their common ancestor. In all, *closely related species are more similar on average than species picked at random*. This means that the species data points are not independent. The situation is like that confronted when pollsters interview more than one person from the same household, or when a bird researcher measures more than one chick from the same nest. However, in the present case, the non-independence is not the fault of the way the species were sampled by the researcher. It is generated by the process of evolution. This is a problem unique to biology.

The preceding example is about discrete variables, but the same issue arises when species data are numerical. The following extreme case makes the point. The branching tree in the figure on page 511 shows a phylogeny for 10 hypothetical species. In this example, the species fall into two lineages that split from a common ancestor a long time ago

Species

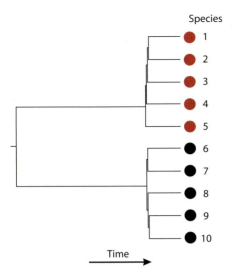

1
2
3
4
5
6
7
8
9
10

Time

(at the first branching point at the far left of the figure). The lineages have been evolving separately every since. More recently, each lineage split into five new species more or less simultaneously. We've used distinct colors to represent species in the two groups.

Now consider two numerical variables, *X* and *Y*, measured on all 10 hypothetical species. It will often be the case that species in the same group (indicated by red or black) will be more similar to each other in these traits than to species in the alternative group, just because they share a more recent common ancestor. The scatter plot below shows example measurements for the 10 species. The symbols

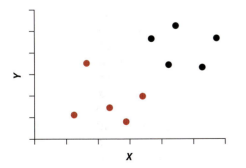

Y

X

in the scatter plot indicate the main lineage to which the species belong.

If we paid no attention to the evolutionary relationships among the 10 species represented by the different symbols, and if we assumed that the species data were independent, we would conclude that *X* and *Y* were positively correlated ($r = 0.69$, $df = 8$, $P = 0.016$). However, it is clear from the scatter plot that closely related species have similar values of *X* and of *Y*, a likely outcome of the common history they shared up until very recently. Within the two groups there appears to be no correlation between *X* and *Y*.

What can be done about the non-independence of species data resulting from shared evolutionary history? Happily, methods have been developed that correct for the problem. They make it possible to test whether an association is present and to put confidence limits on the strength of the association. The most widely used method for analyzing associations between continuously varying species traits is known as **phylogenetically independent contrasts**, invented by Felsenstein (1985). His explanation of how and why it works is very clear, and we refer you to his original paper for details. Analogous methods have been developed for discrete species traits.[1]

None of the methods that correct for the problem of non-independence of species traits are fool-proof, because all make assumptions that can be difficult to verify. For example, they all assume that the process of trait evolution

[1] Specialized computer programs are available to apply the method of phylogenetically independent contrasts to continuous data, such as COMPARE (Martins 2004) and MESQUITE (Maddison and Maddison 2007). Pagel (1994) invented the most widely used method for analyzing associations between discrete traits. Specialized programs available to carry out this method include MESQUITE (Maddison and Maddison 2007) and BAYESTRAITS (Pagel and Meade 2007).

through time can be adequately mimicked by a simple mathematical model of a "random walk." If the mathematical model badly describes the process of evolution in a specific instance, then using the method can be even worse than ignoring the problem of shared history altogether and just using conventional statistical methods. Biologists nowadays tend to cover all bases and analyze their data both

ways. Another strategy is to begin an analysis of species data by examining whether closely related species really are more similar on average than species picked at random. If not, then conventional statistical methods are adequate. If so, then the more specialized methods are used, often along with the results from conventional statistical methods, so that the outcomes can be compared.

18

Multiple explanatory variables

Up to this point in the book, we have discussed methods to predict a response variable from at most one explanatory variable. Many biological studies, however, investigate more than one explanatory variable at the same time. One reason for this is efficiency: for the same amount of work, or just a little more, we can obtain answers to more than one question if we include more than one explanatory variable in a study. For example, a study of the causes of nearsightedness in children might want to examine the role of both genes *and* environment rather than just one of them alone. The same sample of subjects can be used to address both questions.

The analysis of data having more than one explanatory variable follows the purpose and design of the study. We considered several study designs in Chapter 14, and three of these will serve as the basis of the present chapter. The first design is an experiment that includes *blocking* (Section 14.4) to improve the detection of treatment effects. The second design is the *factorial* experiment (Section 14.5), which is carried out to investigate the effects of two or more treatment variables (often called "factors") and their interaction. The third design adjusts for the effects of known confounding variables (also called "*covariates*"; see Interleaf 4 and Section 14.1) when comparing two or more groups.

In this chapter, we introduce an all-purpose method that can be used to analyze data from designs fulfilling all of these objectives and more. The method is called *general linear models*. General linear models is a large topic. Our purpose in this chapter is to introduce the basic elements by example using visual displays of data and models. Each analysis begins with a model statement. We will show you how model statements are constructed, and how the results are interpreted, when analyzing block and factorial designs, and when adjusting for a covariate. Our goal is to give you an overview of what is possible when there are multiple explanatory variables, but we don't have space here to describe all of the many important complications that arise when analyzing these kinds of study designs. To go further or to apply this approach to your own data, consult more specialized books, such as Grafen and Hails (2002).

18.1   From linear regression to general linear models

The basic idea behind general linear models is that a response variable Y can be represented by a linear model plus random error. By **model**, we mean a mathematical representation of the relationship between a response variable Y and one or more explanatory variables. The scatter of Y measurements around the model is random error. Linear regression is an example of a linear model (Chapter 17). Let's quickly review key elements of regression and then see how they can be generalized.

> A *model* is a mathematical representation of the relationship between a response variable Y and one or more explanatory variables.

Modeling with linear regression

The model of linear regression is a straight line,

$$Y = \alpha + \beta X,$$

where α and β are the intercept and slope, respectively. Individual values of the response variable are scattered above and below the regression line, representing random error. In Example 17.3, for example, we fit a linear regression to the relationship between the number of "dee" notes in the alarm calls of chickadees and the body mass of predator species presented to them. If we call the response variable DEES and the explanatory variable PREDMASS, then the equation with the variable names is

$$\text{DEES} = \alpha + \beta\,(\text{PREDMASS}).$$

Least squares yielded the best fit of this model to the data, which is drawn in the right panel of Figure 18.1-1. The data points are scattered above and below the best-fit line. The vertical line segments indicate the residuals (i.e., the vertical deviations between the data points and the predicted values on the regression line), which represent random error.

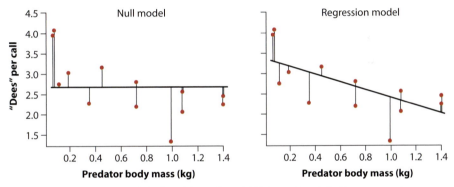

Figure 18.1-1 Comparison of the fits of the null model (*left panel*) and the linear regression model (*right panel*) to the chickadee data of Example 17.3. Vertical lines are the residuals, the deviations between the observed and predicted values from each model.

When we tested the null hypothesis that $\beta = 0$, we were really comparing the fit of the linear regression model to the data with the fit of a second model, the *null model*, in which the slope β is set to zero. The null model is a simplified linear model because setting β equal to zero removes the variable predator mass (PREDMASS) from the equation, yielding the model statement

$$\text{DEES} = \alpha',$$

where α' is just a constant like the intercept in the full regression model. Fitting this null model to the data using least squares yields the flat line depicted in the left panel of Figure 18.1-1; the best estimate of the constant is the estimated sample mean number of "dees" per call. Again, we've illustrated the residuals from the null model as vertical line segments between the data points and the predicted values on the line.

By visually comparing the two panels of Figure 18.1-1 you can see that the full regression model—the one that includes the explanatory variable PREDMASS—is the superior fit. The residuals are smaller than for the null model. Even if the null hypothesis were true, though, we expect the data to have a slope that departs from zero just by chance. The real question is whether the regression model fits the data *significantly* better than the null model. The *F*-ratio (Section 17.3) is used to evaluate this: if *F* is large, then the reduction in the magnitudes of the residuals represents a significant improvement in fit, and the null model is rejected.

Generalizing linear regression

The method of general linear models takes this regression approach and extends it in two key ways. First, it allows more explanatory variables to be included in the model. Second, the method incorporates *categorical* explanatory variables in the same framework, not just numerical explanatory variables as in regression.

For example, the linear model for single-factor ANOVA (one categorical variable) is

$$Y = \mu + A.$$

The constant μ is the grand mean, the average of all the observations combined, and A stands for the group or treatment effect. For each observation, A is the difference between the mean of its group and the grand mean. The sum of μ and A for an observation yields its corresponding group mean. The data themselves are scattered above and below the group means, representing random error.

The resemblance between the model for a single categorical variable and the model for linear regression highlights their fundamental similarity. Both models include a response variable and an explanatory variable. Both also have a constant term—namely, the intercept in linear regression, and the grand mean in the case of a categorical variable. The only real difference is that the explanatory variable is categorical for one model and numerical for the other.

A unified model encompassing both linear regression and single-factor ANOVA can therefore be stated using the following generic format:

$$\text{RESPONSE} = \text{CONSTANT} + \text{VARIABLE}.$$

RESPONSE is the response variable, which is numerical. CONSTANT refers to a constant, and can represent the intercept or mean, depending on whether the explanatory variable (VARIABLE) is categorical or numerical. To keep the format as generic as possible, the model statement leaves out the coefficients for the explanatory variable—namely, the slope in the case of a numerical explanatory variable, and group effects in the case of a categorical explanatory variable. Henceforth, we will use simple word statements like this to describe all types of general linear models. Model statements will be recognizably different mainly by the explanatory variables that are, or are not, included.

Analyzing a categorical treatment variable

Example 18.1 illustrates how categorical data are analyzed in the general linear-model framework.

Example 18.1   **I feel your pain**

Langford et al. (2006) discovered that lab mice experiencing pain appear to "empathize" with familiar mice also in pain. Their conclusions were based in part on the results of an experiment in which individual mice experienced a very mild pain as a result of injection of 0.9% acetic acid into the abdomen. These mice were placed in one of three different treatments: (1) isolation, (2) with a companion
mouse that was not injected, or (3) with a companion mouse also injected and exhibiting behaviors associated with pain. The response variable was the percentage of time that each treated mouse exhibited a characteristic "stretching" behavior (measured by abdominal constriction) when experiencing discomfort. The companion mice used in the second and third treatments were familiar (cage mates) to the target mouse. The data from males are summarized in Table 18.1-1. The results suggest that treated mice stretch the most when a companion mouse is also experiencing mild pain.

Table 18.1-1 The percentage of time that male mice experiencing mild pain spent "stretching" in different companion treatments.

Companion treatment	*n*	Mean percent time stretching $\overline{Y}$	Standard deviation *s*
Isolation	17	37.2	18.3
Companion mouse (not injected)	13	35.4	20.3
Companion mouse (injected)	12	58.1	11.5

We can analyze these data with a general linear model as follows. The percentage of time spent stretching (call it STRETCHING) is our response variable, and the companion treatment (COMPANION) is the explanatory variable of interest. The model statement can be written as

$$\text{STRETCHING} = \text{CONSTANT} + \text{COMPANION}.$$

This word statement is similar to how most statistical packages on the computer require you to enter it if you want to analyze your data with a general linear model.[1] The CONSTANT term stands for the grand mean. COMPANION represents the treatment variable, indicating the group to which individual mice belong. If there is a treatment effect, the COMPANION group mean will differ. The hypotheses are

H_0: Companion treatment means are the same.
H_A: Companion treatment means are not the same.

As in linear regression, the significance of a treatment variable in a general linear model is tested by comparing the fit of *two* models to the data using an *F*-test. The two models correspond to the null and alternative hypotheses. The model that includes the COMPANION term represents the alternative hypothesis, that there is an effect of the COMPANION treatment on stretching behavior. The other model is the null hypothesis. Under the null hypothesis, all the group means are the same (i.e., there is no treatment effect). In this case, the COMPANION variable contributes nothing and is removed from the model, yielding

STRETCHING = CONSTANT.

Figure 18.1-2 shows the fits of the null (left panel) and alternative (right panel) models to the mouse data. Horizontal lines give the predicted values (analogous to the $\hat{Y}$ in linear regression) under the two models. Vertical lines are the residuals, the deviations between the observed and the predicted values for the response variable[2] (equivalent to the residuals in regression).

Even if the null hypothesis were true, we expect the full model—the one including the treatment variable COMPANION—to fit the data best, because the means of the different groups will not be identical simply by chance. The real question is whether the model that includes the COMPANION variable fits the data *significantly* better than the null model.

The *F*-ratio is used to test whether including the treatment variable (COMPAN-ION) in the model results in a significant *improvement* in the fit of the model to the data, compared with the fit of the null model lacking the treatment variable. The test is summarized in Table 18.1-2. The *P*-value (0.0032) indicates whether the improvement in fit is sufficiently large to warrant rejection of H_0 in favor of the alternative model.

---

[1] You would provide the real variable names from your data spreadsheet when generating the model statement on the computer. For example, if the companion treatment variable was named "Companion.Males.Familiar" in your data set, then this is what you should use in the model statement instead of COMPANION. Some packages might not require you to type the CONSTANT term in the word statement for the sake of brevity, but a constant is nevertheless included in the analysis.

[2] In single-factor ANOVA, the predicted value for every data point under the null model is just the grand mean. The predicted value for an observation under the alternative model is the mean for its group.

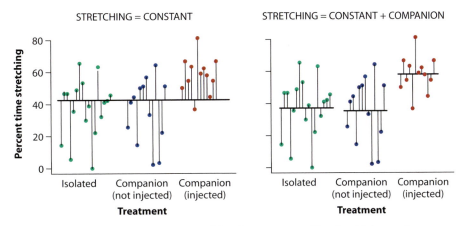

Figure 18.1-2 The fit of a general linear model to the mouse stretching data (*right panel*) compared with the fit of the null model (*left panel*). Groups (i.e., isolated, companion, and injected companion) are indicated with different colors. Points within each group are spread out horizontally to increase visibility. Vertical lines are residuals between the observed points and the values predicted by the model (*horizontal lines*).

Table 18.1-2 ANOVA results for the mouse data (Example 18.1).

Source of variation	Sum of squares	df	Mean squares	F	P
COMPANION	4,040.92	2	2020.46	6.67	0.0032
Residual	11,807.43	39	302.75		
Total	15,848.35	41			

In this example, H_0 is soundly rejected ($F = 6.67$, $df = 2,39$, $P = 0.0032$). We conclude, therefore, that the model including the treatment variable (COMPANION) fits the data better than the null model of no treatment effect.

> *F* measures the improvement in the fit of a general linear model to the data when the term describing a given explanatory variable is included in the model, compared with the null model in which the term is absent.

18.2 Analyzing experiments with blocking

Here we show how to analyze an experiment designed with one treatment variable plus a blocking variable. We use general linear models because the method is readily extended to this type of design. Blocking (Section 14.4) results in an additional variable (block) that must be included in the analysis, which requires an analytic approach that goes beyond a single-factor ANOVA.

Analyzing data from a randomized block design

A randomized block design is like a paired design but for more than two treatments. In the typical randomized block design, such as Example 18.2, every treatment is replicated exactly once within each block.

Example 18.2 ## Zooplankton depredation

Svanbäck and Bolnick (2007) set up a field experiment in the shallows of a small lake on Vancouver Island (British Columbia) that allowed them to measure how fish abundance affects the abundance and diversity of prey species. They used a randomized block design to minimize noise caused by background variation in prey availability between locations in the lake. Three fish abundance treatments, Low, High, and Control, were set up at each of five locations in the lake. In the Low treatment, 30 small fish were added to a mesh cage having a 3 m × 3 m surface area. In the High treatment, 90 fish were added to a second cage nearby. An unenclosed space of equal area adjacent to each pair of enclosures served as the Control. Table 18.2-1 shows the diversity of zooplankton prey in each treatment at the five locations after 13 days. Diversity is measured using an index called Levins' D, which takes both the number of species and their rarity into account (e.g., common species count more than rare species). Does treatment affect zooplankton diversity?

To analyze these data, we can't simply combine all 15 measurements into three treatment groups, because the three measurements made at each location in the lake are not independent. We ran into the same issue in Section 12.2 when analyzing paired measurements. Instead, we need to designate each location as a distinct category of a blocking variable and include it in the analysis. If the blocking variable accounts for some of the variation in the data, then it can improve our ability to detect an effect of the treatment of interest. As in a paired t-test, treatment effects are assessed by the differences in response to different treatments within each block.

Table 18.2-1 Zooplankton diversity D in three fish abundance treatments.

Abundance treatment	Location (block)				
	1	2	3	4	5
Control	4.1	3.2	3.0	2.3	2.5
Low	2.2	2.4	1.5	1.3	2.6
High	1.3	2.0	1.0	1.0	1.6

Model formula

Let's call the abundance treatment ABUNDANCE in our statement of the general linear model. BLOCK is the individual location. Finally, let's call the response variable DIVERSITY. The full linear model, including all terms, is

$$DIVERSITY = CONSTANT + BLOCK + ABUNDANCE.$$

This linear model resembles one for single-factor ANOVA, except that we've added a blocking variable. Fish abundance treatment is the factor we are interested in, so the null and alternative hypotheses are

H_0: Mean zooplankton diversity is the same in every abundance treatment.
H_A: Mean zooplankton diversity is not the same in every abundance treatment.

The linear model, including all of the terms, represents the alternative hypothesis, whereas the model representing the null hypothesis (the null model) leaves out ABUNDANCE, the term specified by the null hypothesis:

$$DIVERSITY = CONSTANT + BLOCK.$$

Fitting the model to data

The graphs in Figure 18.2-1 visually compare the fits of the null (left panel) and alternative (right panel) models. (To minimize clutter we've left out the vertical lines for the residuals.) The predicted values under the full model (right panel) suggest that the diversity of zooplankton was lowest in the High treatment and highest in the Control. The predicted values under the null model (left panel) lie on five horizontal lines rather than one line, as in Figure 18.1-2, because the null model for a test of the effect of ABUNDANCE still includes the blocking variable, and there are five blocks.

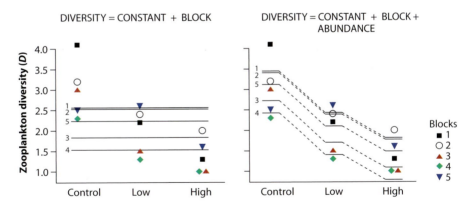

Figure 18.2-1 Comparison of the fits of two linear models to the zooplankton diversity data. Symbols indicate blocks (locations). Horizontal lines are predicted values, with numbers indicating blocks. The left panel shows the fit of the null model, including the CONSTANT and BLOCK terms. The right panel shows the fit of the "full" model, which includes the ABUNDANCE treatment variable.

The ANOVA table provides the results of the test (Table 18.2-2). Adding ABUNDANCE significantly improves the fit of the linear model ($F = 16.37$, $df = 2,14$, $P = 0.001$).

Table 18.2-2 ANOVA results of fitting the linear model to the zooplankton diversity data.

Source of variation	Sum of squares	df	Mean square	F	P
BLOCK	2.3400	4	0.5850	2.79	0.101
ABUNDANCE	6.8573	2	3.4287	16.37	0.001
Residual	1.6760	8	0.2095		
Total	10.8733	14			

The ANOVA table also tests the effect of BLOCK. This result is of less interest than that for ABUNDANCE, the treatment variable. Blocking is included in the analysis only to reduce the effect of variation between locations in zooplankton diversity when testing the effect of the fish abundance treatment. The F-ratio in the ANOVA table corresponding to BLOCK (i.e., 2.79) represents the improvement in the fit of the general linear model when the BLOCK term is included in the model, compared with the fit when the BLOCK term is deleted from the model (but all other terms, including ABUNDANCE, are retained). Whether significant or not, BLOCK should still be retained in the analysis because it is there by design, and because it can still improve the ability to detect a treatment effect.

18.3 Analyzing factorial designs

Here we illustrate the analysis of data from a factorial experiment, an experiment in which all combinations of the values of two (or more) explanatory variables are investigated. Factorial experiments investigate the effects of the factors and their interaction on a response variable (Section 14.5). The explanatory variables are called "factors" because they represent treatments of direct interest. (In contrast, a blocking variable is not considered a factor because it is not of direct interest—a block is only included to improve the detection of treatment effects.)

We restrict ourselves to an example involving two *fixed* factors. As discussed in Section 15.5, the different categories of a fixed factor are predetermined, of direct interest, and repeatable. The analysis is different when one or more of the factors is *random*, instead. A random factor is a variable whose groups are not predetermined, but instead are randomly sampled from a "population" of groups. In our example, the study design is balanced. Special issues arise when sample sizes are unequal between treatment combinations, but these are beyond the scope of this chapter.

Analysis of two fixed factors

Here we examine the analysis of a two-factor experiment in which both factors are fixed. The data in Example 18.3 are from a field experiment that examined the interaction of two environmental factors on plant growth in an ocean habitat.

Example 18.3	Interaction zone

Harley (2003) investigated how herbivores affect the abundance of plants living in the intertidal habitat of coastal Washington using field transplants of a red alga, *Mazzaella parksii*. The experiment also examined whether the effect of herbivores on the algae depended on where in the intertidal zone the plants were growing. Thirty-two study plots were established

just above the low tide mark, and another 32 plots were set at mid-height between the low and high tide marks. The plots were cleared of all organisms, and then a constant amount of new algae was glued onto the rock surface in the center of each plot. Using copper fencing, herbivores (mainly limpets and snails) were excluded from a randomly chosen half of the plots at each height. The remaining plots were left accessible to herbivores. In the photo on the right, clumps of algae are shown placed in the centers of two plots, one with a copper ring and the other without a ring. The design was balanced and included every combination of the height and herbivore treatments (the data are summarized in Table 18.3-1). At the end of the experiment, the surface area covered by the algae (in cm²) was measured in each plot. The data were square-root transformed to improve the fit to the assumptions of normal distributions with equal variance. Means and standard errors are shown in Figure 18.3-1.

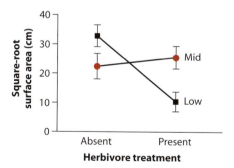

Figure 18.3-1 "Interaction plot" showing the mean surface area of algae at every combination of treatments. Lines connect the means of plots in the different herbivory treatments, separately for each intertidal height category. Error bars represent one standard error above and below each mean. Original units are cm², and $n = 16$ in each treatment combination.

The experiment had two factors, herbivory treatment and height. Figure 18.3-1 suggests that an interaction between these factors might be present, because herbivore treatment had a strong effect on algal cover at low height, but little, or even an opposite effect, at mid-height. We can fit a general linear model to the data to decide which effects are statistically significant.

Model formula

We'll use ALGAE to indicate the response variable, the square-root-transformed algal cover. HERBIVORY and HEIGHT refer to the two factors of interest. Finally, HERBIVORY*HEIGHT will indicate the interaction between HERBIVORY and HEIGHT. The full general linear model is then written as

ALGAE = CONSTANT + HERBIVORY + HEIGHT + HERBIVORY*HEIGHT.

This formula captures every effect that can be examined from the data. A significant HERBIVORY term would indicate that the growth rate of algae differs between HERBIVORY treatments, when averaged over HEIGHT categories. A significant HEIGHT term would indicate that algal growth differs between HEIGHT levels, when averaged over the HERBIVORY treatments. HEIGHT and HERBIVORY are known as the **main effects** because each represents the effects of that factor alone, when averaged over the categories of the other factor. The HERBIVORY*HEIGHT term in the model represents the differences in slope between line segments in the interaction plot (Figure 18.3-1). The interaction term will be different from zero if the effect of HERBIVORY on algal growth is different for different values of HEIGHT.

Fitting the model to data

An *F*-test from an analysis of variance is used to examine the contribution of each main effect and their interaction to the fit of the model to the data. There are three sets of null and alternative hypotheses to be tested:

HERBIVORY (main effect):

> H_0: There is no difference between herbivory treatments in mean algal cover.
> H_A: There is a difference between herbivory treatments in mean algal cover.

HEIGHT (main effect):

> H_0: There is no difference between height treatments in mean algal cover.
> H_A: There is a difference between height treatments in mean algal cover.

HERBIVORY*HEIGHT (interaction effect):

> H_0: The effect of herbivory on algal cover does not depend on height in the intertidal zone.
> H_A: The effect of herbivory on algal cover depends on height in the intertidal zone.

Each of these sets of hypotheses is tested by comparing the fit of the full model to the data with the fit when the term of interest is deleted from the model. To test the interaction term, for example, the fit of the full model is compared with a null model in which the main effects are present but the interaction is absent:

$$ALGAE = CONSTANT + HERBIVORY + HEIGHT.$$

The fits of these two models are depicted in Figure 18.3-2.

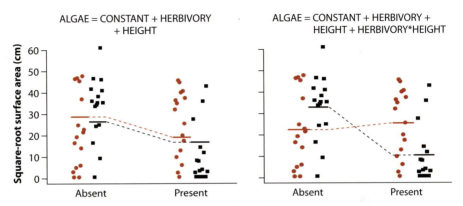

Figure 18.3-2 Visual depiction of the fit of the full model (*right panel*) compared with the fit of the null model (*left panel*), which lacks HERBIVORY*HEIGHT, the interaction term. Horizontal lines are predicted values under each model. Red circles indicate mid-height above low tide, whereas black squares indicate low height. Points in each combination of herbivory and height treatments are spread out to reduce overlap.

Table 18.3-1 ANOVA results of fitting the two-factor model to the herbivory data.

Source of variation	Sum of squares	df	Mean square	F	P
HERBIVORY	1512.18	1	1512.18	6.36	0.014
HEIGHT	88.97	1	88.97	0.37	0.543
HERBIVORY*HEIGHT	2616.96	1	2616.96	11.00	0.002
Residual	14,270.52	60	237.842		
Total	18,488.63	63			

Each combination of herbivory and height treatment represents a separate group in a panel. Under the full model (Figure 18.3-2, right panel), which includes the HERBIVORY*HEIGHT interaction term, the predicted values are the group sample means. The null model (Figure 18.3-2, left panel) lacks the interaction term, which constrains the difference between predicted values at low and mid-heights to be the same in both herbivory treatments. This leads to greater residuals (i.e., greater vertical distances between points and predicted values) under the null model than under the full model. Yet, is the fit of the full model, which includes the interaction term, *significantly* better than the fit of the null model? According to an *F*-test (Table 18.3-2), the interaction term is indeed significant, so the null hypothesis of no interaction is rejected ($F = 11.00$, $df = 1,60$, $P = 0.002$).

The other two *F*-ratios in Table 18.3-1 test the significance of the main effects, HERBIVORY and HEIGHT. Each test compares the fit of the full model with that of a null model in which the term of interest is removed (but all other terms are still included). The results show that the null hypothesis of no HERBIVORY main effect is also rejected ($F = 6.36$, $df = 1,60$, $P = 0.014$). This effect is evident in the interaction plot by a higher overall mean algal cover in the absence of herbivores than in their presence (Figure 18.3-1). No significant main effect of HEIGHT was detected ($F = 0.37$, $df = 1,60$, $P = 0.543$). This is just the main effect, however; HEIGHT still interacted with HERBIVORY to influence algal growth (Figure 18.3-1). Height in the intertidal zone affects algal cover, but only by affecting the way herbivory acts on algal cover.

The importance of distinguishing fixed and random factors

F-tests to compare the fits of null and alternative models to the data are straightforward when all factors are fixed, as in Example 18.3. When one or more factors are random, however, a subtle change occurs that affects how F-ratios are calculated. The change is required because the groups are randomly sampled in the case of the random factor, which contributes sampling error to the design. This sampling error adds noise to the measurement of differences between group means for other factors that interact with the random factor. The F-ratios must be calculated differently to compensate. Most statistics packages assume that all factors are fixed until instructed otherwise. Designating factors as random takes some extra work on your part (you might need to consult the manual of your statistics package to figure out how to do this). If factors are not properly identified as fixed or random, the results given by the computer program will be wrong. Consult more advanced statistics references, such as Sokal and Rohlf (2008), for details on how random factors influence the expected mean squares for treatment effects.

18.4 Adjusting for the effects of a covariate

Our last application uses general linear models to adjust treatment effects to account for a known confounding variable. Confounding variables bias estimates of treatment effects, as we described in Interleaf 4 and Chapter 14. Experimental studies eliminate such biases by randomly assigning treatments to experimental units, but often experiments are not feasible. The best strategy in an observational study is to include known confounding variables in the analysis and to "correct" for their distorting influence on the estimation of the treatment effect.

Here we show the simplest case involving a single treatment variable and a single numerical confounding variable, often called a "covariate." This is the first time in this chapter that we've included both a categorical and a numerical variable in the same model. The approach involves two rounds of model fitting. In the first round, the interaction between the treatment and the covariate is tested. If no interaction is detected, then the interaction term is dropped from the model in the second round, in which the treatment effect is tested. The method is also called "analysis of covariance" or ANCOVA.

Example 18.4   ### Mole-rat layabouts

Mole-rats are the only known mammals with distinct social castes. A single queen and a small number of males are the only reproducing individuals in a colony. Remaining individuals, called workers, gather food, defend the colony, care for the young, and maintain the burrows. Recently, it was discovered that there might be two worker castes in the Damaraland mole-rat (*Cryptomys damarensis*). "Frequent workers" do almost all of the

work in the colony, whereas "infrequent workers" do little work except on rare occasions after rains, when they extend the burrow system. To assess the physiological differences between the two types of workers, Scantlebury et al. (2006) compared daily energy expenditures of wild mole-rats during a dry season. Energy expenditure appears to vary with body mass in both groups (Figure 18.4-1), but

infrequent workers are heavier than frequent workers. How different is mean daily energy expenditure between the two groups when adjusted for differences in body mass?

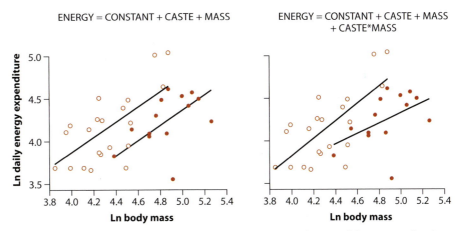

Figure 18.4-1 Log-transformed daily energy expenditure and body mass of "frequent workers" (*open circles*, $n_1 = 21$) and "infrequent workers" (*red-filled circles*, $n_2 = 14$) of Damaraland mole-rats in a dry season. Original units for the two measurements are kJ/day and g. Predicted values in the right panel include the CASTE*MASS interaction term, whereas the null model (the left panel) lacks the interaction term.

Testing interaction

To analyze these data, we used the log transformation of daily energy expenditure and body mass to improve the fit to the assumptions of general linear models. We'll call the log-transformed response variable ENERGY and the log-transformed body mass MASS. The factor of interest is CASTE, whereas MASS is the covariate. All together, the full general linear model is

$$\text{ENERGY} = \text{CONSTANT} + \text{CASTE} + \text{MASS} + \text{CASTE*MASS}.$$

CASTE*MASS indicates the interaction. The MASS variable is numerical, whereas the CASTE variable is categorical. The model describes a linear regression of ENERGY on MASS, separately for each category of CASTE, as shown by the predicted values for each worker caste in Figure 18.4-1 (right panel).

The question of interest asked whether energy expenditure differed between worker castes after adjusting for their differences in size. The easiest way to answer this question is to assume that the regression lines predicting energy expenditure from body mass have the same slope in the two castes. Therefore, the first step when adjusting for a covariate is a test of equal slopes. This is a test of the interaction term in the general linear model. The hypotheses for the interaction term are

H_0: There is no interaction between caste and body mass.
H_A: There is an interaction between caste and body mass.

To test the hypotheses, we compared the fit of the full model, which contains the CASTE*MASS interaction term, with that of the null model (ENERGY = CONSTANT + CASTE + MASS), which lacks the interaction.

The fit of the null model to the data is illustrated in the left panel of Figure 18.4-1. Without an interaction term, the regression lines for worker castes have the same slope. To compare the two models, we focus exclusively on the F-test of the interaction term (CASTE*MASS) in Table 18.4-1, the ANOVA table of results. The interaction term is not statistically significant ($F = 1.02$, $df = 1,32$, $P = 0.321$). The null hypothesis of equal slopes (no interaction) is therefore *not* rejected by the data.

Table 18.4-1 ANOVA table for the general linear model fitted to the mole-rat data. We test only the interaction term in this round.

Source of variation	Sum of squares	df	Mean square	F	P
CASTE	0.0570	1	0.0570		
MASS	1.3618	1	1.3618		
CASTE*MASS	0.0896	1	0.0896	1.02	0.321
Residual	2.7249	31	0.0879		
Total	4.2333	34			

This does not mean that the slopes are truly equal—it is not wise to "accept" a null hypothesis just because you have failed to reject it. But the *assumption* that the slopes are equal seems reasonable to make at this point, and the data do not contradict it.

> A general linear model with one numerical and one categorical explanatory variable fits separate regression lines to each group of the categorical variable. An interaction between the variables means that the regression slopes differ among the groups.

Dropping the interaction term

If we can assume that there really is no difference between the regression slopes, then we can drop the interaction term from the general linear model, yielding

$$ENERGY = CONSTANT + CASTE + MASS.$$

The fit of this model is illustrated in Figure 18.4-1 (left panel). Now, for the second round of the analysis, the hypotheses are

H_0: Castes do not differ in energy expenditure.
H_A: Castes differ in energy expenditure.

The test involves comparing the fit of the "full" model (i.e., ENERGY = CONSTANT + CASTE + MASS) with that of the null model lacking the CASTE term (i.e., ENERGY = CONSTANT + MASS).

This null model is fitted by a single linear regression of energy on mass, calculated on all of the data combined. Both models include MASS, so the test of differences between castes is "adjusted" for mass differences. The ANOVA results are listed in Table 18.4-2.

The F-ratio for CASTE is significant ($F = 7.25$, $df = 1,32$, $P = 0.011$), confirming that the two worker castes differ in their mean daily energy expenditure after adjusting for body mass. The magnitude of the difference is reflected by the vertical gap between the regression lines for the two castes (Figure 18.4-1, left panel). The results suggest that the infrequent workers tend to expend less energy than the frequent workers during the dry season. The results for MASS are also statistically significant ($F = 21.39$, $df = 1,32$, $P < 0.001$), indicating that energy expenditure changes with body mass—the regression slope in Figure 18.4-1 is significantly different from zero.

Keep in mind that this was an observational study. That is, energy expenditure was "adjusted" using naturally occurring variation in body mass, rather than experimentally induced variation in mass, which might yield a different regression slope. The analysis of covariance is still prone to bias resulting from *other* confounding variables not included in the model. The justification for using the method in observational studies is that bias is reduced by including one (or more) important covariates in the model, but it is not necessarily eliminated.

Table 18.4-2 ANOVA table for the general linear model without an interaction term fitted to the mole-rat data.

Source of variation	Sum of squares	df	Mean square	F	P
MASS	1.8815	1	1.8815	21.39	<0.001
CASTE	0.6375	1	0.6375	7.25	0.011
Residual	2.8145	32	0.0880		
Total	5.3335	34			

How would we have proceeded if the interaction term *was* statistically significant, and therefore could *not* be dropped from the model? In this event, the difference between castes is not a constant but changes with mass (as illustrated in the right panel of Figure 18.4-1). Adjusting for mass would then require that we specify a value of body mass at which to estimate caste differences.

Dropping an interaction term to adjust for a confounding variable does not imply that we should usually drop non-significant terms from general linear models fitted to data. A sensible rule is that the analysis should follow the design and purpose of the study. In general, variables that are part of the study design should usually be retained in the general linear model fitted to the data.

18.5   Assumptions of general linear models

The assumptions of general linear models are the same as those for regression and ANOVA.

▶ The measurements at every combination of values for the explanatory variables (e.g., every block and treatment combination) are a random sample from the population of possible measurements.
▶ The measurements for every combination of values for the explanatory variables have a normal distribution in the corresponding population.
▶ The variance of the response variable is the same for all combinations of the explanatory variables.

As in linear regression, the residual plot is a useful technique to evaluate these assumptions in general linear models. In Figure 18.5-1 we display the residual plot

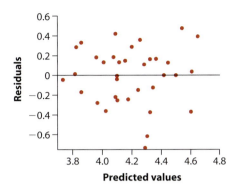

Figure 18.5-1 Residual plot for the general linear model without an interaction term fitted to the mole-rat data in Example 18.4.

for the mole-rat data fitted with the general linear model after dropping the interaction term. Residuals in a general linear model have the same interpretation as in linear regression. Each residual is the difference between an observed Y-value and the value of Y predicted by the model. The residual plot has the predicted values along the horizontal axis and the residuals along the vertical axis. Statistical packages on the computer can compute predicted values and residuals for you.

If the assumptions of general linear models are met, then the residual plot should have the following features:

▶ a roughly symmetric cloud of points above and below the horizontal line at zero, with a higher density of points close to the line than away from the line,
▶ little noticeable curvature as we move from left to right along the horizontal axis, and
▶ approximately equal variance of points above and below the horizontal line at all predicted values.

The residual plot for the mole-rat example meets these criteria reasonably well (Figure 18.5-1), although one data point with a ln body mass of 4.9 has a very low residual and might represent an outlier (see Figure 18.4-1). In general, if assumptions are violated, then a log or other transformation of the response variable can sometimes improve the situation, just as in single-factor analysis of variance. When an outlier is present, it is advisable to determine how its presence influences the results. In the present example, deleting the outlier with a ln body mass of 4.9 did not perceptibly change the results.

18.6 **Summary**

- Some experiments and observational studies have more than one explanatory variable. These usually can be analyzed using a general linear model approach, in which the response variable is represented by a linear model plus random error.
- A model is a mathematical representation of the relationship between a response variable Y and one (or more) explanatory variables.
- Linear regression is an example of a linear model. General linear models expand the regression approach to include multiple explanatory variables that may be numerical or categorical.
- The general linear model approach begins with a statement of the model to be fitted to the data.
- The F-ratio is used to test whether including a term of interest in the general linear model results in a significant improvement in the fit of the model to the data, compared with the fit of the null model lacking the term.
- General linear models assume that every combination of values of the explanatory variables has a random sample of Y-values from a population having a normal distribution with equal variance.
- After fitting a model to the data, a plot of the residuals against the predicted values (i.e., a residual plot) is a useful method to evaluate whether the assumptions of general linear models are met.
- The analysis of data with more than one explanatory variable follows the design and purpose of the study.
- Experiments with blocking include the block as an explanatory variable in the general linear model statement.
- Model statements for factorial designs include the main effects of the factors and their interaction.
- F-ratios are calculated differently when a factorial design includes one or more random factors, compared with that when only fixed factors are present.
- Analyses of treatment effects that adjust for known confounding variables, or covariates, are usually analyzed in two steps. First, the interaction between the treatment and the covariate is tested. Second, if the interaction is not significant, then a revised model that drops the interaction is fitted to the data in the final test of the treatment effect.
- A general linear model with one numerical and one categorical explanatory variable (also called analysis of covariance, or ANCOVA) fits separate regression lines to each group of the categorical variable. An interaction between the variables means that the regression slopes differ among the groups.

PRACTICE PROBLEMS

1. Rest in fruit flies, *Drosophila melanogaster*, has many features in common with mammalian sleep, and its study might lead to a better understanding of sleep in mammals, including humans. Hendricks et al. (2001) examined the role of the signaling molecule cyclic AMP (cAMP) by comparing the mean number of hours of resting in six different lines of mutant or transgenic fly lines having different levels of cAMP expression. Measurements are hours per 24 hours divided by the mean of "wild type" flies. Means ($\pm$ SE) are shown for the different fly lines in the following graph:

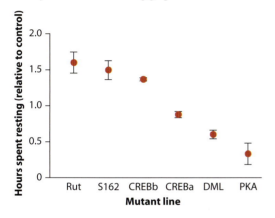

a. Write the statement of the general linear model to be fit to these data to compare means between groups. Indicate what each term in the model represents.
b. Write the corresponding statement for the null hypothesis of no differences between mutant lines.
c. Using a ruler, add the predicted values for each model to the figure (approximate positions will suffice).
d. What test statistic should be used to test whether the null model should be rejected in favor of the alternative?

2. A study of the Magellanic penguin (*Spheniscus magellanicus*) measured stress-induced levels of the hormone corticosterone in chicks living in either tourist-visited areas or undisturbed areas of a breeding colony in Argentina (Walker et al. 2005). Chicks at three stages of development were included in the study—namely, recently hatched, midway through growth, and close to fledging. Penguin chicks were stressed (captured) by the researchers and their corticosterone concentrations were measured 30 minutes later. The following graph diagrams the mean hormone concentrations for the three age groups of chicks from tourist-visited (*filled circles*) and undisturbed (*open circles*) areas of the colony:

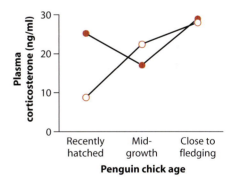

a. What is the response variable?
b. What are the explanatory variables?
c. The line segments in this plot are not parallel. What does this suggest?
d. Is this an observational or experimental study? Explain.

e. Did the study use a factorial design? Explain.

3. Refer to Practice Problem 2.
   a. Write a complete model statement for a general linear model to fit to the penguin data. Indicate what each term in the model represents.
   b. What are the null hypotheses tested in the ANOVA table for the general linear model?
   c. What are the assumptions of your analysis?

4. Give three reasons why studies in biology sometimes have more than one explanatory variable.

5. For each of the following scenarios, draw an interaction plot (like that in Figure 18.3-1) showing the results of a hypothetical experiment having two factors, A and B, each having two groups, in which there is
   a. a main effect of A, no main effect of B, and no interaction between A and B;
   b. a main effect of A, a main effect of B, and an interaction between A and B;
   c. no main effect of A or B, and an interaction between A and B.

6. Evidence is mounting that a part of the brain known as the hippocampus is crucial for tasks that depend on relating information from multiple sources, such as tasks requiring spatial memory. Broadbent et al. (2004) tested this experimentally by surgically inducing lesions of different extent in the hippocampus of rats and measuring the subsequent memory performance of the rats in a maze. Their data are plotted in the following graph:

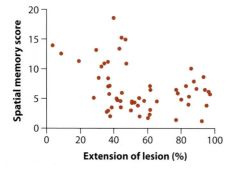

a. Write a model statement for a general linear model to fit to these data. Indicate what each term in the model represents.
b. Write the corresponding statement for the null model.
c. Using a ruler, add the predicted values for each model to the figure (approximate positions will suffice).

7. The foraging gene (*for*) has been found to underlie variation in foraging behavior in several insect species. Ben-Shahar et al. (2002) examined whether the gene might influence behavioral differences in the honey bee (*Apis mellifera*). Worker bees perform tasks in the hive such as brood care ("nursing") when they are young, but switch to foraging for nectar and pollen outside the hive as they age. The authors compared *for* gene expression in nurse and foraging worker bees in three bee colonies. The results are compiled in the accompanying table. Gene expression is measured in arbitrary units.

Worker type	Colony	*for* gene expression
Nurse	1	0.99
Forager	1	1.93
Nurse	2	1.00
Forager	2	2.36
Nurse	3	0.24
Forager	3	1.96

a. Draw an interaction plot for these data.
b. Treating COLONY as a blocking variable, write the statement of a general linear model to fit to these data. Indicate what each term in the model represents.
c. Write the corresponding null model for a test of whether worker types differ in their mean gene expression.
d. Is worker type a random effect or a fixed effect? Explain.
e. What is the purpose of a blocking variable in experimental design?

ASSIGNMENT PROBLEMS

8. The following table lists the ANOVA results for the general linear model fit to the data in Practice Problem 7.

Source of variation	Sum of squares	df	Mean square	F	P
BLOCK	0.342	2	0.171	2.25	0.31
WORKERTYPE	2.693	1	2.693	35.34	0.03
Residual	0.152	2	0.076		
Total	3.187				

a. Explain in words what the *F*-ratio for WORKERTYPE measures.
b. Explain in words what the *F*-ratio for BLOCK measures.
c. The term for BLOCK is not statistically significant. Should it be dropped from the general linear model? Explain.
d. Explain in words what the residuals are.
e. Explain what is plotted along each axis in the following residual plot for the bee data.

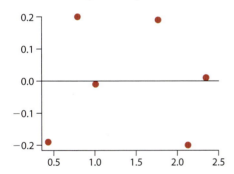

9. Were Neanderthals smaller-brained than modern humans? Estimates of cranial capacity from fossils indicate that Neanderthals had large brains, but also that they had a large body size. The accompanying graph shows the data from Ruff et al. (1977) on estimated log-transformed brain and body sizes of Neanderthal specimens (*filled circles*) and early modern humans (*open circles*). The goal of the analysis was to determine whether humans and Neanderthals have different brain sizes once their differences in body size are taken into account.

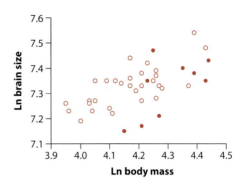

The ANOVA results of the model are listed in the following table:

Source of variation	Sum of squares	df	Mean square	F	P
SPECIES	0.00547	1	0.00547	1.24	0.274
MASS	0.09810	1	0.09810	22.15	<0.001
SPECIES *MASS	0.00485	1	0.00485	1.09	0.303
Residual	0.15503	35	0.00443		
Total	0.26345	38			

a. Examine the ANOVA table and the graph and write the statement of the general linear model that was fit to these data.
b. What does the SPECIES*MASS term represent?
c. What does the *F*-ratio corresponding to the SPECIES*MASS term represent?
d. What null and alternative hypotheses are tested with the SPECIES*MASS term?
e. What can you conclude from the *F*-ratio and *P*-value listed in the table for the SPECIES*MASS term?
f. In view of the goals of the study, what steps would you recommend next in order to test whether the brain sizes of Neanderthal and early modern humans differ after adjusting for differences in body size?

10. Using a ruler, add the predicted values from the analysis recommended in part (f) of Assignment Problem 9 to the scatter plot (approximate positions will suffice).

11. In a study of the effects of commercial fishing on fish populations, Hsieh et al. (2006) measured the year-to-year coefficient of variation (CV) of larval population sizes of exploited and unexploited fish species in the California current system. They compared the two groups of fish species using a general linear model that adjusted for differences between exploited and unexploited species in the age of maturation (MATURATION), which also seemed to influence the coefficient of variation. Data for 13 exploited species and 15 unexploited species are shown in the following graph, along with the predicted values of the model for the two groups of fish:

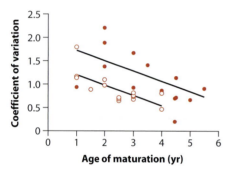

The model was

$$CV = CONSTANT + MATURATION + EXPLOITATION,$$

where EXPLOITATION is the explanatory variable representing the two groups of fish species (exploited and unexploited). The ANOVA results are listed in the table at the top of right column of this page.

Source of variation	Sum of squares	df	Mean square	F	P
MATURATION	1.7313	1	1.7313	17.65	0.003
EXPLOITATION	1.5924	1	1.5924	16.23	0.005
Residual	2.4518	25	0.0981		
Total	5.7755	27			

a. Explain the steps that likely led to the authors using the model analyzed above.
b. What are the assumptions of this analysis?
c. Is there a significant difference between exploited and unexploited fish in their year-to-year coefficients of variation? Explain the basis for your conclusion.

12. The tortoise beetle *Deloyala guttata* feeds and lays eggs on leaves of the two morning glory species *Ipomea pandurata* and *I. purpurea*. Rausher (1984) investigated whether there was genetic variation in the population in the relative abilities of beetles to exploit the two plant species. To test this, he randomly sampled six beetle families from a local population by crossing randomly sampled males and females. He then raised half the offspring from each family on leaves of *I. pandurata* and the other half on *I. purpurea*. Here we analyze development time (i.e., the days from hatching to formation of the pupa) of female offspring from each family. If genetic variation is present in the relative abilities to exploit the two plant species, then there should be an interaction between family and plant species. Means are given in the table below; standard errors were about 0.7.

a. Draw an interaction plot for these results. Briefly describe (in words) the patterns

Family number	Morning glory species	Development time (days)	Morning glory species	Development time (days)
#18	*I. pandurata*	15.1	*I. purpurea*	14.1
#19	*I. pandurata*	14.8	*I. purpurea*	14.5
#25	*I. pandurata*	15.9	*I. purpurea*	14.0
#50	*I. pandurata*	16.9	*I. purpurea*	17.1
#65	*I. pandurata*	14.7	*I. purpurea*	14.7
#66	*I. pandurata*	15.6	*I. purpurea*	14.4

revealed. Is an interaction present? How can you tell?

b. Write a model statement for a "full" general linear model to fit to these data. Explain what each term in the model represents.

c. Which factors in the general linear model are random and which are fixed? Explain.

d. What are the null hypotheses to test in the corresponding ANOVA table?

e. What assumptions are required to test these hypotheses?

f. What does the F-statistic for any given term in the model signify?

13. Females of the yellow dung fly, *Scatophaga stercoraria*, mate with multiple males, and the sperm of different males "compete" to fertilize her eggs. The last male to mate usually gains a disproportionate number of fertilizations. In a laboratory experiment on male and female dung flies from two populations, one in Switzerland and the other in the United Kingdom (UK), Hosken et al. (2002) found that fertilizations by the last male (assessed by DNA fingerprinting) depended on the population of origin of both the male and female. Females were mated to two

males in turn, one from each population, and the father of each egg laid was then determined. On average, males from the same population as the female fared worse than foreign males. The following graph, redrawn from Hosken et al. (2002), shows the mean percentage of offspring sired by the second male $\pm$ SE:

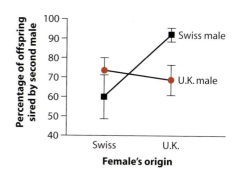

a. What do we call the type of experimental design that was carried out?

b. Write the statement of a general linear model to fit to these data.

c. Based on the graph, which F-ratios are likely to be greater than one?

19

Computer-intensive methods

The advent of fast and cheap computers has changed the way statistics is done. Most obviously, graphics and the tedious calculations required for classical statistical techniques can be done at the touch of a button, so that the large amount of time that was once spent on such tasks could be used to collect more data or drink more coffee. But the computer has allowed more than just speedier calculations of what could already be done. Computers have also made possible new approaches for analyzing data that were not feasible before. This chapter describes three of these methods: simulation, randomization, and bootstrapping.

Simulation and randomization are primarily methods for hypothesis testing, whereas bootstrapping is a method designed to calculate the precision of estimates. These methods are particularly useful when the assumptions of standard statistical methods cannot be met or when no standard method exists. Simulation, randomization, and bootstrapping all require a large number of calculations, and they are all impractical except with the aid of a computer. The value of these techniques is that they can be applied to almost any type of statistical problem.

In this chapter, we describe each of these three methods using relatively simple examples. We emphasize the conceptual basis of each method without assuming that the reader is adept at computer programming.

19.1 Hypothesis testing using simulation

Simulation is a computer-intensive method used in hypothesis testing, where the major challenge is to determine the null distribution—that is, the probability distribution of the test statistic when the null hypothesis is true. In many situations, this distribution can be calculated from probability theory and simulation is not needed. In some situations, however, the null distribution is too difficult to calculate from theory. Computer simulation of the sampling process can be an excellent way of getting an approximation of the null distribution in these cases. **Simulation** uses a computer to mimic—or "simulate"—sampling from a population under the null hypothesis.

With simulation, we use a computer to create an imaginary population whose parameter values are those specified by the null hypothesis. We then use the computer to simulate sampling from this imaginary population, using the same protocol as was used to collect the real data. Each time we take a simulated sample, we use it to calculate the test statistic. We repeat this simulation process a large number of times. The frequency distribution of values obtained for the test statistic from all of the simulations is an approximate null distribution for our hypothesis test.

> *Simulation* uses a computer to imitate the process of repeated sampling from a population to approximate the null distribution of a test statistic.

We used simulation once already without saying so. In Example 6.2, we used computer simulation to generate the null distribution for the number of right-handed toads in a random sample of 18 toads under the null hypothesis that the proportion of right-handed toads in the population was 0.5. Simulation was unnecessary in that case because, as we now know, the binomial test is faster, easier, and gives an exact *P*-value. For Example 19.1, however, simulation is the best or only option.

Example 19.1   **How did he know? The nonrandomness of haphazard choice**

Some stage performers claim to have real telepathic powers—that is, the ability to read minds. However, these powers can be convincingly faked. In one example, the performer asks every member of the audience to think of a two-digit number. After a show of pretending to read their minds, the performer states a number that a surprisingly large fraction of the audience was thinking of. This feat would be surprising if the people thought of all two-digit numbers with equal probability, but not if people everywhere tend to pick the same few numbers. Figure 19.1-1 shows the numbers chosen independently by 350 volunteers (Marks 2000). Are all two-digit numbers selected with equal probability?

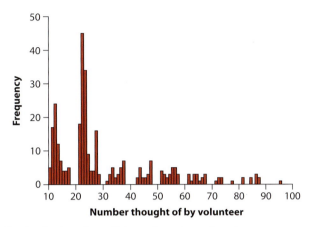

Figure 19.1-1 The distribution of two-digit numbers chosen by volunteers.

According to the histogram in Figure 19.1-1, the two-digit numbers chosen don't seem to occur with equal probability at all. A large number of people chose numbers in the teens and twenties, with few choosing larger numbers. Almost nobody chose multiples of ten. Let's use these data to test the null hypothesis that every two-digit number occurs with equal probability. The hypotheses are

H_0: Two-digit numbers are chosen with equal probability.
H_A: Two-digit numbers are not chosen with equal probability.

To analyze this problem, our first thought might be to use a χ^2 goodness-of-fit test (Chapter 8). There are 90 categories of outcome (each of the 90 integers between 10 and 99). The expected frequency of occurrence of each category is 350/90 = 3.89, and the observed frequencies are those shown in Figure 19.1-1. The resulting value of the test statistic is $\chi^2 = 1111.4$. If this number exceeds the critical value for the null distribution at $\alpha = 0.05$, then we can reject H_0.

But here we run into a problem. The expected frequency of 3.89 for each category violates the requirements of the χ^2 test that no more than 20% of the categories should have expected values less than five. As a result, the null distribution of the test statistic χ^2 is not a χ^2 distribution, so we cannot use that distribution to calculate a P-value. How do we determine the null distribution of our test statistic so that we can get a P-value?

One possible solution is to use computer simulation to generate the null distribution for the test statistic. Here's how it's done, in five steps:

1. *Use a computer to create and sample an imaginary population whose parameter values are those specified by the null hypothesis.* In the case of the mentalist's numbers, simulating a single sample involves drawing

Table 19.1-1 A subset of the results of a simulation. Each row has the first 12 of 350 numbers randomly sampled from an imaginary population in which all numbers between 10 and 99 occur with equal probability. The last column has the χ^2 statistic calculated on each simulated sample of 350 numbers.

Simulation number	Simulated samples of 350 numbers (the first 12 numbers of each sample are shown)													Test statistic χ^2
1	32	42	27	74	78	86	98	71	28	50	41	54	...	86.1
2	38	63	98	36	88	10	74	35	62	52	90	48	...	80.5
3	52	45	59	44	44	25	94	29	27	64	24	47	...	95.4
4	69	55	31	42	35	48	52	37	91	40	67	67	...	78.4
5	11	66	32	11	76	96	73	20	64	40	37	49	...	87.7
6	46	15	87	23	92	18	91	26	31	23	40	51	...	96.9
7	61	35	58	33	58	82	67	95	16	64	59	64	...	95.4
8	47	54	16	39	91	68	49	57	10	21	79	51	...	106.7
9	21	84	30	66	81	13	16	18	81	91	52	95	...	92.3
10	88	26	48	44	34	72	89	14	98	35	99	54	...	100.5
...														...

350 two-digit numbers between 10 and 99 at random and with equal probability. Each simulated sample must have 350 numbers because 350 is the sample size of the real data. The first row of Table 19.1-1 lists the first 12 numbers of our first simulated sample of 350 numbers.

2. *Calculate the test statistic on the simulated sample.* For the number-choosing example, we have decided to use χ^2 as the test statistic. This statistic does not necessarily have a χ^2 distribution under H_0, because of the small expected frequencies, but χ^2 is still a suitable measure of the fit between the data and the null hypothesis. In our first simulated sample, the χ^2 value turned out to be 86.1 (Table 19.1-1).

3. *Repeat steps 1 and 2 a large number of times.* We repeated the simulated sampling process 10,000 times, calculating χ^2 each time. (Typically a simulation should involve at least 1000 replicate samples.) Table 19.1-1 shows a subset of outcomes for the first 10 of our simulated samples.

4. *Gather all of the simulated values for the test statistic to form the null distribution.* The distribution of simulated test statistics can be used as the null distribution of the estimate. The frequency distribution of all 10,000 values for χ^2 that we obtained from our example simulations is plotted in Figure 19.1-2. This is our approximate null distribution for the χ^2 statistic.

5. *Compare the test statistic from the data to the null distribution.* We use the simulated null distribution to get an approximate P-value. In an ordinary χ^2 goodness-of-fit test, the P-value for the χ^2 statistic is the probability under the

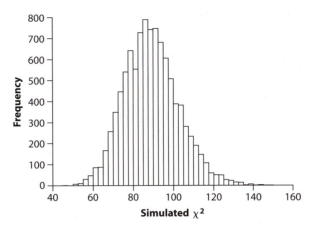

Figure 19.1-2 The null distribution for the χ^2 statistic based on 10,000 simulated random samples from an imaginary population conforming to the null hypothesis (Example 19.1). The test statistic calculated from the data, $\chi^2 = 1111.4$, is far greater than for any of the simulated samples.

null distribution of obtaining a χ^2 statistic as large or larger than the observed value of the test statistic. The same is true with a simulated null distribution for χ^2: the P-value is approximated by the fraction of simulated values for χ^2 that equal or exceed the observed value of χ^2 (in our case, $\chi^2 = 1111.4$). According to Figure 19.1-2, *none* of the 10,000 simulated χ^2 values exceeded the observed χ^2 statistic. This means that the approximate P-value is less than 1 in 10,000 (i.e., $P < 0.0001$).[1] To be more precise, we would have to run more simulations.

These results show that, when people choose numbers haphazardly, the outcome is highly non-random. This can make mentalists appear to have telepathic powers when the only power they possess is that of statistics.

Simulation is more work than a "canned" statistical test. It would have been much easier to analyze the data from Example 19.1 if we could have used the χ^2 distribution or some other known null distribution instead. This is the reason that simulation is usually reserved for cases in which the null distribution is unknown.

19.2 Randomization test

Randomization tests are used to test the hypothesis that two variables are associated. Randomization tests can be used to test for an association between two categorical variables (like a contingency analysis), between a categorical variable and a numerical variable (like a two-sample t-test), or between two numerical variables (like correlation).

---

[1] We can't say $P = 0$, because we might find a more extreme value from the null distribution if we ran more simulations.

We can use a randomization test in place of these standard approaches when the assumptions of these standard methods are not met or when the null distribution is otherwise unknown. Randomization tests make fewer assumptions than most standard tests and so are very versatile. Randomization tests are also usually more powerful than nonparametric tests based on ranks. They do not require that we know the null distribution of the measure of association between the two variables, so randomization tests can be used in an even broader range of circumstances than most parametric tests of association.

In a **randomization test**, a test statistic is chosen that measures the association between the two variables in the data (for example, χ^2 for a contingency analysis, the difference between sample means for two groups, or a correlation coefficient for two numerical variables). The assignment to individuals of the values of one of the variables is scrambled. This yields a "randomized" data set in which every individual keeps its original measurement for one variable but has a randomly reassigned value for the second variable. Any association that was present between the two variables in the original data is broken up in the new data set (except that produced by chance). This randomization procedure is repeated many times, and the test statistic of association is calculated for each randomized data set. The frequency distribution of the test statistic calculated on the randomized data sets is used as an approximate null distribution for the test statistic. If the observed value of the test statistic is unusual compared to the null distribution, we reject the null hypothesis of no association between the variables.

For example, if we measured the lifespans of a random sample of recently deceased people, we probably would find that lifespan was related to sex—that is, men don't live as long as women. Imagine that we took the lifespan data from this study and mixed it up, randomly reassigning each measured value of lifespan to a male or female. With this mixed-up data set, we would not expect any relationship between sex and lifespan to remain. By randomizing the assignment of the variables to individuals, we get a snapshot of what the data would look like if the null hypothesis of no association were true. Randomization repeats this process many times to get the full null distribution. If lifespan and sex really are associated, then we would tend to find a larger difference in lifespan between men and women in the real data than we would typically find in the randomized data sets.

> A *randomization test* generates a null distribution for the association between two variables by repeatedly and randomly rearranging the values of one of the two variables in the data.

Do not confuse "randomization tests" with "randomization," the random assignment of treatments to individuals as a part of an experimental design.[2]

---

[2] Randomization tests are sometimes called "permutation tests" in part to prevent this confusing similarity of terms.

We illustrate randomization with Example 19.2, in which we compare the means of two groups. A difference between groups in a response variable is equivalent to an association between the response variable and the group variable. Randomization tests could equally be used to compare medians between groups, the correlation between numerical variables, and a host of other types of associations between variables.

Example 19.2	**Girls just wanna have genetic diversity**

Pseudoscorpions of the species *Cordylochernes scorpioides* live in tropical forests where they ride on the backs of harlequin beetles to reach the rotting figs where they feed. Females of the species are promiscuous and mate with multiple males over their short lifetimes. It is unclear what advantages there are for a female to mate multiple times, because the males don't help care for her young, and mating just once provides all the sperm she needs to fertilize her eggs.

One possible advantage is that the sperm of some males is genetically incompatible with a given female and, by mating multiple times, a female increases the chances of mating with at least one male whose sperm is compatible with her. To investigate this idea, Newcomer et al. (1999) recorded the number of successful broods by female pseudoscorpions randomly assigned to one of two treatments. One group of females was each mated to two different males (DM), whereas females in the other group were each mated twice to the same male (SM). By mating each female twice, the same total amount of sperm was provided in both treatments, but DM females received genetically more diverse sperm than SM females. The researchers compared the mean number of successful broods in each treatment group. The data[3] are listed in Table 19.2-1 and are displayed in Figure 19.2-1.

These data are not normally distributed, as can be seen from the histograms in Figure 19.2-1. As a result, we cannot use a two-sample *t*-test. Using the Mann–Whitney *U*-test (Section 13.5) to compare the means would be risky, moreover, because the distributions seem to have different shapes, violating an assumption of the nonparametric test. A randomization test also assumes that the distributions have the same

---

[3] The sample sizes were larger in the original paper. We took a subsample of their data to make the calculations easier to follow.

Table 19.2-1 The number of successful broods of pseudoscorpion females that were mated twice to either a single male (SM) or two different males (DM). Data are presented for 20 SM females and 16 DM females. Data from the different treatments are color-coded to more easily show the origin of each value later in Table 19.2-2.

Mating treatment	Number of successful broods	Mating treatment	Number of successful broods
SM	4	DM	2
SM	0	DM	0
SM	3	DM	2
SM	1	DM	6
SM	2	DM	4
SM	3	DM	3
SM	4	DM	4
SM	2	DM	4
SM	4	DM	2
SM	2	DM	7
SM	0	DM	4
SM	2	DM	1
SM	0	DM	6
SM	1	DM	3
SM	2	DM	6
SM	6	DM	4
SM	0		
SM	2		
SM	3		
SM	3		

shape, but it is much more robust to departures from this assumption than the Mann–Whitney U-test and is more powerful, too. A randomization test, therefore, is the more appropriate choice.

There are a total of 36 data points, 20 from the SM group (where the females mated to the same male twice) and 16 from the DM group (where the females mated with two different males). The hypotheses we are testing are

H_0: There is no difference in the mean number of successful broods between these two groups.

H_A: There is some difference in the mean number of successful broods between these two groups.

This is a two-sided test, so we reject H_0 if the difference in the number of successful broods is much greater than zero or if it is much less than zero.

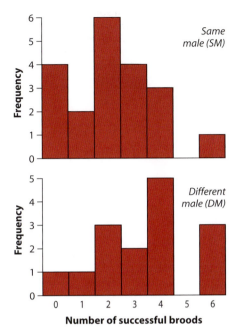

Figure 19.2-1 The number of successful broods for female pseudoscorpions that mated either twice with the same male or once with each of two different males.

To carry out a randomization test of the hypotheses, we need to decide on a test statistic to describe the difference between the groups. The simplest test statistic is the observed difference between the group sample means, $\overline{Y}_{SM} - \overline{Y}_{DM}$. (We could convert this quantity to a t-statistic by dividing by the standard error of the difference, but there is no need to, because we won't be using the t-distribution.) From the data, the difference in the mean number of successful broods is

$$\overline{Y}_{SM} - \overline{Y}_{DM} = 2.2 - 3.625 = -1.425.$$

To generate the null distribution of possible values of $\overline{Y}_{SM} - \overline{Y}_{DM}$, follow these three steps:

1. *Create a randomized set of data in which the response variables are randomly reordered.* To do this, list all of the observations, as in Table 19.2-1. Now take all of the data values for the response variable and randomly rearrange them among individuals, while leaving the other variable (the group, or explanatory, variable) unchanged. Let's call the result a "randomized sample." Table 19.2-2 shows one such randomized sample. Your computer program might do this randomization for you, but it can also be done as follows. First, choose one of the 36 response measurements at random with equal probability and place it in the first row of the "*Number of successful*

Table 19.2-2 Outcome of a single randomization. The response measurements (the number of successful broods) are color coded as in Table 19.2-1 to indicate their original groups.

Mating treatment	Number of successful broods	Mating treatment	Number of successful broods
SM	4	DM	2
SM	0	DM	2
SM	7	DM	3
SM	4	DM	3
SM	2	DM	2
SM	2	DM	3
SM	2	DM	3
SM	1	DM	4
SM	4	DM	1
SM	4	DM	2
SM	0	DM	1
SM	3	DM	4
SM	3	DM	6
SM	4	DM	6
SM	6	DM	2
SM	2	DM	0
SM	4		
SM	6		
SM	0		
SM	0		

broods" column. In our first randomized sample, this first site had measurement 4. Next, choose one of the remaining 35 response measurements at random and place it in the second row; in our randomized sample the result was 0 (see Table 19.2-2). Continue with this process, eliminating each real data point from the sampling pool as we use it, until all 36 response measurements have been used up.[4]

The results of our first randomized sample are shown in Table 19.2-2. Each response measurement from the data is present exactly once, but by chance it has sometimes been assigned to a different group than the group from which it came. The size of each group in the randomized sample is the same as its size in the original data.

---

[4] In statistics jargon this process is called "sampling without replacement." Once a value is sampled, it is "removed" from the population, so the same value cannot be chosen again for the same randomized sample.

2. *Calculate the measure of association for the randomized sample.* For the single randomized sample shown in Table 19.2-2, the mean of the SM group is 2.9 and the mean of the DM group is 2.75. The difference is thus

$$\overline{Y}_{SM} - \overline{Y}_{DM} = 2.9 - 2.75 = 0.15.$$

This is the result from the first replicate of the randomization process.

3. *Repeat the randomization process many times.* We repeated this randomization process a total of 10,000 times, recording the value of $\overline{Y}_{SM} - \overline{Y}_{DM}$ each time. Figure 19.2-2 shows the resulting distribution of values for $\overline{Y}_{SM} - \overline{Y}_{DM}$ from all of these randomizations. This distribution approximates the null distribution of the difference between the two groups.

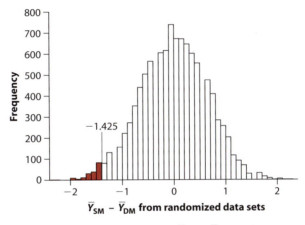

Figure 19.2-2 The null distribution of the test statistic $\overline{Y}_{SM} - \overline{Y}_{DM}$ from 10,000 replicates of the randomization process applied to the pseudoscorpion data. Of these 10,000 randomizations, only 176 had a value of $\overline{Y}_{SM} - \overline{Y}_{DM}$ equal to or less than the observed value, -1.425.

To determine an approximate *P*-value for the test, we use the simulated null distribution for $\overline{Y}_{SM} - \overline{Y}_{DM}$ in the same way that we would use a theoretical null distribution (had one been available). To calculate the *P*-value, we begin by finding the proportion of values in the null distribution that equal or lie farther in the tail than the observed value of the test statistic from the data. For our example, this is the fraction of test statistics from all of the randomized samples that are less than or equal to the observed value, $\overline{Y}_{SM} - \overline{Y}_{DM} = -1.425$. Only 176 of the 10,000 randomized data sets, a proportion 0.0176, yielded such an outcome (Figure 19.2-2). To obtain *P*, we multiply this proportion by two to take into account the equally extreme outcomes at the other tail of the null distribution (remember: this is a two-tailed test). Therefore, the *P*-value is approximately 2(0.0176) = 0.0352, and we reject the null hypothesis. Female pseudoscorpions have more successful broods when they mate with different males than when they mate with the same male repeatedly. In this species, there is a reproductive advantage to females that are promiscuous and mate with more than one male.

Because the randomization samples are generated by a random process, the null distribution would not be exactly the same if this test were repeated. As a result, the *P*-value would vary slightly from one test to another. Nevertheless, with a large enough number of randomized data sets, the *P*-value can be estimated with good precision.

Assumptions of randomization tests

Randomization tests make few assumptions and can be applied in a wide variety of circumstances, but some assumptions are required. First of all, the data must be a random sample from the population.

Secondly, for randomization tests that compare means or medians between groups, the distribution of the variable must have the same shape in every population. Randomization tests are robust to this assumption when sample sizes are large.

Randomization tests have lower power (i.e., lower ability to reject a false null hypothesis) than parametric tests when the sample size is small. They have similar power to parametric tests when sample sizes are large.

19.3 Bootstrap standard errors and confidence intervals

The **bootstrap** is a computer-intensive procedure used to approximate the sampling distribution of an estimate. Bootstrapping creates this sampling distribution by taking new samples randomly and repeatedly *from the data themselves*. Unlike simulation and randomization, the bootstrap is not directly intended for testing hypotheses. Instead, the bootstrap is used to find a standard error or confidence interval for a parameter estimate. The bootstrap is especially useful when no formula is available for the standard error or when the sampling distribution of the estimate of interest is unknown.

Recall from Section 4.1 that the sampling distribution is the probability distribution of sample estimates when a population is sampled repeatedly in the same way. The standard error is the standard deviation of this sampling distribution. In principle, therefore, we might obtain a standard error of an estimate by taking repeated samples from the population, calculating the sample estimate each time, and then taking the standard deviation of the sample estimates. In reality, however, we can't really do this repeated sampling; collecting data is expensive, and it is best to put all individuals collected into one sample if we had them. Bootstrapping does something like this repeated sampling, but using the computer. If the size of our sample from the population is large, then we do have easy access to a part of the population—namely, the part that was already sampled. Bootstrapping is a kind of repeated sampling, but, instead of taking individuals from the population directly, we use a computer to draw the samples from the *data*, a procedure called "resampling." If the data set is large enough, then bootstrap samples drawn in this way will have statistical properties very similar to the distribution of possible sample estimates obtained from the population itself.

> *Bootstrapping* uses resampling from the data to approximate the sampling distribution of an estimate.

The bootstrap is therefore a bit strange: we resample from the data itself to generate many new data sets, and from these we infer the sampling distribution of the estimate. If you think about it, this is almost cheating, because we use the one and only data set to infer the distribution of estimates from all possible data sets. Hence the name, "bootstrap," coming from the idea of picking yourself up by your own bootstraps.[5] The method was proposed by Bradley Efron in 1979, when desktop computers started to become available. The bootstrap is now commonly used in biology and other sciences.

Example 19.3 shows how to calculate a standard error and a confidence interval using the bootstrap. This particular example estimates a median, a simple quantity for which it is otherwise difficult to calculate a sampling distribution. Bootstrapping, however, can be applied to essentially any type of estimate.[6]

Example 19.3   ## The language center in chimps' brains

One of the things that makes humans different from other organisms is our well-developed capacity for complex speech. Chimps and gorillas can learn some rudimentary language, but with a capacity far below that of humans. Speech production in humans is associated with a part of the brain called "Brodmann's area 44," which is part of Broca's area. In humans, this area is larger in the left hemisphere of the brain  than in the right, and this asymmetry has been shown to be important for language development. With the advent of magnetic resonance imaging (MRI), it is possible to ask whether this area is asymmetric in apes' brains as well. A sample of 20 chimpanzees was

---

[5]   The term derives from the adventures of Baron Munchausen, who found himself at the bottom of a hole in the ground, lost until he had the idea of picking himself up by his own bootstraps. (From *The Travels and Surprising Adventures of Baron Munchausen Illustrated with Thirty-seven Curious Engravings from the Baron's Own Designs and Five Illustrations*, by G. Cruikshank.)

[6]   The most common use of the bootstrap in biology is to calculate the uncertainty of estimates of phylogenies—that is, evolutionary relationships between species or other taxa (see Practice Problem 8 for an example).

scanned with MRI, and the asymmetry of their Brodmann's area 44 was recorded (Cantalupo and Hopkins 2001). This asymmetry score is the proportional difference between the areas of Brodmann's area 44 on the left and right sides. (The proportional difference is the left measurement minus the right, divided by the average of the two sides.) The raw data are listed in Table 19.3-1. The sample median asymmetry score was 0.14. We want to quantify the uncertainty of this estimate of the population median.

Table 19.3-1 Asymmetry scores for Brodmann's area 44 in 20 chimpanzees.

Name of chimp	Asymmetry score
Austin	0.30
Carmichael	0.16
Chuck	−0.24
Dobbs	−0.25
Donald	0.36
Hoboh	0.17
Jimmy Carter	0.11
Lazarus	0.12
Merv	0.34
Storer	0.32
Ada	0.71
Anna	0.09
Atlanta	1.12
Cheri	−0.22
Jeannie	1.19
Kengee	0.01
Lana	−0.24
Lulu	0.24
Mary	−0.30
Panzee	−0.16

The frequency distribution of asymmetry scores shown in Figure 19.3-1 is skewed to the right and might even be bimodal. A transformation of these data would be difficult to find because the range of values includes negative numbers. What to do? Bootstrapping provides a suitable approach.

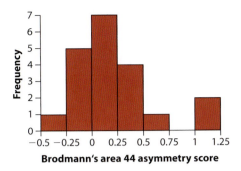

Figure 19.3-1 The frequency distribution of asymmetry scores for Brodmann's area 44 in 20 chimpanzees. A negative score indicates that the area is larger on the right side of the chimp's brain, while chimps with positive scores show a larger area in the left hemisphere.

Bootstrap standard error

To generate a bootstrap standard error, there are four steps to follow. First, we list the steps all at once here, and then we go through them again with the data.

1. *Use the computer to take a random sample of individuals from the original data.* Each individual in the data has an equal chance of being sampled. The bootstrap sample should contain the same number of individuals as the original data. Each time an observation is chosen, it is left available in the data set to be sampled again, so the probability of it being sampled remains unchanged.[7]
2. *Calculate the estimate using the measurements in the bootstrap sample from step 1.* This is the first **bootstrap replicate estimate**.
3. *Repeat steps 1 and 2 a large number of times* (10,000 times is reasonable). The frequency distribution of all bootstrap replicate estimates approximates the sampling distribution of the estimate.
4. *Calculate the sample standard deviation of all the bootstrap replicate estimates obtained in steps 1–3.* The resulting quantity is called the **bootstrap standard error**.

The last point is worth repeating: the standard error is the standard deviation of the sampling distribution of estimates.[8]

We can now apply these four steps to the chimp data. There are 20 data points in the sample, so each bootstrap sample must also have 20 measurements. Each of the

---

[7] In statistics jargon, this is called "sampling with replacement."

[8] A common mistake is to calculate the standard error of the bootstrap estimates by dividing the standard deviation by the square root of the number of bootstrap replicates, by analogy with the standard error of the mean. This results in a standard error that is much too small.

20 measurements in the bootstrap sample is chosen with equal probability from the values in the original data. Applying step 1, the following is the first bootstrap replicate that we obtained:

0.24	0.36	0.30	0.16	0.34	−0.24	0.30	1.19	0.32	0.32
0.36	0.01	0.01	0.11	0.11	−0.25	0.12	0.32	−0.24	0.17

Each of the measurements in this first bootstrap sample was present in the original data set. By chance, some of the original data points are present more than once in the bootstrap sample. For example, the score 0.32 (from the chimp named Storer) is present twice. Also by chance, some of the original data points are absent in this first bootstrap sample. For example, the score 0.71 (from the chimp named Ada) was not sampled. The sample median of this bootstrap sample is 0.205, so this is our first bootstrap replicate estimate of the median asymmetry score (step 2).

We repeated this process 10,000 times, calculating the sample median of the measurements each time (step 3). Figure 19.3-2 shows the frequency distribution of the bootstrap replicate estimates of the sample median.

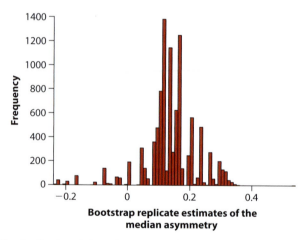

Figure 19.3-2 The distribution of 10,000 bootstrap replicate estimates for the median asymmetry of Brodmann's area 44 in chimpanzees.

The mean of the bootstrap replicate estimates is 0.142, which is very close to the estimated median from the original data (0.14). Remember that the bootstrap procedure is calculating a sampling distribution for an estimate, not a null distribution for a hypothesis test. As such, the overall mean of the bootstrap replicate estimates should be close to the estimate first calculated on the original data.[9]

---

[9] The exception is when the estimate itself is biased. The bootstrap can be used to estimate and correct for bias, but we don't present the method here.

The standard deviation of these bootstrap replicate estimates is 0.085 (step 4). This is the bootstrap standard error of our sample median: SE = 0.085.

> The *bootstrap standard error* is the standard deviation of the bootstrap replicate estimates obtained from resampling the data.

Because the bootstrap samples come from the data, which generally do not represent the full population, the bootstrap standard error tends to be slightly smaller than the true standard error. This effect is negligible when the sample size is large.

Confidence intervals by bootstrapping

The approximate sampling distribution generated by the bootstrap can also be used to calculate an approximate confidence interval for the population parameter. We present the most commonly used method here.[10] The bootstrap $1 - \alpha$ confidence interval can be obtained from the bootstrap sampling distribution by finding the points that separate $\alpha/2$ of the distribution into each of the left and right tails. In other words, a 95% confidence interval ranges from the 0.025 quantile to the 0.975 quantile of the bootstrap sampling distribution.

For example, let's compute a 95% confidence interval for the population median asymmetry using the bootstrap sampling distribution displayed in Figure 19.3-2. To determine the lower bound of the 95% confidence interval, we must find the 0.025 quantile—that is, the value that was greater than or equal to 250 (i.e., 0.025 × 10,000) of the bootstrap replicate estimates. In our example, the 250th smallest bootstrap replicate estimate was −0.075. To determine the upper bound of the confidence interval, we must find the value that is less than or equal to only 2.5% of the bootstrap estimates—in other words, the 9751st measurement in the ordered bootstrap estimates (9750 is 0.975 × 10,000). For the chimp brain data, this number is 0.31. As a result, the 95% bootstrap confidence interval for the median asymmetry of Brodmann's area 44 in chimps is

$$-0.075 < \mu < 0.31.$$

This is a relatively wide confidence interval, indicating that the data are consistent with a broad range of possible values for the population median asymmetry of Brodmann's area 44 in chimps. This asymmetry of brain structure, thought to be so important for language development in humans, might have a median as low as zero (or

---

[10]   Consult the excellent book *An Introduction to the Bootstrap*, by Efron and Tibshirani (1993), for more details and other options.

very slightly larger on the left side than on the right side of the brain) or as large as 0.31 in our closest relative.[11]

Bootstrapping data sets with multiple samples

The bootstrap can also be used to obtain standard errors and confidence intervals for comparisons between parameters of two or more populations (e.g., the difference between two means). When using the bootstrap to compare groups, the data from each group is resampled separately, and the bootstrap sample uses the original sample size for that group. The bootstrap procedure generates new data sets that mimic sampling from the original populations repeatedly; therefore, the bootstrap samples have to be drawn from the groups that they represent. For example, if the original data had a sample of 100 males that were compared to 200 females, each bootstrap data set would also have 100 males and 200 females. The males in the bootstrap sample would be drawn from the males in the real data set, and the females in each bootstrap sample would be drawn from the real females.

Assumptions and limitations of the bootstrap

The main assumption of the bootstrap is that each sample is a random sample from its corresponding population. Additionally, each sample must be large enough that the frequency distribution of the measurements in the sample is a good approximation of the frequency distribution in the population. The larger the sample, the greater the resemblance between the frequency distribution of measurements in the sample and that in the population. Bootstrap analyses based on small samples will, on average, produce standard errors that are too small and confidence intervals that are too narrow, overestimating the precision of the estimate.

19.4 Summary

▶ Simulation is a method for hypothesis testing in which a computer is used to mimic repeated sampling from an imaginary population whose properties conform to those stated in the null hypothesis. The frequency distribution of test statistics calculated on the simulated samples gives a null distribution of the test statistic.

---

[11]   The 95% confidence interval for *mean* asymmetry is similarly broad, but it doesn't overlap zero.

▶ The simulated null distribution of a test statistic is used to calculate *P*-values for hypothesis testing.

▶ A randomization test is a method used to generate the null distribution for a measure of association between two variables by randomly rearranging the observed values for one of the variables. The frequency distribution of test statistics calculated on many randomized data sets gives the null distribution of the test statistic.

▶ The null distribution of a test statistic generated using randomization is used to calculate *P*-values for hypothesis testing.

▶ Bootstrapping is a method for calculating standard errors of estimates and confidence intervals for parameters. It uses resampling from the data to approximate the sampling distribution for an estimate.

▶ The standard deviation of the bootstrap sampling distribution for an estimate is the bootstrap standard error of the estimate.

▶ Bootstrap confidence intervals can be calculated from quantiles of the distribution of bootstrap replicate estimates.

▶ Bootstrapping requires large sample sizes to generate reliable estimates of the sampling distribution.

PRACTICE PROBLEMS

1. The following are very small data sets of birth weights (in kg) of either singleton births or individuals born with a twin.

   Singleton: 3.5, 2.7, 2.6, 4.4   Twin: 3.4, 4.2, 1.7

   We are interested in the difference in mean weight between singleton babies and twin babies.
   a. Construct a valid bootstrap sample from these data for this difference.
   b. Construct a valid randomization sample from these data for this difference.
   c. In what ways are your answers for parts (a) and (b) different?
   d. Assume that we wanted to estimate and test the difference in *medians* between these two groups. Would that change the way in which the bootstrap sample or the randomization would be created?

2. Using the data in Practice Problem 1, state whether each of the following sets of numbers (a through f) is a possible randomization sample for use in testing the difference between the means of singleton and twin birth weights. If not, explain why.

	Singletons	Twins
a.	3.5, 2.7, 2.6, 4.4	3.4, 4.2, 1.7
b.	3.4, 4.2, 1.7, 3.5	2.7, 2.6, 4.4
c.	2.7, 2.6, 4.4	3.4, 4.2, 1.7, 3.4
d.	3.5, 3.5, 3.5, 3.5	3.5, 3.5, 3.5
e.	3.8, 3.8, 3.8, 3.8	3.8, 3.8, 3.8
f.	3.4, 3.5, 4.4, 3.4	4.4, 2.7, 2.6

3. Using the data from Practice Problem 1, state whether each of the following data sets (a through f) is a possible bootstrap replicate sample for use in determining a bootstrap confidence interval for the difference in mean birth weight. If not, explain why.

	Singletons	Twins
a.	3.5, 2.7, 2.6, 4.4	3.4, 4.2, 1.7
b.	3.4, 4.2, 1.7, 3.5	2.7, 2.6, 4.4
c.	2.7, 2.6, 4.4	3.4, 4.2, 1.7, 3.4
d.	3.5, 3.5, 3.5, 3.5	3.4, 3.4, 3.4
e.	3.8, 3.8, 3.8, 3.8	3.4, 4.2, 1.7
f.	3.5, 3.5, 4.4, 2.7	4.2, 1.7, 4.2

4. The following table lists 100 bootstrap replicate estimates for an estimate of the mean length (in cm) of timber wolf jaws. (The values have been sorted into ascending order for your convenience.) Use the numbers to approximate a 90% confidence interval for the mean length.

10.15	10.17	10.17	10.18	10.20
10.20	10.20	10.20	10.21	10.21
10.21	10.21	10.22	10.22	10.23
10.23	10.23	10.23	10.23	10.24
10.24	10.24	10.24	10.24	10.25
10.25	10.25	10.25	10.25	10.26
10.26	10.26	10.26	10.26	10.26
10.27	10.27	10.27	10.27	10.27
10.27	10.27	10.28	10.28	10.28
10.28	10.28	10.29	10.29	10.29
10.29	10.29	10.29	10.30	10.30
10.30	10.30	10.31	10.31	10.31
10.31	10.32	10.32	10.32	10.32
10.33	10.33	10.33	10.34	10.34
10.34	10.35	10.35	10.35	10.35
10.35	10.35	10.35	10.36	10.36
10.36	10.37	10.37	10.37	10.37
10.37	10.38	10.38	10.38	10.38
10.38	10.39	10.40	10.40	10.40
10.40	10.41	10.41	10.44	10.48

5. Of the two scatter plots provided, one represents the real relationship between body size and brain size for 29 species of dinosaur (Hurlburt 1996) and the other shows the first randomized data set created from that same data. Which do you think is the real data, and why?

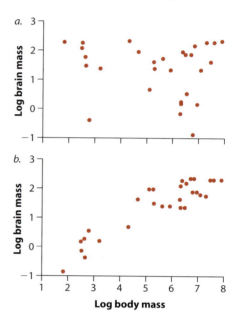

6. Prairie voles are monogamous and social, whereas their close relative the meadow voles are polygamous and solitary. The species differ in their expression of a vasopressin receptor gene (*V1aR*) in their forebrains (expression measures the rate of protein production by a gene). The scientific hypothesis is that this receptor of vasopressin, an important neurotransmitter, might influence the voles' social behavior. Geneticists were able to experimentally increase the expression of the *V1aR* gene in a sample of meadow voles, and they compared the behavior of the resulting individuals to that of control individuals without excess *V1aR* (Lim et al. 2004). They measured the time each vole spent huddling with a partner, under the assumption that greater huddling time is indicative of a more social animal, as described in Assignment Problem 11 in Chapter 3.

The two graphs below show distributions of the difference in median in huddling time between the two groups, each with 1000 replicates. One of the graphs shows bootstrap replicate estimates, and the other graph was produced by randomizations.

a. Which distribution comes from the bootstrap and which comes from randomization?

b. Estimate by eye the bootstrap standard error of the mean difference in median huddling time.

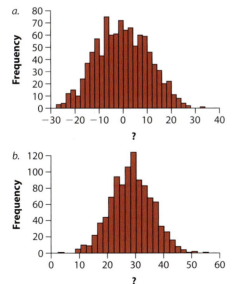

7. The "broken stick" model is often used in ecology to represent how species should divide up space or resources if these are divided randomly. For example, Waldron (2007) compared the geographic range sizes of closely related bird species in North America. For each of 65 pairs of bird species, he used maps to calculate the total range sizes of the two species (in km^2), and then he calculated the ratio of the size of the smaller of the two ranges over the larger range size. This ratio can range from nearly zero, if one species of a pair has a much larger range than the other, to one, if both species have the same range sizes. He observed that the average ratio for all the pairs was 0.48. He then used simulation to test the null hypothesis that the mean of this ratio was no different than that expected if species divided the region "randomly" according to a randomly broken stick, with one species of every pair getting the short end and the other species the long end. Using a computer, he took a stick of unit length, randomly broke it into two parts, and then calculated the ratio of the length of the smaller piece to the length of the larger piece. He did this 65 times, once for every pair of closely related species, and then took the average ratio. This process was repeated 10,000 times. The resulting 10,000 values for the ratio are shown in the following graph:

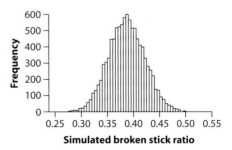

a. What does this frequency distribution of simulated values for the ratio estimate?

b. Only 42 of the 10,000 simulated values for the ratio were greater than or equal to 0.48, the observed ratio in the bird data. Using this information, test the null hypothesis that the observed mean ratio of the bird range sizes is no different than expected if the species divided the region randomly as a broken stick.

8. The most common use of the bootstrap in biology is to estimate the uncertainty in trees of evolutionary relationships (phylogenies) estimated from DNA sequence data. For example, the tree provided for this problem is a phylogeny of the carnivores, a group of mammals, based on their gene sequences (Flynn et al. 2000). The arrangement of branches on the tree indicates how species are related by descent. Pick any two named species on the right (representing the present time), and follow their branches backward in time (to the left) until the branches meet at a node representing their common ancestor.

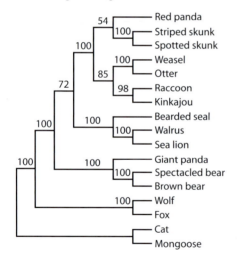

Species are closely related if they share a recent common ancestor. In this tree, for example, the cat is more closely related to the mongoose than either of these two species is to the wolf. The

tree is not necessarily the true phylogeny, but an estimate based on the assumption that species sharing a recent common ancestor will have had less time to become different in their DNA sequences than more distantly related species. The bootstrap is used to calculate the uncertainty in the arrangement of the branches of the estimated evolutionary tree. To do this, the bootstrapping procedure resamples the DNA sequence data and recalculates a phylogenetic tree from each new bootstrap sample (we'll spare you the details). This is repeated a large number of times. The method then examines every bootstrap replicate tree and compares it to the tree estimated from the original data. Every node of the carnivore tree shown here gives a number between 0 and 100 that refers to the percentage of bootstrap replicate trees that have exactly the same set of named species descended from that node. For example, the number "100" on the branch leading to the two skunks means that 100% of the bootstrap replicate trees had a node just like it leading to the same two skunks. The number "54" leading to the trio of red panda (pictured on the first page of this chapter), striped skunk, and spotted skunk indicates that just 54% of the bootstrap replicate trees had a node containing only those same three species. The remaining 46% of the bootstrap replicate trees had different arrangements (e.g., the red panda might have been grouped with the weasel or raccoon).

a. What percentage of the bootstrap replicate trees showed the raccoon as the species most closely related to the kinkajou?

b. What percentage of the bootstrap replicate trees grouped the giant panda with the two bear species?

c. From the results of this bootstrap, which of the following statements can be made with greater confidence, that "red pandas are most closely related to skunks" or that "giant pandas are most closely related to bears"?

9. Use a six-sided die to generate 10 bootstrap replicates of the following data set, which has three numbers from each of two groups:[12]

Group A: 2.1, 4.5, 7.8
Group B: 8.9, 10.8, 12.4

a. Write out the bootstrap replicate data sets.
b. Use these 10 bootstrap replicates to get an approximate 80% confidence interval for the difference between the medians of Group A and Group B.

10. Use a six-sided die to make 10 randomized data sets suitable for testing the difference between the medians of the following two groups:

Group A: 2.1, 4.5, 7.8
Group B: 8.9, 10.8, 12.4

a. Write out the 10 randomized data sets.
b. Using just these 10 randomized data sets, test the null hypothesis that the two groups have the same median, using $\alpha = 0.2$ for the significance level.

11. Suppose we want to calculate the bootstrap standard errors for the difference between two groups in their medians. We would also like to do the same for the difference between the two groups in their interquartile ranges. Could we use the same procedure to generate the bootstrap replicate samples for both standard errors?

12. Suppose we want to use the randomization test to test the difference between two groups in their medians. We would also like to do the same for the difference between the two groups in their interquartile ranges. Could we use the same procedure to generate the randomization samples for both tests?

13. In a hypothetical study of a predatory fish, the bootstrap was used to help generate confidence intervals for the mean waiting time between meals. A waiting-time measurement was obtained for 79 fish in a random sample from the predatory fish population. The mean waiting time was 109 seconds. Ten thousand bootstrap replicates were used to calculate a sampling distribution for the estimate. The mean of the bootstrap replicates was 108.6 seconds, and the standard deviation of the bootstrap replicate estimates was 10.4 seconds. What is the bootstrap standard error of the estimated mean waiting time?

14. Assume that you have two small samples, both drawn from separate populations that are normally distributed but with unequal variances. You want to compare the means of these two populations. What technique would you use to test the null hypothesis that the means are equal?

15. The snail *Helix aspersa* has an unusual mating system. Each individual is a hermaphrodite, producing both sperm and eggs, but the truly bizarre part is that these snails have evolved a "love dart" with which they try to stab their mate. (If you look closely at the photo [on the next page], you can see the love dart protruding through the head of the lower snail.) This love dart is coated with a drug that enhances the amount of sperm that the stabbed snail receives and stores. This storage effect is expected to be greater when the recipient is smaller, because the relative dose would therefore be larger. Rogers and Chase (2001) measured the size of stabbed snails (i.e., the volume of their shells) and the number of stabber sperm that were stored. Is the size of the recipient snail correlated with the amount of sperm it stores? The data are listed in the table provided. The table also gives two computer generated data sets. One of the computer generated data sets was created by a bootstrap procedure that resampled the observed data, and the other was made by a randomization procedure.

---

[12] In this question and in ones that follow, the sample sizes and numbers of bootstrap replicates are often too small to provide accurate bootstrap calculations or powerful tests and are presented as exercises only. This question calls for the 80% confidence interval only to prevent you from having to roll the die too many times. In general, results based on bootstraps use the same levels of confidence as more traditional statistics.

a. Was "computer generated data A" created for a bootstrap estimate or for a randomization test? Describe how you determined this.

b. Was "computer generated data B" created for a bootstrap estimate or for a randomization test? Describe how you determined this.

c. What test statistic should be used on these data to test whether snail size is correlated with the number of sperm stored?

Original data		Computer generated data A		Computer generated data B	
Shell volume (cm³)	**Number of sperm stored**	**Shell volume (cm³)**	**Number of sperm stored**	**Shell volume (cm³)**	**Number of sperm stored**
2.2	2474	2.2	1807	6.0	260
2.5	2897	2.5	2897	3.7	2516
2.6	2658	2.6	1843	2.5	2897
2.6	2471	2.6	2474	3.3	2009
2.7	2250	2.7	260	3.3	2009
2.9	2606	2.9	440	2.5	1843
3.2	2978	3.2	2332	3.0	1552
3.7	2516	3.7	2009	2.5	2897
3.3	2332	3.3	2250	4.0	1138
3.3	2009	3.3	1158	2.5	1843
2.9	1807	2.9	2516	2.7	2250
3.0	1552	3.0	2606	4.2	440
2.5	1843	2.5	1138	2.5	2897
3.5	1158	3.5	2978	2.5	2897
4.2	440	4.2	2658	4.2	440
6.0	260	6.0	1552	3.7	2516
5.9	788	5.9	788	2.7	2250
4.0	1138	4.0	2471	3.3	2009

ASSIGNMENT PROBLEMS

16. Using the data in Practice Problem 15 on shell volume and the number of sperm stored, we carried out a bootstrap analysis of the correlation coefficient r between the two variables. The cumulative frequency distribution provided here summarizes the 1000 bootstrap replicates of the estimate of the correlation coefficient. The correlation coefficient calculated from the original data was -0.78. From this graph, determine the approximate 95% bootstrap confidence interval for the correlation between shell volume and the number of sperm stored.

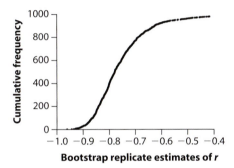

17. The cane toad (*Bufo marinus*), a large, toxic toad introduced to Australia in 1935, has been rapidly spreading across the continent at over 50 km/yr. Are the toads that are moving fastest at the leading edge of the expanding range in Australia morphologically different from the toads in the middle of the range? Phillips et al. (2006) compared the leg lengths of individually marked toads to the distances that they moved in a three-day period. In the following pair of graphs, the relative length of the leg is compared to the movement distance (each measurement is displayed as a deviation from the mean). One of the following two graphs is the original data, and the other shows the first randomized data set. Which is more likely to be the randomized data set? Explain how you obtained your answer.

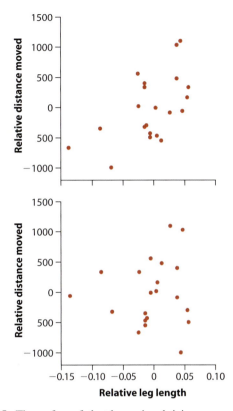

18. The surface of plant leaves is a thriving ecosystem, home to many microorganisms such as bacteria. These bacteria can affect the health of the plant, because some cause disease and some serve as ice-nucleation sites, thus causing frost damage. Hirano et al. (1982) were interested in the probability distribution of the number of bacteria on corn plant leaves. The frequency distribution of the number of bacteria on a random sample of single leaves is shown in the following figure. The number of bacteria is calculated as the number per gram of leaf.

a. Imagine that we wished to estimate the mean number of bacteria per gram of leaf and the uncertainty of our estimate. Would a confidence interval for the mean based on the

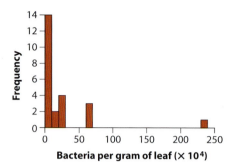

Bacteria per gram of leaf (× 10⁴)

t-distribution be appropriate? Explain your answer.

b. Name two appropriate methods to calculate a confidence interval for the mean with these data.

c. The authors also measured the frequency distribution of the number of bacteria on soybean leaves. The distribution was shaped similarly to the one shown here for corn leaves. List two valid methods that could be used to test for a difference in the median numbers of bacteria on corn and soybean leaves.

19. In Example 13.1, we described the results of a study that measured the change in biomass in marine areas after an increase in protection. The distribution of biomass ratios (i.e., the biomass with protection divided by the biomass without protection) was highly skewed, so we performed a log transformation to overcome this problem when estimating the mean. Here we take a different approach, and use the bootstrap to estimate the *median* biomass ratio. The distribution

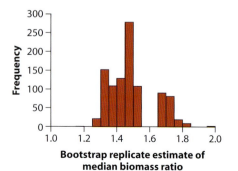

Bootstrap replicate estimate of median biomass ratio

of bootstrap replicate estimates of the median are shown in the accompanying histogram. The mean and standard deviation of this distribution are 1.484 and 0.134, respectively, and there are 1000 bootstrap replicates.

a. What does this frequency distribution estimate?

b. What is the bootstrap standard error of the median biomass ratio?

c. What is the 95% confidence interval for the median biomass ratio? Interpret the histogram to give an approximate answer.

20. Outliers can occur in a sample of data for several reasons, including instrument failure, measurement error, and real variation in the population. The "trimmed mean" is a method developed for estimating a population mean when a measurement is prone to producing outliers. A trimmed mean is an ordinary sample mean calculated on data after the most extreme measurements have been dropped according to a percentile criterion. A 5% trimmed mean drops the measurements below the 5th percentile (0.05 quantile) and above the 95th percentile. In 1882, Simon Newcomb estimated the speed of light by measuring the time it took for light to return to his lab after being bounced off a mirror, a round trip of 7442 meters. The following is the frequency distribution of 66 measurements Newcomb made of this time, in microseconds (i.e., in millionths of a second) (Stigler 1977).

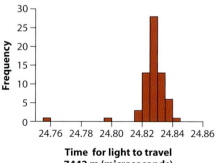

**Time for light to travel
7442 m (microseconds)**

The distribution includes outliers, and the trimmed mean is an objective way to increase the precision of the estimate. For the light data, the 5% trimmed mean drops the three smallest and three largest values, yielding the value 24.8274 for the mean of the remaining measurements.

Bootstrapping is an excellent way to calculate the uncertainty of the trimmed mean. The results of 1000 bootstrap replicate estimates of the trimmed mean are shown in the following histogram. Of the bootstrap estimates, 2.5% were below 24.8254, 5% were below 24.8260, 95% were below 24.8285, and 97.5% were below 24.8288.

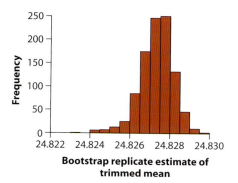

Bootstrap replicate estimate of trimmed mean

a. Calculate a 95% confidence interval for the trimmed mean.

b. The ordinary, "untrimmed" mean of the 66 data points, including the outliers, is 24.8262. Is the ordinary sample mean contained in the 95% confidence interval for the trimmed mean?

21. The variance-to-mean ratio is a useful measurement of how clumped or dispersed events are in space or time relative to the random expectation (see Section 8.6). A high ratio indicates that events are clumped, whereas a low ratio indicates "overdispersion" of events. Davis et al. (in preparation) used this approach to examine the dispersion of "compensatory mutations" affect-

ing protein sequences. Most mutations to genes are harmful, but compensatory mutations occasionally occur that counteract some of the damage caused by other mutations. The authors gathered data on 77 harmful mutations of varying lengths and a total of 328 compensatory mutations that have been discovered in the same genes. These data come from organisms ranging from viruses to fruit flies. The authors recorded the number of compensatory mutations at each amino acid position in the genes. They calculated the mean and variance in the number of compensatory mutations per amino acid position and calculated variance/mean = 2.64. They used simulation to test the null hypothesis that mutations were randomly and independently located in the genes. Compensatory mutations were placed at random and independently. After each simulation, the computer calculated the resulting variance-to-mean ratio in the number of compensatory mutations per amino acid position. The results of 10,000 such simulations are plotted in the following histogram:

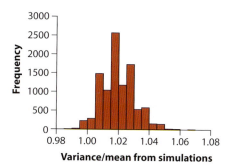

Variance/mean from simulations

a. What does this frequency distribution estimate?

b. Using the results shown in the frequency distribution and the observed variance/mean ratio of 2.64, test the null hypothesis that the true variance/mean ratio is as expected from the random placement of compensatory mutations.

20

Likelihood

Biologists gather data in hopes of discovering the correct value of a population parameter among its many possible values. Consider, for example, the relationship between the metabolic rate of organisms and their body mass. According to one theory, the rate should increase in proportion to body mass raised to the power of 3/4, whereas another theory predicts the power should be 2/3, instead (Savage et al. 2004). Which of these alternatives is correct? We must look to the data to find out.

Likelihood measures how well alternative values of a parameter fit the data. The approach is based on the idea that the best choice of a parameter value, among all the possibilities, is the one with the highest likelihood—the one for which the data have the highest probability of occurring. If the probability of obtaining the data is much higher for one possible value of the parameter than for another, then the first is probably closer to the truth.

Likelihood is a very general approach that can be applied to every type of problem encountered so far in this book, though we have not made much use of it until now. One advantage of likelihood techniques is that they can be applied to data that are not normally distributed.

Here we introduce the concept of likelihood and identify some of its applications in biology. Our goals are to review the key features of the approach so that its uses in the scientific literature may be more readily understood.

20.1 What is likelihood?

Likelihood measures how well a set of data supports a particular value for a parameter. The likelihood of a specific value for a parameter is the probability of obtaining the observed data if the parameter were equal to that specific value. Using the likelihood method involves calculating the probability of obtaining the observed data for each possible value of the parameter, and then comparing this probability between different possible values.

> Likelihood measures how well the data support a particular value for a parameter. It is the probability of obtaining the observed data if the parameter equaled that value.

The likelihood of a particular value tells us little by itself. Rather, the likelihood of a value gains meaning when compared with the likelihoods of other possible values. The likelihood should be relatively high for those values close to the true population parameter and relatively low for values that are far from it. The parameter value gaining greatest support among all possible values is called the **maximum likelihood estimate**. It is the value for which the probability of obtaining the observed data is highest. With the likelihood approach, the maximum likelihood estimate is our best estimate of the true value of the parameter.

> The *maximum likelihood estimate* is the value of the parameter for which the likelihood is highest. It is our best estimate of the parameter.

Nearly all of the estimation techniques we have learned about in this book (for example, the mean and proportion) yield maximum likelihood estimates, although we haven't referred to them in that way.

20.2 Two uses of likelihood in biology

Here we showcase two of the most frequent uses of likelihood in biology—namely, phylogeny estimation and gene mapping. We don't present any of the computations or details. Our goal is to illustrate how likelihood is applied.

Phylogeny estimation

Thomas Henry Huxley[1] was right when he said that either the chimpanzee or the gorilla was the closest living relative of the human species. But which ape is our closest cousin? At least three possible relationships between humans and other apes have been proposed, as shown by the tree diagrams in Figure 20.2-1. According to the hypothesis represented by the leftmost tree, we share our most recent common ancestor with the chimpanzees (the bonobo is the pygmy chimp). According to the tree on the right, however, we are closest to the gorilla, instead. Finally, the tree in the middle represents the hypothesis that we stand apart from the rabble, equally related to chimp and gorilla, who are each other's closest relatives.

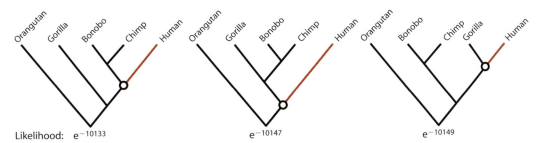

Likelihood: e^{-10133} e^{-10147} e^{-10149}

Figure 20.2-1 Three proposed trees of ancestor–descendant relationships between humans and the other great apes. The human branch and our shared ancestor with the other apes is highlighted. Numbers at the bottom are the likelihoods of each proposal based on gene sequence data (Rannala and Yang 1996). The likelihood of the leftmost tree is the highest at e^{-10133}, where e is the base of the natural logarithm.

Likelihood is frequently used to estimate trees of ancestor–descendant (phylogenetic) relationships using DNA sequence data. Rannala and Yang (1996) determined the likelihood of each of the three proposed trees in Figure 20.2-1 using gene sequence data from all five apes. The likelihood of a given tree refers to the probability of obtaining the observed gene sequences for the five species if that tree were the correct tree. This probability is based on a probability model of gene sequence evolution. In this model, the gene sequences for any pair of species were identical at the moment they split from their most recent common ancestor, after which each species accumulated differences gradually and randomly over time. Under this model, the most closely related species today should have the most similar gene sequences, and the most distantly related species should have the most different sequences.

The tree on the left in Figure 20.2-1 had the highest likelihood, making it the maximum likelihood estimate. Humans are most closely related to the chimps, justifying our nickname, the "third chimpanzee" (Diamond 1992).

---

[1]   Huxley was known as "Darwin's bulldog" for his vigorous support of Darwin's theory of evolution.

Gene mapping

Huge efforts are underway to find the genes that underlie inherited forms of human diseases. One approach tests whether genetic "markers" in the human genome differ between individuals having the disease of interest and those not having the disease. A marker is a unique, easily identifiable site in the genome whose gene sequence ("state") tends to vary among individuals in the population. Many markers throughout the human genome are available for gene-mapping studies.

Hamshere et al. (2005) used this approach to find a gene responsible for human schizoaffective disorder. This debilitating mental illness is known to run in families, and it is therefore likely to have a genetic component. The data were the marker states of a sample of individuals afflicted with the disease and their healthy family members. If a gene were present, then the states of markers closest to the affected gene should show differences in frequency between diseased and healthy individuals.

Their approach is illustrated in Figure 20.2-2 with results from chromosome 1. At each marker along the chromosome, the researchers calculated *two* likelihoods, one for each of two hypotheses. One hypothesis was that a gene increasing risk of schizoaffective disorder was really present at that site. The second hypothesis was the null hypothesis that no gene affecting the disorder was present at that site. The likelihood in each case is the probability of obtaining the observed data (the differences between healthy and diseased individuals in the frequency of marker states) if the given hypothesis is correct. The log of the ratio of these two likelihoods (the first likelihood divided by the second likelihood) measures the strength of evidence that a gene is located at that site.[2] This procedure was repeated at every marker along the chromosome, yielding the curve shown in Figure 20.2-2.

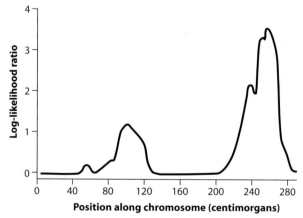

Figure 20.2-2 Evidence for a gene affecting schizoaffective disorder on human chromosome 1. Redrawn from Hamshere et al. (2005), with permission.

---

[2] The arbitrary convention in gene-mapping studies is to use the base-10 logarithm rather than the natural log. The resulting quantity is called the "LOD score."

The highest point on the curve is located approximately at position 260, which indicates where, along chromosome 1, a gene for schizoaffective disorder is most likely to be present.

20.3 Maximum likelihood estimation

Likelihood is a probability. The likelihood L of a particular value for a parameter is the probability of obtaining the data if the parameter equaled that value. Using a vertical bar to represent *given* or *conditional upon* (Chapter 5), this definition of L can be written as

$$L[value \mid data] = \Pr[data \mid parameter = value].$$

Maximum likelihood estimation involves finding the parameter value that has the largest L. Here we show how this is accomplished using Example 20.3, which comes from a study that investigated a population proportion.

Example 20.3	**Unruly passengers**

The tiny wasp, *Trichogramma brassicae*, parasitizes eggs of the cabbage white butterfly, *Pieris brassicae*. The wasp rides on a female butterfly (the arrow in the photo at the right points to a small wasp on a butterfly's leg). When the butterfly lays her eggs on a Brussels sprout or other cabbage, the wasp climbs down and parasitizes the freshly laid eggs. Fatouros et al. (2005) tested whether the wasps could distinguish mated female butterflies (with fertilized eggs) from unmated females. They carried out a series of trials in which a single wasp was presented simultaneously with two female cabbage white butterflies, one of them a virgin female and the other recently mated. Of the  32 wasps that rode on females, 23 chose the mated female, whereas nine chose the unmated female. We can use these data to provide an interval estimate of the population proportion.

You learned in Chapter 7 how to estimate a population proportion and to calculate a confidence interval. The probability model is relatively simple for this case, however, so we can use it to demonstrate how to estimate a parameter using maximum likelihood.

Probability model

Maximum likelihood estimation requires a probability model that specifies the probabilities of different outcomes of the data-sampling process depending on the parameter being estimated. In the wasp study (Example 20.3), the outcome that was measured was the number of wasps that chose the mated female butterfly. The parameter of interest is p, the unknown proportion of wasps in the population that would choose the mated female. Let's assume that each of the n wasps tested represented a single random trial and that separate trials were independent. Under these conditions, the number of wasps choosing the mated female should follow a binomial distribution with the probability of "success" in any one trial equal to p. In this case, the probability that exactly Y females choose mated females depends on p as follows:

$$\Pr[Y \text{ choose mated} \mid p] = \binom{n}{Y} p^Y (1 - p)^{n-Y}.$$

This formula for the binomial distribution[3] was first introduced in Chapter 7, except that here we have written the probability of Y successes as conditional on the unknown p. We use this formulation because we need to vary p to see how this affects the calculated probability of obtaining the observed data.

The likelihood formula

We use the symbols $L[p \mid Y$ chose mated$]$ to indicate "the likelihood of a particular value of p, given that Y wasps chose the mated female." This likelihood is defined as "the probability that Y wasps choose the mated female in n trials, given that the population parameter equals the particular value for p." Thus, the formula for the likelihood is

$$L[p \mid Y \text{ choose mated}] = \binom{n}{Y} p^Y (1 - p)^{n-Y}.$$

To calculate the likelihood of a specific value for p, set Y equal to 23, the observed number of wasps choosing mated females. Also, plug in the total number of trials, $n = 32$, yielding

$$L[p \mid 23 \text{ choose mated}] = \binom{32}{23} p^{23} (1 - p)^9.$$

The likelihood of $p = 0.5$ is

$$L[p = 0.5 \mid 23 \text{ choose mated}] = \binom{32}{23} (0.5)^{23} (1 - 0.5)^9$$
$$= 0.00653.$$

This likelihood represents the support for the possibility that exactly half of the wasps in the population would choose the mated female, given that 23 of 32 sampled

---

[3] Remember that $\binom{n}{Y}$ is called "n choose Y" and is shorthand for $\dfrac{n!}{Y!(n-Y)!}$. The symbol $n!$ represents "n factorial."

wasps did so. We cannot interpret this number in isolation, however, because likelihoods are only informative when compared with the likelihoods of other values for the parameter.

It is usually easier to work with the log of the likelihood rather than the likelihood itself.[4] The **log-likelihood** is the natural log of the likelihood. The formula for the log-likelihood of p, given the observed Y, is

$$\ln L[p \mid Y \ choose \ mated] = \ln\left[\binom{n}{Y}\right] + Y \ln[p] + (n - Y)\ln[1 - p].$$

To understand how to use this formula, plug in the values $n = 32$ and $Y = 23$. The log-likelihood of $p = 0.5$, given the data, is

$$\ln L[p \mid 23 \ choose \ mated] = \ln\left[\binom{32}{23}\right] + 23 \ln[0.5] + 9 \ln[1 - 0.5]$$
$$= -5.03125.$$

This value is the same as the natural log of $L[0.5 \mid 23$ choose mated$] = 0.00653$ calculated previously (except for rounding errors).

> The *log-likelihood* of a value for the population parameter is the natural log of its likelihood.

The maximum likelihood estimate

The maximum likelihood estimate of a parameter is the specific value having the highest likelihood, given the data. The value of the parameter that maximizes the likelihood is also the one that maximizes the log-likelihood, so we can work with the log-likelihood to find the maximum likelihood estimate.

A straightforward way to find the maximum likelihood estimate is to use the computer to calculate the log-likelihood across the range of possible parameter values and then pick the highest one. This is the approach we use here. Alternatively, we could use calculus to find the maximum.

The work of finding the maximum likelihood value for p can be much reduced by using a spreadsheet program on the computer, rather than your calculator. In either case, start by evaluating the log-likelihood for several values of p over a broad range to see its overall shape. In Figure 20.3-1, for example, we calculate the log-likelihood of values of p between 0.1 and 0.9 in increments of 0.1. You can see that the log-likelihood is highest when the proportion p is about 0.7 and that it declines at larger and smaller values.

---

[4]   One reason is that some likelihoods can be so small as to push the lower limits of your calculator and even your computer. Using logs also makes it easier to evaluate quantities such as $\binom{32}{23}$, which might push the upper limits of computer memory. With logs, multiplication becomes addition: $\ln[A \times B] = \ln[A] + \ln[B]$, and powers become multiples: $\ln[A^B] = B \ln[A]$.

	A	B
	Arial ... 10 ... B *I* U ... \$ % , ...	
	SUM ... X ✓ *fx* =LN(COMBIN(32,23))+23*LN(A2)+9*LN(1-A2)	
	A	**B**
1	**proportion *p***	**log-likelihood**
2	0.1	=LN(COMBIN(32,23))+23*LN(A2)+9*LN(1-A2)
3	0.2	-21.876
4	0.3	-13.752
5	0.4	-8.523
6	0.5	-5.031
7	0.6	-2.846
8	0.7	-1.890
9	0.8	-2.468
10	0.9	-5.997

Figure 20.3-1 The log-likelihood calculated for a range of values of *p* using a spreadsheet program on the computer. Column A contains a range of possible values for *p*. Column B shows the corresponding log-likelihood. The first cell in column B shows the formula for the log-likelihood of a proportion in the language of the spreadsheet program we used.

Next, narrow the search using a finer sequence of values for *p*. Figure 20.3-2 shows the spreadsheet results for values of *p* between 0.52 and 0.88 in increments of 0.01. The log-likelihood reaches its maximum when *p* is 0.72. This value is therefore the maximum likelihood estimate.

In Figure 20.3-3, we plot the log-likelihood of all possible values of *p* between 0.4 and 0.9. The resulting curve is called the **log-likelihood curve**. The maximum[5] of the log-likelihood curve occurs at the value $\hat{p} = 0.72$, so this is the maximum likelihood estimate of the population proportion. We give this value the symbol $\hat{p}$ to indicate that 0.72 is our maximum likelihood estimate.[6]

Likelihood-based confidence intervals

The log-likelihood curve also allows us to calculate an interval estimate for the population parameter. The range of values for *p* whose log-likelihood lies within $\dfrac{\chi^2_{1,\,\alpha}}{2}$

---

[5] Log-likelihood curves based on simple probability models usually have only one peak, but there can be more than one peak in more complex models. It is important to check the full range of possible values for the parameter to be sure you find the highest peak.

[6] Notice that the maximum likelihood estimate of $\hat{p} = 0.72$ is the same as the conventional estimate for the proportion, $Y/n = 23/32 = 0.72$. This is no coincidence: Y/n is the formula for the maximum likelihood estimate of a population proportion. This shortcut could have spared us work, except that we wanted to show the general approach for finding a maximum likelihood estimate.

	A	B	C
1	proportion *p*	log-likelihood	log-likelihood - maximum
4	0.52	-4.497	-2.634
5	0.53	-4.248	-2.385
6	0.54	-4.012	-2.149
7	0.55	-3.787	-1.925
8	0.56	-3.575	-1.712
9	0.57	-3.375	-1.512
10	0.58	-3.187	-1.324
11	0.59	-3.010	-1.148
12	0.60	-2.846	-0.983
13	0.61	-2.694	-0.831
14	0.62	-2.554	-0.691
15	0.63	-2.426	-0.563
16	0.64	-2.310	-0.447
17	0.65	-2.207	-0.344
18	0.66	-2.117	-0.254
19	0.67	-2.039	-0.177
20	0.68	-1.976	-0.113
21	0.69	-1.926	-0.063
22	0.70	-1.890	-0.027
23	0.71	-1.869	-0.006
24	0.72	-1.863	0.000
25	0.73	-1.873	-0.010
26	0.74	-1.900	-0.037
27	0.75	-1.944	-0.081
28	0.76	-2.007	-0.144
29	0.77	-2.089	-0.226
30	0.78	-2.192	-0.329
31	0.79	-2.318	-0.455
32	0.80	-2.468	-0.605
33	0.81	-2.644	-0.781
34	0.82	-2.848	-0.985
35	0.83	-3.084	-1.221
36	0.84	-3.354	-1.491
37	0.85	-3.663	-1.800
38	0.86	-4.014	-2.152
39	0.87	-4.416	-2.553
40	0.88	-4.873	-3.010

Figure 20.3-2 The log-likelihood calculated for a narrower range of values for *p* using a spreadsheet program. The log-likelihood is maximized at $\hat{p} = 0.72$. Column C calculates the difference between each log-likelihood and the maximum likelihood.

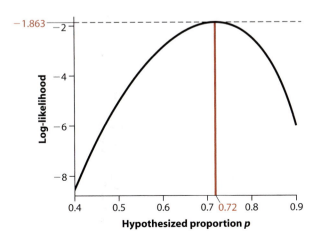

Figure 20.3-3 The log-likelihood curve for *p*, the estimated proportion of wasps choosing the mated female butterfly. The log-likelihood is maximized at $\hat{p} = 0.72$.

units of the maximum constitutes the $1 - \alpha$ **likelihood-based confidence interval** (Meeker and Escobar 1995). For example, an approximate 95% confidence interval for p is the range of values whose log-likelihood lies within 1.92 units of the maximum, since $\chi^2_{1,0.05}/2 = 3.84/2 = 1.92$. The 95% confidence interval is therefore determined directly from the log-likelihood curve. Figure 20.3-4 shows that the limits of the 95% confidence interval for the proportion p are $0.55 < p < 0.86$. A spreadsheet program (e.g., Figure 20.3-2) can also help to find these limits.

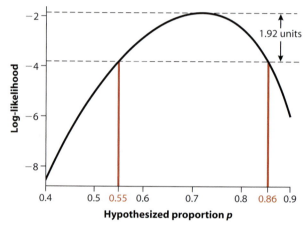

Figure 20.3-4 The likelihood-based 95% confidence interval for p. The top horizontal line indicates the highest log-likelihood (-1.863). The line immediately below corresponds to 1.92 units below the maximum. The 95% confidence interval, indicated by the red lines, is the range of values for the parameter whose log-likelihoods fall within 1.92 units of the maximum. This interval ranges from 0.55 to 0.86 for the wasp–butterfly data.

Based on these data, we conclude that the proportion of wasps in the population that would choose the mated female butterfly lies between 0.55 and 0.86. This most-plausible range for the population proportion is relatively broad and includes relatively weak preference values (but still greater than 0.5) as well as relatively strong preference values. A larger sample size would be needed to narrow the range further.

This confidence-interval method based on the likelihood curve is more accurate than other methods used to generate confidence intervals for the maximum likelihood estimates (Meeker and Escobar 1995). It also requires no formula other than that needed to calculate the log-likelihood curve.

> The *likelihood-based confidence interval* for a population parameter is calculated directly from the log-likelihood curve.

20.4 **Versatility of maximum likelihood estimation**

The great advantage of maximum likelihood estimation is its versatility. It can be applied to almost any situation in which it is possible to write down a model describing the probability of different outcomes. Using Example 20.4, we demonstrate a less familiar problem than estimating a proportion, in which likelihood methods have proved invaluable.

Example 20.4	**Conservation scoop**

Counting elephants is more challenging than you might think, at least when they live in dense forest and feed at night. Eggert et al. (2003) used "capture–recapture" to estimate the total number of forest elephants inhabiting Kakum National Park in Ghana without having to see a single one. The researchers spent about two weeks in the park collecting elephant dung (see image at right), from which they were  able to extract pure elephant DNA. Using five genes, the researchers generated a unique DNA fingerprint for every elephant "encountered" via dung deposits. By using this fingerprint method, they identified 27 elephant individuals over the first seven days of sampling. We call this the first sample, and refer to these 27 elephants as "marked." Over the next eight days, they sampled 74 individuals, of which 15 had already been detected in the first sample. We'll refer to these 15 elephants as "recaptured." Based on the number of recaptures in the second sample, what is the total population size of elephants in the park?

This kind of data can be used to estimate population size because the first sample tells us how many have been marked, and the second sample tells us the proportion of the total population that has been marked. By dividing the number marked by the proportion marked, we can estimate the total number of individuals in the population, N, which is the parameter of interest.

Probability model

The simplest probability model for population size estimation makes the following assumptions:

▶ The population of elephants in the park is constant—that is, there were no births, deaths, immigrants, or emigrants while the study was being carried out.

▶ Random sampling—that is, the dung of every elephant, marked or unmarked, had an equal and independent chance of being sampled.

Our data are the number of "recaptures" in the second sample—that is, individuals that had already been marked in the first sample. We know that 15 individuals were recaptured in the data, but, if we could go back and take another "second" sample of 74 individuals under identical conditions, we would probably obtain a different value for the number recaptured. In other words, the number recaptured has a probability distribution that depends on how many individuals there are in the population (N) and on the sample sizes.

To use the likelihood approach, we need to know the probability of obtaining the observed number of recaptures for different possible values of the parameter N. Finding this probability can be one of the biggest challenges of using the likelihood method. Often, though, someone has already investigated similar data and has published the needed formula in the scientific literature. For mark–recapture data, the formula is already available—we found it in Gazey and Staley (1986). Use n_1 to represent the number of individuals captured and marked in the first seven days of study. In this case, $n_1 = 27$. Let n_2 equal the size of the second sample of individuals obtained over the next eight days. In this case, $n_2 = 74$. The probability of Y recaptures, given the population size N, is[7]

$$\Pr[number\ recaptures = Y \mid N] = \frac{\binom{n_1}{Y}\binom{N - n_1}{n_2 - Y}}{\binom{N}{n_2}}.$$

The term on the right of the equals sign represents the proportion of all possible random samples of n_2 elephants that include exactly Y recaptures (previously marked) and $n_2 - Y$ new individuals (not previously marked).

The likelihood formula

From this point on, the steps are the same as those we followed in Example 20.3 when estimating a proportion. Let $L[N \mid Y\ recaptures]$ indicate the likelihood of a particular value for N, given that there were Y recaptures. The likelihood is defined as the probability of obtaining Y recaptures if N equals the hypothesized value—namely, $\Pr[number\ of\ recaptures = Y \mid N]$. Thus, the formula for the likelihood is

$$L[N \mid Y\ recaptures] = \frac{\binom{n_1}{Y}\binom{N - n_1}{n_2 - Y}}{\binom{N}{n_2}}.$$

[7] This probability distribution is called the hypergeometric distribution, which gives the probability of a given number of individuals of a particular type from a sample of known size and a population of known size, assuming that the individuals are sampled without replacement.

It is easier to work with the log-likelihood rather than the likelihood itself:

$$\ln L[N \mid Y \; recaptures] = \ln\left[\binom{n_1}{Y}\right] + \ln\left[\binom{N - n_1}{n_2 - Y}\right] - \ln\left[\binom{N}{n_2}\right].$$

To calculate the log-likelihood of a specific value of N, substitute $Y = 15$ into the preceding equation. Plugging in the fixed sample sizes, $n_1 = 27$ and $n_2 = 74$, as well, yields

$$\ln L[N \mid 15 \; recaptures] = \ln\left[\binom{27}{15}\right] + \ln\left[\binom{N - 27}{74 - 15}\right] - \ln\left[\binom{N}{74}\right].$$

We are now ready to calculate the log-likelihood for specific values of N.

The maximum likelihood estimate

We plugged the formula for the log-likelihood into a spreadsheet program and calculated the log-likelihood of all possible values of N between 90 and 250. Try this yourself to see if you obtain the same answers we did. Our resulting log-likelihood curve is shown in Figure 20.4-1. The maximum of the log-likelihood curve occurs at the value $\hat{N} = 133$, so this is the maximum likelihood estimate of the elephant population size.

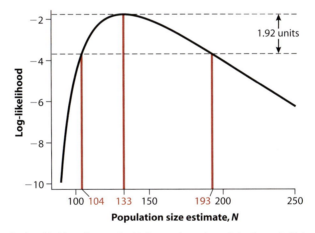

Figure 20.4-1 The log-likelihood curve for N, the total number of elephants in Kakum National Park in Ghana. The log-likelihood is maximized at $\hat{N} = 133$. The values 104 and 193 are the limits of the likelihood-based 95% confidence interval for N.

We also used the spreadsheet program to determine the likelihood-based 95% confidence interval for N. This confidence interval includes all parameter values whose log-likelihood is within $\dfrac{\chi^2_{1,0.05}}{2} = 3.84/2 = 1.92$ units below the maximum log-likelihood. These limits occurred at the N values 104 and 193, as shown in Figure 20.4-1. Based on these data, the population size of elephants in Kakum National Park, Ghana, is likely to be in the range

$$104 < N < 193.$$

Such a broad interval estimate for the population size estimation is not unusual for capture–recapture methods.

Bias

The maximum likelihood estimate should be relatively close to the true population parameter, compared with other values having lower likelihood. Nevertheless, maximum likelihood estimates are often biased. On average, that is, a maximum likelihood estimate tends to fall to one side of the population parameter that it is intended to estimate. For example, the maximum likelihood method for estimating population size using capture–recapture techniques tends to underestimate population size, on average, even when all the assumptions are met (Krebs 1999). In some cases, corrections are known that will compensate for bias. The bias is usually small, however, compared with the breadth of the 95% confidence interval. Bias diminishes as sample size increases (Edwards 1992).

20.5   Log-likelihood ratio test

The log-likelihood ratio test uses likelihood to compare how well two probability models fit the data. In one of the models, the parameter or parameters of interest are constrained to match that specified by a null hypothesis. The constraint is relaxed in the second probability model, which corresponds to the alternative hypothesis. This second model is fit using parameter values that best match the data. If the likelihood of the second model, in which the parameters best match the data, is significantly higher than the likelihood of the first model, in which the parameters are constrained to match H_0, then we reject the null hypothesis.

In this section, we introduce the log-likelihood ratio test. We analyze the wasp choice data from Example 20.3 (Fatouros et al. 2005) and test the null hypothesis that the population proportion p is 0.5. We use the binomial distribution to calculate the log-likelihood of $p = 0.5$, and we compare it to the log-likelihood of the alternative model, in which the parameter p is set to the maximum likelihood estimate.

Likelihood ratio test statistic

The test statistic for the likelihood ratio test, G, is twice the natural log of the ratio of the likelihoods of the two hypotheses. For the simplest case of only a single parameter,

$$G = 2 \ln\left(\frac{L[\text{maximum likelihood value of parameter} \mid \text{data}]}{L[\text{parameter value under } H_0 \mid \text{data}]}\right).$$

The formula is similar when there is more than one parameter to be estimated. The remarkable feature of the log-likelihood ratio is that, if H_0 is true, then G is approximately χ^2 distributed. This means that we can use the χ^2 distribution to calculate a P-value and thus decide whether the null hypothesis should be rejected. The approximation is reliable only when the sample sizes are large.[8]

The degrees of freedom for G equal the difference between the two hypotheses in the number of parameters that require estimation using the data. In the case of only a single parameter, there is one degree of freedom.

Testing a population proportion

In the wasp experiment of Example 20.3, more wasps chose the mated female butterfly (23) than the unmated female (9). Is this by itself evidence that wasps in the population prefer mated female butterflies? We will use the log-likelihood ratio test to answer this question. The hypotheses are

H_0: Wasps choose mated and unmated females with equal probability ($p = 0.5$).
H_A: Wasps prefer one type of female over the other ($p \neq 0.5$).

We could use the binomial test (Chapter 7) or the χ^2 goodness-of-fit test (Chapter 8) to test these hypotheses. The probability models are relatively simple for this case, though, so we will use it to demonstrate the log-likelihood ratio test. This is a two-tailed test because a preference for either the mated or the unmated female is possible if the null hypothesis is false.

Our test statistic is the log-likelihood ratio,

$$G = 2 \ln\left(\frac{L[\hat{p} \mid Y \text{ chose mated female}]}{L[p_0 \mid Y \text{ chose mated female}]}\right),$$

where $L[p_0 \mid Y \text{ chose mated female}]$ is the likelihood of p_0, the value of p specified by the null hypothesis. $L[\hat{p} \mid Y \text{ chose mated female}]$ is the likelihood of $\hat{p}$, the maximum likelihood estimate of p. It is generally easiest to work with log-likelihoods, so we rewrite the previous formula as

$$G = 2(\ln L[\hat{p} \mid Y \text{ chose mated female}] - \ln L[p_0 \mid Y \text{ chose mated female}]).$$

Under the null hypothesis, $p_0 = 0.5$, and in the data $Y = 23$ wasps chose the mated butterflies. Plugging these values into the likelihood formula for p, we get

$$\ln L[p_0 \mid 23 \text{ chose mated female}] = \ln\left[\binom{32}{23}(0.5)^{23}(1 - 0.5)^9\right]$$

$$= \ln\left[\binom{32}{23}\right] + 23 \ln(0.5) + 9 \ln(0.5)$$

$$= -5.031.$$

---

[8]   For small samples, simulation can be used to find the null distribution of G (see Chapter 19).

Under the alternative hypothesis, p is unknown and is estimated using maximum likelihood. We already determined in Section 20.3 that the maximum likelihood estimate is $\hat{p} = 0.72$. The corresponding log-likelihood of this $\hat{p}$ was found to be -1.863 (Figure 20.3-2).

We now have what we need to calculate the test statistic:

$$G = 2(\ln L[\hat{p} \mid 23 \ chose\ mated\ female] - \ln L[p_0 \mid 23 \ chose\ mated\ female]$$
$$= 2[-1.863 - (-5.031)]$$
$$= 6.336.$$

Under the null hypothesis, the test statistic G is approximately χ^2 distributed. The difference between the two hypotheses in the number of parameters needing estimation from the data was one (p). The test statistic G therefore has $df = 1$.

The critical value for the χ^2 distribution having one degree of freedom and $\alpha = 0.05$ is available in Statistical Table A:

$$\chi^2_{1,\,0.05} = 3.84.$$

Because $G > 3.84$, $P < 0.05$. Using a computer to calculate the exact probability for the χ^2 distribution, we find that $P = 0.012$. We reject the null hypothesis based on these data. The wasps indeed prefer mated female cabbage whites over unmated females.[9]

20.6 Summary

▶ Likelihood measures the level of support provided by data for a particular value of a population parameter. The likelihood of a specified value is the probability of obtaining the observed data if the parameter equaled that value.

▶ The maximum likelihood estimate of a parameter is the value having highest support among all possible values. It is the parameter value for which the probability of obtaining the observed data (the likelihood) is highest.

▶ Maximum likelihood estimation depends on a probability model that specifies the probabilities of different outcomes from the process that produced the data.

▶ The log-likelihood is the natural log of the likelihood. Maximizing the log-likelihood is the same as maximizing the likelihood.

▶ The log-likelihood curve describes the log-likelihood of a range of values of the parameter. The maximum likelihood estimate can be found from the value that gives the highest point on the curve.

▶ The likelihood-based confidence interval for a single parameter is the range of values of the parameter whose log-likelihoods lie within $\dfrac{\chi^2_{1,\,\alpha}}{2}$ units of the maximum. The range for an approximate 95% confidence interval includes all values for the parameter whose log-likelihoods are within $3.841/2 = 1.92$ units of the maximum.

---

[9] The researchers determined that the chemical that the wasps use to distinguish mated from unmated females is benzyl cyanide, which the male butterfly passes to the female during mating. The compound is an "anti-aphrodisiac," rendering the mated female less attractive to other male butterflies (Fatouros et al. 2005).

▶ The log-likelihood ratio test uses likelihood to compare the fits of two probability models to the data. In the case of a single parameter, one of the models constrains the parameter to equal that specified by a null hypothesis. The alternative probability model is fit using the maximum likelihood estimate of the parameter.

20.7   Quick Formula Summary

Likelihood

Formula: $L[value \mid data] = \Pr[data \mid parameter = value]$.

Likelihood-based confidence interval

What is it for? To obtain a confidence interval estimate for a parameter.

What does it assume? Data are randomly sampled from a population. The specific assumptions of the probability model are correct.

Formula: Obtained directly from the log-likelihood curve. To do so, find the range of parameter values whose log-likelihood lies within $\dfrac{\chi^2_{1,\alpha}}{2}$ units of the maximum.

Log-likelihood ratio test for a single parameter

What is it for? To compare the fit of two probability models to data using likelihood. In one model, the parameter is fixed to that specified by the null hypothesis. The other model is fit using the maximum likelihood estimate of the parameter.

What does it assume? Data are randomly sampled from a population. The specific assumptions of the probability model are correct.

Test statistic: G

Distribution under H₀: Approximately a χ^2 distribution with degrees of freedom equal to the difference between the two hypotheses in the number of parameters requiring estimation from the data.

Formula: $G = 2 \ln\left(\dfrac{L[maximum\ likelihood\ value\ of\ parameter \mid data]}{L[parameter\ value\ under\ H_0 \mid data]}\right)$

PRACTICE PROBLEMS

1. Albino animals are well known, but what about plants? Apiron and Zohary (1961) found a recessive mutation in the orchard grass (*Dactylis glomerata*) that causes a chlorophyll deficiency. Individuals with two deficient copies of the gene lack the green pigment entirely (and die), but "heterozygous" individuals, which have one normal copy of the gene and one deficient copy, appear to produce normal levels of chlorophyll. Using progeny testing, the authors found that 12 of 47 plants sampled from the Matzleva Valley in Israel were heterozygotes. Assume that the plants were randomly sampled.
   a. What probability distribution should we use to calculate the probability of obtaining a specific number of heterozygous individuals (p) in a sample of size 47?
   b. Write the formula for the likelihood of p, given the data. What does the likelihood measure?
   c. Write the formula for the log-likelihood of p.
   d. Calculate the log-likelihood of the value $p = 0.50$, given the data.

2. Refer to Practice Problem 1. The spreadsheet at the bottom of the page lists log-likelihoods of values of p between 0.1 and 0.4, in increments of 0.02.
   a. Using only the spreadsheet calculations, identify the maximum likelihood estimate of p to within 0.02 units.

   b. Using these same calculations, generate a likelihood-based 95% confidence interval for p.

3. Noonan et al. (2006) used DNA sequence data and the method of maximum likelihood to add the extinct Neanderthal species to the tree of humans and other apes (see Figure 20.2-1). The following graph is a likelihood curve[10] for the time of the split between humans and Neanderthals, which is more recent than that between humans and the other living apes.

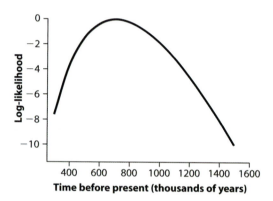

Time before present (thousands of years)

[10] The variable on the vertical axis is the log-likelihood after adding a constant to each log-likelihood value, so that the maximum is zero. This doesn't change any of the ensuing calculations.

Spreadsheet for Problem 2.

	A proportion p	B log-likelihood	C log-likelihood - maximum
1	proportion p	log-likelihood	log-likelihood - maximum
2	0.10	-6.639	-4.615
3	0.12	-5.238	-3.214
4	0.14	-4.193	-2.169
5	0.16	-3.414	-1.390
6	0.18	-2.844	-0.820
7	0.20	-2.444	-0.420
8	0.22	-2.186	-0.162
9	0.24	-2.051	-0.027
10	0.26	-2.024	0.000
11	0.28	-2.094	-0.070
12	0.30	-2.252	-0.228
13	0.32	-2.492	-0.468
14	0.34	-2.809	-0.785
15	0.36	-3.201	-1.176
16	0.38	-3.663	-1.639
17	0.40	-4.195	-2.171

a. What does the likelihood curve measure in this example?

b. With the aid of a ruler, use the likelihood curve to find the maximum likelihood estimate of divergence time.

c. Using this same curve, approximate the likelihood-based 95% confidence interval for divergence time.

d. What does the interval in part (c) measure? Explain.

4. Yashin et al. (2000) compared a sample of 197 centenarians (people 100 years old or older) to a group of 465 younger people (aged between 5 and 80) to examine whether the two groups differed in the frequencies of genetic markers. As part of an attempt to estimate which genetic markers are associated with mortality, the researchers tested for the presence of "hidden heterogeneity," intrinsic differences in mortality rate among individuals within groups. To accomplish this, they generated a likelihood model for a heterogeneity parameter, σ^2, and calculated the log-likelihood for all values of σ^2 between 0 and 0.8, given the data. Their methods are too complicated to describe fully here, but their results are summarized in the graph below. The maximum likelihood estimate for σ^2 was 0.66. The null and alternative hypotheses are

H_0: Heterogeneity in mortality is absent $(\sigma^2 = 0)$.

H_A: Heterogeneity in mortality is present $(\sigma^2 \neq 0)$.

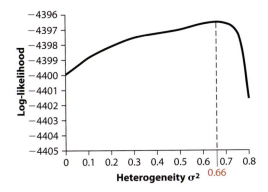

a. Using the log-likelihood curve, find the approximate value of the log-likelihood under the null hypothesis that heterogeneity is zero. Assume that the probability model used depends only on the single parameter shown.

b. With the aid of a ruler, use the log-likelihood curve to find the log-likelihood corresponding to the maximum likelihood estimate (representing the best-case scenario under the alternative hypothesis).

c. Using your results from parts (a) and (b), calculate the test statistic for a log-likelihood ratio test.

d. Under H_0, what is the approximate null distribution for the log-likelihood ratio statistic?

e. Using your results from parts (a)–(d), carry out the log-likelihood ratio test. Report your conclusion.

5. "Have you ever stolen something worth more than $10?" Anyone asked this question in a survey might be reluctant to answer truthfully, especially if he or she did not wish to make known a past misbehavior. The respondent might be more willing to tell the truth if there was a random component to the answer (Warner 1965). This approach was used to estimate the true fraction of thieves among the third-year biology undergraduate population on a university campus in 2006. A total of 185 students participated. Each student was instructed to flip a coin and conceal the outcome. He or she was to respond with a "yes" if the outcome was heads. If the outcome was tails, the student was to answer the theft question truthfully with a "yes" or a "no." The result: 113 of the 185 students responded with a "yes," whereas the remaining 72 answered "no." Assume that students answered truthfully and independently, that the probability of heads was 0.5, and that the sample of students was a random sample. Use likelihood to estimate the fraction of thieves in the student population.

a. Construct a probability tree (Chapter 5) to show that the probability of a student answering "yes" is $(1+s)/2$, where s is the fraction of thieves in the population.

b. If the assumptions are met, the number of "yes" answers in the survey of n students, Y, should follow a binomial distribution with the probability of a "yes" equal to $(1+s)/2$:

$$\Pr[Y \text{ yeses} \mid s] = \binom{n}{Y}\left(\frac{1+s}{2}\right)^Y\left(\frac{1-s}{2}\right)^{n-Y}.$$

Write the formula for the likelihood of s, given the data.

c. Write the formula for the log-likelihood of s, given the data.

d. Calculate the log-likelihood that the fraction of thieves s is zero.

6. Refer to Practice Problem 5.

a. Using a spreadsheet program, calculate the log-likelihood of values of s between 0.0 and 0.5 in increments of 0.01. Using this information, determine the maximum likelihood estimate of the fraction of thieves.

b. Using the same approach as in part (a), calculate the likelihood-based 95% confidence interval for the parameter s.

c. Are there truly thieves among us? Using the values for the likelihood calculated in your answers to Practice Problem 5, use the log-likelihood ratio test to test the null hypothesis that s is zero.

7. A regulatory gene controls the expression of other genes—that is, it turns them on and off. Many of these targets are themselves regulatory genes, instructing genes to turn on or off. Guelzim et al. (2002) determined the number of regulatory genes controlled by a sample of genes in yeast. Their data are listed in the following table.

Number of regulatory genes controlled i	Frequency f_i
0	72
1	18
2	10
3	5
4	1
5	3
Total	109

The shape of this frequency distribution suggests that it might be approximated by a proba-

bility distribution known as the *geometric distribution*. Under a geometric distribution, the fraction of genes controlling no regulatory genes (here, $i = 0$) is p. The fraction controlling exactly one regulatory gene ($i = 1$) is $(1 - p)p$. The fraction controlling exactly two regulatory genes ($i = 2$) is $(1 - p)^2 p$, and so on, yielding

$$\Pr[i] = (1 - p)^i p,$$

where i is 0, 1, 2, or more. Maximum likelihood methods can be used to estimate the parameter p, the fraction controlling no regulatory genes, from data. The log-likelihood formula for p is

$$\ln[L[p \mid data]] = \ln[(1 - p)]\left(\sum i f_i\right) + n \ln[p],$$

where f_i is the frequency of observations corresponding to $i = 0, 1, 2$, and so on, and n is the total sample size.

a. Using the data in the table and the preceding formula, calculate the log-likelihood of values of p between 0.1 and 0.9 in increments of 0.01. Use a spreadsheet program. Draw the relationship between the log-likelihood and p with a log-likelihood curve.

b. What is the maximum likelihood estimate $\hat{p}$, to an accuracy of two decimal places?

c. What is the value of the log-likelihood at the maximum likelihood estimate $\hat{p}$?

d. Using the formula for the geometric distribution, $\Pr[i] = (1 - p)^i p$, calculate the predicted proportion of genes regulating $i = 0$ to 5 genes based on $\hat{p}$. Plot these values on a histogram with the observed proportions. Based on the result, do the data appear to follow a geometric distribution?

e. Identify one method you might use to test the null hypothesis that a geometric distribution fits the data.

8. Phylogenetic trees like those in Figure 20.2-1, if dated, permit us to estimate the rate at which new species have formed over time. For example, the following tree indicates the timing of events in the history of the living species of *Dubautia* of the Hawaiian silverswords (Baldwin and Sanderson 1998), a famously diverse group of plants found only on the Hawaiian islands.

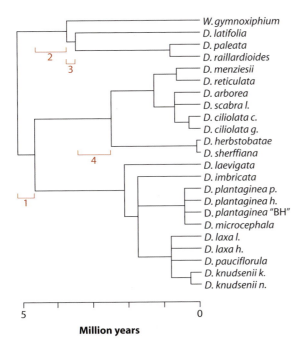

W. gymnoxiphium
D. latifolia
D. paleata
D. raillardioides
D. menziesii
D. reticulata
D. arborea
D. scabra l.
D. ciliolata c.
D. ciliolata g.
D. herbstobatae
D. sherffiana
D. laevigata
D. imbricata
D. plantaginea p.
D. plantaginea h.
D. plantaginea "BH"
D. microcephala
D. laxa l.
D. laxa h.
D. pauciflorula
D. knudsenii k.
D. knudsenii n.

Million years

Assume that the number of *Dubautia* species has increased exponentially over time and has suffered no extinctions. In this case, the time interval between successive branching events on the tree provides information about the rate of increase of the number of species, λ ("lambda"). The earliest four intervals are indicated in red on the tree; with 23 species, there are $n = 21$ such intervals in total. Let Y_i measure the duration of each interval i multiplied by the number of species alive at that time. We'll call these Y_i the "waiting times." The $n = 21$ values of Y_i from the above tree are

1.0, 3.0, 0.8, 5.0, 2.4, 2.1, 0.8, 3.6, 5.0, 0.0, 0.0, 0.0, 1.4, 0.0, 1.6, 3.4, 0.0, 0.0, 2, 0.0, 4.4

Under the assumptions given, the Y_i values should follow an *exponential* distribution (Hey 1992), with probability highest at zero and declining smoothly with increasing Y. To illustrate, an exponential distribution is superimposed on the following histogram of the Y_i values.

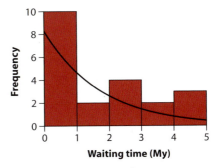

Use the waiting-time data to estimate the rate λ. The formula for the log-likelihood of λ, given the Y_i, is

$$\ln L[\lambda \mid observed\ waiting\ times]$$
$$= n \ln[\lambda] - \lambda \sum Y_i.$$

a. Using a spreadsheet program, calculate the log-likelihood of values of λ between 0.1 and 0.9 in increments of 0.01. Draw the relationship between the log-likelihood and λ with a log-likelihood curve.

b. Find the maximum likelihood estimate of λ to two decimal places. This estimates the number of new species produced per species per million years in *Dubautia*.

c. Using a similar approach as in part (b), find the likelihood-based 95% confidence interval for λ.

ASSIGNMENT PROBLEMS

9. Huntington disease is an inherited neurodegenerative disease caused by a mutation in the huntingtin gene. When did this mutation arise? García-Planells et al. (2005) found that most Spanish cases of the disease share the same ancestral mutation. Using gene sequence data from patients at genetic loci linked to the mutation, the authors applied a likelihood method similar to that used to date events in the human–ape lineage (Figure 20.2-1). Their calculations produced the following log-likelihood curve for the date of origin of the huntingtin mutation, measured as the number of generations before the present.

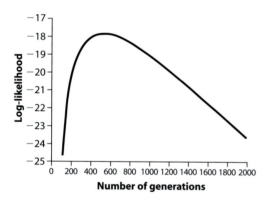

Number of generations

a. With the aid of a ruler, use this curve to find the maximum likelihood estimate for the date of origin of the mutation shared by most Spanish cases of Huntington disease.

b. Using this same curve, approximate a 95% confidence interval for the date of origin.

c. What does the interval in part (b) measure? Explain.

10. The Mediterranean shrub *Thymelaea hirsute* has five sexual types, the most curious of which is "gender labile." Such individuals change their predominant sex from year to year. Ramadan et al. (1994) found 13 gender-labile individuals in a sample of 68 shrubs from a single habitat in Egypt.

a. What probability distribution would we use to calculate the probability of a specific number of gender-labile individuals in a sample of size 68?

b. Write the formula for the likelihood of p, given the data. What does this likelihood measure?

c. What are your assumptions in part (b)?

d. Write the formula for the log-likelihood of p.

e. Calculate the log-likelihood of the value $p = 0.40$, given the data.

Spreadsheet for Problem 11.

	A proportion p	B log-likelihood	C log-likelihood - maximum
1	proportion p	log-likelihood	log-likelihood - maximum
2	0.10	-4.652	-2.549
3	0.11	-4.027	-1.925
4	0.12	-3.517	-1.415
5	0.13	-3.105	-1.003
6	0.14	-2.778	-0.676
7	0.15	-2.524	-0.422
8	0.16	-2.336	-0.234
9	0.17	-2.207	-0.104
10	0.18	-2.130	-0.028
11	0.19	-2.102	0.000
12	0.20	-2.119	-0.016
13	0.21	-2.176	-0.074
14	0.22	-2.272	-0.170
15	0.23	-2.404	-0.302
16	0.24	-2.570	-0.467
17	0.25	-2.768	-0.665
18	0.26	-2.996	-0.894
19	0.27	-3.254	-1.151
20	0.28	-3.539	-1.437
21	0.29	-3.853	-1.750
22	0.30	-4.192	-2.090

11. Refer to Assignment Problem 10. The spreadsheet on the previous page lists log-likelihoods for a range of values of p between 0.1 and 0.3, in increments of 0.01.
    a. Using only our calculations based on these increments, identify the maximum likelihood estimate.
    b. Using these same calculations, generate a likelihood-based 95% confidence interval for p.

12. Sacktor et al. (2000) measured the neuropsychological performance of 33 HIV-positive patients undergoing antiretroviral therapy. Hand-use performance improved in 23 of the 33 patients but deteriorated in the remaining 10 patients.

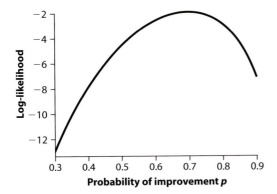

The preceding graph shows the log-likelihood curve for the population proportion p of patients whose hand-use performance improved after antiretroviral therapy.
    a. Using only the information in the log-likelihood curve, test the null hypothesis that the probability p is 0.5 using the log-likelihood ratio test.
    b. Using the same curve, provide an approximate 95% confidence interval for the proportion of patients improving in the population.

13. Ants are capable of chemically discriminating between nestmates and non-nestmates. Ozaki et al. (2005) discovered sensory cells in the antennae of ants that respond only to the cuticular hydrocarbons (CHCs) of non-nestmates. They showed that 42 of 48 head-attached antenna preparations responded to chemical extracts of CHC compounds obtained from non-nestmates. Use likelihood to estimate the proportion of preparations p responding to non-nestmate CHC extracts.
    a. Using a spreadsheet program, calculate the log-likelihood of a series of values of p between 0.50 and 0.99 in increments of 0.01. Draw the relationship between the log-likelihood and p with a log-likelihood curve.
    b. Find the maximum likelihood estimate for p to two decimal places.
    c. Calculate the likelihood-based 95% confidence interval for the parameter p.

14. Although we used the log-likelihood ratio test to analyze the data on the preference of parasitic wasps for female butterflies (Example 20.3), we could have used the binomial test (Chapter 7) or the χ^2 goodness-of-fit test (Chapter 8). The data are reproduced in the following frequency table.

Preference	Observed frequency
Chose mated female	23
Chose virgin female	9
Total	32

Analyze the data again, but this time use the G-test, a goodness-of-fit test that we learned about in Chapter 9. The formula for the G-statistic is given in the following equation, where $Observed_i$ and $Expected_i$ refer to the observed and expected frequencies in category i, respectively. Under the null hypothesis, G is approximately χ^2 distributed with $k - 1$ degrees of freedom, where k is the number of categories in total.

$$G = 2\sum Observed_i \ln\left[\frac{Observed_i}{Expected_i}\right]$$

    a. Restate the hypotheses for the parasitic wasp and butterfly data, and provide the expected frequencies under the null hypotheses.
    b. Calculate the G-statistic using the formula provided.
    c. Compare your result in part (b) with the critical value from the χ^2 distribution with the

appropriate degrees of freedom and state your conclusion.

d. Compare the value of G that you calculated in part (b) with the value of the log-likelihood ratio statistic that we calculated in Section 20.3. Based on your comparison, what can you infer about the G-test?

15. Dispersal and the movement distance of organisms are sometimes described using a probability distribution known as the geometric distribution (see Practice Problem 7). For example, the following histograms show the number of home ranges separating the locations where individual male and female field voles, *Microtus agrestis*, were first trapped, and the locations they were caught in a subsequent trapping period (Sandell et al. 1991). Most individuals didn't move at all, a few moved just one home range away, and a small fraction moved two home ranges.

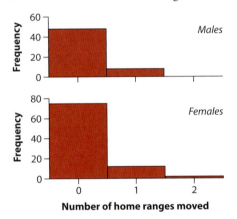

Number of home ranges moved

If these frequencies are described by a geometric distribution, then the fraction of observations in the first category (here, $i = 0$) is given by the parameter p. The fraction in the second category ($i = 1$) is $(1 - p)p$. The fraction in the third category ($i = 2$) is $(1 - p)^2 p$, and so on, yielding

$$\Pr[i] = (1 - p)^i p,$$

where i is 0, 1, 2, or more. Maximum likelihood methods can be used to estimate the parameter p from the data. The log-likelihood formula for p is

$$\ln L[p \mid data] = \ln[1 - p]\left(\sum i f_i\right) + n \ln[p],$$

where f_i is the frequency of observations corresponding to $i = 0$, 1, 2, and so on.

a. Forty-eight of 56 male voles stayed put ($i = 0$), whereas the remaining eight moved one home range. Using a spreadsheet program, calculate the log-likelihood of values of p between 0.70 and 0.99 using intervals of 0.01. Draw the relationship between the log-likelihood and p with a log-likelihood curve.

b. Find the maximum likelihood estimate $\hat{p}$ for male voles to two decimal places.

c. Obtain a likelihood-based 95% confidence interval for p.

16. Refer to Assignment Problem 15. Seventy-five of 89 female voles remained where they were first caught ($i = 0$), 12 moved just one home range away ($i = 1$), and two moved two home ranges away ($i = 2$).

a. Repeating the procedures described in Assignment Problem 15, calculate the maximum likelihood estimate of p for females.

b. Similarly, find a likelihood-based 95% confidence interval for p. Does it overlap that for males? (See Assignment Problem 15.)

17. Is the perception of eye-gazing in humans acquired through experience or is it innate? To test this, Farroni et al. (2002) presented 17 infants (3–5 days old) with paired photographs of faces. The face in one photograph had a direct gaze, whereas the other had an averted gaze (see image). Fifteen of 17 infants spent more time gazing at the face with the direct gaze. Use the

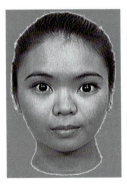
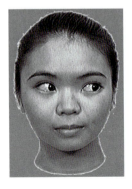

log-likelihood ratio test to test the null hypothesis of no preference. Show the steps of your calculations.

18. Lifespans of individuals in a population often approximate an exponential distribution, a continuous probability distribution having probability density $f(Y) = \lambda e^{-\lambda Y}$, where λ is the mortality rate. To estimate the mortality rate of foraging honey bees, Visscher and Dukas (1997) recorded the entire foraging lifespan of 33 individual worker bees in a local bee population in a natural setting. The 33 lifespans (in hours) are listed as follows and in a histogram.

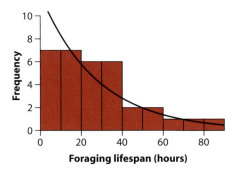

Foraging lifespan (hours)

2.3, 2.3, 3.9, 4, 7.1, 9.5, 9.6, 10.8, 12.8, 13.6, 14.6, 18.2, 18.2, 19, 20.9, 24.3, 25.8, 25.9, 26.5, 27.1, 30, 33.3, 34.8, 34.8, 35.7, 36.9, 41.3, 44.2, 54.2, 55.8, 65.8, 71.1, 84.7

If lifespan follows an exponential distribution, then the log-likelihood of λ, given the data is

$$\ln L[\lambda \mid observed\ lifespans] = n \ln(\lambda) - \lambda \sum Y_i,$$

where Y_i is the lifespan of individual i, and n is the sample size.

a. Estimate λ, the mortality rate of bees per hour. Using a spreadsheet, find the maximum likelihood estimate of λ to two decimal places.

b. What is the value of the log-likelihood at the maximum likelihood estimate $\hat{\lambda}$?

c. What is likelihood-based 95% confidence interval for λ?

21

Meta-analysis: combining information from multiple studies

Most papers in scientific journals present the outcome of just one experiment or of one observational study. Each is typically based on a single random sample or on a small number of related samples. Compelling issues in biology, however, deserve more than one study.

You will rarely find only one published study on a topic, unique in addressing an interesting problem. The first test of a truly original idea is almost always followed by studies that duplicate it to some extent. Most new studies typically address a question that has been asked and answered before, often many times. For example, by 1985 at least 118 data sets had been published testing whether the phase of the moon affects human behavior (Rotton and Kelly 1985). Twenty-four studies have investigated whether acupuncture can help you stop smoking (White et al. 2006). Studies on a topic accumulate in the literature, and at some point these must be summarized to

Photo: Copyright 2007, Fred Espenek, www.MrEclipse.com

yield an overall conclusion. How do we combine the information that we get from multiple studies?

The review article is the traditional answer to this question. An expert in the field assembles the studies published on a topic, thinks about them carefully and (hopefully) fairly, and then writes a review article summarizing the overall conclusions reached. There is much to be said for this approach. A first-rate review article advances a field far beyond a mere summary. Such a review will propose new hypotheses, uncover previously unnoticed relationships, and point to new paths of research. A strong review provides a solid indication of the state of thought and knowledge about a particular topic.

But the traditional review usually lacks a quantitative methodology, which might lead to two problems. First, the conclusions of one reviewer are often partly subjective, perhaps weighing studies that support the author's preferences more heavily than studies with opposing views. An extreme example of this is Linus Pauling's (1986) book *How to Live Longer and Feel Better*, in which the author cited 30 studies supporting his own idea that taking large daily doses of vitamin C reduces the risk of contracting the common cold. Pauling[1] cited no studies opposing this idea, even though a number of such studies had been published (Knipschild 1994). Not all reviews are blatantly biased, but complete objectivity is difficult for any reviewer to achieve when evaluating a diverse set of published studies.

The second problem is that it is extremely difficult to balance multiple studies by intuition alone without quantitative tools. The combined strength of statistical support and the average magnitude of an effect simply cannot be determined without some calculations. "Vote-counting" is a step in the right direction—that is, count the number of published studies that have rejected the null hypothesis of zero effect and compare this count to the number of studies failing to reject the null hypothesis. Yet,

---

[1] Linus Pauling is the only person to have won two unshared Nobel Prizes, Chemistry in 1954 and Peace in 1962, so his behavior does not represent that of a crank.

even this approach ignores important information on the size of the study, the strength of an effect, and the magnitude of our uncertainty.

A set of techniques called *meta-analysis* provide quantitative tools for the synthesis of information from multiple studies. In this chapter, we give an overview of the approach using examples, and we point out advantages and limitations. We also indicate how best to report the results of your own studies so that your results will be useful to future meta-analyses.

21.1   What is meta-analysis?

Meta-analysis is the "analysis of analyses." It is not a single statistical test but rather a set of techniques that allows us to address statistical questions about collections of data obtained from the published literature and other sources. The researcher begins a meta-analysis by gathering all known studies about a particular topic. Typically, these studies report a broad range of results, with some rejecting the null hypothesis and others not rejecting the null hypothesis. Given such a broad range of results, how can we decide what the truth is? Meta-analysis provides ways to combine information from independent studies.

With meta-analysis, we can address questions such as:

- ▶ How strong is the effect we are studying, on average?
- ▶ Does the collection of studies reject a null hypothesis?
- ▶ How variable is the effect? What factors might explain this variability?
- ▶ Is there any bias resulting from the publication of some studies and not others? Are there enough unpublished studies to change our conclusions?

The quantitative details of meta-analysis lie beyond the scope of this book. If you want to know more about the techniques, have a look at Rosenthal (1991), Cooper and Hedges (1993), Hedges and Olkin (1985), or Green and Higgins (2006).

> A *meta-analysis* compiles all known scientific studies estimating or testing an effect and quantitatively combines them to give an overall estimate of the effect.

Why repeat a study?

Why would anyone bother to test an idea that has already been addressed in one or more previous studies that have already provided answers? Repetition is necessary in science, because errors—both statistical errors and methodological flaws—can be made in individual studies. A single study might yield a significant result when the null hypothesis is true (a Type I error) or it may fail to detect a true effect (a Type II error). Moreover, what is true in one circumstance may not be true in another. For example, one kind of animal might tend to evolve to larger size on islands (think of the three-meter Komodo dragon on the Indonesian islands of Komodo and Flores), whereas another kind of animal may do the opposite (for example, the five-foot pygmy elephants on Flores, thought to be the original diet of the dragons). Moreover, we all make mistakes sometimes, so experimental designs can be flawed or errors can be made in recording data. In short, we need to have more than one study on any phenomenon that we care about.

21.2 The power of meta-analysis

One of the most useful aspects of meta-analysis is that it can greatly increase the statistical power of our collective statistical tests. Any one study will have limitations, sometimes severe, on the amount of time and money available to do research. As a result, the sample size of any given study is limited, yet the power of a study depends on sample size. We can combine the power of individual studies when we address the scientific question with meta-analysis. As a result, weak but important effects can sometimes be detected. This power can be particularly important in medical studies, when even a small reduction in mortality rates from an inexpensive and relatively harmless treatment can result in thousands of lives saved.

Example 21.2	**Aspirin and myocardial infarction**

People at risk of stroke or myocardial infarction (a "heart attack") are often recommended to take aspirin or another antiplatelet medication. These kinds of medication are known to reduce the risk of future stroke, myocardial infarction, and death (all of which we will refer to as "vascular events"). But how do we know this is true? It turns out that the effect of aspirin and other antiplatelet agents was confirmed by a large meta-analysis, conducted by a large collaboration of researchers (Antiplatelet Trialists' Collaboration 1994). They combined information from 142 randomized trials of antiplatelet treatments on patients who had previously had a stroke or similar disease, yielding a total of more than 70,000 patients. The results of these trials are summarized in Figure 21.2-1.

Before the meta-analysis, whether aspirin helped prevent vascular events was far from clear. Of the 142 studies reviewed, only 87 showed a better result for patients on antiplatelet therapy than for the control patients. Moreover, only 19 of these 87 showed significantly better results for the treated patients than the control patients at

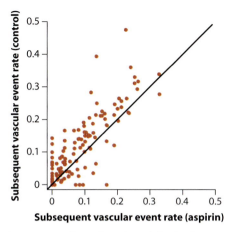

Figure 21.2-1 The vascular event rate for patients in antiplatelet therapy compared with patients in the corresponding control group. Each dot is a separate study, except at the origin, where 28 studies overlap. The one-to-one line marks equal vascular event rates in the two treatments.

$\alpha = 0.05$, while 68 had non-significant results. Two of the 142 studies even showed a significantly worse rate of vascular events with aspirin treatment! A simple "vote-counting" procedure does not give overwhelming support for aspirin.

When the results are combined in a meta-analysis, however, the answer is surprisingly robust. It was relatively straightforward to combine the results because all of the experiments involved the same kind of data: frequencies of control and treated patients who either did or did not have subsequent vascular events. Each of these studies had roughly the same numbers of treated individuals as control individuals. For display purposes only, we added up the frequencies of patients in each category and produced the contingency table shown in Table 21.2-1.

Table 21.2-1 Numbers of patients in each treatment group and outcome category in 142 studies combined.

	Treated with antiplatelet therapy	Control (no antiplatelet therapy)	Row totals
Vascular event (stroke, infarction, death)	4,183	5,400	9,583
No vascular event	32,353	31,311	63,664
Column Totals	36,536	36,711	73,247

In total, 14.7% (5,400/36,711) of the patients in the control groups had subsequent vascular events, compared with 11.4% (4,183/36,536) in the treated group. This difference may seem small, but it corresponds to the control patients having a 30% higher chance of serious problems than the treated patients. Put another way, if just the individuals in the controls of these experiments had been given antiplatelet treatment, there would have been about 1200 fewer strokes, infarctions, and deaths.

Since the side effects associated with antiplatelet treatment are low, this result (if statistically supported) would justify using aspirin. We cannot do a simple odds-ratio (OR) calculation on the combined table, because the data points are not all independent. We can, however, combine the odds ratios for each study using the Mantel–Haenszel confidence interval (see the Quick Formula Summary in Section 21.8). The Mantel–Haenszel procedure is specifically designed to combine information across multiple contingency analyses. Doing so, the researchers found that taking aspirin had a clear, beneficial effect ($0.71 < OR < 0.75$).

Without the analysis of the data across all available studies, there would have been little chance of confirming the value of this relatively important treatment. In fact, the meta-analysis came 10 years after it would have been possible to show an effect by combining across studies that already existed at the time. It took another six years before the results of this meta-analysis were accepted by the medical community, perhaps in part because of the relative newness of meta-analysis techniques (Hunt 1997). The number of extra deaths and suffering in the intervening years that could have been avoided by earlier application and acceptance of these meta-analysis techniques is sobering to think about.

21.3 Meta-analysis can give a balanced view

Studies that find dramatic results are sometimes given more attention in the press and scientific literature than better studies with less "newsworthy" results. As a consequence, we require fair summaries of the breadth of knowledge on a question to give us a balanced view. Meta-analysis attempts to cover all studies fairly.

Example 21.3	**The Transylvania effect**

Many people believe that a full moon can affect human behavior. The word lunacy is derived from the Latin *luna*, moon, and legends of strange happenings, such as werewolves and vampires, have been connected to full moons for centuries. Some studies have shown that abnormal human behavior increases during full moons, as measured by such variables as homicide rates, psychiatric hospital admissions, suicide rates, crisis calls, etc. This phenomenon is called the "Transylvania effect." Other studies, however, have found no significant effects. A meta-analysis of these studies (Rotton and Kelly 1985) found that there was no statistically significant change in "lunacy" during the full moon. Moreover, the best estimate of the average effect was that less than 1% of the variation in these events was explained by moon phase. Even if there were an effect of the moon on human abnormal behavior, the average effect is so small as to be unimportant. The results are shown in the funnel plot (Interleaf 10) in Figure 21.3-1.

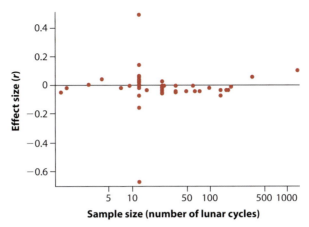

Figure 21.3-1 A funnel plot of the effect sizes from the meta-analysis of the "Transylvania effect." The horizontal line marks zero correlation between moon phase and unusual behavior.

The literature on the "Transylvania effect" includes several studies with apparently very strong relationships between moon phase and human behavior. The most extraordinary of these papers have gotten a great deal of media attention and word-of-mouth among police and hospital workers. Thus, there is a tendency to believe that the sensational studies are typical of all those performed. This meta-analysis shows that these are just the tips of the sampling iceberg; when all data sets are examined, there is little or no effect of moon phase on aberrant human behavior.

21.4   The steps of a meta-analysis

In this section, we describe the process of combining information using meta-analysis. We focus on some of the issues that arise when summarizing data, but we don't provide mathematical details.

Define the question

A key step in meta-analysis is defining the question and the breadth of studies to which it applies. Some meta-analyses apply a question to a very narrow set of studies, all of which are expected to estimate the same true value. Meta-analyses in medical research are usually of this type. For example, in the aspirin/heart attack meta-analysis of Example 21.2, the question was whether aspirin affected the risk of future myocardial events. To answer this question, the authors reviewed studies that had very similar properties. They included only studies that were all done on human patients who had suffered a stroke or heart attack. All of the studies were randomized clinical trials comparing an aspirin treatment to a control treatment, and all measured the future risks to the patients in terms of significant health problems—namely, heart attacks, strokes, or death. Different studies, therefore, estimated very similar effects, and all should give more or less the same answer, except for sampling error. The goal of the meta-analysis was to combine the multiple studies—in effect, to create one large study with more power than any of the single studies. A meta-analysis of such a homogeneous set of studies is called a "fixed-effect" meta-analysis.

At the other extreme, a meta-analysis might address a question whose answer requires reviewing a broad and heterogeneous mix of studies. Meta-analyses in non-medical areas of biology are often of this type. For example, Kingsolver et al. (2001) conducted a meta-analysis to estimate the average strength of natural selection in the wild. Answering this question required combining studies carried out on different species living in different kinds of environments. The variables measured were not the same from study to study. A meta-analysis of such a heterogeneous set of studies is called "random-effect" meta-analysis. In this case, the goal is to obtain an average over the separate effects of all the studies. If the studies in the meta-analysis are similar enough, and if we can standardize their responses in some way, then the average effect may be informative.

Example 21.4 describes a meta-analysis of a heterogeneous mix of studies, even though all were carried out on the same species.

Example 21.4   **Testosterone and aggression**

Using a meta-analysis, Book et al. (2001) asked the question, "Are testosterone levels and aggression correlated in humans?" The studies reviewed for this meta-analysis were all done on humans, but the ways aggression was measured were extremely diverse. Some compared the levels of testosterone in prisoners convicted of violent crimes to those of prisoners convicted of property crimes. One correlated the levels of testosterone in university students with their answers to questionnaires that asked them for levels of agreement to statements like "If somebody hits me, I hit back." Others were less prosaic. One measured the levels of aggression in !Kung San males by counting "their scars and sometimes still open wounds in the head region." Another compared drunken Finnish spouse-abusers to drunken Finns drinking quietly in a bar. Another compared members of "rambunctious" fraternities to "responsible" frats.[2]

In this example, "aggression" was defined in a variety of ways. They tested the null hypothesis: "Testosterone has no average effect on human aggression levels."

---

[2] One of the "rambunctious" frat houses was described as "only standing because it was made out of steel and concrete." At one of the "responsible" frats, they "talked a lot about computers and calculus."

The authors wanted to ask a rather broad question with a variety of types of people and types of aggression, so many studies were relevant. These studies do not repeat themselves; they each address different *specific* questions, but they all address the same *general* question. Meta-analyses in biology can be even broader, often including results from a large number of species. This is appropriate, if the question is general enough, but it should always be remembered that the different sub-groups in the meta-analysis may actually have different responses.

Review the literature

Once the question is defined, the hardest part of a meta-analysis ensues. We must collect all of the available information that pertains to the question at hand. In principle, this is simple, but, in practice, collecting even a fraction of the pertinent literature can be a Herculean task. Meta-analysts often take years to collect the information for their reviews, although computerized databases are making the task easier. Most libraries do not carry subscriptions to all journals, so getting copies of some articles can be difficult. More importantly, many good studies are not published in journals at all and therefore are not in computer databases. Sometimes they are reported in the "gray literature"—monographs and in-house publications of various foundations, institutes, and government agencies. Lots of science is only reported in doctoral dissertations and Master's theses. Finally, many studies are not published at all, so an effort has to be made to contact researchers in the field to discover what information exists but that is not widely available.

It is crucial to find everything because of various forms of "publication bias." As discussed in Interleaf 10, studies that find large and significant effects are more likely to be published, more likely to be in "first-rate" journals, and more likely to be referenced in other articles. All of this means that the studies that we can find easily are *different* from those that we cannot so easily find. Some statistical techniques exist to partially account for such biases (see the discussion of funnel plots in Interleaf 10 and the "file-drawer problem" in Section 21.5), but they are not nearly as reliable as a proper literature review.

Meta-analyses contain from as few as a handful of studies to as many as thousands, depending on the amount of information available, the amount of effort expended by the meta-analysts themselves, and the breadth of the question asked. The more information there is available for analysis, the more reliable the final results are likely to be.

One key issue that arises when compiling studies is whether to include studies of apparently poor quality. An unfortunate fact of science is that not all studies are done well; in spite of long training, large amounts of money spent on research, and the peer review process, some bad science is still published. What do we do about these studies? We could just ignore them (and thus adhere to the maxim "garbage-in, garbage-out"). But this raises a thorny point: we may be more likely to reject a study as "bad" if its conclusions do not agree with our own opinions about the scientific question at hand. Good meta-analyses circumvent this problem by having the methods (alone, separated from the results) of each study rated by independent scientists and then weighting the study's contribution to the meta-analysis by the judges' scores. Others set standards a priori, such as collecting only studies that experimentally and randomly assigned individuals to treatments and controls. This question of how to deal with varying quality of science is not fully resolved, and it represents one of the greatest challenges for the future of meta-analysis.

Compute effect sizes

The core quantity of meta-analysis is the **effect size**, which measures how strong the association is between the explanatory variable X and the response variable Y. The magnitude is important—for example, we might need to know whether the effect of a drug is large enough to exceed any negative consequences or costs the treatment might have.

> The *effect size* is a standardized measure of the magnitude of the relationship between two variables.

In meta-analysis, we are combining across multiple studies, and usually these studies do not measure their results on the same yardstick. Some may even be measuring completely different things. How do we combine measures of "aggression" that range from the number of scars (a numerical variable) to whether or not someone is imprisoned for murder (a categorical variable)? How do we combine information that comes from a correlation analysis with results from ANOVAs, t-tests, and goodness-of-fit tests? Fortunately, many methods have been developed for putting these kinds of answers on a common, standardized scale. These standardized answers are called effect sizes.

Most meta-analyses use one of three common ways to derive an effect size. These are the odds ratio (OR; see Section 9.2), the correlation coefficient (r; see Section 16.1), and the standardized mean difference (SMD; see below). Which measure of effect size is used depends on the variables. The odds ratio compares the frequencies of "success" and "failure" between two groups or treatments. The correlation coefficient typically measures the relationship between two numerical variables, although it can also be generalized to describe many kinds of data (Rosenthal 1991). The standardized mean difference (SMD) is useful when most studies being combined compare the mean of a numerical response variable between two groups. SMD is defined as

$$\text{SMD} = \frac{\overline{Y}_1 - \overline{Y}_2}{s_{\text{pooled}}},$$

where s_{pooled} is the square root of the pooled sample variance. In other words, SMD is the difference in the sample means of two groups divided by the pooled estimate of their standard deviation (the square root of the pooled sample variance). SMD measures the difference between groups in units of standard deviation.[3]

What these effect size measures have in common is that the results of each study can be expressed on a common scale. For any one meta-analysis, only one effect-size measure is used. The goal is to make the result of any one study to be expressed on the same scale as all of the other studies. In all cases, the effect sizes must be calculated in the same way. For example, when using SMD, the mean of the control group should be subtracted from the mean of the treatment group in every study. If a positive effect size indicates improvement in one study, then every other study that showed an improvement associated with the treatment ought to have a positive effect size, too.

Table 21.4-1 lists results from nine of 54 studies in the meta-analysis of testosterone and aggression (Book et al. 2001; only nine are shown here to keep the table brief). Of these nine studies, the original test statistic was either a correlation coeffi-

Table 21.4-1 A subset of the results of a meta-analysis on the relationship between testosterone and aggression in humans.

References	Type of study	Test statistic type	Test statistic	Effect size (r)	df	P
Gray et al. (1991)	Aging men (39–70); questionnaires	r	0.02	0.02	1677	0.21
Houser (1979)	20-something males; questionnaires	r	0.086	0.086	3	0.45
Olweus et al. (1980)	Swedish male adolescents; self reports of aggression	r	0.24	0.24	56	0.035
Banks and Dabbs (1996)	College students compared with "Americans who belonged to a deviant and delinquent urban subculture"	F	5.83	0.30	1,61	0.008
Harris et al. (1996)	University students; questionnaires	r	Men: 0.36 Women: 0.41	0.36 0.41	153 149	1.7×10^{-6} 5.8×10^{-8}
Christiansen and Winkler (1992)	!Kung San men; number of scars from fights	t	0.65	0.06	112	0.27
Lindman et al. (1992)	Drunken Finns arrested for spousal abuse vs. drunken Finns in bars	t	1.06	0.18	34	0.15
Dabbs et al. (1996)	"Rambunctious" frats vs. "responsible" frats	F	6.59	0.26	1,93	0.006

---

[3]   There are other measures similar to SMD that standardize the difference between the means by slightly different ways, including Cohen's *d* (which uses the $\sqrt{s_1^2 + s_2^2}$ in place of s_{pooled}) and Hedges' *g* (which uses the standard deviation from the control group in the denominator).

cient (*r*), a *t*-statistic, or an *F*-ratio. All test statistics were converted to the equivalent correlation coefficient *r* to put them on the same scale. [See Rosenthal (1991) for computational details.] Degrees of freedom (*df*) are also listed in Table 21.4-1, indicating the size of each study, along with the *P*-value.

Several of these nine studies rejected the null hypothesis that there is no association between aggression and testosterone, but not all. The funnel plot in Figure 21.4-1 shows how variable the results are among all 54 studies. Some studies show a strong positive relationship between testosterone and aggression, whereas others show no significant effect. One study even shows a significant negative correlation. This variability stems in part from sampling error, but it also likely reflects differences between studies in the types of variables measured and the populations sampled.

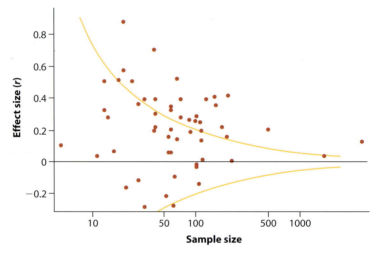

Figure 21.4-1 A funnel plot of studies comparing human aggression to levels of testosterone. The curves show the approximate boundaries of the critical regions that would reject the null hypothesis in any one study with $\alpha = 0.05$.

Determine the average effect size

Once we have all the effect sizes in comparable units, we can calculate the average effect over all of the studies. Typically, however, we do not take just a simple average of the individual effect sizes. Instead, we acknowledge that some studies give us more information than others, and we weight studies according to the precision of their estimates. The most common reason that some studies have more information is simply that their sample sizes were larger. It is also possible to weight studies by their quality; thus, studies done with superior methodology can be weighted more heavily than those with more suspect technique.

Calculate confidence intervals and make hypothesis tests

Meta-analysis techniques make it possible to calculate confidence intervals and test hypotheses for the mean effect size, thus accounting for the input of all studies included. Typically, the confidence intervals for the effect size averaged over all studies is much smaller than that obtained from any one study, and hypothesis tests are much more powerful than any of the individual studies. Different techniques are used for fixed-effects and random-effects meta-analyses. See Gurevitch and Hedges (1999) and Cooper and Hedges (1994) for further discussion.

Our example data show an average correlation between testosterone and aggression of $\bar{r} = 0.096$, with a 95% confidence interval of $0.055 < \bar{\rho} < 0.136$. In other words, the evidence points to a real effect, but it is quite small. Most of the individual studies were not powerful enough to detect an effect this small because their sample sizes were too small.

The mean effect size has to be understood in the context of the studies that are used to calculate it. Averaging over systematically biased estimates will still give a biased estimate. Moreover, the results can only be interpreted relative to the population of individuals that may have been included in the studies. If the studies are all about humans, then the interpretation can only be reasonably applied to humans. If the studies come from a diversity of species, then the conclusions can be applied to a mix of species like those in the studies. It is rare—well, to be honest, impossible— that the group of studies available is a truly random selection of the population that we want to know about. A mixture of studies to determine the effects of hormones on color expression, for example, would be heavily biased towards studies done on birds, so we would have to interpret the results according to this bias. Even in medical studies, the results are drawn disproportionately from large research hospitals, and it is certainly possible that the standard of care in these hospitals is systematically different from other, less-studied, health-care settings.

Look for effects of study quality

Not all studies in a meta-analysis are the same. They vary in sample size and in their methodologies. As part of a meta-analysis, we can and should ask whether these differences among studies matter.

Larger studies are more likely to give a more reliable answer, because the sampling error is likely to be lower. One way to address this issue is to look for a correlation between sample size and effect size. Such a relationship may arise because of publication bias, causing small studies that are published to have larger effect sizes than large studies. For example, in the meta-analysis comparing morphological asymmetry to mating success discussed in Interleaf 10, the funnel plot seemed to show a deficiency of small studies of small effect. If we use a Spearman's correlation to look for an association between effect size and sample size among those studies, we find a negative correlation ($r_S = -0.30$, $P = 0.003$)—in other words, the effect size is much smaller for large studies than for small studies. As a result, we should distrust the smaller studies in that meta-analysis. In contrast, in the meta-analysis of the effects of aspirin on vascular events (Example 21.2), there was no detectable relationship between sample size and effect size. In that meta-analysis, we have no reason to doubt the contribution of the smaller studies.

Moreover, there are often mean differences in effect size between studies of high quality and those of low quality. This difference may reflect the fact that some low-quality methods are biased. For example, studies without blinding (Section 14.3) have systematically larger effect sizes than those studies whose methods included this control (Jüni et al. 2001). Meta-analyses commonly find differences between observational and experimental studies, between randomized and nonrandomized studies, etc. When such differences are found, we should only use the better quality studies to draw our conclusions.

Look for associations

Another advantage of meta-analysis is that the different studies can be used to examine the effects of methodological or other differences between studies. That is, we are looking for **moderator variables**, variables that can explain some of the variation in effect size.

Scientifically interesting factors can be responsible for the heterogeneity of effect size across studies. Meta-analysis can suggest relationships that were not even addressed in any of the component studies. For example, a meta-analysis of the efficacy of homework showed that homework stimulated learning, despite claims to the contrary in the education literature (Cooper and Valentine 2001). Even more surprising, homework seemed to have little benefit for elementary school children, but very large benefits for high-school students. This effect could be shown despite the fact that none of the studies being analyzed treated this variable directly and no one had predicted it theoretically. A potentially important conclusion was reached by combining information across studies for questions that had not been explicitly addressed in any one study.

21.5 File-drawer problem

All of the meta-analysis techniques we have described assume that the studies being reviewed are a random sample of all possible studies on that topic. But there is one very important way in which the studies available for synthesis are not random: they tend to be drawn mainly from the published literature, and this literature usually suffers from publication bias. As mentioned in Interleaf 10, publication bias is the difference in mean effect size of published results from the mean of all studies on the topic.

In meta-analysis, the difficulties caused by publication bias are called the **file-drawer problem**, in reference to the unknown studies sitting unavailable in researchers' file drawers or unfound in obscure journals.

> The *file-drawer problem* is the possible bias in estimates and tests caused by publication bias.

Meta-analysis can increase power and reduce the Type II error rate—that is, combining across studies makes it easier to reject a false null hypothesis. In some cases, however, meta-analysis may have a much higher chance of rejecting a true null hypothesis (Type I error) than expected, if the data available for review are a biased sample.

There are a few methods that partially address these problems. Funnel plots (see Figure 21.4-1 and Interleaf 10), for example, can give some indication of the bias resulting from small studies. (If you haven't done so already, please read Interleaf 10. We won't repeat that material here.)

Another partial solution to the file-drawer problem is to calculate how many missing studies would be needed to change the overall result of the meta-analysis. A standard technique assumes that all of these missing studies failed to reject the null hypothesis of no effect. The method then calculates the number of missing studies required to reach the point at which the null hypothesis is no longer rejected by the meta-analysis. This number is called the **fail-safe number**.[4] If the fail-safe number is small (i.e., roughly the same as the number of published studies included in the initial meta-analysis), then the results of that meta-analysis would be regarded as unreliable. If the fail-safe number is very large (e.g., in the millions), then we can be more certain that the meta-analysis is giving us the right answer—it is simply too unlikely that there are a million unpublished studies out there on the subject.

The fail-safe number calculation assumes that the direction of the effect detected in a study does not influence the likelihood of publication. Unfortunately, studies that detect an effect opposite to that of most published studies might also be lost from publication. For example, studies that detect significant harm to patients taking a drug may be less likely to be published than those that find an advantage to taking the drug. If studies in the opposite direction from that desired also go missing from a meta-analysis database, then the fail-safe number cannot be reliably interpreted.

21.6   How to make your paper accessible to meta-analysis

Many published papers do not report enough information for meta-analysts to extract the numbers that they need. As a result, many otherwise relevant papers have to be

---

[4]   Calculation details of the fail-safe number, also called the "file-drawer method," can be found on p. 405 of Cooper and Hedges (1994).

discarded. This difficulty can be avoided by a few simple changes in the way information is presented. Here are some suggestions.

▶ **Always give estimates of the sizes of the effects and provide their standard errors.** Much too often, the size of the effect (e.g., the odds ratio, the correlation coefficient, etc.) is not given, and an author presents only a *P*-value. Also, give estimates of the mean and standard deviation of the important variables, as both may be needed for some effect-size calculations. It is surprising how often estimates of means and the sizes of effects are not given, even though it is the size of the effect that we care about, not just its statistical significance.

▶ **Always give precise *P*-values.** The convention in most journals has been to report *P*-values relative to some significance level—that is, $P < 0.05$ or even simply NS (not significant). This practice dates back to the days when computers were not readily available and test statistics were compared only to critical values in tables. Modern statistics packages for the computer give much more precise *P*-values, and these are essential for many meta-analysis techniques.

▶ **Give the value of your test statistics and the number of degrees of freedom.** The degrees of freedom are essential for most of meta-analysis. In particular, the degrees of freedom are needed for calculating a weighted average of effect size across studies. Similarly, the values of the test statistics are needed for some methods used to calculate effect sizes.

▶ **Make the data accessible.** Publish the raw data in the paper or on an online archive. If the data are available for scrutiny by others, the information in the study can be faithfully transcribed into future meta-analyses.

21.7 Summary

▶ Most scientific questions have been addressed by more than one published study. Meta-analysis is a general name for methods that quantitatively combine information from multiple studies that address the same question.

▶ Meta-analysis increases power and decreases the Type II error rate.

▶ An effect size is a standardized measure of the results of a study, used for all studies included in a meta-analysis. Common effect sizes include odds ratios, correlation coefficients, and standardized mean differences.

▶ The studies easily found in the literature are not a random sample of all studies done. Because of publication bias, the mean effect size in collections of published studies will often be larger than the true mean. Because meta-analysts have more difficulty finding unpublished studies than published studies, the values estimated via meta-analysis can be biased. This difficulty is called the file-drawer problem.

▶ One measure of the possible effect of publication bias is the fail-safe number. The fail-safe number is the number of hypothetical, unpublished studies failing to reject the null hypothesis that would need to be added to the meta-analysis to change its results.

▶ The confidence intervals of estimates of effect sizes are smaller, usually much smaller, for a meta-analysis than for its component studies, because the meta-analysis uses more information. On the other hand, meta-analysis can give estimates of effect size that are more biased than those from the larger studies they summarize, due to publication bias.

▶ Meta-analysis can also investigate the influence of study methodology and other variables not studied in the original articles.

▶ To make your study amenable to meta-analysis, provide exact calculations of effects, standard errors, standard deviations, P-values, test statistics, and degrees of freedom.

21.8 Quick Formula Summary

Mantel–Haenszel estimate of odds ratios from combined studies

What is it for? To estimate a confidence interval for an odds ratio based on data from multiple studies.

What does it assume? Data in each study are from random samples, there is no publication bias, and the odds ratio in the population is the same for all studies.

Degrees of freedom: 1

Formula: $OR_{MH} = \dfrac{\sum_{i=1}^{S} a_i d_i / n_i}{\sum_{i=1}^{S} b_i c_i / n_i},$

where a, b, c, and d are defined for each study according to the following table, s is the number of studies, and $n_i = a_i + b_i + c_i + d_i$.

		Explanatory variable	
		Treatment	Control
Response	Success	a_i	b_i
variable	Failure	c_i	d_i

The confidence interval for a Mantel–Haenszel estimate should be calculated with the aid of a computer.

Mantel–Haenszel test

What is it for? To test the null hypothesis that the odds ratio estimated in multiple studies equals one.

What does it assume? Data in each study are from random samples, there is no publication bias, and the odds ratio in the population is the same for all studies.

Test statistic: χ^2_{MH}

Distribution under H$_0$: χ^2 distribution with one degree of freedom.

Formula:
$$\chi^2_{MH} = \frac{\left[\left|\sum_{i=1}^{S}\left(a_i - \frac{(a_i + b_i)(a_i + c_i)}{n_i}\right)\right| - 0.5\right]^2}{\sum_{i=1}^{S}\frac{(a_i + b_i)(a_i + c_i)(b_i + d_i)(c_i + d_i)}{n_i^2(n_i - 1)}},$$

where the terms are defined as for the Mantel–Haenszel estimate of odds ratios.

PRACTICE PROBLEMS

Note: Some of these problems address topics related to publication bias, as discussed in Interleaf 10.

1. Give two reasons why researchers are more likely to publish results when $P < 0.05$ than when $P > 0.05$.

2. The accompanying funnel plot shows the results of studies that estimated the heritability of various traits in a large number of species. Heritability is the proportion of the variation among individuals in a population that can be explained by genetic differences among individuals. Because it is a proportion, heritability is a unitless number between zero and one. The methods of estimating heritability, however, allow estimates outside of these limits. The curve on this graph marks the critical values with $\alpha = 0.05$ for the null hypothesis that the population heritability equals zero.
   a. Look carefully at this funnel plot. Is there any evidence for publication bias?
   b. What kinds of decisions made by researchers and editors are behind the pattern?

3. The meta-analysis on the relationship between testosterone and aggression mentioned in Example 21.2 also included several studies on whether success in certain sports, such as tennis, was correlated with testosterone levels. Discuss the value of including these studies in this meta-analysis.

4. The file *Selection.txt* available online[5] contains the effect sizes and sample sizes from a meta-analysis of studies measuring the strength of natural selection on morphological traits in different wild animal and plant populations (Kingsolver et al. 2001). Make and examine a funnel plot of these results. (You don't need to draw the curve showing statistical significance.)

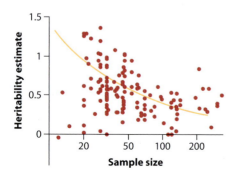

5   http://www.roberts-publishers.com/whitlock/teaching

5. Choose a paper from the scientific literature on a topic that you are interested in. Identify an interesting question that the paper addresses statistically, and attempt to extract from the paper the necessary information for a meta-analysis. What is the sample size? What is the effect size? What test statistic was used, and what was its value? What was the *P*-value of the test?

6. A hypothetical meta-analysis on the effects of a new drug on the reduction in the size of ovarian cancer tumors finds that there are no published studies that failed to find a statistically significant effect of the drug, and the average size of the effect of the drug is small in the largest studies but larger in the smaller studies. The authors found the mean effect size was large, and they interpreted this as evidence for a large benefit of the drug. Why might you doubt the results of this meta-analysis?

7. Why might meta-analysis increase the Type I error rate, relative to individual studies?

ASSIGNMENT PROBLEMS

8. A meta-analysis on 25 studies testing the effects of a new enzymatic treatment on pityriasis (a common skin disease) finds a significant effect of the treatment. The fail-safe number was calculated to be 5.
   a. How much confidence should you give the results of this meta-analysis?
   b. If the fail-safe number had been 1500, how would you feel about the results? Explain your reasoning.

9. In what ways is meta-analysis an improvement over a traditional review of the literature on a research topic?

10. What measure of effect size would most likely be used in each of the following meta-analyses?
   a. A meta-analysis on the effect of the drug ibuprofen on the frequency of individuals contracting kidney disease.
   b. A meta-analysis of whether excluding parasites from a population affects mean growth rate.
   c. A meta-analysis of the relationship between human height and lifespan.
   d. A meta-analysis comparing the survival of parasitized and unparasitized birds.
   e. A meta-analysis comparing the difference in body size of males and females across species.

11. A meta-analysis of the effects of predators on their prey populations collected the results of 45 studies (Salo et al. 2007). The effect size used to describe these studies is Hedge's *d*, similar to the SMD. As one part of the meta-analysis, the researchers discovered that predators that had been artificially introduced by humans ("alien predators") had larger effects on prey than did native predators. None of the 45 individual studies had compared alien and native predators.
   a. What kind of variable is the classification of predators as alien or native? (Answer using the specific language of meta-analysis.)
   b. Australia has had a disproportionately large number of alien species introduced with great ecological damage. As a result, most of the studies compared in this meta-analysis on alien predators were done in Australia. Comment on how this should affect our understanding of the results.
   c. None of the original studies compared alien and native predators. Explain how meta-analysis made it possible to answer a novel question based on these studies, even though none of the individual studies addressed this question.

12. In most meta-analyses, some studies find answers that are opposite to those found by the average of all studies. Give two reasons for this phenomenon.

Answers to Practice Problems

Chapter 1

1. **(a)** Ordinal. **(b)** Nominal. **(c)** Nominal.
   (d) Ordinal.
2. **(a)** Observational (cats were observed to fall a given number of stories—the researchers did not assign the number of stories fallen to the cats).
   (b) The number of floors fallen. **(c)** The number of injuries per cat.
3. **(a)** Discrete. **(b)** Continuous. **(c)** Discrete.
   (d) Continuous.
4. **(a)** Collectors do not usually sample randomly, but prefer rare and unusual specimens. Therefore, the moths are probably not a random sample. **(b)** A sample of convenience. **(c)** Bias (rare color types are likely to be overrepresented in the sample compared with the population).
5. Accuracy.
6. **(a)** Categorical. **(b)** No random sampling procedure was followed, so the survey is not random; the respondents were volunteers. **(c)** Volunteer bias (those most interested in the issue were more likely to respond).
7. **(a)** US army personnel stationed in Iraq. **(b)** Yes. The number of subjects interviewed was a random sample of only 100 from the population; therefore, the responses of the particular individuals who happened to be sampled will differ from the population of interest by chance. **(c)** The advantage of random sampling is that it minimizes bias and allows the precision of estimates of stress levels to be measured.

8. **(a)** The population being estimated is all of the small mammals of Kruger National Park. **(b)** No, the sample is not likely to be random. In a random sample, every individual has the same chance of being selected, but some small mammals might be easier to trap than others (e.g., trapping only at night might miss all of the mammals active only in the daytime). In a random sample, individuals are selected independently. Multiple animals caught in the same trap might not be independent if they are related or live near one another (this is harder to judge). **(c)** The number of species in the sample might underestimate the number in the Park if sampling was not random (e.g., if daytime mammals were missed) or if rare species happened to avoid capture. In this case, m would be a biased estimate of M.
9. Chicks were not sampled randomly. The multiple chicks from the same nest are not independent. Chicks from the same nest are likely to be more similar to one another in their measurements than chicks chosen randomly from the population. Ignoring this problem will lead to erroneous measurements of the precision of survival estimates.

Chapter 2

1. **(a)** 25%. **(b)** 33%. **(c)** 33%.

2.

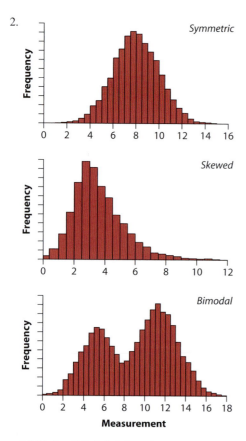

(a) Modes: 7.5–8, 2.5–3, and 11–11.5.
(b) Skewed right.
(c) The bimodal distribution in this example is left-skewed.
3. The lower histogram is correct. Histograms use area, not height, to represent frequency.
4.

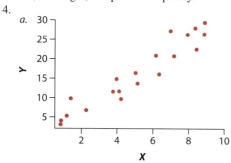

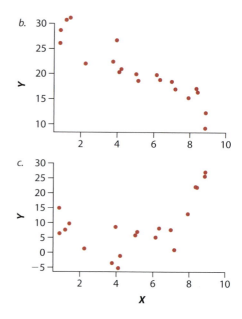

5. Contingency table.

| | Source of fry | | |
	Hatchery	**Wild**	**Total**
Survived	27	51	78
Perished	3973	3949	7922
Total	4000	4000	8000

6. (a) The following table orders the taxa from those with the most endangered species to those with the fewest endangered species. (Other orderings could make sense, such as by phylum.)

Taxon	**Number of species**
Plants	745
Fish	115
Birds	92
Mammals	74
Clams	70
Insects	44
Reptiles	36
Snails	32
Amphibians	22
Crustaceans	21
Arachnids	12

(b) Frequency table.
(c) Bar graph, because data are categorical.

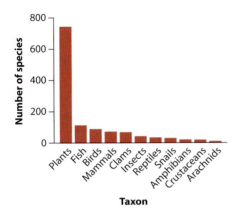

Taxon

(d) The baseline should be zero so that bar height and area correspond to frequency.

7. **(a)** The variables are year (or famine stage) and incidence of schizophrenia. **(b)** Year (and famine stage) is categorical—ordinal. Incidence of schizophrenia is categorical—nominal. **(c)** Contingency table. *(See table at bottom of page.)* **(d)** Proportions: 1956: 0.0082, 1960: 0.0140, and 1965: 0.0083. Other variations of the line graph below are valid (e.g., famine state could be the explanatory variable instead of year).

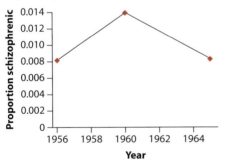

Year

Pattern revealed: incidence of schizophrenia highest in 1960, the famine year.

8. **(a)** Grouped histograms. **(b)** Groups are disease types (diseases transmitted directly from one individual to another and diseases transferred by insect vectors). **(c)** The variable is the virulence of the disease. It is a continuous numerical variable. **(d)** Relative frequency, the fraction of diseases occurring in each interval of virulence. **(e)** Directly transmitted diseases tend to be less virulent than insect-transmitted diseases. Human diseases transmitted directly more frequently have low virulence, and less frequently have high virulence, than diseases transmitted by insect vector.

9. **(a)** Map. **(b)** Date of first appearance of rabies (measured by the number of months following 1 March 1991). **(c)** Geographic location—the township.

10. **(a)** Launch temperature is the explanatory variable, and number of O-ring failures is the response variable, because we want to predict failures from temperature.

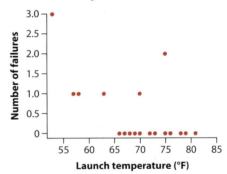

(b) There is a negative association. **(c)** Low temperature is predicted to increase the risk of O-ring failure.

	1956 (pre-famine)	1960 (famine)	1965 (post-famine)	Total
Schizophrenic	483	192	695	1370
Non-schizophrenic	58,605	13,556	82,841	155,002
Total	59,088	13,748	83,536	156,372

11. **(a)** The distribution is left-skewed.

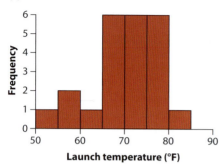

Launch temperature (°F)

(b) The mode is between 65 and 80 °F. (If you used an interval width of 10 °F instead of 5 °F, then the mode is between 70 and 80 °F.)

12. **(a)** Husbands of adopted women resemble a woman's adoptive father more than they resemble the women themselves or the women's adoptive mother. **(b)** Inappropriate baseline. The eye is receptive to bar height and area, but, because of the baseline, height and area do not depict magnitude. A baseline of zero would probably be best (adding a horizontal line at resemblance = 25 might also be a good idea to indicate the random expectation). **(c)** No, the graph is depicting a measurement (resemblance) in each of several categories, not frequency of occurrence.

13. **(a)** Mosaic plot. The explanatory variable is fruit set in previous years. The response variable is fruiting in the given year. Both variables are categorical. **(b)** Line graph. The explanatory variable is year. The response variable is the density of the wood. Both variables are numerical. **(c)** Grouped cumulative frequency distributions. The explanatory variable is the taxonomic group (butterflies, birds, or plants). The response variable is the percent change in range size. Group is a categorical variable. Range change is numerical.

Chapter 3

1. **(a)** The number of doctors studied who sterilized a given percentage of patients under age 25. **(b)** The median would be more informative. The mean would be sensitive to the presence of the outlier (sterilizing more than 30% of female patients under 25), whereas the median is unaffected.

2. **(a)** Box plot. **(b)** Median body mass of the mammals in each group. **(c)** The first and third quartiles of body mass in each group. **(d)** "Extreme values," those lying farther than 1.5 times the interquartile range from the box edge. **(e)** Whiskers. They extend to the smallest and largest values in the data, excluding extreme values (those lying farther than 1.5 times the interquartile range from the box edge). **(f)** Living mammals have the smallest median body size (with a log body mass of about 2), and mammals that went extinct in the last ice age had the largest median size (around 5.5). Mammals that went extinct recently were intermediate in size, with a median of around 3.2. **(g)** The body size distribution of living mammals is right-skewed (long tail toward larger values), whereas the size frequency distribution in mammals that went extinct in the last ice age is left-skewed. The frequency distribution of sizes is nearly symmetric in mammals that went extinct recently. **(h)** One way to answer this is to use the interquartile range as a measure of spread, in which case the mammals that went extinct in the last ice age have the lowest spread. Another way to compare spread is to calculate the standard deviation of log body size in each group, but we are unable to calculate this without the raw data. **(i)** It is likely that extinctions have reduced the median body size of mammals.

3. **(a)** Grouped histograms. **(b)** Explanatory variable: sex. Response variable: number of words per day. **(c)** Men: 8,000−12,000 words per day. Women: 16,000−20,000 words per day. **(d)** Women. **(e)** Men.

4. **(a)** The averages are 7.5 g for the crimson-rumped waxbill (CW), 15.4 g for the cutthroat finch (CF), and 37.9 g for the white-browed sparrow weaver (WS). **(b)** Standard deviations are 0.6 g for CW, 1.2 g for CF, and 3.1 g for WS. The standard deviation is greater when the mean is greater. **(c)** Coefficients of variation: 8.3% for CW, 8.0% for CF, and 8.2% for WS. The coefficients of variation are much more similar than the standard deviations. **(d)** Compare coefficients of variation to

compare variation relative to the mean. Beak length: 3.8%; body mass: 8.2%. Body mass is most variable relative to the mean.

5. Use the cumulative frequency distribution to estimate the median (0.50 quantile, approximately 1.4%), and the first and third quartiles (approximately 0.7% and 2.4%). The interquartile range is about $2.4 - 0.7 = 1.7$% and 1.5 times this amount is about 2.4%, which sets the maximum length of each whisker. The smallest value is about -1.1% and the largest value is about 4.3%. Both of these values are within 1.5 times the interquartile range, so the whiskers extend from -1.1% and 4.3%. The boxplot is:

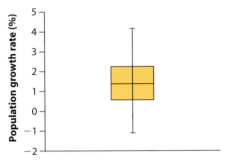

6. **(a)** cm/s.
(b)

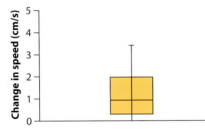

(c) Asymmetric. The range of changes in speed is much greater above the median than below the median. This is true whether we look at the third quartile or the total range. **(d)** The span of the box indicates where the middle 50% of the measurements occur (first quartile to third quartile). **(e)** Mean is 1.19 cm/s, which is greater than the median (0.94 cm/s). The distribution is right-skewed, and the large values influence the mean more than the median. **(f)** 1.15 (cm/s)^2. **(g)** The interval from $\overline{Y} - s$ and $\overline{Y} + s$ is 0.11 to 2.26. This includes 11 of the 16 observations, or 69%.

7. **(a)** Increase the mean by k. **(b)** No effect.
8. **(a)** Increase $\overline{Y}$ by 10 times (11.9 mm/s).
(b) Increase s by 10 times (10.7 mm/s).
(c) Increase median by 10 times (9.4 mm/s).
(d) Increase interquartile range by 10 times (16.8 mm/s). **(e)** No effect. **(f)** Increase s^2 by 100 times (115.3 (mm/s)^2.
9. **(a)** 5.7 hours. **(b)** 2.4 hours. **(c)** $83/114 = 0.73$ (73%). **(d)** Median $= 5$ hours, which is less than the mean. The distribution is right-skewed, and the large values influence the mean more than the median.

Chapter 4

1. **(a)** 0.22 hours. **(b)** The sampling distribution of the estimates of the mean time to rigor mortis. **(c)** Random sampling from the population of corpses.
2. The median will probably be smaller than the mean. The distribution is right-skewed. Extreme values have a greater effect on the mean, pulling it upwards, than on the median.
3. Sample size.
4. **(a)** False ($\hat{p}$ is just an estimate based on a sample). **(b)** True. **(c)** True. **(d)** False. This fraction is a constant (leaving aside the possibility of gene copy number variation between people). **(e)** True.
5. **(a)** The fraction of all Canadians who agree with the statement. **(b)** The sample estimate is 73%. **(c)** The sample size is 1641 people. **(d)** The 95% confidence interval for the population fraction.
6. **(a)** 95.9 ms. The population mean flash duration. **(b)** No, because the calculation was based on a sample rather than on the whole population. By chance, the sample mean will differ from that of the population mean. **(c)** 1.9 ms. **(d)** The spread of the sampling distribution of the sample mean. **(e)** Using the 2SE rule, $95.9 \pm 2(1.9$ ms) yields $92.1 < \mu < 99.7$. **(f)** The interval represents the most plausible values for the population mean μ. In roughly 95% of random samples from the population, when we compute the 95% confidence interval, the interval will include the true population mean.
7. The population mean is likely to be small or zero.

Chapter 5

1. **(a)** 5/8. **(b)** 1/4. **(c)** 7/8 (either in this case means pepperoni or anchovies or both). **(d)** No (some slices have both pepperoni and anchovies). **(e)** Yes. Olives and mushrooms are mutually exclusive. **(f)** No. Pr[mushrooms] = 3/8; Pr[anchovies] = 1/2; if independent, Pr[mushrooms and anchovies] = Pr[mushrooms] × Pr[anchovies] = 3/16. Actual probability = 1/8. Not independent. **(g)** Pr[anchovies | olives] = 1/2 (two slices have olives, and one of these two has anchovies). **(h)** Pr[olives | anchovies] = 1/4 (four slices have anchovies, and one of these has olives). **(i)** Pr[last slice has olives] = 1/4 (two of the eight slices have olives; you still get one slice—it doesn't matter whether your friends pick before you or after you). **(j)** Pr[two slices with olives] = Pr[first slice has olives] × Pr[second slice has olives | first slice has olives] = 2/8 × 1/7 = 1/28. **(k)** Pr[slice without pepperoni] = 1 − Pr[slice with pepperoni] = 3/8 **(l)** Each piece has either one or no topping.

2. Pr[successful hunting] = Pr[find prey] × Pr[captures prey | finds prey] = 0.8 × 0.1 = 0.08.

3. 45 of 273 trees have cavities, so the probability of choosing a tree with a cavity is 45/273 = 0.165.

4. **(a)** Pr[vowel] = Pr[A] + Pr[E] + Pr[I] + Pr[O] + Pr[U] = 8.2% + 12.7% + 7.0% + 7.5% + 2.8% = 38.2%. **(b)** Pr[five randomly chosen letters from an English text would spell "STATS"] = Pr[S] × Pr[T] × Pr[A] × Pr[T] × Pr[S] = 0.063 × 0.091 × 0.082 × 0.091 × 0.063 = 2.7 × 10^{-6}. (Each draw is independent, but all must be successful to satisfy the conditions, so we must multiply the probability of each independent event.) **(c)** Pr[2 letters from an English text = "e"] = 0.127 × 0.127 = 0.016.

5. **(a)** Pr[A_1 or A_4] = Pr[A_1] + Pr[A_4] = 0.1 + 0.05 = 0.15. **(b)** Pr[$A_1 A_1$] = Pr[A_1] × Pr[A_1] = 0.1 × 0.1 = 0.01. (This is only true because the individuals mate at random. If we did not know this, we could not calculate the probability.) **(c)** Pr[not $A_1 A_1$] = 1 − Pr[$A_1 A_1$] = 1 − 0.01 = 0.99. **(d)** Pr[two individuals not $A_1 A_1$] = Pr[not $A_1 A_1$] × Pr[not $A_1 A_1$] = 0.99 × 0.99 = 0.9801. **(e)** Pr[at least one of two individuals chosen at random has genotype $A_1 A_1$] = 1 − Pr[neither individual is $A_1 A_1$] = 1 − 0.9801 = 0.0199.
(f) Pr[three randomly chosen individuals have no A_2 or A_3 alleles] = Pr[six randomly chosen alleles are not A_2 or A_3] = Pr[one allele is neither A_2 nor A_3]^6; Pr[one allele is neither A_2 nor A_3] = 1 − Pr[individual has either A_2 or A_3] = 1 − [Pr[A_2] + Pr[A_3)] = 1 − (0.15 + 0.6) = 0.25. Pr[all three lack these alleles] = 0.25^6 = 0.00024.

6. Pr[sum to 7] = 1/6.

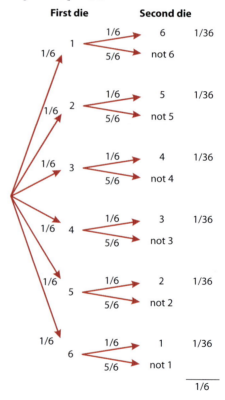

7. **(a)** Pr[no dangerous snakes] = Pr[not dangerous in the left hand] × Pr[not dangerous in the right hand] = 3/8 × 2/7 = 6/56 = 0.107.

(b) Pr[bite]= Pr[bite | 0 dangerous snakes] Pr[0 dangerous snakes] +
Pr[bite | 1 dangerous snake] Pr[1 dangerous snake] +
Pr[bite | 2 dangerous snakes] Pr[2 dangerous snakes].
Pr[0 dangerous snakes] = 0.107 (from part a).
Pr[1 dangerous snake] = (5/8 × 3/7) + (3/8 × 5/7) = 0.536.
Pr[2 dangerous snakes] = 5/8 × 4/7 = 0.357.
Pr[bite | 0 dangerous snakes] = 0.
Pr[bite | 1 dangerous snake] = 0.8.
Pr[bite | 2 dangerous snakes] = $1 - (1 - 0.8)^2 = 0.96$.
Putting these all together:
Pr[bite]= (0 × 0.107) + (0.8 × 0.536) + (0.96 × 0.357) = 0.772.

(c) $\Pr[\text{defanged} \mid \text{no bite}] = \dfrac{\Pr[\text{no bite} \mid \text{defanged}]\Pr[\text{defanged}]}{\Pr[\text{no bite}]}$

Pr[no bite | defanged] = 1; Pr[defanged] = 3/8;
Pr[no bite] = Pr[defanged] Pr[no bite | defanged] + Pr[dangerous] Pr[no bite | dangerous]
= (3/8 × 1) + [5/8 × (1− 0.8)] = 0.5.
So, [defanged | one snake did not bite] = (1.0 × 3/8)/(0.5) = 0.75.

8. **(a)** Pr[all five researchers calculate 95% CI with the true value]? Each one has a 95% chance, all samples are independent, so Pr = $0.95^5 = 0.774$. **(b)** Pr[at least one does not include true parameter] = $1 -$ Pr[all include true parameter] = $1 - 0.774 = 0.226$.

9. **(a)** Pr[cat survives seven days] = Pr[cat not poisoned one day]^7 = $0.99^7 = 0.932$. **(b)** Pr[cat survives a year] = Pr[cat not poisoned one day]^{365} = $0.99^{365} = 0.026$. **(c)** Pr[cat dies within year] = $1 -$ Pr[cat survives year] = $1 - 0.026 = 0.974$.

10. **(a)** Pr[single seed lands in suitable habitat] = 0.3. **(b)** Pr[two independent seeds land in suitable habitat] = $0.3^2 = 0.09$. **(c)** Pr[two of three independent seeds land suitable habitat] = $3 \times (0.7 \times 0.3 \times 0.3) = 0.189$.

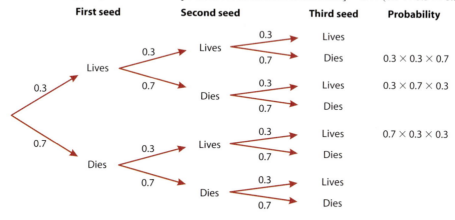

11 **(a)** False positive rate = Pr[positive test | no HIV] = 8 out of 4000 = 0.002. **(b)** False negative rate = Pr[negative test | HIV] = 20 out of 1000 = 0.02. **(c)** 988 tested positive, of which 980 had HIV; 980/988= 0.9919.

12. Each win has probability 0.09, each loss probability 0.91.
(a) Pr[WWLWWW] = $0.09^5 \times 0.91 = 5.4 \times 10^{-6}$. **(b)** Pr[WWWWWL] = $0.09^5 \times 0.91 = 5.4 \times 10^{-6}$.
(c) Pr[LWWWWW] = $0.09^5 \times 0.91 = 5.4 \times 10^{-6}$. **(d)** Pr[WLWLWL] = $0.09^3 \times 0.91^3 = 5.5 \times 10^{-4}$.
(e) Pr[WWWLLL] = $0.09^3 \times 0.91^3 = 5.5 \times 10^{-4}$. **(f)** Pr[WWWWWW] = $0.09^6 = 5.3 \times 10^{-7}$.

13. (a)

First set	Second set	Third set	Number of games	Probability
	1/2 → W	No game	2	1/4
W	1/2 → L	1/2 → W	3	1/8
1/2		1/2 → L	3	1/8
	1/2 → W	1/2 → W	3	1/8
1/2		1/2 → L	3	1/8
L	1/2 → L	No game	2	1/4

The probability that the match lasts two sets is $1/4 + 1/4 = 1/2$. The probability that it lasts three games is $1/8 + 1/8 + 1/8 + 1/8 = 1/2$.

(b)

First set	Second set	Third set	Winner of match	Probability
	0.55 → W	No game	Stronger	0.55^2
W	0.45 → L	0.55 → W	Stronger	$0.55^2 \times 0.45$
0.55		0.45 → L	Weaker	$0.45^2 \times 0.55$
	0.55 → W	0.55 → W	Stronger	$0.55^2 \times 0.45$
0.45		0.45 → L	Weaker	$0.45^2 \times 0.55$
L	0.45 → L	No game	Weaker	0.45^2

The probability of the weaker player winning is $(0.45^2 \times 0.55) + (0.45^2 \times 0.55) + 0.45^2 = 0.425$.

14. Pr[next person will wash his/her hands] $=$

Pr[wash | man] $\times$ Pr[man] + Pr[wash | woman] $\times$ Pr[woman]

$= 0.74 \times 0.4 + 0.83 \times 0.6$

$= 0.794$.

15. Total probability: $4/36 = 1/9 = 0.11$.

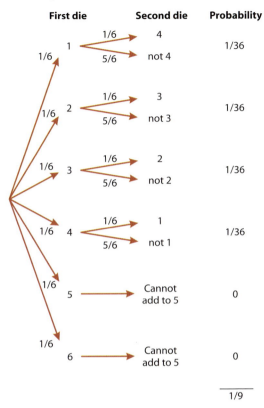

First die	Second die	Probability

16. **(a)** Pr[one person not blinking] $= 1 -$ Pr[person blinks] $= 1 - 0.04 = 0.96$. **(b)** Pr[at least one blink in 10 people] $= 1 -$ Pr[no one blinks] $= 1 - 0.96^{10} = 0.335$.

Chapter 6

1. **(a)** True. **(b)** False.
2. **(a)** H_0: The rate of correct guesses is 1/6.
   (b) H_A: The rate of correct guesses is not 1/6.
3. **(a)** Alternative hypothesis. **(b)** Alternative hypothesis. **(c)** Null hypothesis. **(d)** Alternative hypothesis. **(e)** Null hypothesis.

4. **(a)** Lowers the probability of committing a Type I error. **(b)** Increases the probability of committing a Type II error. **(c)** Lowers power of a test. **(d)** No effect.
5. **(a)** No effect. **(b)** Decreases the probability of committing a Type II error. **(c)** Increases the power of a test. **(d)** No effect.
6. **(a)** $P = 2 \times (\text{Pr}[15] + \text{Pr}[16] + \text{Pr}[17] + \text{Pr}[18]) = 0.0075$. **(b)** $P = 2 \times (\text{Pr}[13] + \text{Pr}[14] + \ldots + \text{Pr}[18]) = 0.096$. **(c)** $P = 2 \times (\text{Pr}[10] + \text{Pr}[11] + \text{Pr}[12] + \ldots + \text{Pr}[18]) = 0.815$. **(d)** $P = 2 \times (\text{Pr}[0] + \text{Pr}[1] + \text{Pr}[2] + \text{Pr}[3] + \ldots + \text{Pr}[7]) = 0.481$.
7. False. The hypotheses aren't variables subject to chance. Rather, the P-value measures how unusual the data are if the null hypothesis is true.
8. Failing to reject H_0 does not mean H_0 is correct, because the power of the test might be limited. The null hypothesis is the default and is either rejected or not rejected.
9. Begin by stating the hypotheses. H_0: Size on islands does not differ in a consistent direction from size on mainlands in Asian large mammals (i.e., $p = 0.5$); H_A: Size on islands differs in a consistent direction from size on mainlands in Asian large mammals (i.e., $p \neq 0.5$), where p is the true fraction of large mammal species that are smaller on islands than on the mainland. Note that this is a two-tailed test. The test statistic is the observed number of mammal species for which size is smaller on islands than mainland: 16. The P-value is the probability of a result as unusual as 16 out of 18 when H_0 is true: $P = 2 \times (\text{Pr}[16] + \text{Pr}[17] + \text{Pr}[18]) = 0.00135$. Since $P < 0.05$, reject H_0. Conclude that size on islands is usually smaller than on mainlands in Asian large mammals.
10. **(a)** Not correct. The P-value does not give the size of the effect. **(b)** Correct. H_0 was rejected, so we conclude that there is indeed an effect. **(c)** Not correct. The probability of committing a Type I error is set by the significance level, 0.05, which is decided beforehand. **(d)** Not correct. The probability of committing a Type II error depended on the effect size, which wasn't known. **(e)** Correct.

Chapter 7

1. **(a)** 91/220 had cancer, so the estimated probability of a cast or crew member developing cancer is 0.414. **(b)** The standard error of the proportion is 0.033. [The square-root of (0.414)(1 − 0.414)/(220 − 1)]. This quantity measures the standard deviation of the sampling distribution of the proportion. **(c)** Using the Agresti–Coull method to generate confidence intervals, we first calculate $p' = (x + 2)/(n + 4) = (91 + 2)/(220 + 4) = 0.415$. The lower bound for the 95% confidence interval is

$$0.415 - 1.96\sqrt{\frac{p'(1 - p')}{n + 4}} = 0.351.$$

The upper bound is

$$0.415 + 1.96\sqrt{\frac{p'(1 - p')}{n + 4}} = 0.480.$$ The confidence intervals do not bracket the typical 14% cancer rate for the age group. It is unlikely that 14% is the true cancer rate for this group.

2. **(a)** 46/50 bills had cocaine, so the estimated proportion is 0.92. **(b)** The 99% confidence interval, calculated using the Agresti–Coull method: $p' = 48/54 = 0.89$. Z for 99% = 2.58. The bounds of the confidence interval are 0.779 and 0.999. **(c)** We are 99% confident that the true proportion of US one-dollar bills that have a measurable cocaine content lies between 0.779 and 0.999.

3. **(a)** No, the probability of drawing a red card changes depending on the cards that have been drawn. **(b)** Yes, the sampling is with replacement, so the probability of success remains constant. **(c)** No, the probability of drawing a red ball will change with each draw. **(d)** No, the number of trials is not known. **(e)** Yes, the number of red-eyed flies in a sample from a *large* population can be described using the binomial distribution. (Because the population sampled is large, we will assume that the sampling of each individual has a negligible effect on the probability of drawing a red fly on the next draw.) **(f)** No, the individuals within different families may

have different probabilities of having red eyes due to shared genetic differences.

4. **(a)** This estimate pools numbers from two different groups. Women with rosacea are not a random sample of the population. **(b)** If the control group women are a random sample, 15/16 = 0.938 have mites. **(c)** $p' = 17/20 = 0.85$. The 95% confidence interval is 0.69 to 1.00. (Proportions cannot be larger than 1, so the confidence interval is truncated at 1.) **(d)** For the women with rosacea, $p' = 18/20 = 0.90$. The 95% confidence interval is $0.769 < p < 1.0$.

5. **(a)** Pr[male is eaten] = 21/52 = 0.404. $p' = 23/56 = 0.411$. The 95% confidence interval is from 0.282 to 0.540. **(b)** This estimate is consistent with a 50% capture of males, but not with a 10% capture of males. **(c)** The larger sample would not change the estimate of the proportion (the fraction is exactly the same), but it would reduce width of the confidence interval. (The sample size is in the denominator of the confidence interval equations.)

6. Estimated probability of moving north is 22/24 = 0.917. To test the null hypothesis that north and south movements of ranges are equally probable, we need to calculate the probability of 22, 23, or 24 species moving northward if $p = 0.5$. We will make this a two-tailed test, so we'll multiply this probability by two.

$$\begin{aligned}\Pr[22] &= \binom{24}{22}0.5^{22}(1 - 0.5)^2 \\ &= 1.6 \times 10^{-5}\end{aligned}$$

Doing the same for 23 and 24, summing all three, and multiplying by two, we find that $P = 3.6 \times 10^{-5}$. We can confidently reject the null hypothesis.

7. **(a)** The first study with the narrower confidence interval probably had the larger sample size. **(b)** The estimate with the smaller confidence interval is the more precise, in the sense that the true value is likely to be somewhere near the estimate. **(c)** The differences are what we might expect due to random sampling: the second confidence interval overlaps the estimate of the first study.

8. **(a)** This is evidence that males' choices may be affected by female positioning of the females: the probability of seeing 19 of 24 choose the 0M

female with the null hypothesis that there is no preference ($p = 0.5$). With a binomial test,

$$P = 2\{\Pr[19]+\Pr[20]+\Pr[21]+\Pr[22]+\Pr[23]+\Pr[24]\}$$

$$= 2\{0.00253+0.0006+0.0001+...\}$$

$$= 0.0066.$$

Therefore $P < 0.05$, and we can reject the null hypothesis that the males have no preference. **(b)** The males may have preferred one female over the other for a number of reasons. If the females were sisters, this would reduce the number of differences due to genes and maternal environment, and so would be more likely to test for fetal position. Ideally, this would be done with 24 sets of sisters so that each trial is independent.

9. **(a)** 30%. **(b)** The binomial distribution. **(c)** The standard deviation of the proportion of A cells is the standard error:

$$\sigma_{\hat{p}} = \sqrt{\frac{0.3(1-0.3)}{15}} = 0.118.$$

(d) 95% of the technicians should construct a confidence interval for the proportion of A cells that includes 0.3.

10. **(a)** 1/6 of the 12 dice should have "3," or 2 dice. **(b)** Pr[no three out of 12 rolls] = $(1 - \Pr[3 \text{ on one roll}])^{12} = (5/6)^{12} = 0.112$. **(c)** $\binom{12}{3} 0.167^3 (1-0.167)^9 = 0.197$. **(d)** Each die has six faces, each with probability 1/6 of showing. The average value for the die will be the sum of each value times its probability, or $(1 \times 1/6) + (2 \times 1/6) + ... (6 \times 1/6)$ or $1/6 \times 21 = 3.5$. For twelve dice, the average sum of the numbers showing will be $12 \times 3.5 = 42$. **(e)** Pr[all dice show 1 or 6] = $(\Pr[1]+\Pr[6])^{12} = 1.88 \times 10^{-6}$.

11. **(a)** 6052 out of 12,028 of the deaths occurred in the week before Christmas, or 0.503. **(b)** Using the Agresti–Coull confidence interval approximation for proportions, $p' = (X + 2)/(n + 4)$, where X is the number of successes (= "died before Christmas") and n is the total number of trials (= "died within a week of Christmas") =

6054/12,032 = 0.503. Then the confidence interval is 0.494 to 0.512. **(c)** This interval includes 50%, which is what is expected by chance. There is no statistical support for the belief that the living hang on until that special day.

12. **(a)** We can calculate the P-value by summing the probabilities of 13, 14, 15, or 16 resin clades having more species, then multiplying by two. $\Pr[13]=\binom{16}{13}0.5^{13}(1-0.5)^3 =0.0085$. Summing the probabilities for 13 through 16, then multiplying by two, $P = 0.021$. **(b)** This is an observational study, not an experimental study. We did not randomly assign clades to have latex/resin. It is possible that clades with resin have higher diversity not due to the resin, but due to some third factor that affects both resin and the number of species.

Chapter 8

1. **(a)** 5. There would be six categories, one for each outcome, and so five degrees of freedom, since no parameters are estimated. **(b)** 10. There are 11 categories (from 0 heads to 10 heads), and 10 degrees of freedom, since no parameters are estimated. **(c)** 9. There are now nine degrees of freedom, since p must be estimated from the data. **(d)** 3. There would be five categories (0 to 4 insect heads per sample), and three degrees of freedom, since the mean number of heads per sample would need to be estimated from the data.

2. No, the tickets purchased on a given day are not all independent; for example, a single person could buy multiple tickets at one purchase. Therefore, this is *not* a random sample as presented.

3. To test whether nematodes are distributed at random among fish, we can see whether the data fit the Poisson distribution. The mean number of parasites per fish is

$\overline{X} = 103(0)+72(1)+44(2)+14(3)+3(4)+1(5)+1(6)=0.94538.$

The expected values from a Poisson distribution with this mean are given below, from the proportion expected from the Poisson distribution times the total sample size, $n = 238$:

Number of parasites	Observed	Expected
0	103	92.47
1	72	87.42
2	44	41.32
3	14	13.02
4	3	3.08
5	1	0.58
6	1	0.09

The expected values for 5 and 6 are below one, so we will combine 4 through 6 into one category.

Number of parasites	Observed	Expected	$\dfrac{(Observed - Expected)^2}{Expected}$
0	103	92.47	1.20
1	72	87.42	2.72
2	44	41.32	0.17
3	14	13.02	0.07
≥ 4	5	3.75	0.42

$\chi^2 = 4.58$. There are five categories, one parameter estimated, so $df = 3$. $\chi^2_{3,0.05} = 7.81$. $4.58 < 7.81$, so we cannot reject the Poisson distribution of parasites on fish. $p = 0.21$.

4. (a) If pine seedlings are distributed randomly, the frequencies should fit the Poisson distribution. (b) The mean number of seedlings per quadrat was 1.325.

Number of seedlings	Observed	Expected
0	47	21.26
1	6	28.18
2	5	18.67
3	8	8.24
≥ 4	14	3.65

The expected counts for 5–7 were below 1, so 4 through 7 were combined. $\chi^2 = 88.0$, with $df = 3$ (five categories, minus one, minus one parameter estimated). $\chi^2_{3,0.001} = 16.27$, and χ^2 is greater than this critical value, so $P < 0.001$. (c) The variance, 3.49, is much greater than the mean here, indicating that some quadrats are much more likely than others to have seedlings: seedlings are clumped.

5. (a) The mean number of goals per side is 1.26. (b) Fitting this to a Poisson and testing with a goodness-of-fit test, we must again combine the four goals and above categories to avoid expected values less than one.

Number of goals	Observed	Expected
0	37	36.3
1	47	45.7
2	27	28.8
3	13	12.1
≥ 4	4	5.0

$\chi^2 = 0.42$, which is less that the critical value $\chi^2_{3, 0.05} = 7.81$, so we cannot reject the null hypothesis of a Poisson distribution of goals per side per game. $p = 0.94$.

6. **(a)** $\Pr[8] = \binom{15}{8} 0.5^8 (1-0.5)^7 = 0.19638$. **(b)** We test the fit of these data to the binomial distribution, combining categories where the expected value is less than one. This leads to the following table:

Total wins	Number of wrestlers	Expected proportion	Expected number	$\dfrac{(\textit{Observed} - \textit{Expected})^2}{\textit{Expected}}$
0–2	32	0.0037	7.85	74.3
3	56	0.0139	29.52	23.8
4	112	0.0417	88.56	6.2
5	190	0.0916	194.84	0.1
6	290	0.1527	324.73	3.7
7	256	0.1964	417.51	62.5
8	549	0.1964	417.51	41.4
9	339	0.1527	324.73	0.6
10	132	0.0916	194.84	20.3
11	73	0.0417	88.56	2.7
12	56	0.0139	29.52	23.8
13–15	47	0.0037	7.85	195.3
	Test statistic:			454.6

There are 12 categories and no estimated parameters, so there are 11 degrees of freedom. The test statistic greatly exceeds the critical value for χ^2 with $df = 11$ for $P = 0.001$, so the data do *not* fit the binomial distribution.

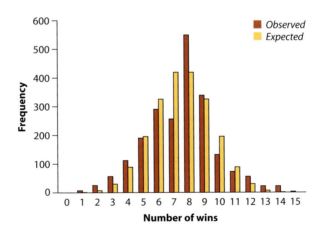

(c) These data show fewer than expected wrestlers with seven wins and more than expected with eight wins. In a match when one wrestler has already won eight rounds, cheating so as to lose to an opponent who has won seven rounds would tend to inflate the number of wins at eight. **(d)** The binomial distribution of wins assumes that each match is independent, and that each wrestler has the same probability of winning each match. Motivation and stamina might well change from match to match. A wrestler with only seven wins would have a lot more motivation to win another match than would one who had already won eight rounds.

7. **(a)**

Number of heads	Observed	Binomial expectation	Expected
0	6	0.0039	3.91
1	32	0.0313	31.25
2	105	0.1094	109.38
3	186	0.2188	218.75
4	236	0.2734	273.44
5	201	0.2188	218.75
6	98	0.1094	109.38
7	33	0.0313	31.25
8	103	0.0039	3.91
Total	1000	1.0	1000

The expected values for 0 and 8 are less than 5, so we combined (0 and 1) and (7 and 8) to produce:

Number of heads	Observed	Expected	$\dfrac{(Observed - Expected)^2}{Expected}$
0 or 1	38	35.16	0.23
2	105	109.38	0.18
3	186	218.75	4.90
4	236	273.44	5.13
5	201	218.75	1.44
6	98	109.38	1.18
7 or 8	136	35.16	289.2
Total	1000	1000	$\chi^2 = 302.27$

A χ^2 goodness-of-fit test to the binomial on these data with the null hypothesis that the proportion of heads is $p = 0.5$, gives $\chi^2 = 302.3$. Reject the null hypothesis with $P < 0.001$ ($df = 6$). **(b)** Graphing this, we can see very easily that there is an excess of coins yielding eight heads. **(c)** We would expect the observed number of coins yielding eight heads to be roughly equal to the number of coins yielding zero heads, or around four. This means that there is an excess of two-headed coins, so roughly $103 - 4 = 99$.

8. We will use a goodness-of-fit test to see whether the data differ from the 0.5 expectation if cancer deaths were randomly distributed around Christmas.

	Observed	Expected	$\dfrac{(Observed - Expected)^2}{Expected}$
Die before Christmas	6052	6014	0.24
Die after Christmas	5976	6014	0.24
Total	12,028	12,028	0.48

$\chi^2 = 0.48$, and $df = 1$. $0.48 < 3.84$, the critical value for $P = 0.05$, so we do *not* reject the hypothesis of a random distribution of cancer deaths around Christmas.

9.

	Observed	14% expected	$\dfrac{(Observed - Expected)^2}{Expected}$
Cancer death	91	30.8	117.67
Noncancer death	129	189.2	19.15
Total	220	220	$\chi^2 = 136.82$

$df = 1$; $\chi^2 = 136.82 \gg 10.83$, the critical value for $P = 0.001$, so we can reject the hypothesis that the cast and crew suffered the population-wide cancer mortality rate, $P < 0.001$.

10. The average admissions are 20 per night. What is the probability of five or fewer admissions? Assume that each admission is independent (no riots, please). Then, they should be Poisson distributed. To find the probability of a quiet night, we sum the probabilities of 0 to 5 admissions.

Admissions	Probability
0	2.06×10^{-9}
1	4.12×10^{-8}
2	4.12×10^{-7}
3	2.75×10^{-6}
4	1.37×10^{-5}
5	5.50×10^{-5}

Total probability $= 7.2 \times 10^{-5}$.

Chapter 9

1. **(a)**

	Observed				Expected				$\dfrac{(Observed - Expected)^2}{Expected}$		
	Capture	Escape	Total		Capture	Escape	Total		Capture	Escape	Total
White	9	92	101	White	50.25	50.75	101	White	33.86	33.53	
Blue	92	10	102	Blue	50.75	51.25	102	Blue	33.53	33.20	
Total	101	102	203	**Total**	101	102	203	**Total**			134.13

$\chi^2 = 134.13 \gg 10.83$ (critical value for $P = 0.001$ for 1 df), so $P < 0.001$. The two types of pigeons differ in their rate of capture.

(b) Odds of capture for white-rumped: $(9/101)/(92/101) = 0.098$.

Odds of capture for blue-rumped: $(92/102)/(10/102) = 9.2$.

Odds ratio: $0.098/9.2 = 0.01$.

For the 95% confidence interval, we use the log odds ratio, $\ln(0.01) = -4.54$.

$$SE = [\ln(\widehat{OR})] = \sqrt{\frac{1}{9} + \frac{1}{92} + \frac{1}{92} + \frac{1}{10}} = 0.48$$

For the 95% confidence interval, we use $Z = 1.96$.

$$-4.54 - 1.96(0.48) < \ln(OR) < -4.54 + 1.96(0.48)$$
$$-5.49 < \ln(OR) < -3.60$$
$$0.004 < OR < 0.027$$

2.

	Observed				Expected				$\dfrac{(Observed - Expected)^2}{Expected}$		
	One	Multi	Total		One	Multi	Total		One	Multi	Total
Malarial	69	20	89	**Malarial**	76.77	12.23	89	**Malarial**	0.79	4.94	
Healthy	157	16	173	**Healthy**	149.23	23.77	173	**Healthy**	0.40	2.54	
Total	226	36	262	**Total**	226	36	203	**Total**			8.67

$\chi^2 = 8.67 > 7.88$, the critical value for $P = 0.005$ for 1 df, (but less than 10.83, the critical value for $P = 0.001$), so $P < 0.005$. Infected and uninfected mosquitoes differ in the probability of multiple blood meals.

3. **(a)** The odds of the second suitor being accepted if the first was eaten are 0.5, while the odds of the second suitor being accepted if the first escaped are 22.0. There is a much higher probability of the second suitor being rejected if the first was eaten. **(b)** There are only four cells, so if any cell has an expected frequency of less than five, the assumptions of the χ^2 contingency test will be violated. In this case, the expected value is much less than five, so it would be necessary to use Fisher's exact test in this case.

4.

	Observed					Expected					$\dfrac{(Observed - Expected)^2}{Expected}$			
	Fire	Reverse fire	White	Total		Fire	Reverse fire	White	Total		Fire	Reverse fire	White	Total
Leave	18	6	0	24	**Leave**	8	8	8	24	**Leave**	12.5	0.5	8	21
Stay	2	14	20	36	**Stay**	12	12	12	36	**Stay**	8.33	0.333	5.33	14
Total	20	20	20	60	**Total**	20	20	20	60	**Total**	20.8	0.833	13.3	35

$\chi^2 = 35.0$ for 2 df [there are two rows and three columns, so $(2 - 1)(3 - 1) = 2\ df$]. $35.0 > 13.82$, the critical value for $P = 0.001$ for 2 df, so $P < 0.001$. Yes, there is evidence that reed frogs react consistently to the sound of fire.

5.

	Observed				Expected		
	Juv male	Juv female	Total		Juv male	Juv female	Total
Adult male	1	11	12	**Adult male**	3.82	8.18	12
Adult female	6	4	10	**Adult female**	3.18	6.82	10
Total	7	15	22	**Total**	7	15	22

(a) Two of the expected values are less than 5, so we cannot use a χ^2 contingency analysis. Fisher's exact test is a good alternative. **(b)** The effects are in the expected direction.

6. **(a)** Proportions of kids in each TV viewing class having violent records eight years later, and confidence intervals (using Agresti–Coull calculations from Chapter 7):

	Record	Total	Proportion	p'	Low CI	High CI
1 hr	5	88	0.057	0.0761	0.022	0.130
1−3 hr	87	386	0.225	0.2282	0.187	0.270
3+ hr	67	233	0.288	0.2911	0.233	0.349

(b) Contingency table for test of independence of TV watching and later violent record.

	Observed				Expected				$\dfrac{(Observed - Expected)^2}{Expected}$		
Class	**No record**	**Record**	**Total**	**Class**	**No record**	**Record**	**Total**	**Class**	**No record**	**Record**	**Total**
1 hr	83	5	88	**1 hr**	68.21	19.79	88	**1 hr**	3.21	11.05	
1−3 hr	299	87	386	**1−3 hr**	299.19	86.81	386	**1−3 hr**	0.0001	0.0004	
3+ hr	166	67	233	**3+ hr**	180.60	52.40	233	**3+ hr**	1.18	4.07	
Total	548	159	707	**Total**	548	159	707	**Total**			19.5

$\chi^2 = 19.5$ for 2 df [there are three rows and two columns, so $(3 - 1)(2 - 1) = 2\ df$]. $35.0 > 13.82$, the critical value for $P = 0.001$ for 2 df, so $P < 0.001$. There is a relationship between watching TV and future violence: those that watched less than one hour of TV were less likely to have a record, while those that watched three or more hours were more likely to have a record.

(c) No, this does not prove that TV watching causes aggression in kids. This was an observational study, not an experimental one, so we do not know if kids watching more TV have other factors in common (e.g., lower parental supervision, etc.).

7.

	Observed				Expected				$\dfrac{(Observed - Expected)^2}{Expected}$		
	Arrest	**Healthy**	**Total**		**Arrest**	**Healthy**	**Total**		**Arrest**	**Healthy**	**Total**
Abstainers	12	197	209	**Abstainers**	10.7	198.3	209	**Abstainers**	0.16	0.01	0.17
Drinkers	9	192	201	**Drinkers**	10.3	190.7	201	**Drinkers**	0.16	0.01	0.17
Total	21	389	410	**Total**	21	389	203	**Total**	0.32	0.02	0.34

$\chi^2 = 0.34$ for 1 $df \ll 3.84$ (critical value for $P = 0.05$), so we cannot reject the null hypothesis that there is no difference in heart attack risk between drinkers and non-drinkers ($P = 0.56$).

(b) No, this does not imply that drinking has no effect on cardiac arrest, only that this study found no evidence for a connection. This study was observational: non-drinkers at the age of 40 may have been heavy drinkers at the age of 20, or may have never consumed alcohol. Drinking may have an effect on heart attacks, but with too small an effect to be detected by this sample size.

8. **(a)** 9.7% of women with vaginal deliveries became depressed, as compared with 12.3% of women with C-sections. **(b)** The odds for vaginal are 0.108 vs. 0.141 for C-section, yielding a ratio of 0.76.

(c)

	Observed				Expected				$\dfrac{(Observed - Expected)^2}{Expected}$		
	Depressed	**Not**	**Total**		**Depressed**	**Not**	**Total**		**Depressed**	**Not**	**Total**
Vaginal	1025	9520	10,545	**Vaginal**	1034.83	9510.17	10545	**Vaginal**	0.09	0.01	0.10
C-section	48	341	389	**C-section**	38.17	350.83	389	**C-section**	2.53	0.28	2.81
Total	1073	9861	10,934	**Total**	1073	9861	203	**Total**	2.62	0.29	2.91

$\chi^2 = 2.91$, for 1 df, which is less than 3.84 (critical value for $P = 0.05$), so we cannot reject the null hypothesis that there is no difference in depression risk for different methods of childbirth ($P = 0.088$).

9. **(a)** The odds of a migraine sufferer having a cardiac shunt are 44:49, or 0.90, while the odds for a non-sufferer are 16:77, or 0.21. The odds ratio is 4.32. **(b)** To find the confidence interval, we first take the log of the odds ratio, $\ln(4.32) = 1.464$ and calculate the SE of the odds ratio,

$\sqrt{1/44 + 1/49 + 1/16 + 1/77}$ = 0.344. The confidence interval for the log odds ratio is 1.464 − 1.96(0.344) to 1.464 + 1.96(0.344), or 0.79 to 2.14. Taking the exponential of these to find the confidence interval of the odds ratio, we get exp(0.79) = 2.2, and exp(2.14) = 8.5. 2.2 < OR < 8.5.

10. **(a)** The proportion of babies born with finger defects is 6522 out of 6,839,854, or 0.001. The 95% confidence interval (using the Agresti–Coull method) is from 0.00093 to 0. 00097. It is very likely that the proportion of babies born with finger defects is between 0.00093 to 0.00097.
(b)

| | Observed | | | | Expected | | | | $\frac{(Observed - Expected)^2}{Expected}$ | | |
	Defect	Normal	Total		Defect	Normal	Total		Defect	Normal	Total
Smoke	805	1280	2085	**Smoke**	695.00	1390.00	2085	**Smoke**	17.41	8.71	26.12
No smoke	4366	9062	13428	**No smoke**	4476.00	8952.00	13428	**No smoke**	2.70	1.35	4.05
Total	5171	10,342	15513	**Total**	5171	10,342	203	**Total**	20.11	10.06	30.17

For 1 df, the test statistic exceeds the critical value for $P = 0.001$, so $P < 0.001$. There is an association between finger defects and smoking during pregnancy. **(c)** The odds of a smoking mother having a child with a finger defect are 0.62, while the odds of a non-smoking mother having a child with a finger defect are 0.48. The odds ratio is 1.3 (smoking mothers are 1.3 times as likely to have a child with a finger defect). For the confidence interval, we first calculate the ln odds ratio = 0.27. We then calculate the SE of the ln odds ratio, the square root of the sum of reciprocals of the frequencies, or 0.05. The confidence interval for the log-odds ratio is ±1.96 × 0.05, so from 0.172 to 0.362. Taking the exponent of the confidence interval for the ln-odds ratio, we find that the 95% confidence interval for the odds ratio is from 1.19 to 1.44.

Chapter 10

1. From Statistical Table B. **(a)** 0.09012. **(b)** 0.09012. **(c)** Pr[$Z > -2.15$] = 1 − Pr[$Z < -2.15$] = 1 − Pr[$Z > 2.15$] = 1 − 0.01578 = 0.98422. **(d)** Pr[$Z < 1.2$] = 1 − Pr[$Z > 1.2$] = 1− 0.11507 = 0.88493. **(e)** Pr[$0.52 < Z < 2.34$] = Pr[$0.52 < Z$] − Pr[$2.34 < Z$], because the first value is the area under the curve from 0.52 to infinity, and the second value is the area under the curve from 2.34 to infinity. The difference will be the area under the curve from 0.52 to 2.34. 0.30153 − 0.00964 = 0.29189.
(f) Pr[$-2.34 < Z < -0.52$] = 0.29189: the normal distribution is symmetrical on either side of 0.
(g) Pr[$Z < -0.93$] = Pr[$Z > 0.93$] = 0.17619. **(h)** Pr[$-1.57 < Z < 0.32$] = (1 − Pr[$Z > 1.57$]) − Pr[$Z > 0.32$] = (1 − 0.05821) − 0.37448 = 0.56731.
2. **(a)** To determine the proportion of men excluded, we convert the height limit into standard normal deviates. (180.3 − 177.0)/7.1 = 0.46. Pr[$Z > 0.46$] = 0.32276, so roughly one-third of British men are excluded from applying. **(b)** (172.7 − 163.3)/6.4 = 1.47. Pr[$Z > 1.47$] = 0.07078, which is the proportion of British women excluded. 1 − 0.0708 = 0.9292 is the proportion of British women acceptable to MI5. **(c)** (183.4 − 180.3)/7.1 = 0.44 standard deviation units above the height limit.
3. **(a)** B is most like the normal distribution. A is bimodal, while C is skewed. **(b)** All three would generate approximately normal distributions of sample means due to the central limit theorem.
4. **(a)** Pr[weight > 5 kg]? Transform to standard normal deviate: (5 − 3.339)/0.573 = 2.90. Pr[$2.90 < Z$] = 0.00187. **(b)** Pr[$3 <$ birth weight < 4]? Transform both to standard normal deviates, and subtract probabilities of the Z-values. (3 − 3.339)/0.573 = −0.59. (4 − 3.339)/0.573 = 1.15. Pr[$-0.59 < Z$] = 1 − Pr[$0.59 < Z$] = 1 − 0.2776 = 0.7224. Pr[$1.15 < Z$] = 0.12507. 0.7224 − 0.12507 = 0.59733.
(c) 0.06681 babies are more than 1.5 standard deviations above, with the same fraction below, so 0.13362 of babies are more than 1.5 standard deviations either way.

(d) First, transform 1.5 kg into normal standard deviates: 1.5/0.573 = 2.62 standard deviations. Pr[2.62 < Z] = 0.0044. Since the distribution is symmetric, we multiply this by two to reflect the probability of being 2.62 standard deviations above or below the mean: (0.0044)(2) = 0.0088. **(e)** The standard error is the same as the standard deviation of the mean. It is equal to the standard deviation divided by the square-root of n, or $0.573/\sqrt{10} = 0.18$ kg. To find the probability that the mean of a sample of 10 babies is greater than 3.5 kg, we transform this mean into a Z-score. (3.5 − 3.339)/0.18 = 0.89 (standard deviations of the mean = 0.18673).

5. **(a)** The lower graph has the higher mean (ca. 20 vs. 10), whereas the upper graph has the higher standard deviation. **(b)** The lower graph has the higher mean (ca. 15 vs. 10) and the higher standard deviation (ca. 5 vs. 2.5).

6. The standard deviation is approximately 10, as the region within one standard deviation from the mean will contain roughly 2/3 of the data points.

7. **(a)** In a normal distribution, the modal value occurs at the mean, so the mode is 35 mm. **(b)** A normal distribution is symmetric, so the middle data point is the mean, 35 mm. **(c)** Twenty per-cent of the distribution is less than 20 mm in size. (Why? Normal distributions are symmetric, so if 20% of the distribution is 15 mm or larger than the mean, 20% must be 15 mm or smaller than the mean.)

8. **(a)** The lower distribution (II) would have sample means that had a more normal distribution, because the initial distribution is closer to normal. Both distributions would converge to a normal distribution if the sample size were sufficiently large. **(b)** The distribution of the sums of samples from a distribution will be normally distributed, given a sufficiently large number of samples.

9. **(a)** The probability of sampling haplotype A is p = 0.3. We expect a sample of 400 fish to have 0.3 × 400 A individuals on average, or 120. The standard deviation is approximately

$\sqrt{n\,p\,(1-p)} = \sqrt{400(0.3)(0.7)} = 9.17$. To find the probability of sampling 130 or more of haplotype A, we convert this into a standard normal deviate: (130 − 120)/9.17 = 1.09; Pr[1.09 < Z] = 0.13786. **(b)** Pr[at least 300 B haplotype out of 400]? Average number of B sampled = (1− p) × 400 = 280. Convert to standard normal: (300 − 280)/9.17 = 2.18; Pr[2.18 < Z] = 0.01463. **(c)** Pr[115 ≤ freq A ≤ 125]? Convert to standard normal deviates, find the probabilities, and calculate the difference. (115 −120)/9.17 = −0.55. (125 − 120)/9.17 = 0.55. Pr[−0.55 < Z] = 1 − Pr[0.55 < Z] = 1 − 0.29116 = 0.70884. Pr[0.55 < Z] = 0.29116. 0.70884 − 0.29116 = 0.41768.

10. Average expected number of cancer victims = 0.14 × 220 = 30.8. Standard deviation = $\sqrt{n\,p(1-p)}$ = $\sqrt{220(0.14)(0.86)}$ = 5.15. Convert actual number of victims, 91, into standard normal deviate: (91 − 30.8)/5.15 = 11.7. This is off the chart: $P < 0.00002$.

11. SE = $s/\sqrt{n}$. Z = (Y − mean)/SE. *(See table at bottom of page.)*

Chapter 11

1. The 99% confidence interval must be *larger* than the 95% confidence interval in order to have a higher probability of capturing the true mean.

2. **(a)** The sample mean is 10.32 cm and the standard error of the mean is 0.056 cm. The standard error estimates the standard deviation of the sampling distribution of sample means. **(b)** The 95% confidence interval for the mean is 10.32 ± $(0.056\, t_{0.05(2),34})$. $t = 2.03$, so the 95% confidence interval is 10.21 cm $< \mu <$ 10.44 cm. **(c)** The variance is 0.11005. There are 35 individuals, so 34 df. The χ^2 critical values are looked up in Statistical Table A for $\alpha = 0.01/2$ and $\alpha = 1 − (0.01/2)$; they are 58.96 and 16.50, respectively.

Mean	SD	Y	SE_{10}	Z_{10}	$Pr(\bar{Y} > Y)$	SE_{30}	Z_{30}	$Pr(\bar{Y} > Y)$
14	5	15	1.58	0.63	0.26435	0.91	1.10	0.13567
15	3	15.5	0.95	0.53	0.29806	0.54	0.91	0.18141
−23	4	−22	1.26	0.79	0.21476	0.73	1.37	0.08534
72	50	45	15.81	−1.71	0.95637	9.18	−2.96	0.99846

We now calculate the lower bound of the confidence interval: $34(0.11005)/58.96 = 0.063$. The upper bound is $34(0.11005)/16.5 = 0.227$. $0.063 < \sigma^2 < 0.227$. **(d)** The 99% confidence interval of the standard deviation of the sample is found by taking the square-root of the variance confidence interval: $0.252 < \sigma < 0.476$, around the estimated standard deviation of 0.332.

3. No, using $n = 70$ would assume that the right distance was independent of the left distance on the same animal, which is not true.

4. **(a)** To test whether the mean fitness changed, we will test whether the mean fitness after 100 generations was significantly different from zero. For this, we will use a t-test. We need the sample mean, the sample standard error, and the sample size. Our hypothesized value from the null hypothesis of no change in fitness, μ_0, is zero. The mean is 0.2724, the standard deviation is 0.600, and the sample size is 5. From this, we calculate the SE $= s/\sqrt{n} = 0.6/\sqrt{5} = 0.268$, and we calculate the t-statistic: $t = (\text{mean} - \mu_0)/\text{SE} = (0.2724 - 0)/0.268 = 1.02$. This is a two-tailed test, so we look up the critical value of $\alpha(2) = 0.05$ for 4 df, 2.78. $t < 2.78$, so we do *not* reject the null hypothesis of no change in fitness. **(b)** The 95% confidence interval for the mean fitness change is $0.2724 \pm 2.78(0.268)$: $-0.47 < \mu < 1.02$. **(c)** We are assuming that fitness is normally distributed and that the five lineages are random and independent.

5. **(a)** The mean testes area for monogamous lines is 0.848, while the mean area for polyandrous lines is 0.950. The standard deviation was 0.031 and 0.034, respectively, for the two sets. **(b)** The standard error of the mean is 0.015 for the monogamous lines and 0.017 for the polyandrous lines. **(c)** The 95% confidence interval for the mean in the polyandrous line is $0.95 \pm 0.017t_{0.05(2),3} = 0.95 \pm 3.18(0.017)$, or $0.90 < \mu < 1.00$. **(d)** The 99% CI for the standard deviation of testes area among monogamous lines requires the sample variance (0.001), the degrees of freedom (3), and the critical values of the χ^2 distribution for the appropriate α and df. The χ^2 critical values are looked up in Statistical Table A for $\alpha = 0.01/2$ and $1 - (0.01/2)$; they are 0.07 and 12.84, respectively. Then, the 99% CI for the

variance is $3(0.001)/0.07$ and $3(0.001)/12.84$, or 0.0002 to 0.04. To get the 99% CI for the standard deviation, we take the square root of each of these, or $0.015 < \sigma < 0.20$.

6. **(a)** The 95% CI for the discontinuity score requires the estimated mean (0.183), the SE (calculated as the s divided by $\sqrt{n}$, or 0.051), and the critical t-value for $\alpha(2) = 0.05$ for 6 df, 2.45. Then the CI is $0.18 \pm (0.051)(2.45)$, or $0.058 < \mu < 0.309$. **(b)** To test whether the sample mean continuity score of 0.183 is consistent with the predicted mean, $\mu_0 = 0$, we use the t-test. $t = (\text{mean} - \mu_0)/\text{SE} = (0.183 - 0)/0.051 = 3.6$. The critical value for $\alpha(2) = 0.05$ for 6 df is 2.45. $t > t_{\text{critical}}$, so we reject the null hypothesis that the mean discontinuity score is zero ($P = 0.01$).

7. In this case, we want to see if rats do better than the chance performance, so we are comparing their scores with $\mu_0 = 50\%$. The mean is 68.4% and the standard deviation is 7.1%. There were seven rats, so SE $= 7.1/\sqrt{7} = 2.7$. $t = (68.4 - 50)/2.7 = 6.82$. For $\alpha(1) = 0.05$, t_{critical} for six $df = 1.94$; $t > t_{\text{critical}}$, so we reject the null hypothesis that rats were doing as expected by chance. (Moreover, we can show that $P < 0.001$.) (For this we assumed that the seven rats were random, independent samples and that their performance scores were normally distributed.) We used a one-tailed test, because there is no reasonable alternative that would have the rats err more than chance.

8. The confidence interval for the mean is symmetric, so the lower bound must be 0.3 kg below the mean, or at 2.9 kg.

9. The confidence interval of the variance is the square of the bounds for the standard deviation.

Standard deviation	Variance
$2.22 < \sigma < 4.78$	$4.93 < \sigma^2 < 22.85$
$20.6 < \sigma < 26.1$	$425.4 < \sigma^2 < 678.8$
$36.4 < \sigma < 59.6$	$1325.0 < \sigma^2 < 3552.2$
$13.63 < \sigma < 16.70$	$185.8 < \sigma^2 < 279.0$

10. In this case, we wish to test whether the relative swimming speed in the syrupy pool is different from $\mu_0 = 1$. We calculate t from the mean (1.0117) and the standard error (0.00997): $t = (1.0117 - 1)/0.00997 = 1.17$.

The critical value for $\alpha(2) = 0.05$ and 17 df is 2.11, so we do *not* reject the null hypothesis that swimming speed remained constant in the two environments. **(b)** The 99% confidence interval for the relative swimming speed in syrup is $1.01 \pm 0.01 \, t_{0.01(2),17} = 1.01 \pm 0.01$ (2.9), or $0.98 < \mu < 1.04$.

Chapter 12

1. **(a)** Paired *t*-test. **(b)** Paired *t*-test. **(c)** Two-sample *t*-test. **(d)** Paired *t*-test. **(e)** Two-sample *t*-test. **(f)** Paired *t*-test. **(g)** Two-sample *t*-test. **(h)** Paired *t*-test.

2. **(a)** On average there are 25 more injuries on No Smoking Day than on the prior Wednesday. **(b)** The 99% confidence interval is calculated as the mean difference: (25) $\pm$ SE of the difference times $t_{0.01(2),9 \, df}$, $t = 3.25$ and SE $= 10.22$, so the confidence interval is $-8.2 < \mu_d < 58.2$. **(c)** If we had data from 100 different samples and we calculated the mean difference and a 99% confidence interval each time, we expect that 99 of these samples would include the true mean difference. **(d)** We can test whether the difference is different from zero by using the paired *t*-test: $t = (25 - 0)/10.22 = 2.45$ for 9 df. $t > 2.26$, the critical value for $\alpha(2) = 0.05$, so we conclude that the accident rate does differ between No Smoking Day and a normal day (and is higher on No Smoking Day). $P < 0.05$.

3. The change in mean length is -5.81 mm, the standard deviation is 19.5 mm, and the sample size is 64. The SE is $19.5/\sqrt{64} = 2.44$. The confidence interval for the difference is the mean difference $\pm 2.44t$, where $t_{0.05(2),63} = 2.0$, so the confidence interval is -10.69 mm $< \mu_d < -0.935$ mm. This assumes that the distribution of change in length is a normal distribution and that the iguanas are a random sample from the population. **(b)** The 95% confidence interval for the SD of the change in length is derived using the formula for the confidence interval of the variance in Chapter 11. The variance of the

difference is 380.25. We need the df (63) and the appropriate χ^2 for 0.025 and 0.9725, which are 86.83 and 42.95, respectively. Then, the lower confidence interval is $(380.25)(63)/86.83 = 275.9$ mm^2, and the high is $(380.25)(63)/42.95 = 557.8$ mm^2. The standard deviation 95% confidence interval is found by taking the square root of the variance confidence interval, or 16.6 mm $< \sigma < 23.6$ mm. **(c)** Highly reasonable. The 95% confidence interval does not include zero. **(d)** $t = -2.38$ for 63 df. We reject the null hypothesis of no mean difference with $P < 0.05$; since $t > 2.0$, the cutoff for $\alpha(2) = 0.05$ for 63 df.

4. **(a)** To calculate the 95% confidence intervals for the mean number of dung beetles captured, we must assume that the number captured is normally distributed. So, for the dung-addition treatment, the confidence interval is $4.8 \pm (2.26 \times 1.03)$: $2.47 < \mu < 7.13$. For the control, the confidence interval is $0.51 \pm (2.26 \times 0.28)$, or $-0.13 < \mu < 1.15$. **(b)** The two-sample *t*-test cannot be used to test for differences in the means, since the standard deviations are more than threefold different between the two groups. Instead, a Welch's approximate *t*-test is appropriate.

5. Because these are not paired samples, we will analyze the difference of the means, not the mean of the differences. The monogamous flies had a mean testes size of 0.8475 mm^2, and the polyandrous of 0.95 mm^2, for a difference of 0.1025 mm^2. The 95% confidence interval for this difference requires finding the $SE_{\bar{Y}_1 - \bar{Y}_2}$, $= \sqrt{[s_p^2(1/n_1 + 1/n_2)]}$, where $n_1 = n_2 = 4$, and $s_p^2 = (df_1 s_1^2 + df_2 s_2^2)/(df_1 + df_2)$, where $df_1 = df_2 = 3$, and s_1^2 and $s_2^2 = 0.0010$ and 0.0011, respectively. Then, $SE_{\bar{Y}_1 - \bar{Y}_2} = 0.023$. The confidence interval for the difference is the standard error times $t_{0.05(2),6} = 2.45$, so $0.1025 \pm (2.45 \times 0.023)$, or 0.046 to 0.159. **(b)** The null hypothesis is that there is no difference between the monogamous and polyandrous treatments in testes size, so $(\mu_1 - \mu_2)_0 = 0$. $t =$ the difference in means $= 0.1025$, over the SE of the difference, 0.023. $t = 4.48$, for 6 df. $t_{critical} = 3.71$ for $P = 0.01$, so $P < 0.01$. The mean testes sizes are significantly different.

6. **(a)** On average, 33% more of the male bodies were covered if they emitted pheromones. $s_p = 26.5\%$; $SE_{\bar{Y}_1 - \bar{Y}_2} = 6.02\%$. $df_1 = 48$ and $df_2 = 31$, so $df = 79$. $t_{0.05(2), 79\,df} = 1.99$, so the confidence interval is $21\% < \mu_1 - \mu_2 < 45\%$. **(b)** Using a two-sample t-test, we will assume that the percent coverage is normally distributed, that each snake is independent, and that the standard deviations are not different (they are not more than three-fold different). The null hypothesis is that there is no difference between the males emitting pheromones and those not, so $(\mu_1 - \mu_2)_0 = 0$. $t = 0.33/0.0602 = 5.47 > 3.9$, the critical value for $\alpha(2) = 0.0002$ for 79 df, so we can reject the null hypothesis, with $P < 0.0002$.

7. **(a)** The PLFA levels between the control and addition plots before cicada death were not significantly different (two sample t-test: $t = -0.19$, $df = 42$, and $P > 0.10$). **(b)** After the cicada addition, the PLFA levels differed significantly: $t = 2.89$, $df = 42$, and $P < 0.01$. **(c)** No. We would need the standard deviation of the differences between measurements before and after, and these are not provided.

8. As described, this test assumes that the eight "open-water" samples were independent of the eight "near-shore" samples, because it uses a two-sample t-test (and so would have 14 df). Differences in growth rate could be due to differences between lakes, so the two samples within each lake are not independent. The paired t-test would better reflect this (and would have 7 df).

9. **(a)** Since we assume that the distributions are normal, we can use the F-test to compare the variances. The ratio of the variances (the larger variance is always in the numerator) is $F = (0.1582)^2/(0.0642)^2 = 6.07$. The degrees of freedom are 32 and 19. $F_{0.05(1),32,19} = 2.05$ (between

2.07 for 30 df in the numerator and 2.03 for 40 df), so we reject the null hypotheses that the variances are equal. **(b)** The variances are not equal, but the difference in the standard deviations is not greater than threefold. Therefore, a two-sample t-test could be used: $t = 0.09$, $df = 51$, and $P > 0.05$. You could use Welch's approximate t-test rather than the two-sample t-test. $t = 0.10$ with 46 df, which is not significant.

10. **(a)** Remember to multiply each value of flower length by the number of flowers in that category when calculating the mean and variance. The zeros are dropped in the equation, which is at the bottom of this page. **(b)** For the simplest test, we will assume that distributions are normal. Then we can use the F-test to compare the variances. $42.4/8.6 = 4.93$; 443, 172 df. Looking this up (roughly) we find that the critical value is 1.21 (for 1000,200) and 1.31 (for 200,100) for $\alpha(1) = 0.05$, so we conclude that the variance for the F_2 is significantly greater than the variance for the F_1.

11. Paired t-test: mean difference is 22.3, which is significantly different from 0 ($P < 0.01$, $t = 4.5$, and $df = 7$).

12. **(a)** We would use Welch's approximate t-test for this comparison since the standard deviations differ between *Pteronotus* and the vampires by more than threefold. **(b)** The average strength was *higher* for *Pteronotus*, which is in the opposite direction as predicted by the model.

13. **(a)** We can use a paired t-test in this case, because the body temperature measurements are taken on the same individuals as the brain temperature measurements. To do this, we calculate the difference in temperature between brains and

$$\frac{4(55) + 10(58) + 41(61) + 75(64) + 40(67) + 3(70)}{4 + 10 + 41 + 75 + 40 + 3} = 63.5$$

$$\frac{4(55 - 63.5)^2 + 10(58 - 63.5)^2 + 41(61 - 63.5)^2 + 75(64 - 63.5)^2 + 40(67 - 63.5)^2 + 3(70 - 63.5)^2}{(4 + 10 + 41 + 75 + 40 + 3) - 1} = 8.6$$

bodies for each ostrich, find the mean difference (0.648 °C) and the standard error of the difference (0.116 °C). $\mu_0 = 0$: the null hypothesis is that the brain temperature does not differ from the body temperature. $t = 0.648/0.116 = 5.6$ with 5 df, which is greater than the critical value for $\alpha(2) = 0.01$, 4.03, so $P < 0.01$. We reject the null hypothesis of no difference between brain and body temperature. **(b)** While our test is significant, the deviation is the opposite of that predicted from observations of mammals in similar environments: brains are hotter than bodies in ostriches, not cooler.

14. **(a)** B. **(b)** B. **(c)** From B, we can still be fairly confident that the groups are different, but we need to mentally double the size of the error bars to make this determination. **(d)** With sample sizes of 100, the standard errors will be one-tenth as great as the standard deviations. Graphs A, B, and C will be significantly different.

Chapter 13

1. **(a)** Not normal. The points show strong curvature. **(b)** Not normal. The curve bends steeply at high and low values. **(c)** Approximately normal. **(d)** Not normal. The points follow and S-shape, not a straight line.

2. **(a)** I. No, this is a uniform distribution, not a normal distribution.
II. No, this plot is bimodal, not normal.
III. No, this plot is right-skewed. It looks log-normal rather than normal.
IV. Yes, this is a normal distribution.
(b) I. The sign test would be best for these data, as they are unlikely to transform into anything resembling normal.
II. The sign test would probably be best for this distribution as well. Bimodal distributions are tough to transform into anything else.
III. These data could probably be tested by a one-sample t-test after transformation (probably a log transformation, as it is right-skewed).
IV. These data could be tested by a one-sample t-test, as they are a normal distribution.
(c) I: b. II: d. III: a. IV: c.

3. **(a)** Mean: 2.75; 95% CI: $-3.28 < \mu_{\log[X]} < 8.78$. **(b)** Mean: 1.86; 95% CI: $-0.02 < \mu_{\log[X]} < 3.74$. **(c)** Not possible: cannot use ln transformation on negative values. **(d)** Mean: 4.23; 95% CI: $-2.04 < \mu_{\log[X]} < 10.5$. **(e)** Here, we must use $Y' = \ln(Y + 1)$. Mean: 0.98; 95% CI: $-0.23 < \mu_{\log[X]} < 2.2$.

4. After applying the arcsine to the square root of the raw data, the mean is 1.39 and the SD is 0.09.

5. The simplest analysis is a sign test. We will score each lobster as "+" if it is pointing more towards home than away (i.e., -90 to 90) and "$-$" if it is pointing more away than towards home (i.e., more than 90 or less than -90). Lobsters exactly at 90 or -90 will not be scored. Of the 14 scored lobsters, 11 were "+" and 3 were "$-$".

$$P = 2\left[\sum_{X=11}^{14} \binom{14}{X} 0.5^{14-X} 0.5^X \right] = 0.057.$$

Using the usual standard of $\alpha = 0.05$, we cannot reject the null hypothesis with these data.

6. Mann–Whitney U–test: $U_1 = 39$. $U_2 = 6$, so $U = 39$. The critical value for $n_1 = 5$ and $n_2 = 9$ is 38. U is larger, so we reject the null hypothesis that there is no difference in the time until reproduction due to accidental death or infanticide.

7. A nonparametric approach to this problem is the Mann–Whitney U-test, noting that there will be many ties, so our test will be less powerful than it might be. We assign a rank of "3" to the zeros (the average of $1-5$), a rank of 11.5 to the 1's (the average of $6-17$), a rank of 19.5 to the 2's (the average of $18-21$), and a rank of 23 to the 3's (the average of $22-24$). Then we sum the ranks for the blind group: $R_1 = 140.5$. We then calculated $U_1 = n_1 n_2 + n_1(n_1+1)/2 - R_1 = 81.5$. $U_2 = n_1 n_2 - U_1 = 144 - 81.5 = 62.5$. $U = 81.5$. The critical value for $n_1 = 12$ and $n_2 = 12$ is 107. U is smaller, so we do *not* reject the null hypothesis that there is no difference in the number of gestures used by the blind vs. sighted humans.

8. **(a)** Benton: mean $= 0.67$ and variance $= 0.075$; Warrenton: mean $= 0.24$ and variance $= 0.007$. The standard deviations are more than threefold different, so a two-sample t-test is inappropriate. **(b)** (1) Try a log transformation, followed by a two-sample t-test. (2) Welch's t-test on the non-transformed data, which does not require equal variances. (3) Mann–Whitney U-test. **(c)** A log transformation might be appropriate because the population with the larger mean has the larger variance. The two-sample t-test on the transformed data rejects the null hypothesis: $t = 0.961/0.249 = 3.87$ with 10 df. $P < 0.01$. **(d)** For Benton county, $-0.94 < \mu_{\log[X]} < -0.04$. For Warrenton county, $-1.84 < \mu_{\log[X]} < -1.06$. **(e)** The transformed confidence interval for Benton county is 0.39 to 0.96; for Warrenton county, the interval is 0.16 to 0.35. If we found the means from many samples, 95% of the confidence intervals from these samples would contain the true mean.

9. Two-sample t-tests and confidence intervals are robust to violations of equal standard deviations so long as the sample sizes of the two groups are roughly equal and the standard deviations are within three times of one another.

10. **(a)** Two-sample t-test. **(b)** These distributions are skewed right, so a log-transformation could be tried. **(c)** This one is tough: neither distribution is normal, but they deviate in different ways, so a transformation that would improve the fit for one would not help with the other. Also, the standard deviations will be very different. The differences in shape make the rank-sum test inappropriate. A randomization test (Chapter 19) is best. **(d)** Two-sample t-test. **(e)** Two-sample t-test.

11. **(a)** The null hypothesis was that the spermatids of males and hermaphrodites did not differ in their mean size, while the alternative hypothesis is that they do differ in size. **(b)** $U = 35910$, $n_1 = 211$, $n_2 = 700$. Use the normal approximation:

$$Z = \frac{2U - n_1 n_2}{\sqrt{n_1 n_2 (n_1 + n_2 + 1)/3}} = -11.32.$$

This is highly significant ($P < 0.00002$). **(c)** Because the distributions are not the same shape. Equal shape is an assumption of the U-test when used to compare means or medians.

12. The differences are not normally distributed, but skewed right. We need a sign test. There are 6 "$-$" and 22 "$+$," where a "$+$" means more species in the monoecious group.

$$P = 2\left[\sum_{X=22}^{28} \binom{28}{X} 0.5^{28-X} 0.5^{X}\right] = 0.0037,$$

so we reject the null hypothesis.

13. Mann–Whitney U-tests are insensitive to outliers and may be used appropriately in this case as long as the distributions are otherwise similar.

14. First, we transform the data by taking the natural log of each weight. The mean ln-transformed weights are -1.37 for females and -1.76 for males. Since the transformed weights are normally distributed, we can use the two-sample t-test. $t = 3.51$ with $df = 18$. The critical value for $\alpha(2) = 0.01$ for 18 df is 2.88, so $P < 0.01$, and we reject the null hypothesis. Female mosquitoes weigh more than males.

15. No, the differences are probably not normally distributed: two values less than 500 and three values above 24,000 would not be expected from a normal distribution. **(b)** We can use a sign test: we have five pairs where there are more species feeding on angiosperms than on gymnosperms. The probability of five out of five in the binomial distribution, assuming equal probabilities of either outcome, is 0.5^5, or 0.031. However, for a two-tailed test, we must double this, so $P = 0.062$. We are unable to reject the null hypothesis with the usual significance level. **(c)** The number of species that feed on angiosperms was often two orders of magnitude greater, and all pairs had a higher number of species in the angiosperm group. It is impossible to get a P-value under 5% with only five data points in a sign test, even with all the data in a consistent direction.

Chapter 14

1. **(a)** Limit sampling error. **(b)** Reduce bias. **(c)** Reduce sampling error. **(d)** Reduces bias. **(e)** Reduce bias.

2. **(a)** [Answers will vary.] T, H, H, H, H, T, T, H. **(b)** [Answers will vary.] No. **(c)** $1 - \text{Pr}[\text{exactly 4 heads in 8 tosses}]$
$$= 1 - \binom{8}{4}0.5^8 = 0.727.$$
(d) [Answers will vary.] Assign a random number between zero and one to each unit. Assign the first treatment to the units with the four smallest random numbers and the second treatment to the remaining units.

3. Use a randomized block design, where each block is a position along the moisture gradient. Place three plots in each block, one for each of the three fertilizer treatments (call the fertilizers A, B, and C). Within each block, randomly assign the three fertilizers plots. The figure below illustrates the design for six blocks.

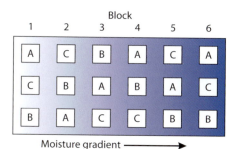

4. The researchers planned their sample size assuming a significance level of 0.05 and an 80% probability of rejecting a false null hypothesis (for a specified difference between treatment means).

5. Observational study. The treatments, presence and absence of brook trout, were not randomly assigned to the units (streams)—the trout were already present in the streams prior to the study. Potential confounding factors (water temperature, stream depth, food supply) might differ between streams with and without brook trout, and randomization was not used to break their association with the treatment variable.

6. **(a)** Decrease bias (reduces effects of confounding variables); decrease sampling error (by grouping similar units into pairs). **(b)** No effect on bias; reduce sampling error. **(c)** Decrease bias (corrects for effects of age, a possible confounding variable); no effect on sampling error. **(d)** No effect on bias; reduce sampling error.

7. **(a)** No. In an experimental study, the experimenter assigns two or more treatments to subjects. Here, there was only one treatment. **(b)** Have two treatments: the salt infusion and a placebo control (e.g., distilled-water infusions). Assign treatments randomly to a sample of severe pneumonia patients. Ensure equal numbers of patients in each treatment. Keep patients unaware of which treatment they are receiving. A clinician unaware of which patient received which treatment should record their subsequent condition.

8. **(a)** Observational study: cancer treatments were not assigned randomly to subjects. **(b)** Yes, it compares marijuana use of cancer patients and non-cancer (control) patients. **(c)** Reducing bias. Age and sex might be confounding variables, affecting both marijuana use and cancer incidence. Using only subjects similar in age and sex reduces the effect of these confounding variables on the association between marijuana use and cancer. **(d)** First, it is not wise to "accept" the null hypothesis, because the study might not have had sufficient power to detect an effect. The 95% confidence interval for the odds ratio ranged from 0.6 to 1.3, which still includes the possibility of a moderate effect. Second, observational studies such as this one cannot decide causation because of confounding variables. Perhaps marijuana use does increase cancer risk, but marijuana users may have lifestyle differences that diminish cancer risk, offsetting any marijuana effect.

9. **(a)** The stings were applied to two volunteers, which means that the reactions to the 40 stings were probably not independent. Treating the sample size as 40 is a case of pseudoreplication.

(b) More conservatively and appropriately, this study should be treated as a paired design with two samples. The mean difference in swelling would be found for each subject, and then the average of the two subjects would be tested to see if it was significantly different from zero. **(c)** Since each subject adds just one data point, there is no need to inflict 20 stings on each subject. Fewer stings per subject would be less cruel, and more subjects would allow more replicates and thus more power.

10. Extreme doses increase power and so enhance the probability of detecting an effect. If an effect is detected, then studies of the effects of more realistic doses would be the next step.

11. **(a)** 13 plots per treatment. The "uncertainty" is $0.4/2 = 0.2$, so $n = 8(0.25/0.2)^2 = 12.5$; round up to 13. **(b)** 50 plots per treatment. The "uncer-tainty" is $0.2/2 = 0.1$, so $n = 8(0.25/0.1)^2 = 50$.

(c) A greater total sample size would be needed if the design were not balanced. For a given total sample size, the expected width of a confidence interval for the difference between two means increases as the design is more imbalanced (because the precision of the treatment mean having the lower sample size is greatly reduced). Achieving the same confidence interval width as a balanced design will therefore require a greater total sample size. **(d)** Because environmental differences between the normal-corn plot and the Bt-corn plot would be confounding variables. In effect, such a design would lack replication because plants in the same plot are not independent.

12. 16 plots per treatment. $n = 16(0.25/0.25)^2 = 16$.

Chapter 15

1. **(a)** Explanatory: Population isolation. Response: Generations persisted. **(b)** Experimental study: treatments were assigned to plants by the experimenters. **(c)** We assume that the variable has a normal distribution. (We also assume random samples.)

Treatment	Mean	95% CI
Isolated	9.25	$5.3 < \mu < 13.2$
Medium	14.50	$11.5 < \mu < 17.5$
Long	10.75	$8.0 < \mu < 13.5$
Continuous	12.75	$8.2 < \mu < 17.3$

(d) (Answers may vary.) In the figure below, open circles are the data (offset where needed to minimize overlap). Means are filled circles. Vertical lines indicate 95% confidence intervals.

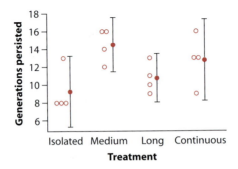

2. **(a)** H_0: The mean persistence times of the four isolation treatments are equal ($\mu_1 = \mu_2 = \mu_3 = \mu_4$). H_A: At least one of the means μ_i is different.

(b)

Source of variation	Sum of squares	df	Mean squares	F-ratio	P
Groups (treatments)	63.188	3	21.063	3.996	0.035
Error	63.250	12	5.271		
Total	126.438	15			

(c) F-distribution with 3 and 12 degrees of freedom. **(d)** The probability of obtaining an F-ratio statistic as large as or larger than the value observed when the null hypothesis is true. **(e)** The total sum of squares is the sum of the deviations squared between each observation and the grand mean. The error sum of squares is the sum of the squared deviations between each observation and its group mean. The group sum of squares is the sum of the squared deviations between the group mean for each individual and the grand mean. **(f)** Use R^2, the ratio of the group sum of squares and the total sum of squares. **(g)** $R^2 = 63.188/126.438 = 0.50$.

3. **(a)** H_0: The mean of group i equals the mean of group j, for all pairs of means i and j. H_A: The mean of group i does not equal the mean of group j, for all pairs of means i and j. **(b)** These are unplanned comparisons, because they are intended to search for differences among all pairs of means. Planned comparisons must be fewer in number and identified as crucial in advance of gathering and analyzing the data. **(c)** Failure to reject a null hypothesis that the difference between a given pair of means is zero does not imply that the means are equal, because power is not necessarily high, especially when the differences are small. If the means of the "Medium" and "Isolated" treatments differ from one another, then one or both of them must differ from the means from the other two groups, but we don't know which.

(d))

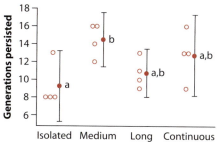

(e) The critical value for a t-test has a Type I error rate of 0.05 when comparing two means. The critical value for the Tukey comparison of all pairs of means is larger so that the probability of making at least one Type I error in *all* of the comparisons is only 0.05. (Note that degrees of freedom are 12 in either case because we are using the error mean square to calculate the standard error of the difference between means.)

4. **(a)** Transformations or a nonparametric Kruskal–Wallis test. **(b)** The transformation should be attempted first, because this would yield the more powerful test.

5. **(a)** H_0: The mean number of shoots is equal in the three treatments ($\mu_1 = \mu_2 = \mu_3$). H_A: At least one of the treatment means μ_i is different.

(b)

Source of variation	Sum of squares	df	Mean squares	F-ratio	P
Groups (treatments)	2952.808	2	1476.40	5.32	0.011
Error	8049.067	29	277.55		
Total	11,001.875	31			

The critical value is $F_{0.05(1),2,29} = 3.33$. Since $F > 3.33$, $P < 0.05$, we reject H_0. Conclude that there are differences between treatments in mean shoot number. **(c)** The variable is normally distributed with equal variance in the three treatment populations. **(d)** Fixed-effects ANOVA. The levels chosen were set by the researcher and are repeatable—they were not randomly sampled from a population of treatments.

6. **(a)** $\bar{Y}_1 - \bar{Y}_2 = (-0.004 - (-0.195)) = 0.191$. $MS_{error} = 0.0345$, $df = 42$, $SE = 0.0679$, $t_{0.05(2),42} = 2.02$, $0.054 < \mu_1 - \mu_2 < 0.328$. **(b)** Yes, because the study was designed mainly to compare PLP1 gene expression in persons with schizophrenia to that of control individuals. It was a single focused comparison, not a broad search for differences between groups. **(c)** The expression measurements are normally distributed in the populations with equal variances. (We also assume random samples.)

7. **(a)** H_0: Mean PLP1 gene expression is equal in the three groups ($\mu_1 = \mu_2 = \mu_3$).
H_A: At least one of the group means μ_i is different.

Source of variation	Sum of squares	df	Mean squares	F-ratio	P
Groups	0.5403	2	0.2701	7.82	0.0013
Error	1.4502	42	0.0345		
Total	1.9905	44			

The critical value is $F_{0.05(1),2,42} = 3.22$. Since $F > 3.22$, $P < 0.05$, we reject H_0. Conclude that mean PLP1 expression differs among groups. **(b)** The expression measurements are normally distributed in the populations with equal standard deviations. **(c)** Fixed-effects ANOVA: we are comparing predetermined and repeatable treatment groups, not a random selection of groups in a population. **(d)** Use R^2 to describe the fraction of the variance explained by group differences: $R^2 = 0.27$. **(e)** Use the Tukey–Kramer method.

8. **(a)** Random-effects ANOVA. The sites were randomly chosen; the goal was to determine whether sites varied in general in Spain, not just whether these four sites were different. **(b)** Because this is a random-effects ANOVA, the hypotheses should be stated in terms of the "population" of sites, from which we have a sample.
H_0: Mean carotenoid plasma concentration in vultures is the same among sites in Spain.
H_A: Mean carotenoid plasma concentration in vultures differs among sites in Spain.

Source of variation	Sum of squares	df	Mean squares	F-ratio	P
Groups (sites)	712.25	3	237.42	30.44	0.0001
Error	1388.26	178	7.80		
Total	2100.51	181			

The critical value is $F_{0.05(1),3,178} = 2.66$. Since $F > 2.66$, $P < 0.05$, we reject H_0. Conclude that sites vary in the mean carotenoid plasma concentration. **(c)** We assume that carotenoid plasma concentration has a normal distribution in every population with equal variances. We also assume that sites were randomly chosen and that site means have a normal distribution. (We also assume random samples.)

9. **(a)** H_0: Mean tested weight is the same in the four treatments ($\mu_1 = \mu_2 = \mu_3 = \mu_4$). H_A: At least one of the treatment means μ_i is different.

Source of variation	Sum of squares	df	Mean squares	F-ratio	P
Groups (treatment)	19,057.5	3	6352.5	3.32	0.025
Error	130,050.0	68	1912.5		
Total	149,107.5	71			

The critical value is $F_{0.05(1),3,68} = 2.74$. Since $F > 2.74$, $P < 0.05$, we reject H_0. Conclude that mean testes weight varies among treatments. **(b)** Testes weight is normally distributed in the four populations with equal variance. (We also assume random samples.) **(c)** Yes, this was an experimental study because the treatments were assigned (randomly) to the subjects (mice) by the researchers. **(d)** $R^2 = 19,057.5/149,107.5 = 0.13$.

10. Need to use a χ^2 contingency test (Chapter 9). Observed and expected frequencies (in parentheses) are shown in the contingency table below.

H_0: The proportion of impregnated females is the same in the four treatments.

H_A: The proportion of impregnated females is not the same in the four treatments.

Number of females

Treatment	Impregnated	Not impregnated	Row sum
Oil	14 (12.75)	4 (5.25)	18
THC	13 (12.75)	5 (5.25)	18
Cannabinol	11 (12.75)	7 (5.25)	18
Cannabidiol	13 (12.75)	5 (5.25)	18
Column sum	51	21	72

$\chi^2 = 1.28$ and $df = 3$. The critical value is $\chi^2_{3,\,0.05} = 7.81$. Since χ^2 is not greater than 7.81, do *not* reject H_0. ($P = 0.73$). We cannot reject the null hypothesis that the proportion of females impregnated is equal between treatments.

11. Random-effects ANOVA: the males were chosen at random from the population. A given male is not a specific, repeatable treatment. The goal is to generalize to the population of males (the hypothesis statements should reflect this). **(b)** H_0: Mean condition of offspring is the same for all males in the population. H_A: Mean condition of offspring differs between males in the population.

Source of variation	Sum of squares	df	Mean squares	F-ratio	P
Groups (males)	9.9401	11	0.9036	4.63	0.0008
Error	4.682	24	0.1951		
Total	14.622	35			

The critical value $F_{0.05(1),11,24} = 2.21$ (between 2.18 and 2.25). Since $F > 2.21$, $P < 0.05$, we reject H_0. Conclude that mean offspring conditions differs among males in the beetle population.

(c) $s^2_A = (0.9036 - 0.1951)/3 = 0.236$. Repeatability $= 0.236/(0.236 + 0.195) = 0.548$.

12. **(a)** ANOVA.

(b) Try transforming the data to better meet the assumptions of normality and equal variances. If this fails, use the Kruskal-Wallis test if the distributions have equal shape.

(c) ANOVA is appropriate if sample size is large enough (appealing to the Central Limit Theorem).

(d) Tukey-Kramer test of all pairs of means.

Chapter 16

1. $r = 0$; $r = -0.8$; $r = 0.5$; $r = 0$.
2. **(a)** 95%: $0.22 < \rho < 0.63$. 99%: $0.14 < \rho < 0.68$. **(b)** 95%: $-0.95 < \rho < -0.76$. 99%: $-0.96 < \rho < -0.70$. **(c)** 95%: $-0.17 < \rho < 0.25$. 99%: $-0.24 < \rho < 0.31$.
3. **(a)** Scatter plot:

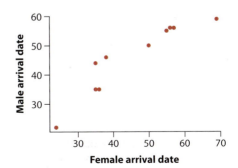

(b) The relationship is linear, positive, and strong. **(c)** $r = 0.93$. SE $= 0.13$. **(d)** The standard error is the standard deviation of the sampling distribution of r. **(e)** $0.72 < \rho < 0.98$.

4. **(a)** No change to the correlation coefficient ($r = 0.93$). Adding a constant to one of the variables does not alter the correlation coefficient. **(b)** No change to the correlation coefficient ($r = 0.93$). Dividing one of the variables by a constant does not alter the correlation coefficient.

5. We can use a paired t-test (Chapter 12).
H_0: Mean arrival date of male and female partners is the same ($\mu_d = 0$).
H_A: Mean arrival date of male and female partners is different ($\mu_d \neq 0$).
$\bar{d} = 0.3$ days (males are slightly earlier on average), SE $= 1.667$, $t = 0.18$, $df = 9$, and $P = 0.86$. $t_{0.05(2),9} = 2.26$; since observed t is less than $t_{0.05(2),9}$, $P > 0.05$, we do not reject H_0. Conclude that we cannot reject null hypothesis of equal mean arrival times of males and females. Assume a normal distribution of differences between arrival rates of males and females.

6. When there is measurement error in one or both of the variables X and Y.

7. A narrower range of values for inbreeding coefficient should lower the correlation with the number of surviving pups compared with a wider range of inbreeding coefficients.

8. Use the Spearman rank correlation:
H_0: The population rank correlation is zero ($\rho_S = 0$).
H_A: The population rank correlation is not zero ($\rho_S \neq 0$).
$r_S = 0.57$. $P = 0.17$. $r_{S\,(0.05(2),7)} = 0.821$. Since r_S is not greater than or equal to $r_{S\,(0.05(2),7)}$, $P > 0.05$, we do not reject H_0. Conclude that we cannot reject the null hypothesis of zero correlation. **(b)** Assume random sample and a linear relationship between the ranks of the two variables.

9. Sampling error in the estimates of earwig density and the proportion of males with forceps means that true density and proportion on an island are measured with error. Measurement error will tend to decrease the estimated correlation. Therefore, the actual correlation is expected to be higher on average than the estimated correlation.

10. **(a)** The assumption of bivariate normality is violated: there is an outlier.
(b) Using a rank correlation would be appropriate.
(c) H_0: The population rank correlation is zero ($\rho_S = 0$).
H_A: The population rank correlation is not zero ($\rho_S \neq 0$).
$r_S = 0.30$. $P = 0.053$. $r_{S\,(0.05(2),41)} = 0.308$. Since r_S is not greater than or equal to $r_{S\,(0.05(2),41)}$, $P > 0.05$, we do *not* reject H_0. Conclude that we cannot reject the null hypothesis of zero correlation.
(d) Random sample, and that there's a linear relationship between the ranks of the two variables.

11. **(a)** The graph shows that on average the number of predators increases as the resemblance to pufferfish decreases. The relationship is fairly strong, though there is some scatter.

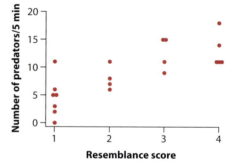

(b) To test the hypothesis, we will need to use the Spearman's rank correlation because the resemblance to pufferfish is a ordinal variable, not a continuous numeric score.
H_0: The rank correlation between resemblance and number of predators is zero ($\rho_S = 0$).
H_A: The rank correlation between resemblance and number of predators is not zero ($\rho_S \neq 0$).
$r_S = 0.79$. $P = 0.00004$.
$r_{S\,(0.01(2),20)} = 0.564$. Since r_S is greater than or equal to $r_{S\,(0.01(2),20)}$, $P < 0.01$, we reject H_0. Conclude that lower resemblance to the pufferfish is correlated with higher number of approaching predators.

Chapter 17

1. (a) $Y = 2/3X + 1$. (b) $Y = X - 1$.
   (c) $Y = -0.5X + 2$. (d) $Y = X - 5$.
2. (a)

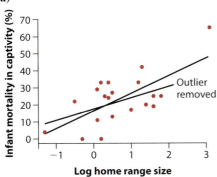

Log home range size

The percent infant mortality increases approximately linearly with the log of the home range size. (b) Mortality $= 16.37 + 10.26$(log home range). (c) H_0: Home range size does not predict infant mortality ($\beta = 0$). H_A: Home range size predicts infant mortality ($\beta \neq 0$).
$b = 10.26$, SE $= 2.69$, $t = 3.81$, $df = 18$, $P = 0.0013$, and $t_{0.05(2),18} = 2.10$. Since $t > 2.10$, $P < 0.05$. Reject H_0. Conclude that home range size predicts infant mortality. (d) Mortality $= 17.51 + 6.60$(log home range). The slope is much lower with the polar bear removed (6.60 instead of 10.26, a reduction of more than a third).

3. (a) The "least squares" regression line is the one the minimizes the sum of squared differences between the predicted Y-values on the regression line for each X and the observed Y-values. (b) Residuals are the differences between the predicted Y-value on the estimated regression line and the observed Y-values. (c) The most conspicuous problem is that the variance of the residuals increases with increasing progesterone concentration, violating the assumption that the variance of Y is the same at all values of X. There are no conspicuous departures from the other two assumptions, linearity and a normal distribution of Y-values at every X.

4. (a)

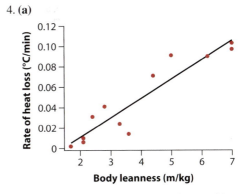

Body leanness (m/kg)

(b) H_0: Rate if heat loss does not change with body leanness ($\beta = 0$). H_A: Rate if heat loss changes with body leanness ($\beta \neq 0$).
$b = 0.0190$, SE $= 0.0023$, $t = 8.29$, $df = 10$, $P = 0.000009$, and $t_{0.05(2),10} = 2.23$. Since $t > 2.23$, $P < 0.05$. (More precisely, $P = 0.023$.) Reject H_0. Conclude that rate of heat loss increases with body leanness. (c) $0.0139 < \beta < 0.0241$. (d) For every X there is a normal distribution of Y-values, of which we have a random sample; the relationship between leanness and heat loss is linear; the variance of Y is the same for all values of X.

5. Caution is warranted because the prediction is based on extrapolation, which is risky because the relationship between winning time and year might not be linear beyond the range of the existing data.

6. (d).

7. (a) The arcsine square root transformation is a good bet for data that are proportions.
   (b) $Y = 0.416 + 7.10X$, where X is genetic distance and Y is arcsine square root transformed proportion sterile.

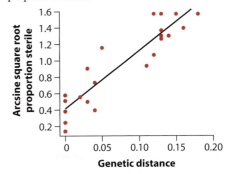

Genetic distance

(c) $5.67 < \beta < 8.52$.

8. **(a)** Such a complicated curve is unwarranted by the data. It would do a poor job of predicting new observations because it does not describe a general trend. A curve fit should be as simple as possible. **(b)** First try to transform the data to make the relationship linear. If that fails, consider nonlinear regression.

9. **(a)** $b = 11.68$ and SE $= 4.85$. **(b)** $-4.08 < \beta < 27.43$. **(c)** The range of most plausible values for the parameter. In 99% of random samples, the confidence interval will bracket the population value for the slope. **(d)** Measurement error in the X-variable (bite force) will tend to lead to an underestimation of the population slope. **(e)** Measurement error in the Y-variable (territory area) will not bias the estimate of slope (though it will increase its uncertainty).

10. **(a)**

Source of variation	Sum of squares	df	Mean squares	F-ratio
Regression	194.3739	1	194.3739	5.8012
Residual	301.5516	9	33.5057	
Total	495.9255	10		

(b) H_0: The slope of the regression of territory area on bite force is zero ($\beta = 0$). H_A: The slope of the regression of territory area on bite force is not zero ($\beta \neq 0$).
$F = 5.8012$, $df = 1,9$, and $P = 0.039$. $F_{0.05(1),1,9} = 5.12$. Since $F > 5.12$, $P < 0.05$. Reject H_0. Conclude that the slope is *not* zero (it is positive).
(c) That the relationship between X and Y is linear; for each X there is a normal distribution of Y-values in the population, of which we have a random sample; the variance of Y is the same at all values of X. **(d)** The variance of the residuals. **(e)** $R^2 = 0.392$. It measures the fraction of the variation in Y that is explained by X.

11. **(a)** First, check the data to ensure this individual was not entered incorrectly. Perform the analysis with and without the outlier included in the data set to determine whether it has an influence on the outcome. If it has a big influence, then it is probably wise to leave it out and limit predictions to the range of X-values between 0 and about 200 (and urge them to obtain more data at the higher X-values). **(b)** Confidence bands give the confidence interval for the predicted *mean* time since death for a given hypoxanthine concentration. **(c)** Confidence bands. **(d)** The prediction interval, because it measures uncertainty when predicting the time of death of a single individual.

12. **(a)**

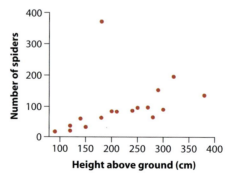

(b) The assumptions of equal variance of residuals, and of a normal distribution of Y-values at each X, are not met because of the presence of an outlier. **(c)** A transformation of the data might improve matters, but success is doubtful. Alternatively, use nonparametric correlation to measure association between the two variables.

Chapter 18

1. **(a)** HOURS = CONSTANT + MUTANT.
    HOURS is hours of resting. CONSTANT is the grand mean. MUTANT is the effect of each line of mutant flies. **(b)** HOURS = CONSTANT. **(c)** Long horizontal line indicates the grand mean, which is the predicted value for the mean of each group under the null hypothesis. The short horizontal lines give the group means, which are the predicted values under the "full" general linear model including the MUTANT term.

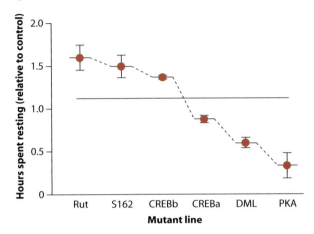

   (d) F.
2. **(a)** Plasma corticosterone concentration. **(b)** Chick age group and disturbance regime. **(c)** That the two explanatory variables interact to affect corticosterone concentrations. **(d)** Observational. The penguins were not assigned by the researcher to age groups or disturbance regimes. **(e)** Yes. Every combination of the two factors (age group and disturbance regime) is included in the design.
3. **(a)** CORTICOSTERONE = CONSTANT + AGE + DISTURBANCE + AGE*DISTURBANCE.
    CORTICOSTERONE is the corticosterone concentration, CONSTANT is the grand mean corticosterone concentration, AGE is the age group of the penguin, DISTURBANCE indicates whether the penguin lived in the undisturbed or tourist-disturbed area, and AGE*DISTURBANCE is the interaction between age and disturbance.
    (b) H_0: Penguin age group has no effect on mean corticosterone concentration. H_0: Disturbance regime has no effect on mean corticosterone concentration. H_0: There is no interaction between penguin age group and disturbance regime.
    (c) Penguins were randomly sampled from each area and each age class. Corticosterone concentrations are normally distributed within age class and disturbance regime. The variance of the corticosterone concentration is the same for each combination of age and disturbance.
4. Three of the following: a. To investigate the effect of more than one variable with the same experimental design; b. to include the effects of blocking in an experimental design; c. to investigate interactions between the effects of different factors; and d. to control for the effects of a confounding variable by including it as a covariate.

5. In these plots, different symbols refer to different groups of factor B. Other answers are possible.

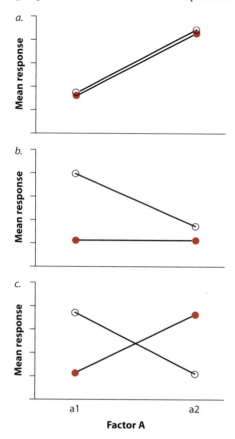

a.

b.

c.

Factor A

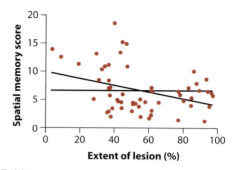

7. **(a)**

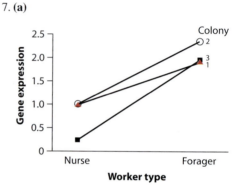

6. **(a)** MEMORY = CONSTANT + LESION. MEMORY is the spatial memory score of a rat, CONSTANT is a constant indicating the value of MEMORY when LESION is zero (*Y*-intercept), and LESION is the extent of the lesion. **(b)** MEMORY = CONSTANT. **(c)** The line having negative slope in the following graph represents the predicted values from the "full" general linear model. The flat (horizontal) line represents the predicted values under the null model.

(b) EXPRESSION = CONSTANT + WORKERTYPE + BLOCK.
EXPRESSION is the *for* gene expression, CONSTANT is the grand mean of *for* gene expression, WORKERTYPE is the worker type, and BLOCK identifies which colony the bee comes from. **(c)** EXPRESSION = CONSTANT + BLOCK. **(d)** Fixed-effect, because the types are repeatable and of direct interest. The worker types in the analysis are not randomly sampled from a population of worker types. **(e)** Blocking variables allow extraneous variation caused by the block variable to be accounted for in the analysis and eliminated. When a block variable is included, the error variance is smaller, making it easier to detect real effects of the factor.

Chapter 19

1. **(a)** For example: Singleton: 3.5, 3.5, 2.6, 4.4; Twin: 3.4, 4.2, 3.4. **(b)** Singleton: 3.5, 4.2, 4.4, 3.4; Twin: 2.7, 2.6, 1.7. **(c)** In a bootstrap replicate, the data for each group are a sample of the data from that same group. Sampling is with replacement (some points might not occur and others might occur more than once). In a randomization replicate, the data for each group are sampled from all of the data without regard to original group. Sampling is without replacement (every data point occurs exactly once). **(d)** No.

2. **(a)** Yes—by chance, the randomization might reassign data to correct groups. **(b)** Yes. **(c)** No. Sample sizes of groups must be the same in each randomization as in the data. **(d)** No. Each data point can occur only once in a randomization. **(e)** No, 3.8 is not in the data. **(f)** No. Each data point can occur only once in a randomization.

3. **(a)** Yes—by chance, bootstrap samples might contain the same observations as the data. **(b)** No. The bootstrap sample for a group can contain only data from that group. **(c)** No. Sample sizes of groups must be the same in each randomization as in the data. **(d)** Yes. **(e)** No, 3.8 is not in the data. **(f)** Yes.

4. $10.20 < \mu < 10.40$.

5. Graph b has the real data. In a, there is little or no relationship between the two variables, whereas b shows a strong relationship. Since randomization tends to break up associations, it is more likely that b is the data and a is a randomization.

6. **(a)** Graph a comes from randomization and b comes from the bootstrap. **(b)** Approximately 10 units. About 2/3 of the bootstraps lie within $\pm$ 10 units of the mean.

7. **(a)** The null distribution for the ratio. **(b)** H_0: Mean ratio of range size equals that expected of a randomly broken stick. H_A: Mean ratio of range size differs from that expected of a randomly broken stick. P is approximately $2 \times (42/10{,}000) = 0.0084$. Since $P < 0.05$, reject H_0. Conclude that mean ratio in birds exceeds that expected from a randomly broken stick.

8. **(a)** 98. **(b)** 100. **(c)** "Giant pandas are most closely related to bears."

9. **(a)** For example:

Group A				Group B		
7.8	4.5	2.1		12.4	8.9	8.9
7.8	7.8	4.5		12.4	12.4	12.4
4.5	2.1	7.8		8.9	8.9	12.4
7.8	2.1	4.5		10.8	12.4	10.8
7.8	2.1	4.5		12.4	8.9	10.8
7.8	4.5	7.8		8.9	10.8	10.8
4.5	4.5	7.8		10.8	12.4	8.9
2.1	7.8	4.5		8.9	8.9	10.8
4.5	7.8	4.5		10.8	10.8	12.4
2.1	7.8	7.8		8.9	12.4	12.4

(b) Difference in median (B minus A) for these 10 bootstrap replicates:
3.0, 4.4, 4.4, 4.4, 4.6, 4.6, 6.3, 6.3, 6.3, 6.3.
$4.4 <$ difference between population medians < 6.3.

10. **(a)** For example:

Group A				Group B		
8.9	2.1	4.5		10.8	7.8	12.4
4.5	8.9	2.1		7.8	10.8	12.4
4.5	12.4	2.1		7.8	8.9	10.8
4.5	8.9	2.1		10.8	12.4	7.8
10.8	4.5	2.1		8.9	12.4	7.8
4.5	10.8	2.1		12.4	8.9	7.8
8.9	12.4	4.5		2.1	7.8	10.8
12.4	10.8	2.1		7.8	4.5	8.9
2.1	10.8	7.8		8.9	12.4	4.5
7.8	4.5	10.8		12.4	2.1	8.9

(b) H_0: The two groups have the same median. H_A: The two groups do not have the same median.
Test statistic: Observed difference between medians (B minus A) = 6.3.
Null distribution: $-3.0, -1.1, 1.1, 1.1, 4.4, 4.4, 4.4, 6.3, 6.3, 6.3$.
3 out of 10 outcomes is greater than or equal to the observed value, 6.3.
P is approximately $2 \times 3/10 = 0.6$. Since $P > 0.20$, do *not* reject H_0.
Conclude that the null hypothesis of no difference between medians is *not* rejected.
11. Yes.
12. Yes.
13. 10.4 seconds.
14. Welch's approximate t-test.
15. **(a)** A randomization test. The variable "shell volume" has been kept fixed and values of the second variable are all included but in random order. **(b)** A bootstrap estimate. Each row of the data set is a different individual sampled from the original sample of individuals. Sampling is with replacement because the same individual sometimes occurs more than once. **(c)** The linear correlation coefficient, r.

Chapter 20

1. **(a)** Binomial distribution with $n = 47$. **(b)** $L[p \mid 12 \text{ heterozygotes}] = \binom{47}{12} p^{12}(1 - p)^{35}$.
   It measures the probability of getting 12 heterozygotes (the data) given that the true value of the proportion of heterozygotes is p.
   (c) $\ln L[p] = \ln\left[\binom{47}{12}\right] + 12 \ln[p] + 35 \ln[1 - p]$.
   (d) -7.90.
2. **(a)** 0.26. **(b)** $0.14 < p < 0.40$.
3. **(a)** The log of the probability of the DNA data given different possible times since the split between Neanderthals and modern humans. **(b)** 710 thousand years. **(c)** $470 < $ time since split $ < 1000$ (in thousands of years). **(d)** This interval is like a 95% confidence interval. It describes the most plausible values of the time since the split between humans and Neanderthals.
4. **(a)** -4400. **(b)** -4396.5. **(c)** $G = 2(-4396.5 - (-4400)) = 7$. **(d)** χ^2 with $df = 1$. **(e)** $\chi^2_{1, 0.05} = 3.84$. Since G is greater than 3.84, $P < 0.05$ (exact $P = 0.008$). Reject H_0. Conclude that heterogeneity is not zero.

5. **(a)** $\Pr[\text{Yes}] = 1/2 + s/2 = (1 + s)/2.$

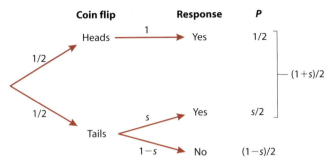

(b) $L[s \mid 113 \text{ yeses}] = \binom{185}{113}\left(\dfrac{1 + s}{2}\right)^{113}\left(\dfrac{1 - s}{2}\right)^{72}.$

(c) $\ln L[s \mid 113 \text{ yeses}] = \ln\left[\binom{185}{113}\right] + 113\ln[(1 + s)/2] + 72\ln[(1 - s)/2].$

(d) $-7.39.$

6. **(a)** The maximum likelihood estimate is $\hat{s} = 0.22$. **(b)** $0.07 < s < 0.36$. **(c)** H_0: The fraction of thieves s is zero. H_A: The fraction of thieves s is not zero.
$G = 2[-2.81 - (-7.39)] = 9.16$. $\chi^2_{1,0.05} = 3.84$. Since $G > 3.84$, $P < 0.05$ (exact $P = 0.0025$). Reject H_0. Conclude that there are thieves among us.

7. **(a)**

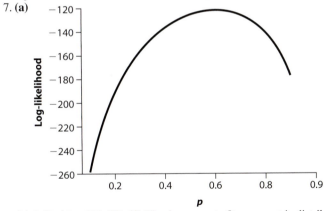

(b) 0.60. **(c)** -121.653. **(d)** The data seem to fit a geometric distribution very well.

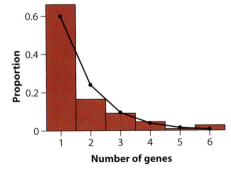

(e) χ^2 Goodness-of-fit test.

8. **(a)**

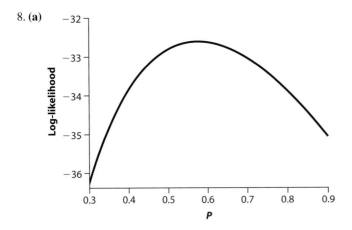

(b) 0.58. **(c)** $0.36 < \lambda < 0.86$.

Chapter 21

1. Results are seen as more interesting or exciting; results are more likely to be accepted for publication; results are more likely to be accepted in a better journal.
2. **(a)** Yes, average heritability declines with increasing sample size. **(b)** Small-sample studies yielding low heritability estimates are unlikely to be submitted for publication (researchers) or accepted for publication (editors).
3. The value of including this type of study is low because we would need to assume that success in tennis is a measure of aggression. Including it might be an act of desperation, resulting from a shortage of human studies that directly measure aggression.
4.

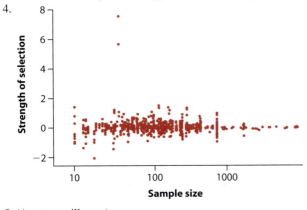

5. (Answers will vary.)
6. The smaller effect size with larger studies suggests that there is a publication bias affecting the estimates: small studies yielding low effect sizes are less likely to be appear in the published literature.
7. Even if H_0 is true, some studies might reject it by chance. If these are the only studies available for review (because of publication bias), then meta-analysis would conclude that H_0 is false.

Literature cited

Abd-El-Al, A. M., A. M. Bayoumy, and E. A. Abou Salem. 1997. A study on *Demodex folliculorum* in rosacea. *Journal of the Egyptian Society of Parasitology* 27: 183–195.

Adolph, S. C., and J. S. Hardin. 2007. Estimating phenotypic correlations: correcting for bias due to intraindividual variability. *Functional Ecology* 21: 178–184.

Agresti, A. 2002. Categorical Data Analysis. Wiley: Hoboken, NJ.

Agresti, A., and B. A. Coull. 1998. Approximate is better than "exact" for interval estimation of binomial proportions. *The American Statistician* 52: 119–126.

Altman, D. G., and J. M. Bland. 1998. Generalisation and extrapolation. *British Medical Journal* 317: 409–410.

Anderson, R. N. 2001. Deaths: Leading causes for 1999. National vital statistics reports, vol. 49, no. 11. Hyattsville, Maryland: National Center for Health Statistics.

Andrade, M. C. B. 1996. Sexual selection for male sacrifice in the Australian redback spider. *Science* 271: 70–72.

Antiplatelet Trialists' Collaboration. 1994. Collaborative overview of randomized trials of antiplatelet therapy–I: Prevention of death, myocardial infarction, and stroke by prolonged antiplatelet therapy in various categories of patients. *British Medical Journal* 308: 81–106.

Aparicio, S., et al. 2002. Whole-genome shotgun assembly and analysis of the genome of *Fugu rubripes*. *Science* 297: 1301–1310.

Apiron, D., and D. Zohary. 1961. Chlorophyll lethal in natural populations of the orchard grass (*Dactylis glomerata* L.). A case of balanced polymorphism in plants. *Genetics* 46: 393–399.

Arnqvist, G., M. Edvardsson, U. Friberg, and T. Nilsson. 2000. Sexual conflict promotes specia-tion in insects. *Proceedings of the National Academy of Sciences (USA)* 97: 10460–10464.

Baldwin, B. G., and M. J. Sanderson. 1998. Age and rate of diversification of the Hawaiian silversword alliance (Compositae). *Proceedings of the National Academy of Sciences (USA)* 95: 9402–9406.

Bandolier. 2002. That old placebo feeling. Bandolier 99-3: http://www.jr2.ox.ac.uk/bandolier/band99/b99-3.html.

Banks, T., and J. M. Dabbs, Jr. 1996. Salivary testosterone and cortisol in a delinquent and violent urban subculture. *Journal of Social Psychology* 136: 49–56.

Barber, V. A., G. P. Juday, and B. P. Finney. 2000. Reduced growth of Alaskan white spruce in the twentieth century from temperature-induced drought stress. *Nature* 405: 668–673.

Barker-Plotkin, A., D. Foster, A. Lezberg, and W. Lyford. 2006. Lyford mapped tree plot. http://harvardforest.fas.harvard.edu/data/p03/hf032/HF032-data.html. Accessed May 26, 2006.

Barss, P. 1984. Injuries due to falling coconuts. *Journal of Trauma* 24: 990–991.

Beall, C. M., et al. 2002. An Ethiopian pattern of human adaptation to high-altitude hypoxia. *Proceedings of the National Academy of Sciences (USA)* 99: 17215–17218.

Beaugrand, G., K. M. Brander, J. A. Lindley, S. Souissi, and P. C. Reid. 2003. Plankton effect on cod recruitment in the North Sea. *Nature* 426: 661–664.

Beaver, L. M., and J. M. Giebultowicz. 2004. Regulation of copulation duration by period and timeless in *Drosophila melanogaster*. *Current Biology* 14: 1492–1497.

Bebarta, V., D. Luyten, and K. Heard. 2003. Emergency medicine animal research: does use of randomization and blinding affect the results? *Academic Emergency Medicine* 10: 684–687.

Bekelman, J. E., Y. Li, and C. P. Gross. 2003. Scope and impact of financial conflicts of interest in biomedical research: a systematic review. *Journal of the American Medical Association* 289: 454–465.

Benjamini, Y., and Y. Hochberg. 1995. Controlling the false discovery rate: a practical and powerful approach to multiple testing. *Journal of the Royal Statistical Society, Series B* 57: 289–300.

Ben-Shahar, Y., A. Robichon, M. B. Sokolowski, and G. E. Robinson. 2002. Influence of gene action across different time scales on behavior. *Science* 296: 741–744.

Bereczkei, T., P. Gyuris, and G. E. Weisfeld. 2004. Sexual imprinting in human mate choice. *Proceedings of the Royal Society of London, Series B, Biological Sciences* 271: 1129–1134.

Berthold, P., and F. Pulido. 1994. Heritability of migratory activity in a natural bird population. *Proceedings of the Royal Society of London, Series B, Biological Sciences* 257: 311–315.

Bisazza, A., C. Cantalupo, A. Robins, L. J. Rogers, and G. Vallortigara. 1996. Right-pawedness in toads. *Nature* 379: 408.

Blackburn, C., et al. 2003. Effect of strategies to reduce exposure of infants to environmental tobacco smoke in the home: cross sectional survey. *British Medical Journal* 327: 257–260.

Blake, E. S., E. N. Rappaport, J. D. Jarrell, and C. W. Landsea. 2005. The deadliest, costliest, and most intense United States tropical cyclones from 1851 to 2004 (and other frequently requested hurricane facts). NOAA Technical Memorandum NWS TPC-4.

Bland, J. M., and D. G. Altman 1994. Matching. *British Medical Journal* 309: 1128.

Bland, M. 2000. An introduction to medical statistics, 3rd ed. Oxford University Press: Oxford, UK.

Blaustein, A. R., J. M. Kiesecker, D. P. Chivers, and R. G. Anthony. 1997. Ambient UV-B radiation causes deformities in amphibian embryos. *Proceedings of the National Academy of Sciences (USA)* 94: 13735–13737.

Blount, J. B., et al. 2003. Carotenoid modulation of immune function and sexual attractiveness in zebra finches. *Science* 300: 125–127.

Boag, P. T., and P. R. Grant. 1984. The classic case of character release: Darwin's finches (*Geozpiza*) on Isla Daphne Major, Galápagos. *Biological Journal of the Linnean Society* 22: 243–287.

Boles, L. C., and K. J. Lohmann. 2003. True navigation and magnetic maps in spiny lobsters. *Nature* 421: 60–63.

Book, A. S., K. B. Starzyk, and V. L. Quinsey. 2001. The relationship between testosterone and aggression: a meta-analysis. *Aggression and Violent Behavior* 6: 579–599.

Borenstein, M. 1997. Hypothesis testing and effect size estimation in clinical trials. *Annals of Allergy, Asthma and Immunology* 78: 5–16.

Bortkiewicz, L. 1898. Das Gesetz der Kleinen Zahlen (Teubner, Leipzig), as cited in Larsen, R. J., and M. L. Marx. 1981. An Introduction to Mathematical Statistics and its Applications. Prentice-Hall: Englewood Cliffs, NJ.

Bouyer, J., M. Pruvot, Z. Bengaly, P. M. Guerin, and R. Lancelot. 2007. Learning influences host choice in tsetse. *Biology Letters* 3: 113–116.

Box, G. E. P., and S. L. Andersen. 1955. Permutation theory in the derivation of robust criteria and the study of departures from assumption. *Journal of the Royal Statistical Society, Series B* 17: 1–34.

Bradley, J. V. 1980. Nonrobustness in one-sample Z and t tests: A large-scale sampling study. *Bulletin of the Psychonomic Society* 15: 29–32.

Brédart, S. and R. M. French. 1999. Do babies resemble their fathers more than their mothers? A failure to replicate Christenfeld and Hill (1995). *Evolution and Human Behavior* 20: 129–135.

Brem, R. B., J. D. Storey, J. Whittle, and L. Kruglyak. 2005. Genetic interactions between polymorphisms that affect gene expression in yeast. *Nature* 436: 701–703.

Brieger, D., et al. 2004. Acute coronary syndromes without chest pain, an underdiagnosed and undertreated high-risk group—Insights from the Global Registry of Acute Coronary Events. *Chest* 126: 461–469.

British Columbia Ministry of Education. 2004. Your surplus, your priorities: building BC together. Government of British Columbia: http://www.gov.bc.ca/bcgov/content/docs/@2NO32_0YQtuW/middle.pdf\par.

Broadbent, N. J., L. R. Squire, and R. E. Clark. 2004. Spatial memory, recognition memory, and the hippocampus. *Proceedings of the National Academy of Sciences (USA)* 101: 14515–14520.

Brooks, R. 2000. Negative genetic correlation between male sexual attractiveness and survival. *Nature* 406: 67–70.

Brown, S. J., et al. 2002. Sequence of the *Tribolium castaneum* homeotic complex: the region corresponding to the *Drosophila melanogaster* Antennapedia complex. *Genetics* 160: 1067–1074.

Brownlee, K. A. 1955. Statistics of the 1954 polio vaccine trials. *Journal of the American Statistical Association* 50: 1005–1013.

Brutsaert, T. D., et al. 2002. Effect of menstrual cycle phase on exercise performance of high-altitude native women at 3600 m. *The Journal of Experimental Biology* 205: 233–239.

Burch, C. L., and L. Chao. 1999. Evolution by small steps and rugged landscapes in the RNA virus ?6. *Genetics* 151: 921–927.

Buri, P. 1956. Gene frequency in small populations of mutant *Drosophila*. *Evolution* 10: 367–402.

Caley, M. J., and D. Schluter. 2003. Predators favour mimicry in a tropical reef fish. *Proceedings of the Royal Society of London, Series B, Biological Sciences* 270: 667–672.

Campbell, S. S., and P. J. Murphy. 1998. Extraocular circadian phototransduction in humans. *Science* 279: 396–399.

Cantalupo, C., and W. D. Hopkins. 2001. Asymmetric Broca's area in great apes. *Nature* 414: 505.

Cattin, M.-F., L.-F. Bersier, C. Banasek-Richter, R. Baltensperger, and J.-P. Gabriel. 2004. Phylogenetic constraints and adaptation explain food-web structure. *Nature* 427: 835–839.

CBC News. November 28, 2005. Politicians facing cynical electorate as campaign opens. CBC News Online (http://www.cbc.ca/news/background/election2005/poll.html).

Chadwick Johnson, J., T. M. Ivy, and S. K. Sakaluk. 1999. Female remating propensity contingent on sexual cannibalism in sagebrush crickets, *Cyphoderis strepitans*: a mechanism of cryptic female choice. *Behavioral Ecology* 10: 227–233.

Chase, J. M., and M. A. Leibold. 2002. Spatial scale dictates the productivity–biodiversity relationship. *Nature* 416: 427-430.

Christenfeld, N., and E. Hill. 1995. Whose baby are you? *Nature* 378: 669.

Christiansen, K., and E. Winkler. 1992. Hormonal, anthropometrical, and behavioral correlates of physical aggression in !Kung San men of Namibia. *Aggressive Behavior* 18: 271–280.

Clemons, T., and M. Pagano. 1999. Are babies normal? *The American Statistician* 53: 298–302.

Clubb, R., and G. Mason. 2003. Captivity effects on wide-ranging carnivores. *Nature* 425: 473–474.

Coello, C. A., et al. 2005. Adverse impact of surgical site infections in English hospitals. *Journal of Hospital Infection* 60: 93–103.

Collins S., and G. Bell. 2004. Phenotypic consequences of 1000 generations of selection at elevated CO_2 in a green alga. *Nature* 431: 566–569.

Colosimo, P. F., et al. 2004. The genetic architecture of parallel armor plate reduction in three-spine sticklebacks. *PLOS Biology* 2: 635–641.

Cook, N. R., et al. 2005. Low-dose aspirin in the primary prevention of cancer: the women's health study: a randomized controlled trial. *Journal of the American Medical Association* 294: 47–55.

Cooper, H. M., and L. V. Hedges, eds. 1993. Handbook of Research Synthesis. Sage Foundation Press: New York.

Cooper, H., and J. C. Valentine. 2001. Using research to answer practical questions about homework. *Educational Psychologist* 36: 143–153.

Craig, J. K., and C. Foote. 2001. Countergradient variation and secondary sexual color: Phenotypic convergence promotes genetic divergence in carotenoid use between sympatric anadromous and nonanadromous morphs of sockeye salmon (*Oncorhynchus nerka*). *Evolution* 55: 380–391.

Cratsley, C. K., and S. M. Lewis. 2003. Female preference for male courtship flashes in *Photinus ignitus* fireflies. *Behavioral Ecology* 14: 135–140.

Dabbs, Jr., J. M., M. F. Hargrove, and C. Huesel. 1996. Testosterone differences among college fraternities: well-behaved vs. rambunctious. *Personality and Individual Differences* 20: 157–161.

Dalal, S. R., E. B. Fowlkes, and B. Hoadley. 1989. Risk analysis of the space shuttle: pre-*Challenger* prediction of failure. *Journal of the American Statistical Association* 408: 945–957.

Dalterio, S., F. Badr, A. Bartke, and D. Mayfield. 1982. Cannabinoids in male mice: effects on fertility and spermatogenesis. *Science* 216: 315–316.

David, P., T. Bjorksten, K. Fowler, and A. Pomiankowski. 2000. Condition-dependent signaling of genetic variation in stalk-eyed flies. *Nature* 406: 186–188.

Dawson, R. J. MacG. 1995. The "unusual episode" data revisited. *Journal of Statistics Education* 3 (3).

Diamond, J. 1992. The Third Chimpanzee. HarperCollins: New York.

Diamond, J. M. 1988. Why cats have nine lives. *Nature* 332: 586-587.

Dickersin, K., Y. I Min, and C. L. Meinert. 1992. Factors influencing publication of research results. *Journal of the American Medical Association* 267: 374–378.

Doll, R., R. Peto, E. Hall, K. Wheatley, and R. Gray. 1994. Mortality in relation to consumption of alcohol: 13 years' observations on male British doctors. *British Medical Journal* 309: 911–918.

Duggan, M., and S. D. Leavitt. 2002. Winning isn't everything: corruption in sumo wrestling. *The American Economic Review* 92: 1594–1605.

Dunkin, R. C., W. A. McLellan, J. E. Blum, and D. A. Pabst. 2005. The ontogenetic changes in the thermal properties of blubber from Atlantic bottlenose dolphin *Tursiops truncates*. *The Journal of Experimental Biology* 208: 1469–1480.

East, E. M. 1916. Studies in size inheritance in *Nicotiana*. *Genetics* 1: 164–176.

Easterbrook, P. J., J. E. Berlin, R. Gopalan, and D. R. Matthews. 1991. Publication bias in clinical research. *The Lancet* 337: 867–872.

Economist Newspaper Limited. 2005. The World in 2006. The Economist Newspaper Limited: London.

Edelaar, P., and C. W. Benkman. 2006. Replicated population divergence caused by localized coevolution? A test of three hypotheses in the red crossbill–lodgepole pine system. *Journal of Evolutionary Biology* 19: 1651–1659.

Edwards, A. W. F. 1992. Likelihood. Johns Hopkins University Press: Baltimore, MD.

Efron, B., and R. J. Tibshirani. 1993. An introduction to the bootstrap. Chapman & Hall: New York.

Eggert, L. S., J. A . Eggert, and D. S. Woodruff. 2003. Estimating population sizes for elusive animals: the forest elephants of Kakum National Park, Ghana. *Molecular Ecology* 12: 1389–1402.

Ehrenberg, A. S. C. 1977. Rudiments of numeracy. *Journal of the Royal Statistical Society, A* 140: 277–297.

Ehrlich, P. R., and P. H. Raven. 1964. Butterflies and plants: a study in coevolution. *Evolution* 18: 586–608.

Elmore, J. G., K. Armstrong, C. D. Lehman, and S. W. Fletcher. 2005. Screening for breast cancer. *Journal of the American Medical Association* 293: 1245–1256.

Elstein, D. J., Ed. 1988. Professional Judgment. A Reader in Clinical Decision Making. Cambridge University Press: Cambridge, UK.

Elton, C., and M. Nicholson. 1942. The ten-year cycle in numbers of the lynx in Canada. *Journal of Animal Ecology* 11: 215–244.

Emlen, D. J. 2001. Costs and the diversification of exaggerated animal structures. *Science* 291: 1534–1536.

Epel, E. S., et al. 2004. Accelerated telomere shortening in response to life stress. *Proceedings of the National Academy of Sciences (USA)* 101: 17312–17315.

Ernst, E., and A. R. White. 1998. Acupuncture for back pain: a meta-analysis of randomized controlled trials. *Archives of Internal Medicine* 158: 2235–2241.

Evason, K., C. Huang, I. Yamben, D. F. Covey, and K. Kornfeld. 2005. Anticonvulsant medications extend worm life-span. *Science* 307: 258–262.

Ewald, P. W. 1993. The evolution of virulence. *Scientific American* 268: 86–93.

Falconer, D. S., and T. F. C. MacKay. 1996. Introduction to Quantitative Genetics. Longmans Green, Harlow: Essex, UK.

Fan, Z., S. R. Shifley, M. A. Spetich, F. R. Thompson III, and D. R. Larsen. 2005. Abundance and size distribution of cavity trees in second-growth and old-growth central hardwood forests. *Northern Journal of Applied Forestry* 22: 162–169.

Farrell, B. D. 1998. "Inordinate fondness" explained: why are there so many beetles? *Science* 281: 555–559.

Farrell, B. D., D. E. Dussord, and C. Mitter. 1991. Escalation of plant defense: do latex and resin canals spur plant diversification? *The American Naturalist* 138: 881–900.

Farren, L., S. Shayler, and A. R. Ennos. 2004. The fracture properties and mechanical design of human fingernails. *The Journal of Experimental Biology* 207: 735–741.

Farrington, D. P. 1994. Cambridge Study in Delinquent Development [Great Britain], 1961–1981. 2nd ICPSR ed. Inter-university Consortium for Political and Social Research: Ann Arbor, MI. Data distributed at http://webapp.icpsr.umich.edu/cocoon/NACJD-STUDY/08488.xml.

Farroni, T., G. Csibra, F. Simion, and M. H. Johnson. 2002. Eye contact detection in humans from birth. *Proceedings of the National Academy of Sciences (USA)* 99: 9602–9605.

Fatouros, N. E., M. E. Huigens, J. J. A. van Loon, M. Dicke, and M. Hilker. 2005. Butterfly anti-aphrodisiac lures parasitic wasps. *Nature* 433: 704.

Faurie, C., and M. Raymond. 2005. Handedness, homicide and negative frequency-dependent selection. *Proceedings of the Royal Society of London, Series B, Biological Sciences* 272: 25–28.

Faurie, C., D. Pontierb, and M. Raymond. 2004. Student athletes claim to have more sexual partners than other students. *Evolution and Human Behavior* 25: 1–8.

Felsenstein, J. 1985. Phylogenies and the comparative method. *The American Naturalist* 125: 1–15.

Fernandes, C. C., J. Podos, and J. G. Lundberg. 2004. Amazonian ecology: tributaries enhance the diversity of electric fishes. *Science* 305: 1960–1962.

Flynn, J. J., M. A. Nedbal, J. W. Dragoo, and R. L. Honeycutt. 2000. Whence the red panda? *Molecular Phylogenetics and Evolution* 17: 190–199.

Focht, D. R., C. Spicer, and M. P. Fairchok. 2002. The efficacy of duct tape vs. cryotherapy in the treatment of verruca vulgaris (the common wart). *Archives of Pediatrics & Adolescent Medicine* 156: 971–974.

Fortin, N. J., K. L. Agster, and H. B. Eichenbaum. 2004. Critical role of the hippocampus in memory for sequences of events. *Nature Neuroscience* 5: 458–462.

Freedman, D., R. Pisani, and R. Purves. 1997. Statistics, 3rd ed. Norton: New York.

Frick, R. W. 1996. The appropriate use of null hypothesis testing. *Psychological Methods* 1: 379–390.

Fritsches, K. A., R. W. Brill, and E. J. Warrant. 2005. Warm eyes provide superior vision in swordfishes. *Current Biology* 15: 55–58.

Fuller, A., P. R. Kamerman, S. K. Maloney, G. Mitchell, and D. Mitchell. 2003. Variability in brain and arterial blood temperatures in free-ranging ostriches in their natural habitat. *The Journal of Experimental Biology* 206: 1171–1181.

Fuller, R. A., K. N. Irvine, P. Devine-Wright, P. H. Warren, and K. J. Gaston. 2007. Psychological benefits of greenspace increase with biodiversity. *Biology Letters* 3: 390–394.

Galton, F. 1894. Note on fitting normal curves to distribution of speeds of old homing pigeons. *Homing News and Pigeon Fanciers' Journal* (April 6): 159–160. http://galton.org.

Galton. F. 1886. Regression towards mediocrity in hereditary stature. *Journal of the Anthropological Institute* 15: 246–263.

García-Planells, J., et al. 2005. Ancient origin of the CAG expansion causing Huntington disease in a Spanish population. *Human Mutation* 25: 453-459.

Gazey, W. J., and M. J. Staley. 1986. Population estimation from mark-recapture experiments using a sequential Bayes algorithm. *Ecology* 67: 941–951.

Geffeney, S., E. D. Brodie, Jr., P. C. Ruben, and E. D. Brodie III. 2002. Mechanisms of adaptation in a predator-prey arms race: TTX-resistant sodium channels. *Science* 297: 1336–1339.

Gettelfinger, B., and E. L. Cussler. 2004. Will humans swim faster or slower in syrup? *AIChE Journal* 50: 2646–2647.

Gianoli, E. 2004. Evolution of a climbing habit promotes diversification in flowering plants. *Proceedings of the Royal Society of London, Series B, Biological Sciences* 271: 2011–2015.

Gilham, C., et al. 2005. Day care in infancy and risk of childhood acute lymphoblastic leukaemia: findings from UK case-control study. *British Medical Journal*, doi:10.1136/bmj.38428.521042.8F.

Glare, P., et al. 2003. A systematic review of physicians' survival predictions in terminally ill cancer patients. *British Medical Journal* 327: 1–6.

Golenda, C. F., V. B. Solberg, R. Burge, J. M. Gambel, and R. A. Wirtz. 1999. Gender-related efficacy difference to an extended duration formulation of topical *N,N*-diethyl-*m*-toluamide (DEET). *American Journal of Tropical Medicine and Hygiene* 60: 654–657.

Gore, S. M., I. G. Jones, and E. C. Rytter. 1977. Misuse of statistical methods: critical assessment of articles in BMJ from January to March 1976. *British Medical Journal* 1: 85–87.

Gøtzsche, P. C. 1987. Reference bias in reports of drug trials. *British Medical Journal* 295: 654–656.

Grafe, T. U., S. Döbler, and K. E. Linsenmair. 2002. Frogs flee from the sound of fire. *Proceedings of the Royal Society of London, Series B, Biological Sciences* 269: 999–1003.

Grafen, A., and R. Hails. 2002. Modern Statistics for the Life Sciences. Oxford University Press: Oxford, UK.

Gray, A., D. N. Jackson, and J. B. McKinlay. 1991. The relation between dominance, anger, and hormones in normally aging men: results from the Massachusetts Male Aging Study. *Psychosomatic Medicine* 53: 375–385.

Gray, E. M. 1997. Female red-winged blackbirds accrue material benefits from copulating with extra-pair males. *Animal Behaviour* 53: 625–639.

Gray, G. M., et al. 2004. Weight of the evidence evaluation of low-dose reproductive and developmental effects of bisphenol A. *Human and Ecological Risk Assessment* 10: 875–921.

Gray, S. M., L. M. Dill, and J. S. McKinnon. 2007. Cuckoldry incites cannibalism: male fish turn to cannibalism when perceived certainty of paternity decreases. *The American Naturalist* 169: 258–263.

Green, J. A., P. J. Butler, A. J. Woakes, I. L. Boyd, and R. L. Holder. 2001. Heart rate and rate of oxygen consumption of exercising macaroni penguins. *The Journal of Experimental Biology* 204: 673–684.

Green, S., and J. Higgins, Eds. 2006. Cochrane Handbook for Systematic Reviews of Interventions 4.2.6.: http://www.cochrane.dk/cochrane/handbook/handbook.htm.

Griffith, S. C., and B. C. Sheldon. 2001. Phenotypic plasticity in the expression of sexually selected traits: neglected components of variation. *Animal Behaviour* 61: 987–993.

Guelzim, N., S. Bottani, P. Bourgine, and F. Képès. 2002. Topological and causal structure of the yeast transcriptional regulatory network. *Nature Genetics* 31: 60–63.

Gundale, M. J., W. M. Jolly, and T. H. Deluca. 2005. Susceptibility of a northern hardwood forest to exotic earthworm invasion. *Conservation Biology* 19: 1075–1083.

Gunnarsson, T. G., J. A. Gill, T. Sigurbjörnsson, and W. J. Sutherland. 2004. Arrival synchrony in migratory birds. *Nature* 431: 646.

Gurevitch, J., and L. V. Hedges. 1999. Statistical issues in ecological meta-analyses. *Ecology* 80: 1142–1149.

Hadfield, J. D., et al. 2006. Direct versus indirect sexual selection: genetic basis of colour, size and recruitment in a wild bird. *Proceedings of the Royal Society of London, Series B, Biological Sciences* 273: 1347–1353.

Haeslery, M. P., and O. Seehausen. 2005. Inheritance of female mating preference in a sympatric sibling species pair of Lake Victoria cichlids: implications for speciation. *Proceedings of the Royal Society of London, Series B, Biological Sciences* 272: 237–245.

Hairston N. G., Jr., et al. 1999. Rapid evolution revealed by dormant eggs. *Nature* 401: 446.

Halpern, B. S. 2003. The impact of marine reserves: do reserves work and does reserve size matter? *Ecological Applications* 13: S117–S137.

Hama, Y., et al. 2001. Sex ratio in the offspring of male radiologists. *Academic Radiology* 8: 421–424.

Hamshere, M. L., et al. 2005. Genomewide linkage scan in schizoaffective disorder: significant evidence for linkage at 1q42 close to disc1, and suggestive evidence at 22q11 and 19p13. *Archives in General Psychiatry* 62: 1081–1088.

Hans, C. N. 2002. On the risk of mortality to primates exposed to anthrax spores. *Risk Analysis* 22: 189–193.

Harley, C. D. G. 2003. Abiotic stress and herbivory interact to set range limits across a two-dimensional stress gradient. *Ecology* 84: 1477–1488.

Harley, M., M. A. Mohammed, S. Hussain, J. Yates, and A. Almasri. 2005. Was Rodney Ledward a statistical outlier? Retrospective analysis using routine hospital data to identify gynaecologists' performance. *British Medical Journal* 330: 929; published online 15 April 2005; doi:10.1136/bmj.38377.675440.8F.

Harpole, W. S., and D. Tilman. 2007. Grassland species loss resulting from reduced niche dimension. *Nature* 446: 791–793.

Harris, J., P. Rushton, E. Hampton, and D. Jackson. 1996. Salivary testosterone and self report aggressive and pro-social personality characteristics in men and women. *Aggressive Behavior* 22: 321–331.

Hasselquist, D., J. A. Marsh, P. W. Sherman, and J. C. Wingfield. 1999. Is avian humoral immunocompetence suppressed by testosterone? *Behavioral Ecology and Sociobiology* 45: 167–175.

Heathcote, J. A. 1995. Why do old men have big ears? *British Medical Journal* 311: 1668.

Hedges, L. V., and I. Olkin. 1985. Statistical Methods for Meta-analysis. Academic Press: Orlando, FL.

Heffner, R. A., M. J. Butler IV, and C. K. Reilly. 1996. Pseudoreplication revisited. *Ecology* 77: 2558–2562.

Heilbuth, J. C. 2000. Lower species richness in dioecious clades. *The American Naturalist* 156: 221–241.

Heiling, A. M., M. E. Herberstein, and L. Chittka. 2003. Crab spiders manipulate flower signals. *Nature* 421: 334.

Hemmingsson, T., and D. Kriebel. 2003. Smoking at age 18–20 and suicide during 26 years of follow-up—how can the association be explained? *International Journal of Epidemiology* 32: 1000–1004.

Hendricks, J. C., et al. 2001. A non-circadian role for cAMP signaling and CREB activity in *Drosophila* rest homeostasis. *Nature Neuroscience* 4: 1108–1115.

Hendry, A. P., O. K. Berg, and T. P. Quinn. 1999. Condition dependence and adaptation-by-time: breeding date, life history, and energy allocation in a population of salmon. *Oikos* 85: 499–514.

Heusner, A. A. 1991. Size and power in mammals. *The Journal of Experimental Biology* 160: 25–54.

Hey, J. 1992. Using phylogenetic trees to study speciation and extinction. *Evolution* 46: 627–640.

Hirano, S. S., E. V. Nordheim, D. C. Arny, and C. D. Upper. 1982. Lognormal distribution of epiphytic bacterial populations on leaf surfaces. *Applied and Environmental Microbiology* 44: 695–700.

Hobbs, J.-P. A., P. L. Munday, and G. P. Jones. 2004. Social induction of maturation and sex determination in a coral reef fish. *Proceedings of the Royal Society of London, Series B, Biological Sciences* 271: 2109–2114.

Hoogland, J. L. 1998. Why do female Gunnison's prairie dogs copulate with more than one male? *Animal Behaviour* 55: 351–359.

Hosken, D. J., and P. I. Ward. 2001. Experimental evidence for testis size evolution via sperm competition. *Ecology Letters* 4: 10–13.

Hosken, D. J., W. U. Blanckenhorn, and T. W. J. Garner. 2002. Heteropopulation males have a fertilization advantage during sperm competition in the yellow dung fly (*Scathophaga stercoraria*). *Proceedings of the Royal Society of London, Series B, Biological Sciences* 269: 1701–1707.

Houser, B. B. 1979. An investigation of the correlation between hormonal levels in males and mood, behavior, and physical discomfort. *Hormones and Behavior* 12: 185–197.

Hovatta, I., et al. 2005. Glyoxalase 1 and glutathione reductase 1 regulate anxiety in mice. *Nature* 438: 662–666.

Hróbjartsson, A., and P. C. Gøtzsche. 2001. Is the placebo powerless?: An analysis of clinical trials comparing placebo with no treatment. *Journal of the American Medical Association* 344: 1594–1602.

Hsieh, C.-h., et al. 2006. Fishing elevates variability in the abundance of exploited species. *Nature* 443: 859–862.

Hubbard, T., et al. 2005. Ensembl 2005. *Nucleic Acids Research* 33: D447–D453.

Huber, R., M. F. Ghilardi, M. Massimini, and G. Tononi. 2004. Local sleep and learning. *Nature* 430: 78–81.

Huey, R. B., and A. E. Dunham. 1987. Repeatability of locomotor performance in natural populations of the lizard *Sceloporus merriami*. *Evolution* 42: 1116–1120.

Huey, R. B., and X. Eguskitza. 2000. Supplemental oxygen and mountaineer death rates on Everest and K2. *Journal of the American Medical Association* 284: 181.

Hunt, M. 1997. How Science Takes Stock. Sage Foundation Press: New York.

Hurlbert, S. H.1984. Pseudoreplication and the design of ecological field experiments. *Ecological Monographs* 54: 187–211.

Hurlbert, S. H., and M. D. White. 1993. Experiments with freshwater invertebrate zooplancti-vores—quality of statistical analyses. *Bulletin of Marine Science* 53: 128–153.

Hurlburt, G. 1996. Relative brain size in recent and fossil amniotes: determination and interpretation. Ph.D. Thesis, Zoology (Vertebrate Paleontology), University of Toronto.

Hyndman, R. J., and Fan, Y. 1996. Sample quantiles in statistical packages. *American Statistician* 50: 361–365.

Inaudi, D., et al. 1995. Chaos: evidence for the butterfly effect. *Annals of Improbable Research* Vol. 1, No. 6; reprinted in M. Abrahams, Ed. 2000. The Best of Annals of Improbable Research. W. H. Freeman: New York.

Ioannidis, J. P. A., J. C. Cappelleri, H. S. Sacks, and J. Lau. 1997. The relationship between study design, results, and reporting of randomized trials of HIV infection. *Controlled Clinical Trials* 18: 431–444.

Iverson, J. M., and S. Goldin-Meadow. 1998. Why people gesture when they speak. *Nature* 396: 228.

James, R. A., P. A. Hoadley, and B. G. Sampson. 1997. Determination of postmortem interval by sampling vitreous humour. *American Journal of Forensic Medicine and Pathology* 18: 158–162.

Jenkins, A. J. 2001. Drug contamination of US paper currency. *Forensic Science International* 121:189–193.

Jensen, H., et al. 2004. Lifetime reproductive success in relation to morphology in the house sparrow *Passer domesticus*. *Journal of Animal Ecology* 73: 599–611.

Jensen, K. H., T. Little, A. Skorping, and D. Ebert. 2006. Empirical support for optimal virulence in a castrating parasite. *PLoS Biology* 4: 1265–1269.

Jerison, H. J. 2006. http://brainmuseum.org/Evolution/paleo/.

Jesson, L. K., and S. C. H. Barrett. 2002. The genetics of mirror-image flowers. *Proceedings of the Royal Society of London, Series B, Biological Sciences* 269: 1835–1839.

Johnson, A. M., et al. 2001. Sexual behaviour in Britain: partnerships, practices, and HIV risk behaviours. *The Lancet* 358: 1835–1842.

Johnson, J. G., et al. 2002. Television viewing and aggressive behavior during adolescence and adulthood. *Science* 295: 2468–2471.

Jules, E. S., and B. J. Rathcke. 1999. Mechanisms of reduced trillium recruitment along edges of old-growth forest fragments. *Conservation Biology* 13: 784–793.

Jüni, P., D. G. Altman, and M. Egger. 2001. Assessing the quality of controlled clinical trials. *British Medical Journal* 323: 42–46.

Kanter, M. H., and J. R. Taylor. 1994. Accuracy of statistical methods in *Transfusion*—a review of articles from July/August 1992 through June 1993. *Transfusion* 34: 697–701.

Kanwisher, J., G. Gabrielsen, and N. Kanwisher. 1981. Free and forced diving in birds. *Science* 211: 717–719.

Keenan, J. P., et al. 2001. Self-recognition and the right hemisphere. *Nature* 409: 305.

Kelly, C., and T. D. Price. 2005. Correcting for regression to the mean in behavior and ecology. *The American Naturalist* 166: 700–707.

Kendall, M., and B. B. Smith. 1939. The problem of *m* rankings. *Annals of Mathematical Statistics* 10: 275–287.

Kim, U., E. Jorgenson, H. Coon, M. Leppert, N. Risch, and D. Drayna. 2003. Positional cloning of the human quantitative trait locus underlying taste sensitivity to phenylthiocarbamide. *Science* 299: 1221–1225.

Kingsolver, J. G., et al. 2001. The strength of phenotypic selection in natural populations. *The American Naturalist* 157: 245–261.

Klem, D., et al. 2004. Effects of window angling, feeder placement, and scavengers on avian mortality at plate glass. *Wilson Bulletin* 116: 69–73.

Knipschild, P. 1994. Systematic reviews: some examples. *British Medical Journal* 309: 719–721.

Kodric-Brown, A., and J. H. Brown. 1993. Highly structured fish communities in Australian desert springs. *Ecology* 74: 1847–1855.

Koella, J.C., F. L. Sørensen, and R. A. Anderson. 1998. The malaria parasite, *Plasmodium falciparum*, increases the frequency of multiple feeding of its mosquito vector, *Anopheles gambiae*. *Proceedings of the Royal Society of London, Series B, Biological Sciences* 265: 763–768.

Kotiaho, J. S., L. W. Simmons, and J. L. Tomkins. 2001. Towards a resolution of the lek paradox. *Nature* 410: 684–686.

Kramer, M. S., and R. Kakuma. 2002. Optimal duration of exclusive breastfeeding. *Cochrane Database of Systematic Reviews*, Issue 1. Art. No.: CD003517. DOI: 10.1002/14651858.CD003517.

Kramer, M. S., et al. 2002. Breastfeeding and infant growth: biology or bias? *Pediatrics* 110: 343–347.

Krebs, C. J. 1999. Ecological Methodology, 2nd ed. Benjamin Cummings: Menlo Park, CA.

Krochmal, A. R., G. S. Bakken, and T. J. LaDuc. 2004. Heat in evolution's kitchen: evolutionary perspectives on the functions and origin of the facial pit of pit vipers (Viperidae: Crotalinae). *The Journal of Experimental Biology* 207, 4231–4238.

LaCroix, A. Z., S. G. Leveille, J. A. Hecht, L. C. Grothaus, and E. H. Wagner. 1996. Does walking decrease the risk of cardiovascular disease hospitalizations and death in older adults? *Journal of the American Geriatric Society* 44: 113–120.

Lafferty, K. D., and A. K. Morris. 1996. Altered behavior of parasitized killifish increases susceptibility to predation by bird final hosts. *Ecology* 77: 1390–1397.

LaMunyon, C. W., and S. Ward. 1998. Larger sperm out compete smaller sperm in the nematode *Caenorhabditis elegans*. *Proceedings of the Royal Society of London, Series B, Biological Sciences* 265: 1997–2002.

Langford, D. J., et al. 2006. Social modulation of pain as evidence for empathy in mice. *Science* 312: 1967–1970.

Lanza, F., et al. 1994. Double-blind comparison of lansoprazole, ranitidine, and placebo in the treatment of acute duodenal ulcer. Lansoprazole study group. *American Journal of Gastroenterology* 89: 1191–1200.

Lappin, A. K., and J. F. Husak. 2005. Weapon performance, not size, determines mating success and potential reproductive output in the collared lizard (*Crotaphytus collaris*). *The American Naturalist* 166: 426–436.

Laurance, W. F., et al. 1997. Biomass collapse in Amazonian forest fragments. *Science* 278: 1117–1118.

Leakey, A. D. B., J. D. Scholes, and M. C. Press. 2005. Physiological and ecological significance of sunflecks for dipterocarp seedlings. *Journal of Experimental Botany* 56: 469–482.

Lehr, R. 1992. Sixteen *s* squared over *d* squared: a relation for crude sample size estimates. *Statistics in Medicine* 11: 1099–1102.

Leopold, S. S., W. J. Warme, E. F. Braunlich, and S. Shott. 2003. Association between funding source and study outcome in orthopedic research. *Clinical Orthopaedics and Related Research* 415: 293–301.

Levey, D. J., R. S. Duncan., and C. F. Levins. 2004. Use of dung as a tool by burrowing owls. *Nature* 431:39.

Levin, P. S., S. Achord, B. E. Fiest, and R. W. Zabel. 2002. Non-indigenous brook trout and the demise of Pacific salmon: a forgotten threat? *Proceedings of the Royal Society of London, Series B, Biological Sciences* 269: 1663–1670.

Lewis, E. B., B. D. Pfeiffer, D. R. Mathog, and S. E. Celniker. 2003. Evolution of the homeobox complex in the Diptera. *Current Biology* 13: R587–R588.

Liang, Y.-L., et al. 2001. Effects of blood pressure, smoking, and their interaction on carotid artery structure and function. *Hypertension* 37: 6–11.

Liberg, O. H., et al. 2005. Severe inbreeding depression in a wild wolf (*Canis lupus*) population. *Biology Letters* 1: 17–20.

Lim, M. M., et al. 2004. Enhanced partner preference in a promiscuous species by manipulating the expression of a single gene. *Nature* 429: 754–757.

Lindman, R., B. von der Pahlen, B. Öst, and P. Eriksson. 1992. Serum testosterone, cortisol, glucose, and ethanol in males arrested for spouse abuse. *Aggressive Behavior* 18: 393–400.

Lounibos, L. P., N. Nishimura, J. Conn, and R. Lourenco-de-Oliveira. 1995. Life history correlates of adult size in the malaria vector *Anopheles darlingi*. *Memórias do Instituto Oswaldo Cruz* 90: 769–774.

Maddison, W. P., and D. R. Maddison. 2007. Mesquite: a modular system for evolutionary analysis. Version 2.0: http://mesquiteproject.org.

Manning, J. T., P. E. Bundred, D. J. Newton, and B. F. Flanagan. 2003. The second to fourth digit ratio and variation in the androgen receptor gene. *Evolution and Human Behavior* 24: 399–405.

Marks, D., 2000. The Psychology of the Psychic. Prometheus Books: Amherst, NY.

Martins, E. P. 2004. COMPARE, version 4.6b. Computer programs for the statistical analysis of comparative data. Distributed by the author at http://compare.bio.indiana.edu/. Department of Biology, Indiana University, Bloomington IN.

Maxwell, E. A. 1976. Analysis of contingency tables and further reasons for not using Yates' correction in 2 × 2 tables. *Canadian Journal of Statistics* 4: 277–290.

McCarthy, C. R. 1994. Historical background of clinical trials involving women and minorities. *Academic Medicine* 69: 695–698.

McGaha, T. L., B. Sorrentino, and J. V. Ravetch. 2005. Restoration of tolerance in lupus by targeted inhibitory receptor expression. *Science* 307: 590–593.

McGraw, K. J., and D. R. Ardia. 2003. Carotenoids, immunocompetence, and the information content of sexual colors: an experimental test. *The American Naturalist* 162: 704–712.

McGuigan, S. M. 1995. The use of statistics in the *British Journal of Psychiatry*. *British Journal of Psychiatry* 167: 683–688.

McKinnon, J. S., et al. 2004. Evidence for ecology's role in speciation. *Nature* 429: 294–298.

Mechelli, A., et al. 2004. Structural plasticity in the bilingual brain. *Nature* 431: 757.

Meeker, W. Q., and W. A. Escobar. 1995. Teaching about confidence regions based on maximum likelihood estimation. *American Statistician* 49: 48–53.

Mehl, M. R., S. Vazire, N. Ramírez-Esparza, R. B. Slatcher, and J. W. Pennebaker. 2007. Are women really more talkative than men? *Science* 317: 82.

Melander, H., J. Ahlqvist-Rastad, G. Meijer, and B. Beermann. 2003. Evidence b(i)ased medicine—elective reporting from studies sponsored by the pharmaceutical industry: review of studies in new drug applications. *British Medical Journal* 326: 1171–1175.

Michalsen, A., et al. 2003. Effectiveness of leech therapy in osteoarthritis of the knee: a randomized, controlled trial. *Annals of Internal Medicine* 139: 724–730.

Miller, L. M., T. Close, and A. R. Kapuscinski. 2004. Lower fitness of hatchery and hybrid rainbow trout compared to naturalized populations in Lake Superior tributaries. *Molecular Ecology* 13: 3379–3388.

Milner-Gulland, E. J., et al. 2003. Reproductive collapse in saiga antelope harems. *Nature* 422: 135.

Miranda, A., O. G. Almeida1, P. C. Hubbard, E. N. Barata and A. V. M. Canário. 2005. Olfactory discrimination of female reproductive status by male tilapia (*Oreochromis mossambicus*). *The Journal of Experimental Biology* 208: 2037–2043.

Modig, A. O. 1996. Effects of body size and harem size on male reproductive behaviour in the southern elephant seal. *Animal Behaviour* 51: 1295–1306.

Moir, W. H., and E. P. Bachelard. 1969. Distribution of fine roots in three *Pinus radiata* plantations near Canberra, Australia. *Ecology* 50: 658–662.

Molofsky, J., and J.-B. Ferdy. 2005. Extinction dynamics in experimental metapopulations. *Proceedings of the National Academy of Sciences (USA)* 102: 3726–3731.

Mood, A. M., 1954. On the asymptotic efficiency of certain nonparametric two-sample tests. *Annals of Mathematical Statistics* 25: 514–522.

Moore, J. L., et al. 2002. The distribution of cultural and biological diversity in Africa. *Proceedings of the Royal Society of London, Series B, Biological Sciences* 269: 1645–1653.

Motulsky, H. J. 1999. Curvefit.com, the complete guide to nonlinear regression. Published by GraphPad Software Inc., San Diego, CA. [Online]. Available from http://www.curvefit.com.

Moyle, L. C., M. S. Olson, and P. Tiffin. 2004. Patterns of reproductive isolation in three angiosperm genera. *Evolution* 58: 1195–1208.

Murphy, B. F., Jr., and J. E. Heath. 1983. Temperature sensitivity in the prothoracic ganglion of the cockroach, *Periplaneta americana*, and its relationship to thermoregulation. *Journal of Experimental Biology* 105: 305–315.

National Aeronautics and Space Administration. 2007. Total Ozone Mapping Spectrometer (TOMS). British Atmospheric Data Centre, 2006–2007. Available from http://badc.nerc.ac.uk/data/toms_cds/.

Negre, B. S., et al. 2005. Conservation of regulatory sequences and gene expression patterns in the disintegrating *Drosophila Hox* gene complex. *Genome Research* 15: 692–700.

Negro, J. J., et al. 2002. An unusual source of essential carotenoids. *Nature* 416: 807–808.

Newberger, D. S. 2000. Down syndrome: prenatal risk assessment and diagnosis. *American Family Physician* 62: 825–832.

Newcomer, S. D., J. A. Zeh, and D. W. Zeh. 1999. Genetic benefits enhance the reproductive success of polyandrous females. *Proceedings of the National Academy of Sciences (USA)* 96: 10236–10241.

Nightingale, F. 1858. Notes on Matters Affecting the Health, Efficiency and Hospital Administration of the British Army. Harrison and Sons: London.

Noonan, J. P., et al. 2006. Sequencing and analysis of Neanderthal genomic DNA. *Science* 314: 1113–1118.

Nosil, P., and B. J. Crespi. 2006. Experimental evidence that predation promotes divergence in adaptive radiation. *Proceedings of the National Academy of Sciences (USA)* 103: 9090–9095.

Nunn, C. L., J. L. Gittleman, and J. Antonovics. 2000. Promiscuity and the primate immune system. *Science* 290: 1168–1170.

Olweus, D., A. Mattsson, D. Schalling, and H. Low. 1980. Testosterone, aggression, physical and personality dimensions in normal adolescent males. *Psychosomatic Medicine* 50: 261–272.

Oppliger, A., P. Christe, and H. Richner. 1996. Clutch size and malaria resistance. *Nature* 381: 565.

Orringer, J. S., et al. 2004. Treatment of acne vulgaris with a pulsed dye laser: a randomized controlled trial. *Journal of the American Medical Association* 291: 2834–2839.

Ozaki, M. A., et al. 2005. Ant nestmate and nonnestmate discrimination by a chemosensory sensillum. *Science* 309: 311–314.

Packer, C., and A. E. Pusey. 1983. Adaptations of female lions to infanticide by incoming males. *The American Naturalist* 121: 716–728.

Pagel, M. 1994. Detecting correlated evolution on phylogenies: a general method for the comparative analysis of discrete characters. *Proceedings of the Royal Society of London, Series B, Biological Sciences* 225: 37–45.

Pagel, M., and A. Meade. 2007. BayesTraits. Version 1.0. Distributed by the authors at http://www.evolution.rdg.ac.uk. School of Biological Sciences, University of Reading, Reading, UK.

Palleroni, A., C. T. Miller, M. Hauser, and P. Marler. 2005. Prey plumage adaptation against falcon attack. *Nature* 434: 973–974.

Palmer, A. R. 1999. Detecting publication bias in meta-analysis: a case study of fluctuating asymmetry and sexual selection. *The American Naturalist* 154: 220–233.

Panel on Scientific Responsibility and the Conduct of Research. 1992. Responsible Science: Ensuring the Integrity of the Research Process, Vol. I. National Academy Press: Washington, DC.

Parmesan, C., et al. 1999. Poleward shifts in geographical ranges of butterfly species associated with regional warming. *Nature* 399: 579–583.

Patel, R. R., D. J. Murphy, and T. J. Peters. 2005. Operative delivery and postnatal depression: a cohort study. *British Medical Journal* 330: 879.

Patterson, T. B., and T. J. Givnish. 2002. Phylogeny, concerted convergence, and phylogenetic niche conservatism in the core Liliales: insights from *rbcL* and *ndhF* sequence data. *Evolution* 56: 233–252.

Pauling, L. 1986. How to Live Longer and Feel Better. W. H. Freeman: New York.

Phillips, B. L., G. P. Brown, J. K. Webb, and R. Shine. 2006. Invasion and the evolution of speed in toads. *Nature* 439: 803.

Pitnick, S., K. E. Jones, and G. S. Wilkinson. 2005. Mating system and brain size in bats. *Proceedings of the Royal Society of London, Series B, Biological Sciences* 273: 719–724.

Pounder, D. J. 1995. Postmortem changes and time of death. University of Dundee. http://www.dundee.ac.uk/forensicmedicine/llb/timedeath.htm.

Provine, R. R. 1989. Faces as releasers of contagious yawning: an approach to face detection using normal human subjects. *Bulletin of the Psychonomic Society* 27: 211–214.

Prugnolle, F., A. Manica, and F. Balloux. 2005. Geography predicts neutral genetic diversity of human populations. *Current Biology* 15: R159–R160.

Ramadan, A. A., A. El-Keblawy, K. H. Shaltout, and J. Lovett-Doust. 1994. Sexual polymorphism, growth, and reproductive effort in Egyptian *Thymelaea hirsuta* (Thymelaeaceae). *American Journal of Botany* 81: 847–857.

Ramos, M., D. J. Irschick, and T. E. Christenson. 2004. Overcoming an evolutionary conflict: removal of a reproductive organ greatly increases locomotor performance. *Proceedings of the National Academy of Sciences (USA)* 101: 4883–4887.

Ramsey, P. H. 1980. Exact Type I error rates for robustness of Student's *t* test with unequal variances. *Journal of Educational Statistics* 5: 337–349.

Rannala, B., and Z. Yang. 2006. Probability distribution of molecular evolutionary trees: a new method of phylogenetic inference. *Journal of Molecular Evolution* 43: 304–311.

Raup, D. M., and J. J. Sepkoski, Jr. 1982. Mass extinctions in the marine fossil record. *Science* 215: 1501–1503.

Rausher, M. D. 1984. Tradeoffs in performance in different hosts: evidence from within- and between-site variation in the beetle *Deloyala guttata. Evolution* 38: 582–595.

Relyea, R. A. 2003. Predator cues and pesticides: a double dose of danger for amphibians. *Ecological Applications* 13: 1515–1521.

Reusch, T. B. H., A. Ehlers, A. Hämmerli, and B. Worm. 2005. Ecosystem recovery after climatic extremes enhanced by genotypic diversity. *Proceedings of the National Academy of Sciences (USA)* 102: 2826–2831.

Ricaurte, G. A., et al. 2002. Severe dopaminergic neurotoxicity in primates after a common recreational dose regimen of MDMA ("ecstasy"). *Science* 297: 2260.

Ricaurte, G. A., et al. 2003. Retraction. *Science* 301: 1479.

Richardson, D. S., J. Komdeur, and T. Burke. 2003. Altruism and infidelity among warblers. *Nature* 422: 580–581.

Richardson, T. E., and A. G. Stephenson. 1991. Effects of parentage, prior fruit set and pollen load on fruit and seed production in *Campanula americana* L. *Oecologia* 87: 80–85.

Ricklefs, R. E., and E. Bermingham. 2001. Non-equilibrium diversity dynamics of the Lesser Antillean avifauna. *Science* 294: 1522–1524.

Riskin, D. K., J. E. A. Bertram, and J,. W. Hermanson. 2005. Testing the hindlimb-strength hypothesis: non-aerial locomotion by Chiroptera is not constrained by the dimensions of the femur or tibia. *The Journal of Experimental Biology* 208: 1309–1319.

Rodgers, J. L., and D. Doughty. 2001. Does having boys or girls run in the family? *Chance Magazine* Fall: 8–13.

Rogers, D. W., and R. Chase. 2001. Dart receipt promotes sperm storage in the garden snail *Helix aspersa. Behavioral Ecology and Sociobiology* 50: 122–127.

Roscoe, J. T., and J. A. Byars. 1971. Sample size restraints commonly imposed on the use of the chi-square statistic. *Journal of the American Statistical Association* 66: 755–759.

Rosenblatt, K. A., J. R. Daling, C. Chen, K. J. Sherman, and S. M. Schwartz. 2004. Marijuana use and risk of oral squamous cell carcinoma. *Cancer Research* 64: 4049–4054.

Rosenthal, R. 1991. Meta-analytic Procedures for Social Research. Sage Foundation Press: New York.

Rossman, A. J. 1994. Televisions, physicians, and life expectancy. *Journal of Statistics Education* 2 (2).

Rotton, J., and L. W. Kelly. 1985. Much ado about the full moon: a meta-analysis of lunar-lunacy research. *Psychological Bulletin* 97: 286–306.

Roubik, D. W. 1978. Competitive interactions between neotropical pollinators and Africanized honey bees. *Science* 201: 1030–1032.

Ruben, D. 2006. Highrise syndrome in cats: is your apartment fall-safe? Accessed 7 January 2006: http://www.petplace.com/article.aspx?id=2570.

Ruff, C. B., E. Trinkaus, and T. W. Holliday. 1997. Body mass and encephalization in Pleistocene *Homo. Nature* 387: 173–176.

Rypstra, A. L. 1979. Foraging flocks of spiders: a study of aggregate behavior in *Cyrtophora citricola* Forskål (Araneae; Araneidae) in West Africa. *Behavioral Ecology and Sociobiology* 5: 291–300.

Sacktor, N. C., et al. 2000. Improvement in HIV-associated motor slowing after antiretroviral therapy including protease inhibitors. *Journal of NeuroVirology* 6: 84–88.

Sakaue, M., et al. 2001. Bisphenol-A affects spermatogenesis in the adult rat even at a low dose. *Journal of Occupational Health* 43: 185–190.

Salo, P., E. Korpimäki, P. B. Banks, M. Nordström, and C. R. Dickman. 2007. Alien predators are more dangerous than native predators to prey populations. *Proceedings of the Royal Society of London, Series B, Biological Sciences* 274: 1237–1243.

Sandell, M., J. Agrell, S. Erlinge, and J. Nelson. 1991. Adult philopatry and dispersal in the field vole *Microtus agrestis. Oecologia* 86: 153–158.

Sandidge, J. S. 2003. Scavenging by brown recluse spiders. *Nature* 426: 30.

Sauer, J. R., J. E. Hines, and J. Fallon. 2003. The North American breeding bird survey, results and analysis 1966–2002. Version 2003.1, USGS Patuxent Wildlife Research Center: Laurel, MD.

Savage, V. M., et al. 2004. The predominance of quarter power scaling in biology. *Functional Ecology* 18: 257–282.

Scantlebury, M., J. R. Speakman, M. K. Oosthuizen, T. J. Roper, and N. C. Bennett. 2006. Energetics reveals physiologically distinct castes in a eusocial mammal. *Nature* 440: 795–797.

Schaal, T., and G. Smith. 2000. Do baseball players regress toward the mean? *American Statistician* 54: 231–235.

Schluter, D. 1988. The evolution of finch communities on islands and continents: Kenya vs. Galápagos. *Ecological Monographs* 58: 229–249.

Schmid, P. E., M. Tokeshi, and J. M. Schmid-Araya. 2000. Relation between population density and body size in stream communities. *Science* 289: 1557–1560.

Schwerzmann, M., et al. 2005. Prevalence and size of directly detected patent foramen ovale in migraine with aura. *Neurology* 65: 1415–1418.

Seaton, E. D., et al. 2003. Pulsed-dye laser treatment for inflammatory acne vulgaris: randomised controlled trial. *The Lancet* 362: 1347–1352.

Shaw, D. J., B. T. Grenfell, and A. P. Dobson. 1998. Patterns of macroparasite aggregation in wildlife host populations. *Parasitology* 117: 597–610.

Shaw, M. R., et al. 2002. Grassland responses to global environmental changes suppressed by elevated CO_2. *Science* 298: 1987–1990.

Sherwood, C. C., et al. 2006. Evolution of increased glia–neuron ratios in the human frontal cortex. *Proceedings of the National Academy of Sciences (USA)* 103: 13606–13611.

Shine, R., B. Phillips, H. Waye, M. LeMaster, and R. T. Mason. 2001. Benefits of female mimicry in snakes. *Nature* 414: 267.

Shoemaker, A. L. 1996. What's normal?—Temperature, gender, and heart rate. *Journal of Statistics Education* 4 (2).

Simard, S. W., D. A. Perry, M. D. Jones, D. D. Myrold, D. M. Durall, and R. Molinak. 1997. Net transfer of carbon between ectomycorrhizal tree species in the field. *Nature* 388: 579–582.

Simmons, L. W., and B. Roberts. 2005. Bacterial immunity traded for sperm viability in male crickets. *Science* 309: 2031.

Simons, D. J., and C. F. Chabris. 1999. Gorillas in our midst: sustained inattentional blindness for dynamic events. *Perception* 28: 1059–1074.

Simpson, S. J., E. Despland, B. F. Hägele, and T. Dodgson. 2001. Gregarious behavior in desert locusts is evoked by touching their back legs. *Proceedings of the National Academy of Sciences (USA)* 98: 3895–3897.

Sloan, R. E. G., and W. R. Keatinge. 1973. Cooling rates of young people swimming in cold water. *Journal of Applied Physiology* 35: 371–375

Smallwood, R. H., D. J. Morgan, G. W. Mihaly, and R. A. Smallwood. 1998. Lack of linear correlation between hepatic ligand uptake rate and unbound ligand concentration does not necessarily imply receptor-mediated uptake. *Journal of Pharmacokinetics and Pharmacodynamics* 16: 397–411.

Smith, D. L., B. Lucey, L. A. Waller, J. E. Childs, and L. A. Real. 2002. Predicting the spatial dynamics of rabies epidemics on heterogeneous landscapes. *Proceedings of the National Academy of Sciences (USA)* 99: 3668–3672.

Smith, F. A., et al. 2003. Body mass of late Quaternary mammals. *Ecology* 84: 3403–3403. Data are from http://www.esapubs.org/archive/ecol/E084/094/metadata.htm.

Smith, T. B. 1993. Disruptive selection and the genetic basis of bill size polymorphism in the African finch, *Pyrenestes*. *Nature* 363: 618–620.

Socha, J. J. 2002. Gliding flight in the paradise tree snake. *Nature* 418: 603–604.

Sokal, R. R., and F. J. Rohlf. 2008. Biometry. W. H. Freeman: New York.

Sokal, R. R., and F. J. Rohlf. 1995. Biometry: The Principles and Practice of Statistics in Biological Research, 3rd ed. W. H. Freeman: New York.

Spalding, K. L., B. A. Buchholz, L.-E. Bergman, H. Druid, and J. Frisén. 2005. Forensics: age written in teeth by nuclear tests. *Nature* 437: 333–334.

Spence, D. S., and E. A. Thompson. 2005. Homeopathic treatment for chronic disease: a 6-year, university-hospital outpatient observational study. *The Journal of Alternative and Complementary Medicine* 11: 793–798.

Srivastava, D. S., and J. H. Lawton. 1998. Why more productive sites have more species: an experimental test of theory using tree-hole communities. *The American Naturalist* 152: 510–529.

St. Clair, D., et al. 2005. Rates of adult schizophrenia following prenatal exposure to the Chinese famine of 1959–1961. *Journal of the American Medical Association* 294: 557–562.

Stafne, G. M., and P. R. Manger. 2004. Predominance of clockwise swimming during rest in

Southern Hemisphere dolphins. *Physiology & Behavior* 82: 919–926.

Stern, J. M., and R. John Simes. 1997. Publication bias: evidence of delayed publication in a cohort study of clinical research projects. *British Medical Journal* 315: 640–645.

Stigler, S. M. 1977. Do robust estimators work with real data? *Annals of Statistics* 5: 1055–1078.

Storey, J. D., and R. Tibshirani. 2003. Statistical significance for genomewide studies. *Proceedings of the National Academy of Sciences (USA)* 100: 9440–9445.

Student. 1931. The Lanarkshire milk experiment. *Biometrika* 23(No. 3/4): 398–406.

Styf, J. R., et al. 1997. Height increase, neuromuscular function, and back pain during 6° head-down tilt with traction. *Aviation, Space, and Environmental Medicine* 68: 24–29.

Sunda, W. G., and Huntsman, S. A. 1997. Interrelated influence of iron, light and cell size on marine phytoplankton growth. *Nature* 390: 389–392.

Svanbäck, R., and D. I. Bolnick. 2007. Intraspecific competition drives increased resource use diversity within a natural population. *Proceedings of the Royal Society of London, Series B, Biological Sciences* 274: 839–844.

Svenson, N. 2006. Blink-free photos, guaranteed. *Velocity*, June 2006, http://velocity.ansto.gov.au/velocity/ans0011/article_06.asp

Tatem, A. J., C. A. Guerra, P. M. Atkinson, and S. I. Hay. 2004. Athletics: momentous sprint at the 2156 Olympics? *Nature* 431: 525.

Tattersall, G. J., W. K. Milsom, A. S. Abe, S. P. Brito, and D. V. Andrade. 2004. The thermogenesis of digestion in rattlesnakes. *The Journal of Experimental Biology* 207: 579–585.

Templer, D. E., D. M. Veleber, and R. K. Brooner. 1982. Geophysical variables and behavior VI. Lunar phase and accident injuries: a difference between day and night. *Perceptual and Motor Skill* 55: 280–282.

Templeton, C. N., E. Greene, and K. Davis. 2005. Allometry of alarm calls: black-capped chickadees encode information about predator size. *Science* 308: 1934–1937.

Thomas, J. A., et al. 2004. Comparative losses of British butterflies, birds, and plants and the global extinction crisis. *Science* 303: 1879–1881.

Tkachev, D., et al. 2003. Oligodendrocyte dysfunction in schizophrenia and bipolar disorder. *The Lancet* 362: 798–805.

Tomkins, J. L., and G. S. Brown. 2004. Population density drives the local evolution of a threshold dimorphism. *Nature* 431: 1099–1103.

Trichopoulou, A., et al. 2005. Modified Mediterranean diet and survival: EPIC-elderly prospective cohort study. *British Medical Journal* 330: 991–997.

Trites, A. W. 1996. Physical growth of northern fur seals (*Callorhinus ursinus*): seasonal fluctuations and migratory influences. *Journal of Zoology, London* 238: 459–482.

Tufte, E. R. 1983. The Visual Display of Quantitative Information. Graphics Press: Cheshire, CT.

Tufte, E. R. 1997. Visual Explanations: Images and Quantities, Evidence and Narrative. Graphics Press: Cheshire, CT.

Turner, D. C. 1975. The vampire bat: a field study in behavior and ecology. Johns Hopkins Press: Baltimore, MD.

United Nations Statistics Division. 2004. Demographic and social statistics. http://unstats.un.org/unsd/demographic/products/socind/population.htm.

US DOT Traffic Safety Facts. 1999. National Center for Statistics and Analysis.

U.S. Fish and Wildlife Service. 2001. Number of U.S. listed species per calendar year. Distributed at http://endangered.fws.gov/wildlife.html.

Van Damme, L., et al. 2002. Effectiveness of COL-1492, a nonoxynol-9 vaginal gel, on HIV-1 transmission in female sex workers: a randomised controlled trial. *The Lancet* 360: 971–977.

Veen, T., et al. 2001. Hybridization and adaptive mate choice in flycatchers. *Nature* 411: 45–50.

Venables, W. N., and B. D. Ripley. 2002. Modern Applied Statistics with S-PLUS. Springer Publishing: New York.

Ventura, S. J., J. A. Martin, S. C. Curtin, F. Menacker, and B. E. Hamilton. 2001. Births: final data for 1999. National Vital Statistics Reports Vol. 49, No. 1.

Visscher, P. K., and R. Dukas. 1997. Survivorship of foraging honey bees. *Insectes Sociaux* 44: 1–5.

Visscher, P. K., R. S. Vetter, and S. Carmazine. 1996. Removing bee stings. *The Lancet* 348: 301–302.

Volkow, N. D., et al. 1997. Relationship between subjective effects of cocaine and dopamine transporter occupancy. *Nature* 386: 827–830.

vom Saal, F. S., and F. H. Bronson. 1980. Sexual characteristics of adult female mice are correlated with their blood testosterone levels during prenatal development. *Science* 208: 597–599.

Wager, T. D., et al. 2004. Placebo-induced changes in fMRI in the anticipation and experience of pain. *Science* 303: 1162–1167.

Waide, R. B., and W. B. Reagan, Eds. 1996. The Food Web of a Tropical Rainforest. University of Chicago Press: Chicago.

Waldron, A. 2007. Geographic range size evolution. *The American Naturalist* 170: 221–231.

Walker, B. G., P. D. Boersma, and J. C. Wingfield. 2005. Physiological and behavioral differences in Magellanic penguin chicks in undisturbed and tourist-visited locations of a colony. *Conservation Biology* 19: 1571–1577.

Walters, C. and J.-J. Maguire. 1996. Lessons for stock assessment from the northern cod collapse. *Reviews in Fish Biology and Fisheries* 6: 125–137.

Wang, P. J., J. R. McCarrey, F. Yang, and D. C. Page. 2001. An abundance of X-linked genes expressed in spermatogonia. *Nature Genetics* 27: 422–426.

Warner, S. L. 1965. Randomized response: a survey technique for eliminating evasive answer bias. *Journal of the American Statistical Association* 60: 63–69.

Waters, A. J., M. J. Jarvis, and S. R. Sutton. 1998. Nicotine withdrawal and accident rates. *Nature* 394: 137.

Waters, J. R., K. S. McKelvey, D. L. Luoam, and C. J. Zabel. 1997. Truffle production in old-growth and mature fir stands in northeastern California. *Forest Ecology and Management* 96: 155–166.

Weisstein, E. W. 2007. Hypothesis testing. From MathWorld—A Wolfram Web Resource: http://mathworld.wolfram.com/HypothesisTesting.html. Accessed February 4, 2007.

Werren, J. H. 1980. Sex ratio adaptations to local mate competition in a parasitic wasp. *Science* 208: 1157–1159.

West, S. A., and B. C. Sheldon. 2002. Constraints in the evolution of sex ratio adjustment. *Science* 295: 1695–1688.

Wheeler, B. 2005. The SuppDists package, version 1.0-13, April 7, 2005. Gnu Public License version 2. http://cran.r-project.org/src/contrib/Descriptions/SuppDists.html.

Wheelwright, N. T., and B. A. Logan. 2004. Previous-year reproduction reduces photosynthetic capacity and slows lifetime growth in females of a neotropical tree. *Proceedings of the National Academy of Sciences (USA)* 101: 8051–8055.

White, A. R., H. Rampes, and J. L. Campbell. 2006. Acupuncture and related interventions for smoking cessation. *Cochrane Database of Systematic Reviews* Issue 1, Art. No.: CD000009. DOI: 10.1002/14651858.CD000009.pub2.

White, S. A., T. Nguyen, and R. D. Fernald. 2002. Social regulation of gonadotropin-releasing hormone. *The Journal of Experimental Biology* 205: 2567–2581.

Whitman, K., A. M. Starfield, H. S. Quadling, and C. Packer. 2004. Sustainable trophy hunting of African lions. *Nature* 428: 175–178.

Whitney, W. O., and C. J. Mehlhaff. 1987. High-rise syndrome in cats. *Journal of the American Veterinary Medical Association* 191: 1399–1403.

Wikelski, M., and C. Thom. 2000. Marine iguanas shrink to survive El Niño. *Nature* 403: 37–38.

Wilkinson, G. S. 1984. Reciprocal food sharing in the vampire bat. *Nature* 308: 181–184.

Willerman, L., R. Schultz, J. N. Rutledge, and E. Bigler. 1991. In vivo brain size and intelligence. *Intelligence* 15:223–228.

Williams, T. M., L. A. Fuiman, M. Horning, and R. W. Davis. 2004. The cost of foraging by a marine predator, the Weddell seal *Leptonychotes weddellii*: pricing by the stroke. *The Journal of Experimental Biology* 207: 973–982.

Winder, B., et al. 2002. Food and drink packaging: who is complaining and who should be complaining. *Applied Ergonomics* 33: 433–438

Wiseman, R., and P. Lamont. 1996. Unravelling the Indian rope-trick. *Nature* 383: 212–213.

Wood, E., et al. 2001. Unsafe injection practices in a cohort of injection drug users in Vancouver: could safer injecting rooms help? *Canadian Medical Association Journal* 165: 405–410.

Wright, Jr., K. P., and C. A. Czeisler 2002. Absence of circadian phase resetting in response to bright light behind the knees. *Science* 297: 571.

Yamamoto, Y., and W. R. Jeffery. 2000. Central role for the lens in cavefish eye degeneration. *Science* 289: 631–633.

Yang, L. H. 2004. Periodical cicadas as resource pulses in North American forests. *Science* 306: 1565–1567.

Yashin, A. I., et al. 2000. Genes and longevity: lessons from studies of centenarians. *Journal of Gerontology* 55A: B319–B328.

Yeh, P. J. 2004. Rapid evolution of a sexually selected trait following population establishment in a novel habitat. *Evolution* 58: 166–174.

Young, A. J., et al. 2006. Stress and the suppression of subordinate reproduction in cooperatively breeding meerkats. *Proceedings of the National Academy of Sciences (USA)* 103: 12005–12010.

Young, D. C., and E. M. Hade. 2004. Holidays, birthdays, and postponement of cancer death. *Journal of the American Medical Association* 292: 3012–3016.

Young, K. V., E. D. Brodie, Jr., and E. D. Brodie III. 2004. How the horned lizard got its horns. *Science* 304: 65.

Zimmerman, D. W. 2003. A warning about the large-sample Wilcoxon-Mann-Whitney test. *Understanding Statistics* 2: 267–280.

Statistical Tables

This appendix gives probabilities and critical values for a few of the most commonly used probability distributions. More can be found in such references as Rohlf and Sokal (1994), *Statistical Tables*.

Using statistical tables

Statistical tables are used to calculate probabilities and to put bounds on the P-value for a test statistic. They are also used to find the critical values for calculating confidence intervals. With easy access to computers becoming more common, statistical tables are becoming less necessary. One advantage of using computer programs is that they provide precise P-values, whereas statistical tables often only bracket a P-value (such as "$0.025 < P < 0.05$"). Moreover, the tables in this book only give critical values corresponding to the most commonly used significance levels, such as 0.05 and 0.01. Statistical tables are useful when computers are unavailable, or when putting bounds on P is satisfactory.

Some probability distributions come in families, with each member identified by its degrees of freedom. Examples include the χ^2 distribution, the Student's t-distribution, and the F-distribution. In such cases, the statistical tables in this book provide quantities for many but not all values of the degrees of freedom. What should you do when the probability distribution corresponding to a specific number of degrees of freedom is missing from the tables? For example, you might be looking for critical values of the t-distribution having 149 degrees of freedom, but Statistical Table C provides such quantities only for t-distributions having 140 and 160 degrees of freedom. In this case, there are two approaches to obtaining the missing critical values, as follows:

1. *The conservative approach*: Find the two critical values in the table corresponding to distributions having degrees of freedom just above and just below the desired degrees of freedom. Use the critical value that makes your confidence interval widest, or that makes it most difficult to reject the null hypothesis of your test.
2. *Interpolation*: Find the two critical values in the table that correspond to the degrees of freedom just above and just below the desired number of degrees of freedom. Use linear interpolation to estimate the critical value for your desired degrees of freedom.

For example, let's say that you want to find the critical value for the F-distribution having five degrees of freedom in the numerator and 132 degrees of freedom in the denominator, $F_{0.05(1),5,132}$. Statistical Table D only gives $F_{0.05(1),5,100} = 2.31$ and $F_{0.05(1),5,200} = 2.26$. Because 132 lies between 100 and 200, the critical value must lie between 2.31 and 2.26. The following figure shows how to find the critical value by linear interpolation.

The desired critical value (C) is obtained as

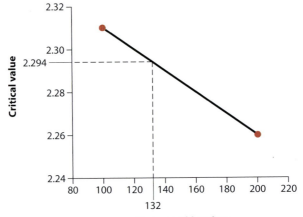

$$C = 2.31 + (2.26 - 2.31)\left(\frac{132 - 100}{200 - 100}\right)$$

$$= 2.294.$$

The general formula for calculating a critical value from interpolation is

$$C = C_{low} + (C_{high} - C_{low})\left(\frac{df - df_{low}}{df_{high} - df_{low}}\right),$$

where df_{low} and df_{high} are the degrees of freedom just below and just above df, the desired number of degrees of freedom, and C_{low} and C_{high} are their corresponding critical values.

Statistical Table A: The χ^2 distribution

This table gives critical values of the χ^2 distribution. The critical value $\chi^2_{df,\alpha}$ defines the area α in the right tail of the χ^2 distribution with df degrees of freedom. To find a critical value in the table, select the desired α along the top row and the number of degrees of freedom df in the far left column. For example, if $df = 5$ and $\alpha = 0.05$, then the critical value is 11.07; that is, the probability of a value greater than or equal to 11.07 is 0.05.

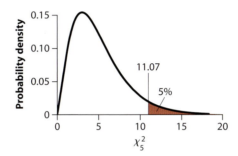

The critical value for a χ^2 distribution having $df = 5$ when $\alpha = 0.05$. The area in red indicates the tail probability corresponding to 0.05, or 5%. The boundary of the red section is $\chi^2 = 11.07$. The probability of a value greater than or equal to 11.07 is 0.05.

df	0.999	0.995	0.99	0.975	0.95	α 0.05	0.025	0.01	0.005	0.001
1	0.0000016	0.000039	0.00016	0.00098	0.00393	3.84	5.02	6.63	7.88	10.83
2	0.002	0.01	0.02	0.05	0.10	5.99	7.38	9.21	10.60	13.82
3	0.02	0.07	0.11	0.22	0.35	7.81	9.35	11.34	12.84	16.27
4	0.09	0.21	0.30	0.48	0.71	9.49	11.14	13.28	14.86	18.47
5	0.21	0.41	0.55	0.83	1.15	11.07	12.83	15.09	16.75	20.52
6	0.38	0.68	0.87	1.24	1.64	12.59	14.45	16.81	18.55	22.46
7	0.60	0.99	1.24	1.69	2.17	14.07	16.01	18.48	20.28	24.32
8	0.86	1.34	1.65	2.18	2.73	15.51	17.53	20.09	21.95	26.12
9	1.15	1.73	2.09	2.70	3.33	16.92	19.02	21.67	23.59	27.88
10	1.48	2.16	2.56	3.25	3.94	18.31	20.48	23.21	25.19	29.59
11	1.83	2.60	3.05	3.82	4.57	19.68	21.92	24.72	26.76	31.26
12	2.21	3.07	3.57	4.40	5.23	21.03	23.34	26.22	28.30	32.91
13	2.62	3.57	4.11	5.01	5.89	22.36	24.74	27.69	29.82	34.53
14	3.04	4.07	4.66	5.63	6.57	23.68	26.12	29.14	31.32	36.12
15	3.48	4.60	5.23	6.26	7.26	25.00	27.49	30.58	32.80	37.70
16	3.94	5.14	5.81	6.91	7.96	26.30	28.85	32.00	34.27	39.25
17	4.42	5.70	6.41	7.56	8.67	27.59	30.19	33.41	35.72	40.79
18	4.90	6.26	7.01	8.23	9.39	28.87	31.53	34.81	37.16	42.31
19	5.41	6.84	7.63	8.91	10.12	30.14	32.85	36.19	38.58	43.82
20	5.92	7.43	8.26	9.59	10.85	31.41	34.17	37.57	40.00	45.31
21	6.45	8.03	8.90	10.28	11.59	32.67	35.48	38.93	41.40	46.80
22	6.98	8.64	9.54	10.98	12.34	33.92	36.78	40.29	42.80	48.27
23	7.53	9.26	10.20	11.69	13.09	35.17	38.08	41.64	44.18	49.73
24	8.08	9.89	10.86	12.40	13.85	36.42	39.36	42.98	45.56	51.18
25	8.65	10.52	11.52	13.12	14.61	37.65	40.65	44.31	46.93	52.62
26	9.22	11.16	12.20	13.84	15.38	38.89	41.92	45.64	48.29	54.05
27	9.80	11.81	12.88	14.57	16.15	40.11	43.19	46.96	49.64	55.48
28	10.39	12.46	13.56	15.31	16.93	41.34	44.46	48.28	50.99	56.89
29	10.99	13.12	14.26	16.05	17.71	42.56	45.72	49.59	52.34	58.30
30	11.59	13.79	14.95	16.79	18.49	43.77	46.98	50.89	53.67	59.70
31	12.20	14.46	15.66	17.54	19.28	44.99	48.23	52.19	55.00	61.10
32	12.81	15.13	16.36	18.29	20.07	46.19	49.48	53.49	56.33	62.49
33	13.43	15.82	17.07	19.05	20.87	47.40	50.73	54.78	57.65	63.87
34	14.06	16.50	17.79	19.81	21.66	48.60	51.97	56.06	58.96	65.25
35	14.69	17.19	18.51	20.57	22.47	49.80	53.20	57.34	60.27	66.62
36	15.32	17.89	19.23	21.34	23.27	51.00	54.44	58.62	61.58	67.99
37	15.97	18.59	19.96	22.11	24.07	52.19	55.67	59.89	62.88	69.35
38	16.61	19.29	20.69	22.88	24.88	53.38	56.90	61.16	64.18	70.70
39	17.26	20.00	21.43	23.65	25.70	54.57	58.12	62.43	65.48	72.05
40	17.92	20.71	22.16	24.43	26.51	55.76	59.34	63.69	66.77	73.40
41	18.58	21.42	22.91	25.21	27.33	56.94	60.56	64.95	68.05	74.74
42	19.24	22.14	23.65	26.00	28.14	58.12	61.78	66.21	69.34	76.08
43	19.91	22.86	24.40	26.79	28.96	59.30	62.99	67.46	70.62	77.42
44	20.58	23.58	25.15	27.57	29.79	60.48	64.20	68.71	71.89	78.75
45	21.25	24.31	25.90	28.37	30.61	61.66	65.41	69.96	73.17	80.08
46	21.93	25.04	26.66	29.16	31.44	62.83	66.62	71.20	74.44	81.40
47	22.61	25.77	27.42	29.96	32.27	64.00	67.82	72.44	75.70	82.72
48	23.29	26.51	28.18	30.75	33.10	65.17	69.02	73.68	76.97	84.04
49	23.98	27.25	28.94	31.55	33.93	66.34	70.22	74.92	78.23	85.35
50	24.67	27.99	29.71	32.36	34.76	67.50	71.42	76.15	79.49	86.66

	α									
df	**0.999**	**0.995**	**0.99**	**0.975**	**0.95**	**0.05**	**0.025**	**0.01**	**0.005**	**0.001**
51	25.37	28.73	30.48	33.16	35.60	68.67	72.62	77.39	80.75	87.97
52	26.07	29.48	31.25	33.97	36.44	69.83	73.81	78.62	82.00	89.27
53	26.76	30.23	32.02	34.78	37.28	70.99	75.00	79.84	83.25	90.57
54	27.47	30.98	32.79	35.59	38.12	72.15	76.19	81.07	84.50	91.87
55	28.17	31.73	33.57	36.40	38.96	73.31	77.38	82.29	85.75	93.17
56	28.88	32.49	34.35	37.21	39.80	74.47	78.57	83.51	86.99	94.46
57	29.59	33.25	35.13	38.03	40.65	75.62	79.75	84.73	88.24	95.75
58	30.30	34.01	35.91	38.84	41.49	76.78	80.94	85.95	89.48	97.04
59	31.02	34.77	36.70	39.66	42.34	77.93	82.12	87.17	90.72	98.32
60	31.74	35.53	37.48	40.48	43.19	79.08	83.30	88.38	91.95	99.61
61	32.46	36.30	38.27	41.30	44.04	80.23	84.48	89.59	93.19	100.89
62	33.18	37.07	39.06	42.13	44.89	81.38	85.65	90.80	94.42	102.17
63	33.91	37.84	39.86	42.95	45.74	82.53	86.83	92.01	95.65	103.44
64	34.63	38.61	40.65	43.78	46.59	83.68	88.00	93.22	96.88	104.72
65	35.36	39.38	41.44	44.60	47.45	84.82	89.18	94.42	98.11	105.99
66	36.09	40.16	42.24	45.43	48.31	85.96	90.35	95.63	99.33	107.26
67	36.83	40.94	43.04	46.26	49.16	87.11	91.52	96.83	100.55	108.53
68	37.56	41.71	43.84	47.09	50.02	88.25	92.69	98.03	101.78	109.79
69	38.30	42.49	44.64	47.92	50.88	89.39	93.86	99.23	103.00	111.06
70	39.04	43.28	45.44	48.76	51.74	90.53	95.02	100.43	104.21	112.32
71	39.78	44.06	46.25	49.59	52.60	91.67	96.19	101.62	105.43	113.58
72	40.52	44.84	47.05	50.43	53.46	92.81	97.35	102.82	106.65	114.84
73	41.26	45.63	47.86	51.26	54.33	93.95	98.52	104.01	107.86	116.09
74	42.01	46.42	48.67	52.10	55.19	95.08	99.68	105.20	109.07	117.35
75	42.76	47.21	49.48	52.94	56.05	96.22	100.84	106.39	110.29	118.60
76	43.51	48.00	50.29	53.78	56.92	97.35	102.00	107.58	111.50	119.85
77	44.26	48.79	51.10	54.62	57.79	98.48	103.16	108.77	112.70	121.10
78	45.01	49.58	51.91	55.47	58.65	99.62	104.32	109.96	113.91	122.35
79	45.76	50.38	52.72	56.31	59.52	100.75	105.47	111.14	115.12	123.59
80	46.52	51.17	53.54	57.15	60.39	101.88	106.63	112.33	116.32	124.84
81	47.28	51.97	54.36	58.00	61.26	103.01	107.78	113.51	117.52	126.08
82	48.04	52.77	55.17	58.84	62.13	104.14	108.94	114.69	118.73	127.32
83	48.80	53.57	55.99	59.69	63.00	105.27	110.09	115.88	119.93	128.56
84	49.56	54.37	56.81	60.54	63.88	106.39	111.24	117.06	121.13	129.80
85	50.32	55.17	57.63	61.39	64.75	107.52	112.39	118.24	122.32	131.04
86	51.08	55.97	58.46	62.24	65.62	108.65	113.54	119.41	123.52	132.28
87	51.85	56.78	59.28	63.09	66.50	109.77	114.69	120.59	124.72	133.51
88	52.62	57.58	60.10	63.94	67.37	110.90	115.84	121.77	125.91	134.75
89	53.39	58.39	60.93	64.79	68.25	112.02	116.99	122.94	127.11	135.98
90	54.16	59.20	61.75	65.65	69.13	113.15	118.14	124.12	128.30	137.21
91	54.93	60.00	62.58	66.50	70.00	114.27	119.28	125.29	129.49	138.44
92	55.70	60.81	63.41	67.36	70.88	115.39	120.43	126.46	130.68	139.67
93	56.47	61.63	64.24	68.21	71.76	116.51	121.57	127.63	131.87	140.89
94	57.25	62.44	65.07	69.07	72.64	117.63	122.72	128.80	133.06	142.12
95	58.02	63.25	65.90	69.92	73.52	118.75	123.86	129.97	134.25	143.34
96	58.80	64.06	66.73	70.78	74.40	119.87	125.00	131.14	135.43	144.57
97	59.58	64.88	67.56	71.64	75.28	120.99	126.14	132.31	136.62	145.79
98	60.36	65.69	68.40	72.50	76.16	122.11	127.28	133.48	137.80	147.01
99	61.14	66.51	69.23	73.36	77.05	123.23	128.42	134.64	138.99	148.23
100	61.92	67.33	70.06	74.22	77.93	124.34	129.56	135.81	140.17	149.45

Statistical Table B: The standard normal (Z) distribution

This table gives probabilities under the right tail of the standard normal distribution. To determine the probability of sampling a value greater than or equal to a specific value of Z, find the first two digits of Z (called a and b) in the far left column of the table, and then find the last digit (c) in the top row of the table. For example, to find the probability of sampling a value greater than or equal to $Z = 1.96$, use $a.b = 1.9$ and $c = 6$. This probability is 0.025, or 2.5%.

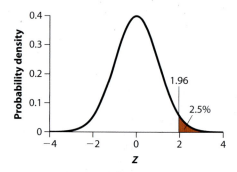

An example of a probability under the standard normal curve: the probability of sampling a value greater than or equal to the value 1.96 is 0.025, or 2.5%.

First two digits of *a.bc*	Second digit after decimal (*c*)									
	0	1	2	3	4	5	6	7	8	9
0.0	0.5	0.49601	0.49202	0.48803	0.48405	0.48006	0.47608	0.47210	0.46812	0.46414
0.1	0.46017	0.45620	0.45224	0.44828	0.44433	0.44038	0.43644	0.43251	0.42858	0.42465
0.2	0.42074	0.41683	0.41294	0.40905	0.40517	0.40129	0.39743	0.39358	0.38974	0.38591
0.3	0.38209	0.37828	0.37448	0.37070	0.36693	0.36317	0.35942	0.35569	0.35197	0.34827
0.4	0.34458	0.34090	0.33724	0.33360	0.32997	0.32636	0.32276	0.31918	0.31561	0.31207
0.5	0.30854	0.30503	0.30153	0.29806	0.29460	0.29116	0.28774	0.28434	0.28096	0.27760
0.6	0.27425	0.27093	0.26763	0.26435	0.26109	0.25785	0.25463	0.25143	0.24825	0.24510
0.7	0.24196	0.23885	0.23576	0.23270	0.22965	0.22663	0.22363	0.22065	0.21770	0.21476
0.8	0.21186	0.20897	0.20611	0.20327	0.20045	0.19766	0.19489	0.19215	0.18943	0.18673
0.9	0.18406	0.18141	0.17879	0.17619	0.17361	0.17106	0.16853	0.16602	0.16354	0.16109
1.0	0.15866	0.15625	0.15386	0.15151	0.14917	0.14686	0.14457	0.14231	0.14007	0.13786
1.1	0.13567	0.13350	0.13136	0.12924	0.12714	0.12507	0.12302	0.12100	0.11900	0.11702
1.2	0.11507	0.11314	0.11123	0.10935	0.10749	0.10565	0.10383	0.10204	0.10027	0.09853
1.3	0.09680	0.09510	0.09342	0.09176	0.09012	0.08851	0.08691	0.08534	0.08379	0.08226
1.4	0.08076	0.07927	0.07780	0.07636	0.07493	0.07353	0.07215	0.07078	0.06944	0.06811
1.5	0.06681	0.06552	0.06426	0.06301	0.06178	0.06057	0.05938	0.05821	0.05705	0.05592
1.6	0.05480	0.05370	0.05262	0.05155	0.05050	0.04947	0.04846	0.04746	0.04648	0.04551
1.7	0.04457	0.04363	0.04272	0.04182	0.04093	0.04006	0.03920	0.03836	0.03754	0.03673
1.8	0.03593	0.03515	0.03438	0.03362	0.03288	0.03216	0.03144	0.03074	0.03005	0.02938
1.9	0.02872	0.02807	0.02743	0.02680	0.02619	0.02559	0.02500	0.02442	0.02385	0.02330
2.0	0.02275	0.02222	0.02169	0.02118	0.02068	0.02018	0.01970	0.01923	0.01876	0.01831
2.1	0.01786	0.01743	0.01700	0.01659	0.01618	0.01578	0.01539	0.01500	0.01463	0.01426
2.2	0.01390	0.01355	0.01321	0.01287	0.01255	0.01222	0.01191	0.01160	0.01130	0.01101
2.3	0.01072	0.01044	0.01017	0.00990	0.00964	0.00939	0.00914	0.00889	0.00866	0.00842
2.4	0.00820	0.00798	0.00776	0.00755	0.00734	0.00714	0.00695	0.00676	0.00657	0.00639
2.5	0.00621	0.00604	0.00587	0.00570	0.00554	0.00539	0.00523	0.00508	0.00494	0.00480
2.6	0.00466	0.00453	0.00440	0.00427	0.00415	0.00402	0.00391	0.00379	0.00368	0.00357
2.7	0.00347	0.00336	0.00326	0.00317	0.00307	0.00298	0.00289	0.00280	0.00272	0.00264
2.8	0.00256	0.00248	0.00240	0.00233	0.00226	0.00219	0.00212	0.00205	0.00199	0.00193
2.9	0.00187	0.00181	0.00175	0.00169	0.00164	0.00159	0.00154	0.00149	0.00144	0.00139
3.0	0.00135	0.00131	0.00126	0.00122	0.00118	0.00114	0.00111	0.00107	0.00104	0.00100
3.1	0.00097	0.00094	0.00090	0.00087	0.00084	0.00082	0.00079	0.00076	0.00074	0.00071
3.2	0.00069	0.00066	0.00064	0.00062	0.00060	0.00058	0.00056	0.00054	0.00052	0.00050
3.3	0.00048	0.00047	0.00045	0.00043	0.00042	0.00040	0.00039	0.00038	0.00036	0.00035
3.4	0.00034	0.00032	0.00031	0.00030	0.00029	0.00028	0.00027	0.00026	0.00025	0.00024
3.5	0.00023	0.00022	0.00022	0.00021	0.00020	0.00019	0.00019	0.00018	0.00017	0.00017
3.6	0.00016	0.00015	0.00015	0.00014	0.00014	0.00013	0.00013	0.00012	0.00012	0.00011
3.7	0.00011	0.00010	0.00010	0.00010	0.00009	0.00009	0.00008	0.00008	0.00008	0.00008
3.8	0.00007	0.00007	0.00007	0.00006	0.00006	0.00006	0.00006	0.00005	0.00005	0.00005
3.9	0.00005	0.00005	0.00004	0.00004	0.00004	0.00004	0.00004	0.00004	0.00003	0.00003
4.0	0.00003	0.00003	0.00003	0.00003	0.00003	0.00003	0.00002	0.00002	0.00002	0.00002

Statistical Table C: Student's *t*-distribution

This table gives critical values of the *t*-distribution. The two-tailed critical value $t_{\alpha(2),df}$ defines the combined area α under both tails of the *t*-distribution having *df* degrees of freedom. To find a critical value in the table, select the desired value of $\alpha(2)$ along the top and the number of degrees of freedom in the far left column. For example, if *df* = 5 and $\alpha(2) = 0.05$, the critical value is 2.57; that is, the probability of a *t*-value greater than or equal to 2.57, or less than or equal to -2.57, is 0.05. The left panel of the following figure illustrates the critical value for this example.

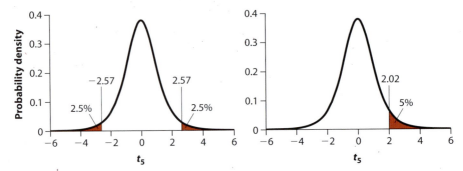

Critical values of the *t*-distribution having five degrees of freedom when $\alpha = 0.05$. The left panel shows the two-tailed case. The area in red shows the combined tail probabilities of 0.05, with 2.5% of the probability in each tail. The boundaries of the areas in red are -2.57 and 2.57. The right panel shows the one-tailed case. The area in red indicates the right tail of the distribution corresponding to 5% of the total probability. The boundary of the red section is 2.02.

The one-tailed critical value $t_{\alpha(1),df}$ defines the area α under the right tail of the *t*-distribution having *df* degrees of freedom. To find a critical value in the table, select the desired value of $\alpha(1)$ along the top and the number of degrees of freedom in the far left column. For example, if *df* = 5 and $\alpha(1) = 0.05$, the critical value is 2.02; that is, the probability of a value greater than or equal to 2.02 is 0.05. The right panel of the preceding figure illustrates the critical value for this example.

df	α(2): α(1):	0.2 0.1	0.10 0.05	0.05 0.025	0.02 0.01	0.01 0.005	0.001 0.0005	0.0001 0.00005
1		3.08	6.31	12.71	31.82	63.66	636.62	6366.20
2		1.89	2.92	4.30	6.96	9.92	31.60	99.99
3		1.64	2.35	3.18	4.54	5.84	12.92	28.00
4		1.53	2.13	2.78	3.75	4.60	8.61	15.54
5		1.48	2.02	2.57	3.36	4.03	6.87	11.18
6		1.44	1.94	2.45	3.14	3.71	5.96	9.08
7		1.41	1.89	2.36	3.00	3.50	5.41	7.88
8		1.40	1.86	2.31	2.90	3.36	5.04	7.12
9		1.38	1.83	2.26	2.82	3.25	4.78	6.59
10		1.37	1.81	2.23	2.76	3.17	4.59	6.21
11		1.36	1.80	2.20	2.72	3.11	4.44	5.92
12		1.36	1.78	2.18	2.68	3.05	4.32	5.69
13		1.35	1.77	2.16	2.65	3.01	4.22	5.51
14		1.35	1.76	2.14	2.62	2.98	4.14	5.36
15		1.34	1.75	2.13	2.60	2.95	4.07	5.24
16		1.34	1.75	2.12	2.58	2.92	4.01	5.13
17		1.33	1.74	2.11	2.57	2.90	3.97	5.04
18		1.33	1.73	2.10	2.55	2.88	3.92	4.97
19		1.33	1.73	2.09	2.54	2.86	3.88	4.90
20		1.33	1.72	2.09	2.53	2.85	3.85	4.84
21		1.32	1.72	2.08	2.52	2.83	3.82	4.78
22		1.32	1.72	2.07	2.51	2.82	3.79	4.74
23		1.32	1.71	2.07	2.50	2.81	3.77	4.69
24		1.32	1.71	2.06	2.49	2.80	3.75	4.65
25		1.32	1.71	2.06	2.49	2.79	3.73	4.62
26		1.31	1.71	2.06	2.48	2.78	3.71	4.59
27		1.31	1.70	2.05	2.47	2.77	3.69	4.56
28		1.31	1.70	2.05	2.47	2.76	3.67	4.53
29		1.31	1.70	2.05	2.46	2.76	3.66	4.51
30		1.31	1.70	2.04	2.46	2.75	3.65	4.48
31		1.31	1.70	2.04	2.45	2.74	3.63	4.46
32		1.31	1.69	2.04	2.45	2.74	3.62	4.44
33		1.31	1.69	2.03	2.44	2.73	3.61	4.42
34		1.31	1.69	2.03	2.44	2.73	3.60	4.41
35		1.31	1.69	2.03	2.44	2.72	3.59	4.39
36		1.31	1.69	2.03	2.43	2.72	3.58	4.37
37		1.30	1.69	2.03	2.43	2.72	3.57	4.36
38		1.30	1.69	2.02	2.43	2.71	3.57	4.35
39		1.30	1.68	2.02	2.43	2.71	3.56	4.33
40		1.30	1.68	2.02	2.42	2.70	3.55	4.32
41		1.30	1.68	2.02	2.42	2.70	3.54	4.31
42		1.30	1.68	2.02	2.42	2.70	3.54	4.30
43		1.30	1.68	2.02	2.42	2.70	3.53	4.29
44		1.30	1.68	2.02	2.41	2.69	3.53	4.28
45		1.30	1.68	2.01	2.41	2.69	3.52	4.27
46		1.30	1.68	2.01	2.41	2.69	3.51	4.26
47		1.30	1.68	2.01	2.41	2.68	3.51	4.25
48		1.30	1.68	2.01	2.41	2.68	3.51	4.24
49		1.30	1.68	2.01	2.40	2.68	3.50	4.24
50		1.30	1.68	2.01	2.40	2.68	3.50	4.23

df	$\alpha(2)$: $\alpha(1)$:	0.2 0.1	0.10 0.05	0.05 0.025	0.02 0.01	0.01 0.005	0.001 0.0005	0.0001 0.00005
51		1.30	1.68	2.01	2.40	2.68	3.49	4.22
52		1.30	1.67	2.01	2.40	2.67	3.49	4.21
53		1.30	1.67	2.01	2.40	2.67	3.48	4.21
54		1.30	1.67	2.00	2.40	2.67	3.48	4.20
55		1.30	1.67	2.00	2.40	2.67	3.48	4.20
56		1.30	1.67	2.00	2.39	2.67	3.47	4.19
57		1.30	1.67	2.00	2.39	2.66	3.47	4.18
58		1.30	1.67	2.00	2.39	2.66	3.47	4.18
59		1.30	1.67	2.00	2.39	2.66	3.46	4.17
60		1.30	1.67	2.00	2.39	2.66	3.46	4.17
61		1.30	1.67	2.00	2.39	2.66	3.46	4.16
62		1.30	1.67	2.00	2.39	2.66	3.45	4.16
63		1.30	1.67	2.00	2.39	2.66	3.45	4.15
64		1.29	1.67	2.00	2.39	2.65	3.45	4.15
65		1.29	1.67	2.00	2.39	2.65	3.45	4.15
66		1.29	1.67	2.00	2.38	2.65	3.44	4.14
67		1.29	1.67	2.00	2.38	2.65	3.44	4.14
68		1.29	1.67	2.00	2.38	2.65	3.44	4.13
69		1.29	1.67	1.99	2.38	2.65	3.44	4.13
70		1.29	1.67	1.99	2.38	2.65	3.44	4.13
71		1.29	1.67	1.99	2.38	2.65	3.43	4.12
72		1.29	1.67	1.99	2.38	2.65	3.43	4.12
73		1.29	1.67	1.99	2.38	2.64	3.43	4.12
74		1.29	1.67	1.99	2.38	2.64	3.43	4.11
75		1.29	1.67	1.99	2.38	2.64	3.43	4.11
76		1.29	1.67	1.99	2.38	2.64	3.42	4.11
77		1.29	1.66	1.99	2.38	2.64	3.42	4.10
78		1.29	1.66	1.99	2.38	2.64	3.42	4.10
79		1.29	1.66	1.99	2.37	2.64	3.42	4.10
80		1.29	1.66	1.99	2.37	2.64	3.42	4.10
81		1.29	1.66	1.99	2.37	2.64	3.41	4.09
82		1.29	1.66	1.99	2.37	2.64	3.41	4.09
83		1.29	1.66	1.99	2.37	2.64	3.41	4.09
84		1.29	1.66	1.99	2.37	2.64	3.41	4.09
85		1.29	1.66	1.99	2.37	2.63	3.41	4.08
86		1.29	1.66	1.99	2.37	2.63	3.41	4.08
87		1.29	1.66	1.99	2.37	2.63	3.41	4.08
88		1.29	1.66	1.99	2.37	2.63	3.40	4.08
89		1.29	1.66	1.99	2.37	2.63	3.40	4.07
90		1.29	1.66	1.99	2.37	2.63	3.40	4.07
100		1.29	1.66	1.98	2.36	2.63	3.39	4.05
120		1.29	1.66	1.98	2.36	2.62	3.37	4.03
140		1.29	1.66	1.98	2.35	2.61	3.36	4.01
160		1.29	1.65	1.97	2.35	2.61	3.35	3.99
180		1.29	1.65	1.97	2.35	2.60	3.35	3.98
200		1.29	1.65	1.97	2.35	2.60	3.34	3.97
400		1.28	1.65	1.97	2.34	2.59	3.32	3.93
1000		1.28	1.65	1.96	2.33	2.58	3.30	3.91

Statistical Table D: The *F*-distribution

These tables give critical values of the *F*-distribution for $\alpha(1) = 0.05$ and $\alpha(1) = 0.01$. The critical value $F_{\alpha(1),df1,df2}$ defines the area α under the right tail of the *F*-distribution having df_1 and df_2 degrees of freedom. To find a critical value in the table, select the numerator degrees of freedom (df_1) listed across the top row and the denominator degrees of freedom (df_2) given in the first column. For example, if $df_1 = 9$, $df_2 = 12$, and $\alpha = 0.05$, then the critical value is 2.80; that is, the probability of a value greater than or equal to 2.80 is 0.05.

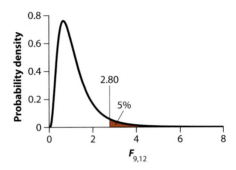

The critical value for the *F*-distribution having nine and 12 degrees of freedom when $\alpha = 0.05$. The area in red indicates the tail probability corresponding to 0.05, or 5%. The boundary of the red section is 2.80.

Critical value of *F*, $\alpha(1) = 0.05$

Denominator *df*	Numerator *df*									
	1	**2**	**3**	**4**	**5**	**6**	**7**	**8**	**9**	**10**
1	161.45	199.50	215.71	224.58	230.16	233.99	236.77	238.88	240.54	241.88
2	18.51	19.00	19.16	19.25	19.30	19.33	19.35	19.37	19.38	19.40
3	10.13	9.55	9.28	9.12	9.01	8.94	8.89	8.85	8.81	8.79
4	7.71	6.94	6.59	6.39	6.26	6.16	6.09	6.04	6.00	5.96
5	6.61	5.79	5.41	5.19	5.05	4.95	4.88	4.82	4.77	4.74
6	5.99	5.14	4.76	4.53	4.39	4.28	4.21	4.15	4.10	4.06
7	5.59	4.74	4.35	4.12	3.97	3.87	3.79	3.73	3.68	3.64
8	5.32	4.46	4.07	3.84	3.69	3.58	3.50	3.44	3.39	3.35
9	5.12	4.26	3.86	3.63	3.48	3.37	3.29	3.23	3.18	3.14
10	4.96	4.10	3.71	3.48	3.33	3.22	3.14	3.07	3.02	2.98
11	4.84	3.98	3.59	3.36	3.20	3.09	3.01	2.95	2.90	2.85
12	4.75	3.89	3.49	3.26	3.11	3.00	2.91	2.85	2.80	2.75
13	4.67	3.81	3.41	3.18	3.03	2.92	2.83	2.77	2.71	2.67
14	4.60	3.74	3.34	3.11	2.96	2.85	2.76	2.70	2.65	2.60
15	4.54	3.68	3.29	3.06	2.90	2.79	2.71	2.64	2.59	2.54
16	4.49	3.63	3.24	3.01	2.85	2.74	2.66	2.59	2.54	2.49
17	4.45	3.59	3.20	2.96	2.81	2.70	2.61	2.55	2.49	2.45
18	4.41	3.55	3.16	2.93	2.77	2.66	2.58	2.51	2.46	2.41
19	4.38	3.52	3.13	2.90	2.74	2.63	2.54	2.48	2.42	2.38
20	4.35	3.49	3.10	2.87	2.71	2.60	2.51	2.45	2.39	2.35
21	4.32	3.47	3.07	2.84	2.68	2.57	2.49	2.42	2.37	2.32
22	4.30	3.44	3.05	2.82	2.66	2.55	2.46	2.40	2.34	2.30
23	4.28	3.42	3.03	2.80	2.64	2.53	2.44	2.37	2.32	2.27
24	4.26	3.40	3.01	2.78	2.62	2.51	2.42	2.36	2.30	2.25
25	4.24	3.39	2.99	2.76	2.60	2.49	2.40	2.34	2.28	2.24
26	4.23	3.37	2.98	2.74	2.59	2.47	2.39	2.32	2.27	2.22
27	4.21	3.35	2.96	2.73	2.57	2.46	2.37	2.31	2.25	2.20
28	4.20	3.34	2.95	2.71	2.56	2.45	2.36	2.29	2.24	2.19
29	4.18	3.33	2.93	2.70	2.55	2.43	2.35	2.28	2.22	2.18
30	4.17	3.32	2.92	2.69	2.53	2.42	2.33	2.27	2.21	2.16
40	4.08	3.23	2.84	2.61	2.45	2.34	2.25	2.18	2.12	2.08
50	4.03	3.18	2.79	2.56	2.40	2.29	2.20	2.13	2.07	2.03
60	4.00	3.15	2.76	2.53	2.37	2.25	2.17	2.10	2.04	1.99
70	3.98	3.13	2.74	2.50	2.35	2.23	2.14	2.07	2.02	1.97
80	3.96	3.11	2.72	2.49	2.33	2.21	2.13	2.06	2.00	1.95
90	3.95	3.10	2.71	2.47	2.32	2.20	2.11	2.04	1.99	1.94
100	3.94	3.09	2.70	2.46	2.31	2.19	2.10	2.03	1.97	1.93
200	3.89	3.04	2.65	2.42	2.26	2.14	2.06	1.98	1.93	1.88
400	3.86	3.02	2.63	2.39	2.24	2.12	2.03	1.96	1.90	1.85

Critical value of *F*, $\alpha(1) = 0.05$ (continued)

Denominator df	Numerator df									
	12	15	20	30	40	60	100	200	400	1000
1	243.91	245.95	248.01	250.10	251.14	252.20	253.04	253.68	254.00	254.19
2	19.41	19.43	19.45	19.46	19.47	19.48	19.49	19.49	19.49	19.49
3	8.74	8.70	8.66	8.62	8.59	8.57	8.55	8.54	8.53	8.53
4	5.91	5.86	5.80	5.75	5.72	5.69	5.66	5.65	5.64	5.63
5	4.68	4.62	4.56	4.50	4.46	4.43	4.41	4.39	4.38	4.37
6	4.00	3.94	3.87	3.81	3.77	3.74	3.71	3.69	3.68	3.67
7	3.57	3.51	3.44	3.38	3.34	3.30	3.27	3.25	3.24	3.23
8	3.28	3.22	3.15	3.08	3.04	3.01	2.97	2.95	2.94	2.93
9	3.07	3.01	2.94	2.86	2.83	2.79	2.76	2.73	2.72	2.71
10	2.91	2.85	2.77	2.70	2.66	2.62	2.59	2.56	2.55	2.54
11	2.79	2.72	2.65	2.57	2.53	2.49	2.46	2.43	2.42	2.41
12	2.69	2.62	2.54	2.47	2.43	2.38	2.35	2.32	2.31	2.30
13	2.60	2.53	2.46	2.38	2.34	2.30	2.26	2.23	2.22	2.21
14	2.53	2.46	2.39	2.31	2.27	2.22	2.19	2.16	2.15	2.14
15	2.48	2.40	2.33	2.25	2.20	2.16	2.12	2.10	2.08	2.07
16	2.42	2.35	2.28	2.19	2.15	2.11	2.07	2.04	2.02	2.02
17	2.38	2.31	2.23	2.15	2.10	2.06	2.02	1.99	1.98	1.97
18	2.34	2.27	2.19	2.11	2.06	2.02	1.98	1.95	1.93	1.92
19	2.31	2.23	2.16	2.07	2.03	1.98	1.94	1.91	1.89	1.88
20	2.28	2.20	2.12	2.04	1.99	1.95	1.91	1.88	1.86	1.85
21	2.25	2.18	2.10	2.01	1.96	1.92	1.88	1.84	1.83	1.82
22	2.23	2.15	2.07	1.98	1.94	1.89	1.85	1.82	1.80	1.79
23	2.20	2.13	2.05	1.96	1.91	1.86	1.82	1.79	1.77	1.76
24	2.18	2.11	2.03	1.94	1.89	1.84	1.80	1.77	1.75	1.74
25	2.16	2.09	2.01	1.92	1.87	1.82	1.78	1.75	1.73	1.72
26	2.15	2.07	1.99	1.90	1.85	1.80	1.76	1.73	1.71	1.70
27	2.13	2.06	1.97	1.88	1.84	1.79	1.74	1.71	1.69	1.68
28	2.12	2.04	1.96	1.87	1.82	1.77	1.73	1.69	1.67	1.66
29	2.10	2.03	1.94	1.85	1.81	1.75	1.71	1.67	1.66	1.65
30	2.09	2.01	1.93	1.84	1.79	1.74	1.70	1.66	1.64	1.63
40	2.00	1.92	1.84	1.74	1.69	1.64	1.59	1.55	1.53	1.52
50	1.95	1.87	1.78	1.69	1.63	1.58	1.52	1.48	1.46	1.45
60	1.92	1.84	1.75	1.65	1.59	1.53	1.48	1.44	1.41	1.40
70	1.89	1.81	1.72	1.62	1.57	1.50	1.45	1.40	1.38	1.36
80	1.88	1.79	1.70	1.60	1.54	1.48	1.43	1.38	1.35	1.34
90	1.86	1.78	1.69	1.59	1.53	1.46	1.41	1.36	1.33	1.31
100	1.85	1.77	1.68	1.57	1.52	1.45	1.39	1.34	1.31	1.30
200	1.80	1.72	1.62	1.52	1.46	1.39	1.32	1.26	1.23	1.21
400	1.78	1.69	1.60	1.49	1.42	1.35	1.28	1.22	1.18	1.15

Critical value of F, $\alpha(1) = 0.01$

Numerator *df*

Denominator df	1	2	3	4	5	6	7	8	9	10
1	4052	4999	5403	5624	5763	5859	5928	5981	6022	6055
2	98.50	99.00	99.17	99.25	99.30	99.33	99.36	99.37	99.39	99.40
3	34.12	30.82	29.46	28.71	28.24	27.91	27.67	27.49	27.35	27.23
4	21.20	18.00	16.69	15.98	15.52	15.21	14.98	14.80	14.66	14.55
5	16.26	13.27	12.06	11.39	10.97	10.67	10.46	10.29	10.16	10.05
6	13.75	10.92	9.78	9.15	8.75	8.47	8.26	8.10	7.98	7.87
7	12.25	9.55	8.45	7.85	7.46	7.19	6.99	6.84	6.72	6.62
8	11.26	8.65	7.59	7.01	6.63	6.37	6.18	6.03	5.91	5.81
9	10.56	8.02	6.99	6.42	6.06	5.80	5.61	5.47	5.35	5.26
10	10.04	7.56	6.55	5.99	5.64	5.39	5.20	5.06	4.94	4.85
11	9.65	7.21	6.22	5.67	5.32	5.07	4.89	4.74	4.63	4.54
12	9.33	6.93	5.95	5.41	5.06	4.82	4.64	4.50	4.39	4.30
13	9.07	6.70	5.74	5.21	4.86	4.62	4.44	4.30	4.19	4.10
14	8.86	6.51	5.56	5.04	4.69	4.46	4.28	4.14	4.03	3.94
15	8.68	6.36	5.42	4.89	4.56	4.32	4.14	4.00	3.89	3.80
16	8.53	6.23	5.29	4.77	4.44	4.20	4.03	3.89	3.78	3.69
17	8.40	6.11	5.18	4.67	4.34	4.10	3.93	3.79	3.68	3.59
18	8.29	6.01	5.09	4.58	4.25	4.01	3.84	3.71	3.60	3.51
19	8.18	5.93	5.01	4.50	4.17	3.94	3.77	3.63	3.52	3.43
20	8.10	5.85	4.94	4.43	4.10	3.87	3.70	3.56	3.46	3.37
21	8.02	5.78	4.87	4.37	4.04	3.81	3.64	3.51	3.40	3.31
22	7.95	5.72	4.82	4.31	3.99	3.76	3.59	3.45	3.35	3.26
23	7.88	5.66	4.76	4.26	3.94	3.71	3.54	3.41	3.30	3.21
24	7.82	5.61	4.72	4.22	3.90	3.67	3.50	3.36	3.26	3.17
25	7.77	5.57	4.68	4.18	3.85	3.63	3.46	3.32	3.22	3.13
26	7.72	5.53	4.64	4.14	3.82	3.59	3.42	3.29	3.18	3.09
27	7.68	5.49	4.60	4.11	3.78	3.56	3.39	3.26	3.15	3.06
28	7.64	5.45	4.57	4.07	3.75	3.53	3.36	3.23	3.12	3.03
29	7.60	5.42	4.54	4.04	3.73	3.50	3.33	3.20	3.09	3.00
30	7.56	5.39	4.51	4.02	3.70	3.47	3.30	3.17	3.07	2.98
40	7.31	5.18	4.31	3.83	3.51	3.29	3.12	2.99	2.89	2.80
50	7.17	5.06	4.20	3.72	3.41	3.19	3.02	2.89	2.78	2.70
60	7.08	4.98	4.13	3.65	3.34	3.12	2.95	2.82	2.72	2.63
70	7.01	4.92	4.07	3.60	3.29	3.07	2.91	2.78	2.67	2.59
80	6.96	4.88	4.04	3.56	3.26	3.04	2.87	2.74	2.64	2.55
90	6.93	4.85	4.01	3.53	3.23	3.01	2.84	2.72	2.61	2.52
100	6.90	4.82	3.98	3.51	3.21	2.99	2.82	2.69	2.59	2.50
200	6.76	4.71	3.88	3.41	3.11	2.89	2.73	2.60	2.50	2.41
400	6.70	4.66	3.83	3.37	3.06	2.85	2.68	2.56	2.45	2.37

Critical value of *F*, $\alpha(1) = 0.01$ (continued)

Denominator df	Numerator df									
	12	15	20	30	40	60	100	200	400	1000
1	6106	6157	6208	6260	6286	6313	6334	6350	6357	6362
2	99.42	99.43	99.45	99.47	99.47	99.48	99.49	99.49	99.50	99.50
3	27.05	26.87	26.69	26.50	26.41	26.32	26.24	26.18	26.15	26.14
4	14.37	14.20	14.02	13.84	13.75	13.65	13.58	13.52	13.49	13.47
5	9.89	9.72	9.55	9.38	9.29	9.20	9.13	9.08	9.05	9.03
6	7.72	7.56	7.40	7.23	7.14	7.06	6.99	6.93	6.91	6.89
7	6.47	6.31	6.16	5.99	5.91	5.82	5.75	5.70	5.68	5.66
8	5.67	5.52	5.36	5.20	5.12	5.03	4.96	4.91	4.89	4.87
9	5.11	4.96	4.81	4.65	4.57	4.48	4.41	4.36	4.34	4.32
10	4.71	4.56	4.41	4.25	4.17	4.08	4.01	3.96	3.94	3.92
11	4.40	4.25	4.10	3.94	3.86	3.78	3.71	3.66	3.63	3.61
12	4.16	4.01	3.86	3.70	3.62	3.54	3.47	3.41	3.39	3.37
13	3.96	3.82	3.66	3.51	3.43	3.34	3.27	3.22	3.19	3.18
14	3.80	3.66	3.51	3.35	3.27	3.18	3.11	3.06	3.03	3.02
15	3.67	3.52	3.37	3.21	3.13	3.05	2.98	2.92	2.90	2.88
16	3.55	3.41	3.26	3.10	3.02	2.93	2.86	2.81	2.78	2.76
17	3.46	3.31	3.16	3.00	2.92	2.83	2.76	2.71	2.68	2.66
18	3.37	3.23	3.08	2.92	2.84	2.75	2.68	2.62	2.59	2.58
19	3.30	3.15	3.00	2.84	2.76	2.67	2.60	2.55	2.52	2.50
20	3.23	3.09	2.94	2.78	2.69	2.61	2.54	2.48	2.45	2.43
21	3.17	3.03	2.88	2.72	2.64	2.55	2.48	2.42	2.39	2.37
22	3.12	2.98	2.83	2.67	2.58	2.50	2.42	2.36	2.34	2.32
23	3.07	2.93	2.78	2.62	2.54	2.45	2.37	2.32	2.29	2.27
24	3.03	2.89	2.74	2.58	2.49	2.40	2.33	2.27	2.24	2.22
25	2.99	2.85	2.70	2.54	2.45	2.36	2.29	2.23	2.20	2.18
26	2.96	2.81	2.66	2.50	2.42	2.33	2.25	2.19	2.16	2.14
27	2.93	2.78	2.63	2.47	2.38	2.29	2.22	2.16	2.13	2.11
28	2.90	2.75	2.60	2.44	2.35	2.26	2.19	2.13	2.10	2.08
29	2.87	2.73	2.57	2.41	2.33	2.23	2.16	2.10	2.07	2.05
30	2.84	2.70	2.55	2.39	2.30	2.21	2.13	2.07	2.04	2.02
40	2.66	2.52	2.37	2.20	2.11	2.02	1.94	1.87	1.84	1.82
50	2.56	2.42	2.27	2.10	2.01	1.91	1.82	1.76	1.72	1.70
60	2.50	2.35	2.20	2.03	1.94	1.84	1.75	1.68	1.64	1.62
70	2.45	2.31	2.15	1.98	1.89	1.78	1.70	1.62	1.58	1.56
80	2.42	2.27	2.12	1.94	1.85	1.75	1.65	1.58	1.54	1.51
90	2.39	2.24	2.09	1.92	1.82	1.72	1.62	1.55	1.50	1.48
100	2.37	2.22	2.07	1.89	1.80	1.69	1.60	1.52	1.47	1.45
200	2.27	2.13	1.97	1.79	1.69	1.58	1.48	1.39	1.34	1.30
400	2.23	2.08	1.92	1.75	1.64	1.53	1.42	1.32	1.26	1.2

Statistical Table E:
Mann–Whitney *U*-distribution

These tables give the two-tailed critical values of the U-distribution for $\alpha = 0.05$ and $\alpha = 0.01$. U is the larger of U_1 and U_2. The critical value $U_{0.05(2),\,n1,n2}$ defines the area α under the right tail of the U-distribution corresponding to sample sizes n_1 and n_2. To find a critical value in the table, select n_1 from the top row and n_2 from the far left column. A "—" means that it is not possible to reject a null hypothesis with that α and those sample sizes. For larger sample sizes, use the Z-approximation from Chapter 13. For example, if $n_1 = 5$, $n_2 = 7$, and $\alpha = 0.05$, then the critical value is 30; that is, the probability of a value greater than or equal to 30 is 0.05 or less (it may be less than 0.05 because the U-distribution is discrete, and no critical value may correspond exactly to 0.05). The following figure illustrates the critical value for this example.

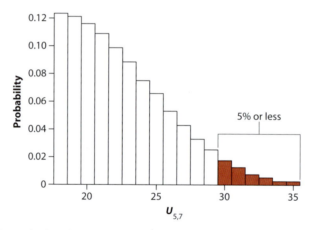

The two-tailed critical value of the Mann–Whitney U-distribution when $n_1 = 5$, $n_2 = 7$, and $\alpha = 0.05$. The area in red indicates the tail probability corresponding to 0.05, or 5%. The boundary of the red section is 30. Thus, the probability of a value greater than or equal to 30 is 0.05 or less. (The distribution gives the two-tailed value, because it measures the probability that either U_1 or U_2 is greater than $U_{0.05(2),\,n1,n2}$.)

$$\alpha(2) = 0.05$$

n_2	3	4	5	6	7	8	9	10	11	12	13	14	15
3	—	—	15	17	20	22	25	27	30	32	35	37	40
4	—	16	19	22	25	28	32	35	38	41	44	47	50
5	15	19	23	27	30	34	38	42	46	49	53	57	61
6	17	22	27	31	36	40	44	49	53	58	62	67	71
7	20	25	30	36	41	46	51	56	61	66	71	76	81
8	22	28	34	40	46	51	57	63	69	74	80	86	91
9	25	32	38	44	51	57	64	70	76	82	89	95	101
10	27	35	42	49	56	63	70	77	84	91	97	104	111
11	30	38	46	53	61	69	76	84	91	99	106	114	121
12	32	41	49	58	66	74	82	91	99	107	115	123	131
13	35	44	53	62	71	80	89	97	106	115	124	132	141
14	37	47	57	67	76	86	95	104	114	123	132	141	151
15	40	50	61	71	81	91	101	111	121	131	141	151	161

$$\alpha(2) = 0.01$$

n_2	3	4	5	6	7	8	9	10	11	12	13	14	15
3	—	—	—	—	—	—	27	30	33	35	38	41	43
4	—	—	—	24	28	31	35	38	42	45	49	52	55
5	—	—	25	29	34	38	42	46	50	54	58	63	67
6	—	24	29	34	39	44	49	54	59	63	68	73	78
7	—	28	34	39	45	50	56	61	67	72	78	83	89
8	—	31	38	44	50	57	63	69	75	81	87	94	100
9	27	35	42	49	56	63	70	77	83	90	97	104	111
10	30	38	46	54	61	69	77	84	92	99	106	114	121
11	33	42	50	59	67	75	83	92	100	108	116	124	132
12	35	45	54	63	72	81	90	99	108	117	125	134	143
13	38	49	58	68	78	87	97	106	116	125	135	144	153
14	41	52	63	73	83	94	104	114	124	134	144	154	164
15	43	53	67	78	89	100	111	121	132	143	153	164	174

Statistical Table F:
Tukey–Kramer q-distribution

This table gives the critical values of the Tukey–Kramer q-distribution for $\alpha = 0.05$. The critical value $q_{0.05,k,N-k}$ defines the area 0.05 under the right tail of the q-distribution having k groups and $N - k$ degrees of freedom, where N is the total sample size of all groups combined. The probability of a q-value greater than or equal to $q_{0.05,k,N-k}$ is 0.05, or 5%.

						Number of groups (k)								
df_{error}	2	3	4	5	6	7	8	9	10	11	12	13	14	15
10	2.23	2.74	3.06	3.29	3.47	3.62	3.75	3.86	3.96	4.05	4.12	4.20	4.26	4.32
11	2.20	2.70	3.01	3.23	3.41	3.56	3.68	3.79	3.88	3.96	4.04	4.11	4.17	4.23
12	2.18	2.67	2.97	3.19	3.36	3.50	3.62	3.72	3.81	3.90	3.97	4.04	4.10	4.16
13	2.16	2.64	2.94	3.15	3.32	3.45	3.57	3.67	3.76	3.84	3.91	3.98	4.04	4.09
14	2.14	2.62	2.91	3.12	3.28	3.41	3.53	3.63	3.72	3.79	3.86	3.93	3.99	4.04
15	2.13	2.60	2.88	3.09	3.25	3.38	3.49	3.59	3.68	3.75	3.82	3.88	3.94	3.99
16	2.12	2.58	2.86	3.06	3.22	3.35	3.46	3.56	3.64	3.72	3.78	3.85	3.90	3.95
17	2.11	2.57	2.84	3.04	3.20	3.33	3.44	3.53	3.61	3.69	3.75	3.81	3.87	3.92
18	2.10	2.55	2.83	3.02	3.18	3.30	3.41	3.50	3.59	3.66	3.72	3.78	3.84	3.89
19	2.09	2.54	2.81	3.01	3.16	3.16	3.39	3.48	3.56	3.63	3.70	3.76	3.81	3.86
20	2.09	2.53	2.80	2.99	3.14	3.27	3.37	3.46	3.54	3.61	3.68	3.73	3.79	3.84
21	2.08	2.52	2.79	2.98	3.13	3.25	3.35	3.44	3.52	3.59	3.66	3.71	3.77	3.82
22	2.07	2.51	2.78	2.97	3.12	3.24	3.34	3.43	3.51	3.57	3.64	3.69	3.75	3.80
23	2.07	2.50	2.77	2.96	3.10	3.22	3.32	3.41	3.49	3.56	3.62	3.68	3.73	3.78
24	2.06	2.50	2.76	2.95	3.09	3.21	3.31	3.40	3.48	3.54	3.61	3.66	3.71	3.76
25	2.06	2.49	2.75	2.94	3.08	3.20	3.30	3.39	3.46	3.53	3.59	3.65	3.70	3.75
26	2.06	2.48	2.74	2.93	3.07	3.19	3.29	3.38	3.45	3.52	3.58	3.63	3.68	3.73
27	2.05	2.48	2.74	2.92	3.06	3.18	3.28	3.36	3.44	3.51	3.57	3.62	3.67	3.72
28	2.05	2.47	2.73	2.91	3.06	3.17	3.27	3.35	3.43	3.50	3.56	3.61	3.66	3.71
29	2.05	2.47	2.72	2.91	3.05	3.16	3.26	3.35	3.42	3.49	3.55	3.60	3.65	3.70
30	2.04	2.46	2.72	2.90	3.04	3.16	3.25	3.34	3.41	3.48	3.54	3.59	3.64	3.68
31	2.04	2.46	2.71	2.89	3.04	3.15	3.25	3.33	3.40	3.47	3.53	3.58	3.63	3.68
32	2.04	2.46	2.71	2.89	3.03	3.14	3.24	3.32	3.40	3.46	3.52	3.57	3.62	3.67
33	2.03	2.45	2.70	2.88	3.02	3.14	3.23	3.32	3.39	3.45	3.51	3.56	3.61	3.66
34	2.03	2.45	2.70	2.88	3.02	3.13	3.23	3.31	3.38	3.45	3.50	3.56	3.60	3.65
35	2.03	2.45	2.70	2.88	3.01	3.13	3.22	3.30	3.37	3.44	3.50	3.55	3.60	3.64
36	2.03	2.44	2.69	2.87	3.01	3.12	3.22	3.30	3.37	3.43	3.49	3.54	3.59	3.64
37	2.03	2.44	2.69	2.87	3.00	3.12	3.21	3.29	3.36	3.43	3.48	3.54	3.58	3.63
38	2.02	2.44	2.69	2.86	3.00	3.11	3.21	3.29	3.36	3.42	3.48	3.53	3.58	3.62
39	2.02	2.44	2.68	2.86	3.00	3.11	3.20	3.28	3.35	3.42	3.47	3.52	3.57	3.62
40	2.02	2.43	2.68	2.86	2.99	3.10	3.20	3.28	3.35	3.41	3.47	3.52	3.57	3.61
41	2.02	2.43	2.68	2.85	2.99	3.10	3.19	3.27	3.34	3.41	3.46	3.51	3.56	3.60
42	2.02	2.43	2.67	2.85	2.99	3.10	3.19	3.27	3.34	3.40	3.46	3.51	3.56	3.60
43	2.02	2.43	2.67	2.85	2.98	3.09	3.18	3.26	3.33	3.40	3.45	3.50	3.55	3.59
44	2.02	2.43	2.67	2.84	2.98	3.09	3.18	3.26	3.33	3.39	3.45	3.50	3.55	3.59
45	2.01	2.42	2.67	2.84	2.98	3.09	3.18	3.26	3.33	3.39	3.44	3.50	3.54	3.59
46	2.01	2.42	2.67	2.84	2.97	3.08	3.17	3.25	3.32	3.39	3.44	3.49	3.54	3.58
47	2.01	2.42	2.66	2.84	2.97	3.08	3.17	3.25	3.32	3.38	3.44	3.49	3.53	3.58
48	2.01	2.42	2.66	2.83	2.97	3.08	3.17	3.25	3.32	3.38	3.43	3.48	3.53	3.57
49	2.01	2.42	2.66	2.83	2.97	3.07	3.17	3.24	3.31	3.37	3.43	3.48	3.53	3.57
50	2.01	2.42	2.66	2.83	2.96	3.07	3.16	3.24	3.31	3.37	3.43	3.48	3.52	3.57

Statistical Table G: Critical values for the Spearman's rank correlation

This table gives two-tailed critical values of the Spearman rank correlation under the null hypothesis that the population correlation is zero. The critical value $r_{S(\alpha,n)}$ defines the combined area α under both tails of the null distribution, where n is the sample size. The probability of a value greater than or equal to $r_{S(\alpha,n)}$ or less than or equal to $-r_{S(\alpha,n)}$ is α. These critical values were obtained using the *SuppDist* package (Wheeler 2005) implemented in R, according to the methods of Kendall and Smith (1939).

n	α = 0.05	α = 0.01	n	α = 0.05	α = 0.01	n	α = 0.05	α = 0.01
5	0.900		37	0.325	0.419	69	0.237	0.308
6	0.943	1.000	38	0.320	0.413	70	0.235	0.306
7	0.821	0.929	39	0.316	0.408	71	0.234	0.304
8	0.762	0.881	40	0.312	0.403	72	0.232	0.302
9	0.700	0.833	41	0.308	0.398	73	0.230	0.300
10	0.648	0.782	42	0.305	0.393	74	0.229	0.298
11	0.618	0.755	43	0.301	0.389	75	0.227	0.296
12	0.587	0.720	44	0.298	0.385	76	0.226	0.294
13	0.560	0.692	45	0.294	0.380	77	0.224	0.292
14	0.538	0.670	46	0.291	0.376	78	0.223	0.290
15	0.521	0.645	47	0.288	0.372	79	0.221	0.288
16	0.503	0.626	48	0.285	0.369	80	0.220	0.286
17	0.485	0.610	49	0.282	0.365	81	0.219	0.285
18	0.472	0.593	50	0.279	0.361	82	0.217	0.283
19	0.458	0.579	51	0.276	0.358	83	0.216	0.281
20	0.447	0.564	52	0.273	0.354	84	0.215	0.280
21	0.435	0.551	53	0.271	0.351	85	0.213	0.278
22	0.425	0.539	54	0.268	0.348	86	0.212	0.276
23	0.415	0.528	55	0.266	0.345	87	0.211	0.275
24	0.406	0.516	56	0.263	0.342	88	0.210	0.273
25	0.398	0.506	57	0.261	0.339	89	0.208	0.272
26	0.389	0.497	58	0.259	0.336	90	0.207	0.270
27	0.382	0.488	59	0.256	0.333	91	0.206	0.269
28	0.375	0.479	60	0.254	0.330	92	0.205	0.267
29	0.368	0.471	61	0.252	0.327	93	0.204	0.266
30	0.362	0.464	62	0.250	0.325	94	0.203	0.264
31	0.356	0.456	63	0.248	0.322	95	0.202	0.263
32	0.350	0.449	64	0.246	0.320	96	0.201	0.262
33	0.345	0.443	65	0.244	0.317	97	0.200	0.260
34	0.339	0.436	66	0.242	0.315	98	0.199	0.259
35	0.334	0.430	67	0.241	0.313	99	0.198	0.258
36	0.329	0.424	68	0.239	0.310	100	0.197	0.257

Photo Credits

Chapter 1
Page 1: Double-crested cormorant © Martin B. Withers / FLPA. Page 3: *Fat Cat Capsizing* © photos courtesy of Richard Watherwax / watherwax.com. Page 14: Large-beaked ground finch, Dolph Schluter.

Interleaf 1
Page 21: R. A. Fisher © Antony Barrington Brown, reproduced with permission of the Fisher Memorial Trust.

Chapter 2
Page 27: White-winged dove © photo courtesy of Jeff Whitlock. Page 31: Sockeye salmon © Fletcher & Baylis / Photo Researchers, Inc. Page 35: Great tit © Andrew Howe / iStockphoto. Page 41: Guppies © photo courtesy of Andrew Hendry. Page 43: Total Ozone Mapping, image courtesy of NASA. Page 46: Diversity in Africa © reproduced with permission from Moore, *et al.* (2002). Page 53: Black-bellied seedcracker © photo courtesy of Thomas B. Smith.

Chapter 3
Page 59: Saiga © Igor Shpilenok / Nature Picture Library. Page 60: Paradise tree snake © Jake Socha. Page 67: *Tidarren* spider reproduced with permission from Ramos, *et al.* (2004), Page 70: Threespine sticklebacks, reproduced with permission from Marchinko, K.B., and D. Schluter (2007) [*Evolution* 61:1084–1090, Wiley-Blackwell Publishing Ltd.] Page 78: Cutthroat finch © John Kormendy.

Chapter 4
Page 83: DNA Crystal © Molecular Expressions.

Chapter 5
Page 99: White-breasted nuthatch © photo courtesy of Jeff Whitlock. Page 113: Jewel wasp © photo courtesy of John H. Werren. Page 126: Blackjack © Matthew Scherf / iStockphoto.com.

Chapter 6
Page 127: *Cyanella alba* © photo courtesy of Lawrence Harder. Page 130: *Bufo bufo* © Wayne Hutchinson / FLPA. Page 139: *Heteranthera multiflora* © photo courtesy of Spencer C. H. Barrett.

Chapter 7
Page 151: *Drosophila* © Oliver Meckes, Nicole Ottawa / Photo Researchers, Inc. Page 158: Mouse chromosomes © photo courtesy of Chromosome Science Laboratory, Inc. Page 167: Follicle mites © Andrew Syred / Photo Researchers, Inc. Page 170: Gorilla © photo courtesy of Daniel Simons [Simons and Chabris (1999)].

Chapter 8
Page 175: Ammonite fossil © Francois Gohier / Photo Researchers, Inc. Page 185: Human X chromosome © Andrew Syred / Photo Researchers, Inc. Page 193: Marine diatom fossil © Stephen Nagy. Page 203: Hurricane Frances, photo courtesy of NASA.

Chapter 9
Page 207: Goby (*Gobiodon erythrospilus*) © photo courtesy of Philip L. Munday. Page 213: California killifish © photo courtesy of Drew Talley Page 214: Great blue heron © photo courtesy of Jeff Whitlock. Page 220: Vampire bat © Ron Austing / FLPA. Page 227: Collared flycatcher © Stefan Johnson.

Chapter 10
Page 231: Crab spider *Thomisus spectabilis* © Ed Nieuwenhuys. Page 248: Brown recluse spider © S. Camazine, K. Visscher / Photo Researchers, Inc. Page 255: European earwig © B. Borrell Casals / FLPA.

Chapter 11
Pages 259 and 261: Stalk-eyed flies *Cyrtodiopsis dalmanni* © photos courtesy of Sam Cotton. Page 266: Human body thermograph © Dr. Ray Clark FRPS and Mervyn de Calcina-Goff FRPS / Photo Researchers, Inc.

Chapter 12
Page 279: Galapagos marine iguana © David Hosking / FLPA. Page 282: Red-winged blackbird © Paul Bannick/www.PaulBannick.com. Page 287 (left and middle): Horned lizard © photos courtesy of Butch Brodie. Page 287 (right): Loggerhead shrike © Paul Wolf / iStockphoto. Page 294: Brook trout © prepared by Ellen Edmondson as part of the 1927–1940 New York Biological Survey. Permission for use in

this book is granted by the New York State Department of Environmental Conservation. Page 307: Burrowing owl © Ronald G. Wolff. Page 310: Ostrich © ra-photos / iStockphoto. Page 313: Tilapia Peter Davey / FLPA. Page 314: Weddell seals © Paul Ward.

Chapter 13

Page 319: Sun star *Solaster stimpsoni* © photo courtesy of Colin Bates. Page 323: Sally Lightfoot crab © photo courtesy of Sally Otto. Page 334: Mating dragonflies © Gary Cox. Page 338: Sagebrush crickets © David H. Funk. Page 341: Cartoon © Danny Shanahan / The Cartoon Bank. Page 348: Male lion committing infanticide © photo courtesy of Tim Caro. Page 351: Zebra finch © photo courtesy of Jeff Whitlock.

Chapter 14

Page 357: Frog deformities © Stanley K. Sessions. Page 360: HIV-1 virus reproduced by permission from Los Alamos National Laboratory. This information has been authored by an employee or employees of the University of California, operator of the Los Alamos National Laboratory under Contract No. W-7405-ENG-36 with the U.S. Department of Energy. Page 370: Experimental tree-holes © photo courtesy of Diane Srivastava. Page 371: Rat © Sergey Goruppa / iStockphoto. Page 373: Bullfrog tadpole © Ron Brancato / iStockphoto. Page 385: Monarch caterpillar © Mark Plonsky. Page 388: Meerkat © Gordana Sermek / iStockphoto.

Chapter 15

Page 393: Egyptian vulture © photo courtesy of Diana Hromish / LunarEye Photography. Page 394: Clock © Alan Crosthwaite / Absolute Stock Photo. Page 407: Mycorrhizae © reproduced by permission of the publisher from Read, David (1997) [*Nature* 388 (6642), Figure 1.] Page 412: Walking stick *Timema cristinae* © photo courtesy of Patrick Nosil. Page 423: *Onthophagus taurus* beetle © Frank Kohler.

Chapter 16

Page 431: Western trillium, Dolph Schluter. Page 438: Wolf © Karel Broz / iStockphoto. Page 444: Indian rope trick © image courtesy of The Nielsen Magic Poster Collection, Page 452: Black-tailed godwit © photo courtesy of Steve Gantlett. Page 454: Filefish © photo courtesy of John Randall.

Chapter 17

Page 463: Red campion *Silene dioica* © Richard Becker / FLPA. Page 465: Male lion © Mark Wilson / iStockphoto. Page 475: Big-eared baby, Mike Whit-

lock. Page 476: Black-capped chickadee © photo courtesy of Jeff Whitlock. Page 483: Dark-eyed junco © photo courtesy of Janine Russell. Page 490: Northern fur seal © photo courtesy of Andrew Trites. Page 501: *Crotaphytus collaris* © reproduced by permission of the publisher from Lappin and Husak (2005), The University of Chicago Press. Page 506: Macaroni penguins © Geno Sajko / iStockphoto. Page 508: Atomic bomb test © photo courtesy of AJ Software & Multimedia.

Interleaf 11

Page 509 (left): *Clintonia borealis* © photo courtesy of Harvey Kirsch Rivernen. Page 509 (right): *Lillium superbum* © Michael Hare / iStockphoto.

Chapter 18

Page 513: Zooplankton © Wim van Egmond / Visuals Unlimited. Page 517: Mice © Emilia Stasiak / iStockphoto. Page 523: Study plots © photo courtesy of Chris Harley. Page 527: Damaraland mole rat © Wendy Dennis / FLPA. Page 533: Magellanic penguins © Alexander Hafemann / iStockphoto. Page 537 (left): Tortoise beetle © photo courtesy of Edward Trammel. Page 537 (right): Yellow dung fly © photo courtesy of Alan J. Silverside, University of the West of Scotland.

Chapter 19

Page 539: Red panda © photo courtesy of Jeff Whitlock. Page 545: Pseudoscorpion with babies © photo courtesy of Jeanne Zeh. Page 551: Chimpanzee and teacher © Frans Lanting. Page 559: Prairie vole © photo courtesy of Michael R. Jeffords. Page 562: *Helix aspera* mating © reproduced by permission of Chase, Ronald, and Katrina C. Blanchard (2006) [*Proceedings of the Royal Society B* 273:1471–1475, Figure 1].

Chapter 20

Page 567: Chimpanzee © Kitch Bain / iStockphoto. Page 571: *Trichgramma* wasp on *Pieris brassicae* © reproduced by permission of the publisher from Fatouros *et al.* (2005), Figure 1, Nature Publishing Group. Page 577: Lori Eggert © photo courtesy of Lori S. Eggert. Page 590: Face images © Farroni (2002).

Chapter 21

Page 593: Blood red moon © 2007 by Fred Espenak, www.MrEclipse.com. Page 600: Testosterone crystal © Molecular Expressions.

Index